DORLING KINDERSLEY HANDBOOKS

MAMMALS

MAMMALS

Editorial Consultant

JULIET CLUTTON-BROCK

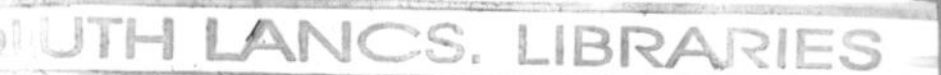

LONDON, NEW YORK, MUNICH,
MELBOURNE, AND DELHI

DK DELHI
Senior Project Editor Sheema Mookherjee
Senior Art Editor Aparna Sharma
Editor Kajori Aikat
Designers Kavita Dutta, Sabyasachi Kundu
DTP Designers Pankaj Sharma, Balwant Singh
Managing Editor Ira Pande
Managing Art Editor Shuka Jain

DK LONDON
Project Editor David Summers
Project Art Editor Kirsten Cashman
DTP Designer Rajen Shah
Production Controller Elizabeth Dodd
Picture Researcher Cheryl Dubyk-Yates
Senior Editor Angeles Gavira
Senior Art Editor Ina Stradins
Category Publisher Jonathan Metcalf
Special Consultant and Writer Steve Parker

First published in Great Britain in 2002
by Dorling Kindersley Limited
80 Strand, London WC2R 0RL

A CIP catalogue record for this book
is available from the British Library

ISBN 0-7513-3374-3
Reproduced by Colourscan, Singapore
Printed and bound by
Kyodo Printing Co., Singapore

CONTENTS

INTRODUCTION • *6*
HOW THIS BOOK WORKS *9*
WHAT IS A MAMMAL *10*
EVOLUTION *12*
DIVERSITY *14*
ANATOMY *16*
REPRODUCTION *20*
SOCIAL GROUPS *24*
SENSES AND COMMUNICATION *26*
LOCOMOTION *28*
FEEDING *30*
DESERT MAMMALS *34*
GRASSLAND MAMMALS *36*
FOREST MAMMALS *38*
POLAR AND MOUNTAIN MAMMALS *40*
AQUATIC MAMMALS *42*
WATCHING MAMMALS *44*

THREATENED MAMMALS *46*
CONSERVATION *48*
MAMMALS AND HUMANS *50*
CLASSIFICATION *52*

CATALOGUE • *54*
EGG-LAYING MAMMALS *54*
MARSUPIALS *56*
INSECTIVORES *78*
BATS *84*

ELEPHANT-SHREWS *94*
FLYING LEMURS *95*
TREE SHREWS *96*

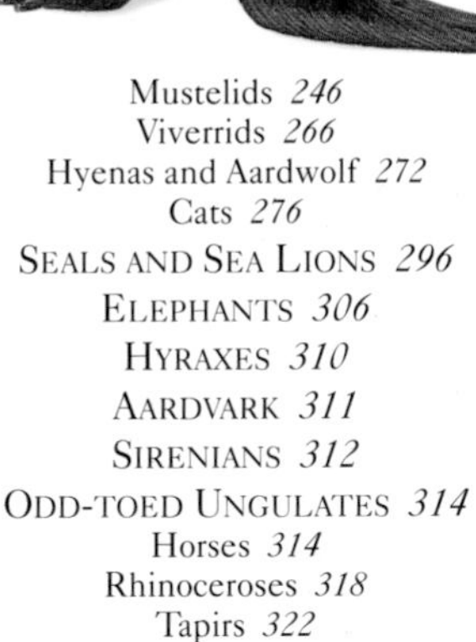

Mustelids *246*
Viverrids *266*
Hyenas and Aardwolf *272*
Cats *276*
SEALS AND SEA LIONS *296*
ELEPHANTS *306*
HYRAXES *310*
AARDVARK *311*
SIRENIANS *312*
ODD-TOED UNGULATES *314*
Horses *314*
Rhinoceroses *318*
Tapirs *322*
EVEN-TOED UNGULATES *324*
Wild Pigs *324*
Hippopotamuses *328*
Camels and Relatives *330*
Deer *334*
Pronghorn *344*

PRIMATES *97*
Prosimians *97*
Monkeys *107*
Apes *122*
SLOTHS, ANTEATERS, ARMADILLOS *130*
PANGOLINS *135*
RABBITS, HARES, AND PIKAS *136*
RODENTS *143*
CETACEANS *182*
Toothed Whales and Dolphins *182*

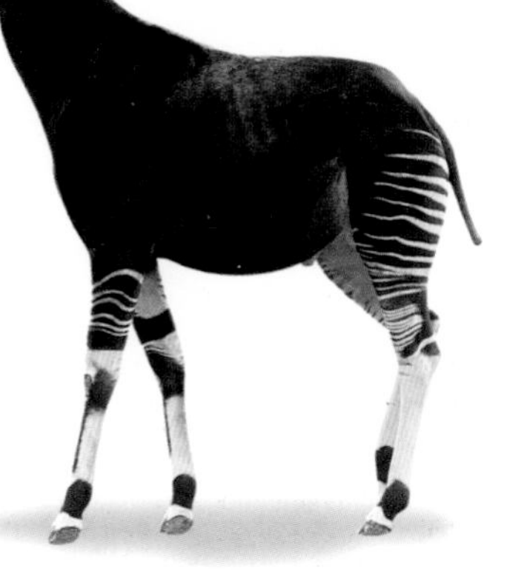

Giraffe and Okapi *345*
Cattle, Antelope, and Relatives *348*

Baleen Whales *204*
CARNIVORES *216*
Dogs and Foxes *216*
Bears *234*
Raccoons *242*
Glossary *382*
Index *386*
Acknowledgments *398*

INTRODUCTION

Mammals are the most familiar group of vertebrates. They are also the most varied and adaptable, having found ways to survive in the broadest range of habitats, from the oceans to the poles. Within this variety, mammals share some fundamental characteristics: they are warm-blooded, give birth to live young, feed their young on milk produced in mammary glands, and all but a few have a covering of hair on their bodies.

LIFE ON EARTH has changed over millions of years through the process of evolution. This has produced a vast array of living things, including untold millions of animal species, from worms, scorpions, and flies, to fish, frogs, reptiles, birds – and mammals. In such lists, the mammal group is usually put at the very top, as though natural processes have been aiming to produce them as an end-point. Mammals are certainly successful in terms of being widespread over a variety of habitats, and they are relatively large in body size and numbers. However, they should not be looked upon as an evolutionary peak. Other animal groups, such as birds in the air, insects on land, and fish and crustaceans in the sea, are equally successful in evolutionary terms, and they outnumber mammals both in terms of species as well as in the total count of individuals. But mammals do possess some unique and fascinating qualities.

◁ A LONG HISTORY
Fossil remains show that small, shrew-like mammals such as this Megazostrodon appeared on Earth over 200 million years ago. This was the time when early dinosaurs were also spreading across the land.

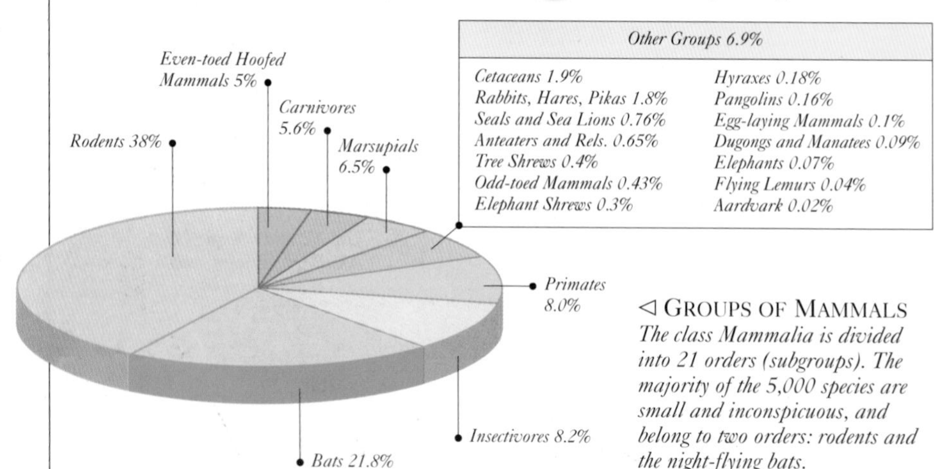

◁ GROUPS OF MAMMALS
The class Mammalia is divided into 21 orders (subgroups). The majority of the 5,000 species are small and inconspicuous, and belong to two orders: rodents and the night-flying bats.

SUCCESS STORY

Mammals have a complex body chemistry, which maintains a constant raised body temperature. This requires them to eat large amounts of food, in comparison to cold-blooded species such as reptiles and insects. However, it also allows them to remain active in cold conditions, and mammals can survive high on mountains during winter, and in the far north and south near the poles. This means mammals are among the few animal groups to thrive in such harsh habitats. The mammal group also displays some of the most complicated behaviours in the animal kingdom. Most species can learn from experience, and the longest-lived, such as elephants, are able to accumulate and pass on invaluable skills and knowledge to their offspring. Some form complex societies with sophisticated methods of communication, in which individuals help each other to survive.

△ INTELLIGENT
Apes have the capacity for problem-solving and the use of tools. This chimpanzee has learned to use a stick to get a meal from a termite mound.

◁ ADAPTABLE
Mammals have colonized almost every habitat, from hyraxes in barren, rocky mountains (left), to whales that can swim huge distances through the sea.

▽ SOCIABLE
Because female mammals suckle their young, there is an extended period of family care. Many species carry social patterns into adulthood, forming groups such as prides, herds, troops, and schools.

Humans and Other Mammals

People have a history of long and close associations with other mammals – we can identify with their warmth, furriness, alertness, facial expressions, active movements, and complex behaviour. Throughout history, mammals have even been objects of worship, condemnation, or sacrifice. They have also long been domesticated for various uses, including companionship, provision of food, and as beasts of burden. This process of selective breeding for desirable qualities, such as temperament or strength, has often given rise to considerable change and variety – take for example the huge number of domestic dog breeds, few of which greatly resemble their ancestor, the wolf. This book focuses on the species that still inhabit the wild and live mostly independently of, though certainly not unaffected by, humans. They, more than ever, need our consideration and care.

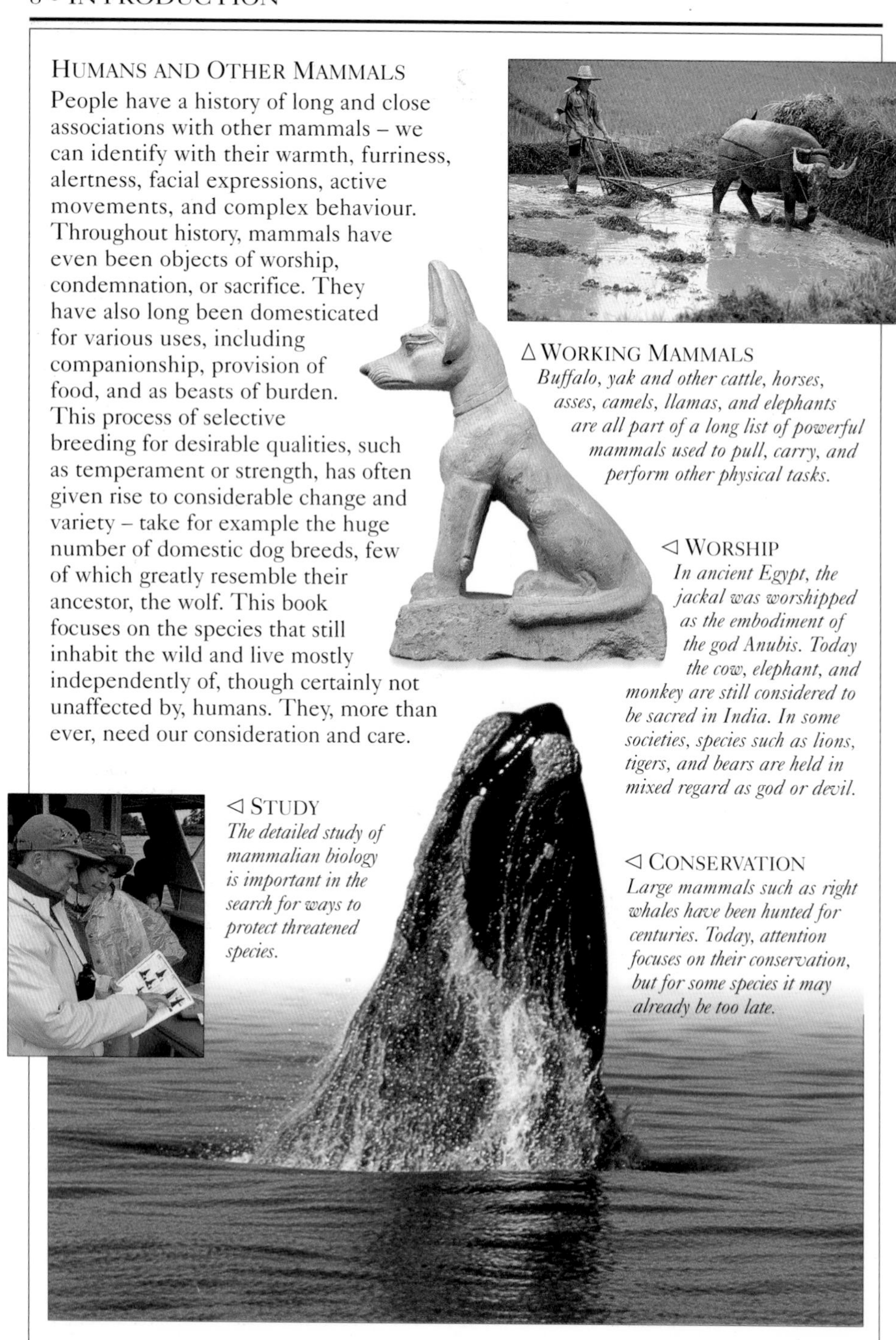

△ Working Mammals
Buffalo, yak and other cattle, horses, asses, camels, llamas, and elephants are all part of a long list of powerful mammals used to pull, carry, and perform other physical tasks.

◁ Worship
In ancient Egypt, the jackal was worshipped as the embodiment of the god Anubis. Today the cow, elephant, and monkey are still considered to be sacred in India. In some societies, species such as lions, tigers, and bears are held in mixed regard as god or devil.

◁ Study
The detailed study of mammalian biology is important in the search for ways to protect threatened species.

◁ Conservation
Large mammals such as right whales have been hunted for centuries. Today, attention focuses on their conservation, but for some species it may already be too late.

How This Book Works

THIS BOOK COVERS the 21 different orders that make up the class Mammalia. Within each of these orders, individual species entries are arranged according to the family they belong to. The sample page below shows a typical entry, with information organized into text, bands, and symbols.

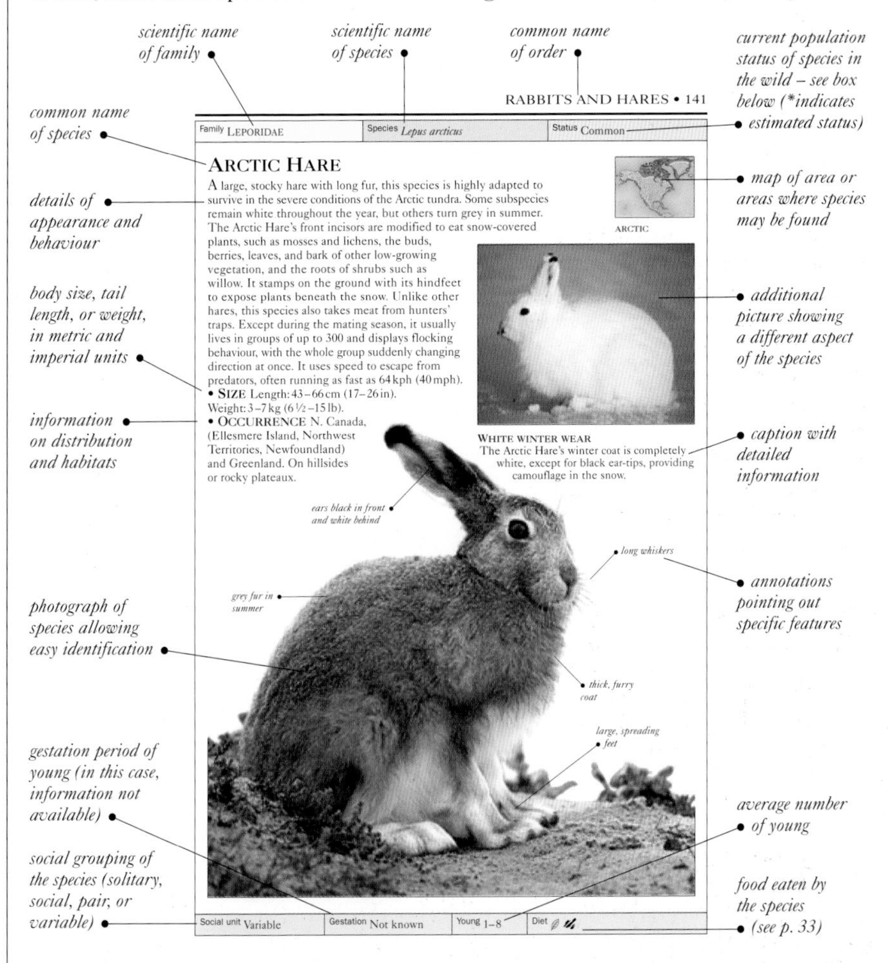

RABBITS AND HARES • 141

Family LEPORIDAE	Species *Lepus arcticus*	Status Common

ARCTIC HARE

A large, stocky hare with long fur, this species is highly adapted to survive in the severe conditions of the Arctic tundra. Some subspecies remain white throughout the year, but others turn grey in summer. The Arctic Hare's front incisors are modified to eat snow-covered plants, such as mosses and lichens, the buds, berries, leaves, and bark of other low-growing vegetation, and the roots of shrubs such as willow. It stamps on the ground with its hindfeet to expose plants beneath the snow. Unlike other hares, this species also takes meat from hunters' traps. Except during the mating season, it usually lives in groups of up to 300 and displays flocking behaviour, with the whole group suddenly changing direction at once. It uses speed to escape from predators, often running as fast as 64 kph (40 mph).

• SIZE Length: 43–66 cm (17–26 in). Weight: 3–7 kg (6½–15 lb).

• OCCURRENCE N. Canada, (Ellesmere Island, Northwest Territories, Newfoundland) and Greenland. On hillsides or rocky plateaux.

WHITE WINTER WEAR The Arctic Hare's winter coat is completely white, except for black ear-tips, providing camouflage in the snow.

Social unit Variable	Gestation Not known	Young 1–8	Diet

IN THIS BOOK, the population status of each species is mainly based on the IUCN Red List of Threatened Species (see p. 46), as follows:

Extinct in the wild Exists in captivity only.
Critically endangered Faces an extremely high risk of extinction.
Endangered Faces a very high risk of extinction.
Vulnerable Faces a high risk of extinction.
Lower risk Dependent on conservation not to qualify for any of the above categories.
Common Found in relatively high densities over a wide range.
Locally common Found in relatively high densities within a restricted area or areas.

WHAT IS A MAMMAL?

MAMMALS ARE THE most familiar kinds of animals in our daily lives. Most of our domesticated animals and pets are mammals. So are the creatures that people like to watch in wildlife parks and zoos. As mammals, humans too have all the typical features of the group. In fact, our closest relatives belonging to the animal world – apes and monkeys – are mammals.

MAMMAL CHARACTERISTICS

Mammals have three main features that set them apart from other classes of animals, including other vertebrates (animals with an internal skeleton – fish and reptiles, as well as mammals). First, mammals are warm-blooded, or to be more accurate, they are endothermic and homeothermic. This means they generate heat within their body to keep themselves at a fairly high temperature, usually 35–40°C (95–104°F), and they maintain this temperature at a constant level despite variations in the surroundings. Second, mammals have hair or fur. Even mammals that seem to be hair-less, such as whales and dolphins, have a few hairs here and there on their bodies. Third, female mammals feed their newborn offsprings on milk that is produced by glands in their body. These are known as mammary glands, and give the name Mammalia to the whole group.

△ MAMMALS AND MILK
A female mammal (as in the Mona Monkey shown here) feeds her newborn on milk from her mammary glands. These are specialized types of sweat glands in the skin, usually located on the female's chest or abdomen.

◁ THE MAMMALIAN BODY
A typical mammal such as the wolf has a large head with keen eyes, ears, and nose. The strong jaws are equipped with several kinds of teeth (see p.18). The long body, four limbs, and tail (see p.16) are mostly covered by hair. However, some mammal groups have evolved very different body shapes (see opposite).

THREE GROUPS OF MAMMALS

Mammals are often divided into three groups, depending on the way they reproduce. In the monotremes or egg-laying mammals, the female lays eggs that hatch into babies. In the marsupial or pouched mammals, the babies develop for a short time in the mother's uterus or womb. They are born at a very early stage of development – tiny, unable to see or hear, with hardly formed limbs, and no fur. The newborn babies then crawl to a pocket-like flap on the female's body, the marsupium, where they feed on the mother's milk and continue to develop. The third and largest group comprises the placental mammals (see p. 22). Here, the offspring develop to a more advanced stage in the uterus, where they receive nourishment from the placenta. Hence, the young of placental mammals are born in a more developed state than marsupial mammals.

△ EGG-LAYING MAMMALS
The five species of monotremes are the four echidnas (seen here is the Short-nosed Echidna) and the Duck-billed Platypus.

◁ POUCHED MAMMALS
The 292 or so species of marsupials include kangaroos (such as the Western Grey Kangaroo shown here), wallabies, koalas, possums, bandicoots, and smaller kinds that resemble rats, mice, and shrews.

SPECIALIZED MAMMALS

Although most mammals live on the ground, several groups are adapted to living in different environments. These mammals have evolved very different body shapes and limb structures, according to the demands of their particular surroundings.

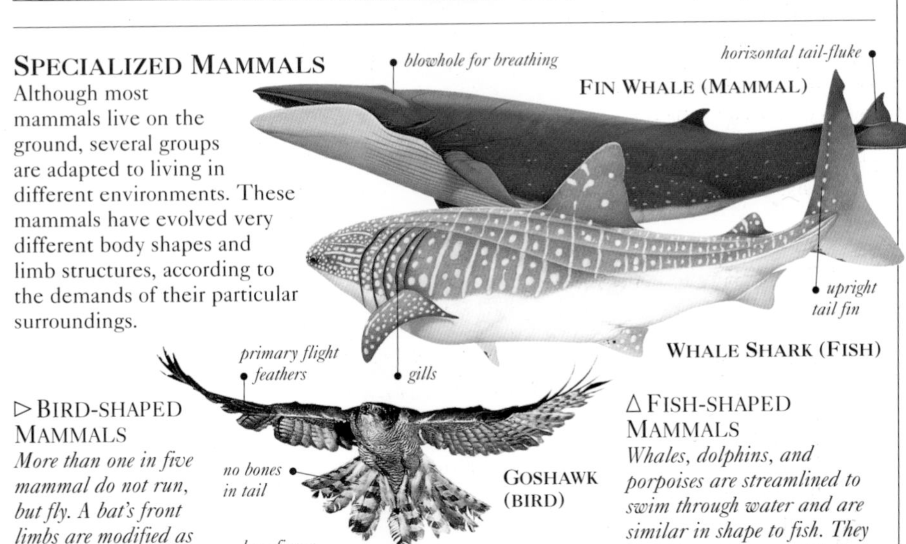

▷ BIRD-SHAPED MAMMALS
More than one in five mammal do not run, but fly. A bat's front limbs are modified as wings that are made of a membrane held by finger bones, rather than feathers as in a bird.

△ FISH-SHAPED MAMMALS
Whales, dolphins, and porpoises are streamlined to swim through water and are similar in shape to fish. They have front limbs modified as flippers and tail-flukes at the rear end. They breathe with their lungs, unlike fish that breathe through gills.

EVOLUTION

LIKE ALL LIVING things, mammals have evolved over long periods of time. The 4,475 or so mammal species alive today are only a small proportion of all the mammals that have lived on earth. Fossil evidence suggests that the first mammals appeared around the same time as the early dinosaurs, over 200 million years ago. However mammalian features such as feeding young on milk, were not preserved as fossils. Prehistoric mammals must be identified from their fossilized remains, particularly the teeth and skull (see opposite).

MAMMAL ANCESTORS

The ancestors of early mammals were small, active predators – a sub-group of mammal-like reptiles known as therapsids. Fossils suggest that some therapsids had developed fur, and were possibly endothermic (warm-blooded), on the way to becoming true mammals. These earliest mammals date from the Middle Triassic Period. They had new features in the skull (see panel, opposite), a lighter and more flexible skeleton, and erect limbs aligned under the body, rather than sprawling out by the sides, as in reptiles. By 200 million years ago, early mammals hunted in the undergrowth for small prey. For the next 135 million years, dinosaurs dominated the land. However, mammals persisted, although none was larger than a pet cat. They probably came out at night, when it was too cool for the great reptiles to be active.

△ EARLY MAMMAL
The remains of one of the earliest mammals, Megazostrodon, discovered in Lesotho, Africa, date from the Late Triassic Period. They indicate a creature 12 cm (4¾ in) in length, outwardly resembling today's shrews or tree shrews.

MOERITHERIUM (50–35 MYA) PHIOMIA (35 MYA) GOMPHOTHERIUM (20 MYA)

TRIASSIC	JURASSIC	CRETACEOUS
Mammal-like reptiles common in this period. First mammals and first dinosaurs appear.	Dinosaurs dominate the land as giant plant-eaters and fierce meat-eaters. Mammals are small, nocturnal insect-eaters that probably laid eggs (monotremes).	Dinosaurs continue to diversify into many groups. Mammals continue as small nocturnal
YEARS AGO (MILLIONS) 205	142	

MESOZOIC ERA

RAPID EVOLUTION

Dinosaurs died out 65 million years ago. Soon after, in the early Tertiary Period, mammals underwent great change. Even those that are highly evolved from their four-legged ancestors, appeared early in this period. Mammals now developed into hundreds of new kinds. Some of these disappeared, but others persisted and established the mammalian groups of today. By 50 million years ago, the first whales swam in the oceans and early bats flew in the air.

▽ EVOLVING FAMILIES

Some mammal groups were once much more common than today. Although the elephant group includes only three surviving species, it has a long and diverse history that has included over 160 species.

DEINOTHERIUM
(22–2 MYA)

ASIAN ELEPHANT
(PRESENT)

hinge at back of skull

uniform teeth

EARLY REPTILE

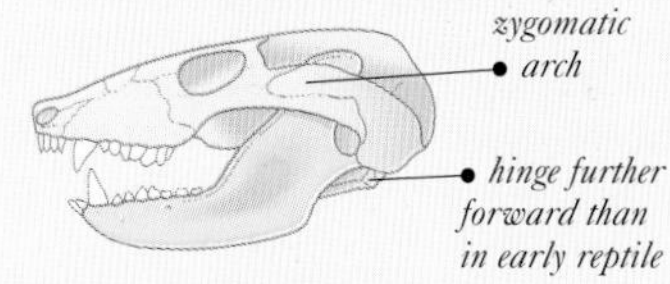

TRIASSIC MAMMAL

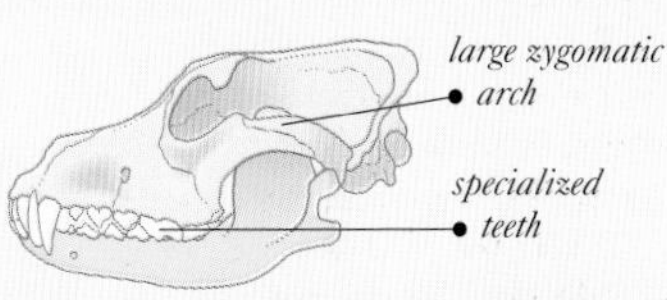

MODERN MAMMAL

DENTARY BONE

EVOLVING SKULL

The mammal's reptilian ancestors had teeth that were all much the same, and also several bones in the lower jaw. As mammals evolved, the lower jaw was reduced to one dentary bone, and the teeth became different shapes. The zygomatic arch developed to anchor stronger chewing muscles.

▽ TIMELINE

Mammals survived with little change, as small predators, through most of the Mesozoic Era. At the start of the Cenozoic Era, with the dinosaurs gone, mammals (and birds) evolved rapidly and soon came to dominate the land.

	TERTIARY	QUARTERNARY
predators. Marsupial and placental mammal groups probably appear.	Mammals undergo rapid evolution. By 40 million years ago, most present-day groups are established.	Series of Ice Ages greatly alter the habits of various mammals Appearance of modern humans coincides with disappearance of many large mammals–mammoths, giant elk, and others.
65	1.8	PRESENT

CENOZOIC ERA

Diversity

Mammals are the most widespread and diverse of all animal groups. They live in more habitats, and in more regions of the world, than any other major group of animals. This is partly due to the mammalian feature of endothermy (warm-bloodedness), which allows mammals to stay active in the coldest places, such as polar seas.

Extreme Size and Shape

The size range of mammals is unsurpassed by all other animal groups. The largest mammal is the Blue Whale, which is not only the biggest creature alive today, but also one of the largest animals ever to have lived on earth. It is more than 70 million times heavier than the smallest mammals, such as the Hog-nosed Bat or Pygmy Shrew, which are smaller than a human thumb. Between these two extremes are mammals of almost every imaginable size and shape, including mice that are smaller than some insects, otters and dolphins that can out-swim fish, fruit bats with wingspans greater than that of most birds, and massive buffaloes with horns longer than adult human arms.

◁ Largest Mammal
A well-fed female Blue Whale weighs over 150 tonnes (150 tons) and is up to 30 m (98 ft) long. This is about the same as the size of the largest dinosaurs, such as Argentinosaurus.

▷ Reaching High
A giraffe reaches leaves 6 m (20 ft) above ground. Everything about this mammal is elongated: its immense neck, drawn-out snout, and stilt-like legs. Even its tongue extends 45 cm (18 in) from its mouth.

△ Smallest Mammal
The Hog-nosed or Bumblebee Bat is nearly as light as a feather – just around 2 g (1/16 oz). It has a body that is 30 mm (1¼ in) long, wings that are 15 cm (6 in) across, and is found in caves in south-west Thailand. Pygmy White-toothed Shrews are known to be almost as light.

MODES OF LIFE

Mammals live and travel in almost every habitat on earth. Different species are adapted to dwelling in all kinds of terrestrial habitats and to cope with a huge range of temperatures and terrains, from the icy tundra and cold mountain peaks, to tropical rainforest, coniferous and deciduous woodland, grassland and scrub, and even the most barren and arid deserts. Mammals are also found in the air, in both fresh as well as salt water, in the soil, and in underground caves. Despite the need to breathe air, some mammals, such as the sperm whales and beaked whales, regularly visit the depths of the sea.

△ ON LAND
The fastest terrestrial animal is the Cheetah, which can reach 100 kph (63 mph). The Pronghorn of North America is almost as rapid.

△ IN BRANCHES
Koalas, possums and gliders, certain lemurs, and many monkeys are arboreal – they spend their lives in trees.

△ IN THE AIR
Although bats are the only true fliers among mammals, several others, such as flying lemurs, are expert gliders.

△ IN WATER
Mammals such as otters visit water often. Others such as seals stay in it for long periods. Whales, dolphins, and porpoises never leave their aquatic habitat.

SIMILAR BUT DIFFERENT

During millions of years of evolution, closely related mammals changed as they adapted to different environments. Tree-rats have large eyes and run nimbly through branches, but their close cousins the mole-rats are almost blind and burrow underground. Conversely, mammals that are only distant relations have evolved to become very similar due to their common lifestyles and habitats.

△ EUROPEAN MOLE
Although this mole is an insectivore, it resembles the Marsupial Mole in size and appearance. It also has spade-like front claws, a stocky body, and tiny eyes and ears – adaptations for living underground.

▷ MARSUPIAL MOLE
This outwardly resembles the European Mole in almost every respect, except colour, yet it belongs to an entirely different order: the marsupials.

ANATOMY

MOST MAMMALS HAVE an anatomical structure that includes a readily identifiable head and neck, a long body, four limbs which end in five digits (fingers or toes) each, and a tail. This basic structure was evident in the earliest mammals, 200 million years ago. However, over the years evolution has modified the design into a huge variety of sizes and shapes.

ENDOSKELETON

The bones of the skeleton are the body's internal supporting framework (endoskeleton). They are light, yet strong and stiff, and linked at flexible joints to permit movement. The same bones occur in similar positions in most mammals, although in the horse, the finger and toe bones have been lost.

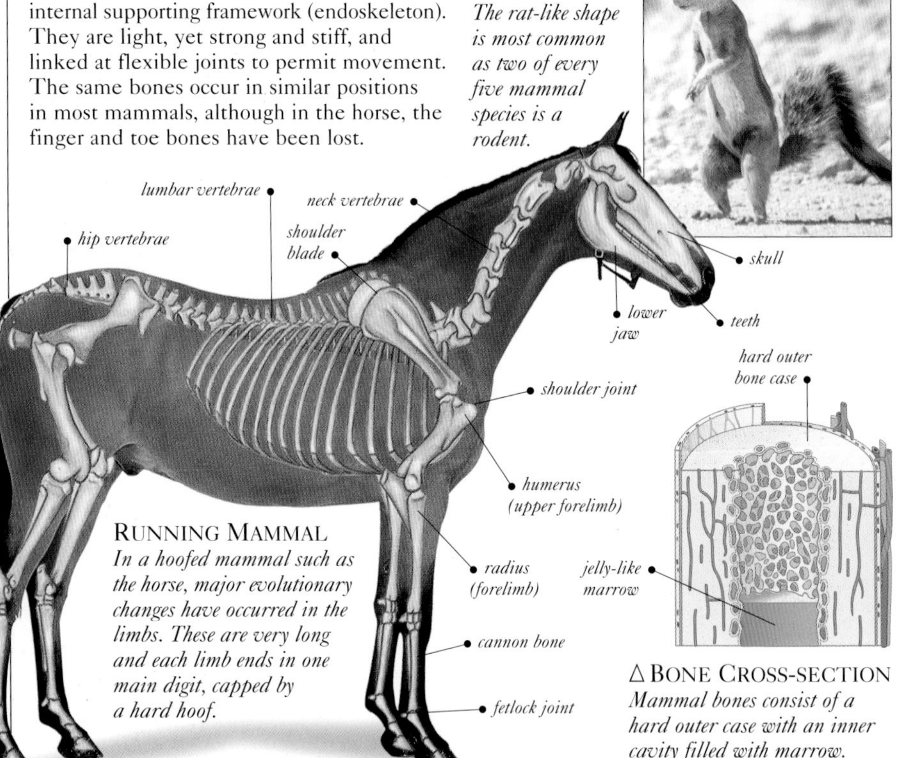

▷ MOST COMMON
The rat-like shape is most common as two of every five mammal species is a rodent.

RUNNING MAMMAL
In a hoofed mammal such as the horse, major evolutionary changes have occurred in the limbs. These are very long and each limb ends in one main digit, capped by a hard hoof.

△ BONE CROSS-SECTION
Mammal bones consist of a hard outer case with an inner cavity filled with marrow.

MAMMAL LIMBS

The outer shape of mammalian limbs, and the bones inside, vary greatly. They depend on how the mammal moves: legs on the ground, wings in air, and flippers in water, so its overall design is specialized. In many species, the limb has other tasks in addition to movement, such as catching prey or grooming fur.

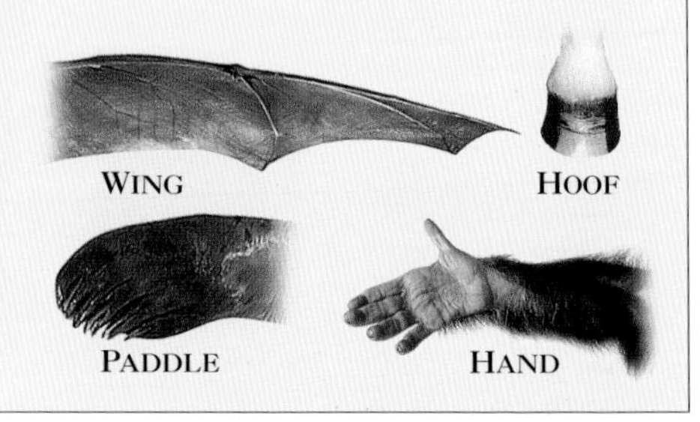

AQUATIC ADAPTATIONS

A cetacean such as a dolphin has one of the most highly evolved anatomies – its body shape and structure are very different from the original or primitive mammal design. As adaptations to pushing through water, the dolphin's head, neck, and body are streamlined and tapering, with hardly any hairs on the smooth skin.

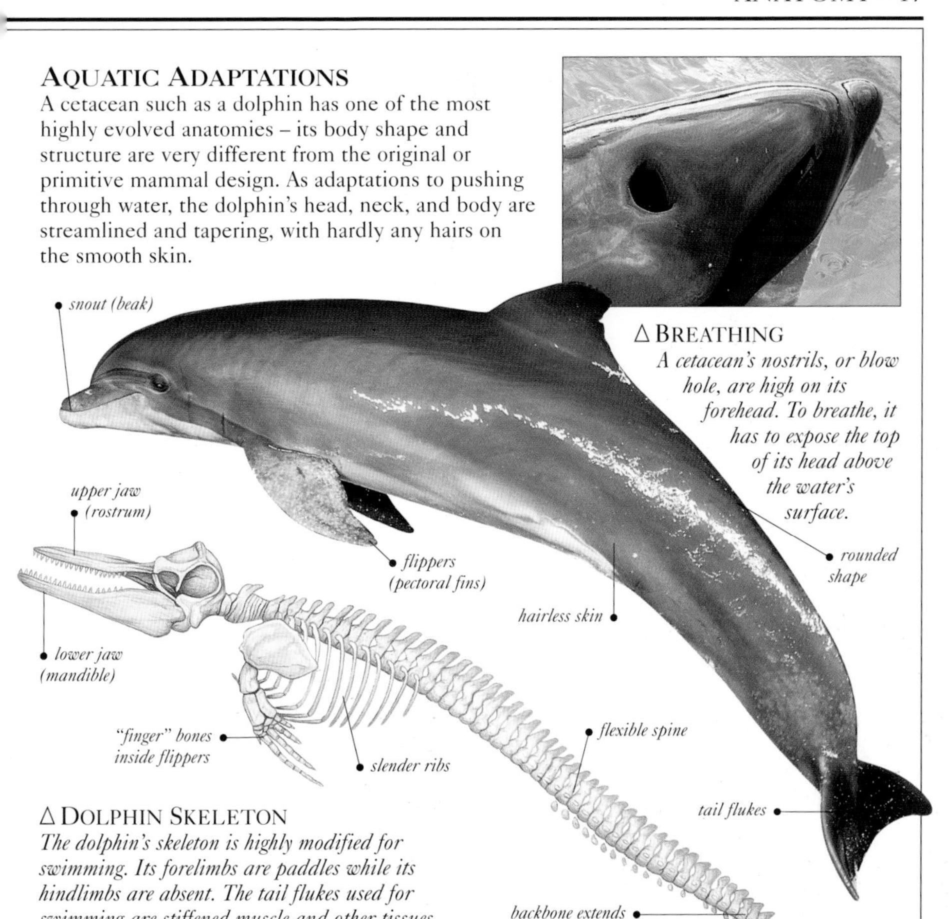

△ BREATHING
A cetacean's nostrils, or blow hole, are high on its forehead. To breathe, it has to expose the top of its head above the water's surface.

△ DOLPHIN SKELETON
The dolphin's skeleton is highly modified for swimming. Its forelimbs are paddles while its hindlimbs are absent. The tail flukes used for swimming are stiffened muscle and other tissues, unsupported by bones.

ADAPTATIONS FOR FLYING

A bat's hindlimbs are similar in overall structure to the limbs of most other mammals. However, its forelimbs are highly evolved as wings. The upper arm bone is short and stout, and the forearm bones are longer. The main part of the wing membrane, or patagium, is held out by the extremely elongated digits or fingers. The membrane itself is a thin layer of muscles and elastic fibres sandwiched between two layers of skin. It is derived from the skin webs that are present between the fingers of most other mammals (including humans). The wings are flapped by powerful muscles in the shoulders and chest.

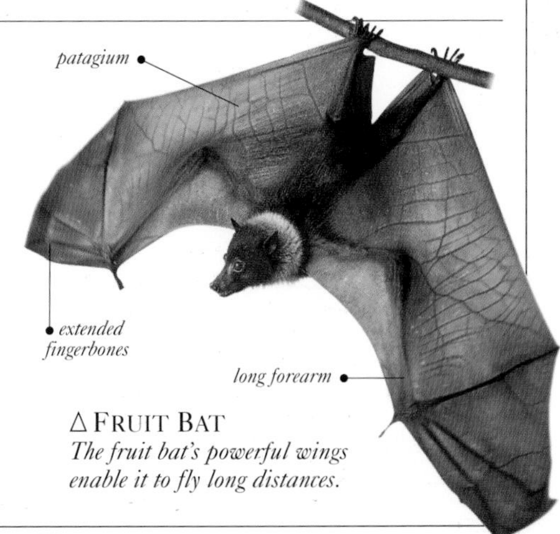

△ FRUIT BAT
The fruit bat's powerful wings enable it to fly long distances.

TEETH

Mammals are unique in having a lower jaw directly hinged to the skull, making the jaw a powerful tool. Unlike other vertebrates, mammals also have teeth that are varied in shape, for specialized tasks. This is known as heterodont dentition. There are four main tooth types. Incisors at the front have straight, sharp edges for biting and gnawing. Canines (eye teeth) are long and pointed to rip and tear. Premolars and molars (cheek teeth) are broad-topped for crushing, or sharp-edged for shearing. These teeth are developed differently in various species. Carnivores have long canines for stabbing and tearing animal prey. Grazers have very small or no canines, but their premolars and molars are massive and ridged for powerful chewing.

sharp cutting edge

CARNIVORE CHEEK TOOTH

broad, grinding surface

HERBIVORE CHEEK TOOTH

multi-cusped chewing surface

OMNIVORE CHEEK TOOTH

molars · *premolars* · *canine* · *incisors*

△ JAWFUL OF TEETH
The mongoose is an omnivorous mammal with a very wide diet. Its generalized dentition includes all four types of teeth, numbering 40 in total.

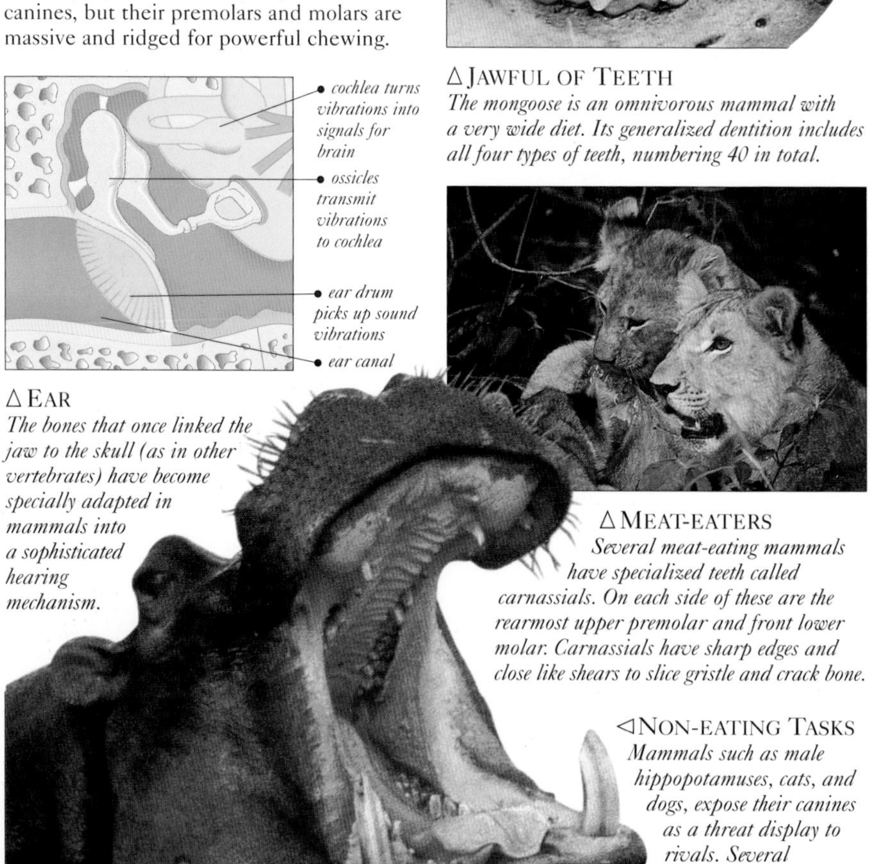

△ EAR
The bones that once linked the jaw to the skull (as in other vertebrates) have become specially adapted in mammals into a sophisticated hearing mechanism.

△ MEAT-EATERS
Several meat-eating mammals have specialized teeth called carnassials. On each side of these are the rearmost upper premolar and front lower molar. Carnassials have sharp edges and close like shears to slice gristle and crack bone.

◁ NON-EATING TASKS
Mammals such as male hippopotamuses, cats, and dogs, expose their canines as a threat display to rivals. Several mammals also groom their fur with their teeth.

SKIN AND HAIR

Mammalian skin has several important functions. It encases and protects the body's delicate internal parts. It provides animals with the sense of touch. Sweat glands in it release watery sweat, which draws heat from the body as it evaporates, to keep it at a fairly constant temperature. Its sebaceous glands make natural oils or waxes to keep the skin supple and waterproof. Skin also produces hair or fur, which is unique to mammals. The skin of large herbivores, such as elephants, is more than 3cm (1¼in) thick, giving protection against predators such as lions and also against pests such as biting flies.

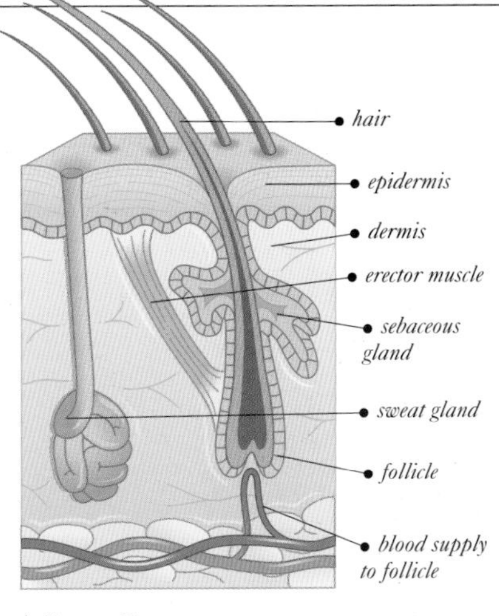

△ SKIN CROSS-SECTION
A mammal's skin consists of two layers: the outer epidermis that is dead and tough, and the inner dermis that contains glands and touch sensors.

▷ WHISKERS
Whiskers (vibrissae) are sensitive hairs around the nose: contacting the surroundings as the mammal moves along.

◁ SPINES
Mammals such as porcupines and hedgehogs have enlarged, stiff, sharp-tipped hairs known as spines, prickles, or quills, used mainly for protection.

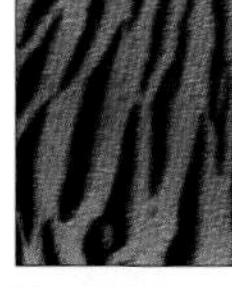

TIGER

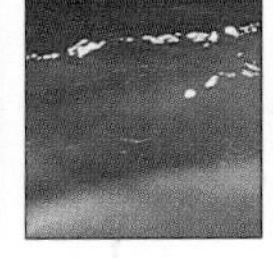

DOLPHIN

ARMADILLO

△ TYPES OF SKIN
The tiger's skin has stripes for camouflage in grass. The dolphin's skin is hairless to glide through water. The armadillo's skin grows horny external plates for protection.

TEMPERATURE CONTROL

Mammals are homeothermic (their body temperature remains at a fairly constant high level). They can regulate their temperature through behaviour, such as wallowing in mud or resting in shade, constricting or widening blood vessels that carry heat to the skin, shivering, raising or lowering their metabolic rate, or sweating.

▷ PANTING
Mammals with thick coats do not cool down by sweating. They pant instead, to get rid of body heat in the warm breath, and by the evaporating saliva.

Reproduction

Mammalian reproduction is sexual. This means that an egg cell from a female joins with a sperm cell from a male, to produce a fertilized egg that grows into a new individual. Unlike some simpler living beings, mammals cannot reproduce asexually, where young are produced by just one parent. Mammalian reproduction has several unique features, especially the growth of the young inside a specialized body part, the uterus or womb. This has a lining called the placenta from which the unborn foetus gets its nourishment.

Courtship and Mating

Courtship ensures that a female and male mammal of the same species come together to breed healthy, viable offspring. Courting may involve sound, scent, and sight. Calls, usually by males, attract potential partners from far away. Many female mammals produce scents that inform males that they are in breeding condition. Displays of body postures and movements enable each partner to assess the other's fitness as a suitable mate.

△ Breeding Competition
Some male mammals battle each other for the chance to mate with females. The biggest and strongest win: they are most likely to father the healthiest offspring.

◁ Mating
In mammals, fertilization is internal: the male sperm cells pass into the female's body, ready to fertilize her eggs.

▽ Breeding Season
Most mammals, such as seals, gather to breed only at a certain time of the year. The young are raised during the most favourable season, usually spring or summer, when there is plenty of food available.

EGG-LAYING MAMMALS

Monotremes form a very small order that do not give birth to formed young. Instead, the young develop within eggs inside the mother's body. After the eggs are laid, the babies hatch and feed on their mother's milk as in other mammals. The five species of monotremes are the platypus and four kinds of echidnas. They are from S.E. Asia and Australia. In the echidna, the eggs, and later the young, are carried in the female's pouch.

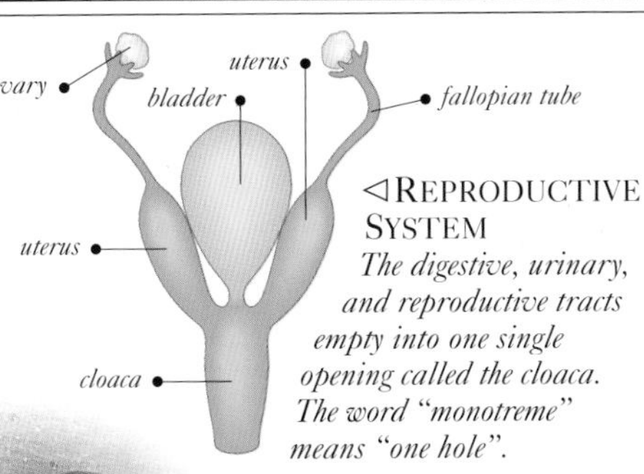

◁ REPRODUCTIVE SYSTEM
The digestive, urinary, and reproductive tracts empty into one single opening called the cloaca. The word "monotreme" means "one hole".

◁ PLATYPUS
The mother Platypus makes her breeding nest inside a riverbank tunnel. She lays her eggs and then keeps them warm and protected for an incubation period of 10 days, until the babies hatch. She has no teats (nipples). Milk oozes from her mammary glands onto her belly, for the babies to lap.

MARSUPIALS

The name marsupial derives from the Latin term for "leathery pouch". In most of the 292 marsupials, the female has a pouch – the marsupium – on her abdomen. Her babies are born at a very early stage of development compared to the young of placental mammals (see p. 22). Then they crawl into their mother's pouch, where they continue to develop, safe and protected.

◁ REPRODUCTIVE SYSTEM
A female marsupial has two vaginas, where the male's sperm enter to fertilize the eggs, and two uteri (wombs). The young is born through a central birth canal, which may form only temporarily.

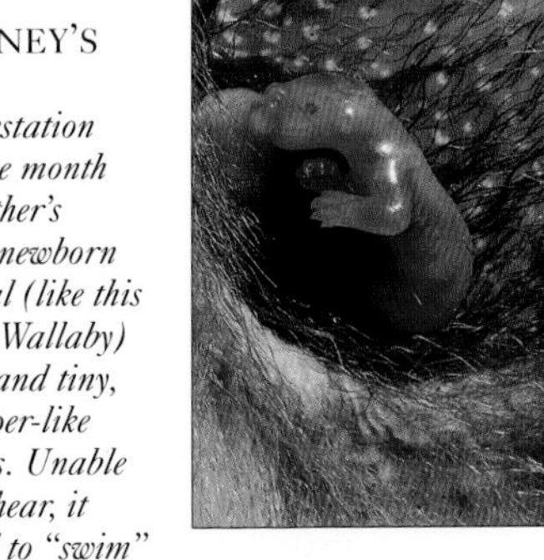

▷ JOURNEY'S END
After a gestation of just one month in its mother's womb, a newborn marsupial (like this Tammar Wallaby) is naked and tiny, with flipper-like limb buds. Unable to see or hear, it uses smell to "swim" through its mother's fur, from the birth opening to the pouch. It then attaches itself to a teat in the pouch to feed on milk, growing for another six months, before venturing out.

▷ JOEY
Older marsupial young (joeys) leave the pouch for short periods, returning to suckle and rest, or if threatened. The pouch can open forwards (as in the kangaroo) or backwards (as in the Koala).

Placental Mammals

All mammal species, except for monotremes and marsupials, are placental mammals. The placenta is a body part that develops inside the female's uterus (womb), alongside the unborn young. It is specialized to pass oxygen and nutrients from the mother's blood to the offspring, and to remove waste substances, allowing the unborn young to grow to an advanced stage in the uterus. The name "placental mammal", however, is not very accurate, as a rudimentary placenta also develops in the uterus of a marsupial mother.

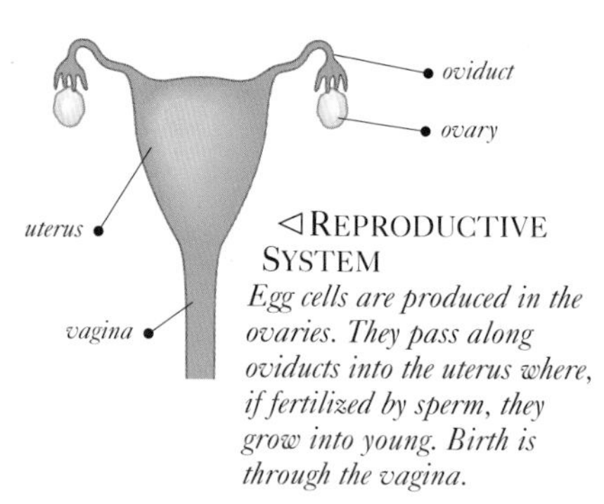

◁ Reproductive System
Egg cells are produced in the ovaries. They pass along oviducts into the uterus where, if fertilized by sperm, they grow into young. Birth is through the vagina.

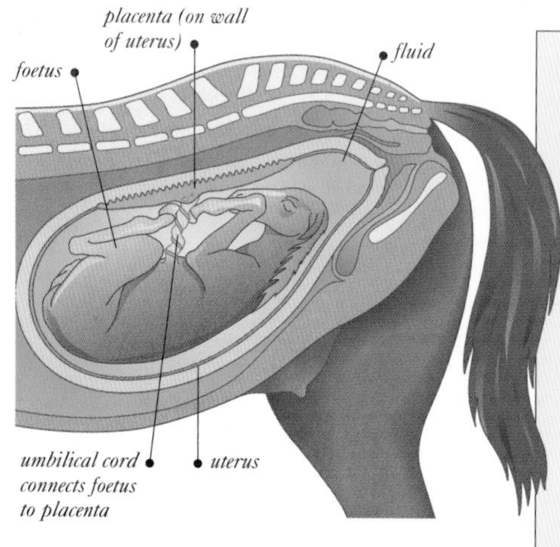

△ Inside the Uterus
The placenta is embedded in the inner lining of the uterus and is linked to the developing young (foetus) by blood vessels that form the umbilical cord. Fluid cushions the foetus from knocks and jolts. The placenta is expelled shortly after the baby.

▷ Just Born
Some newborn mammals are able to walk and run within minutes of birth. These are usually species such as Wildebeest, born in the open and at risk from predators. Babies born in a breeding nest are usually less well developed.

First Breaths

The baby in the womb receives oxygen from its mother's blood through the placenta and does not have to breathe. However, as it is born, the placenta detaches from the uterus and the baby immediately needs air. The births of mammals such as dolphins, porpoises, and whales (cetaceans), and Dugongs and Manatees (sirenians), occur underwater. After birth, the baby must get to the surface quickly to breathe in air. Often, the mother or another adult gently pushes and supports the newborn on its journey to the surface of the water.

△ Underwater Birth
Land mammals usually emerge head-first to make their passage through the birth canal easier. However, baby dolphins and other cetaceans emerge tail-first as their streamlined bodies slip smoothly through the birth canal.

Parental Care

A feature of mammal reproduction, which is uncommon among animals in general, is the long period of parental care. Offspring are fed on the mother's milk for weeks, or even months. They are also kept warm, safe, and protected. This is invariably by the mother, but in some species also by the father, and, in some social mammals, by other members of the group too. The longest period of parental care is ten years or more, in large mammals such as elephants, great apes, and humans.

◁ FAST BREEDERS
Small rodents such as mice are regular food for a range of predators. Rapid breeding is used as self-defence for the whole species. In one year, a pair of House Mice can produce more than 1,000 offspring of 2–3 generations, if all survive.

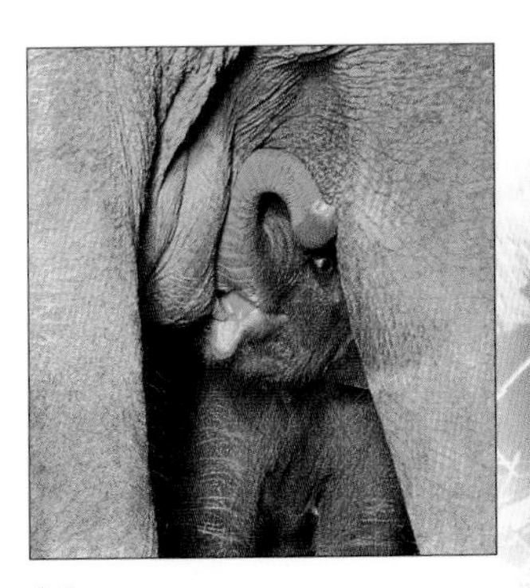

△ LACTATION
The time during which a mother mammal suckles (feeds) her offspring on her milk is known as the lactation period. It may vary from 10–14 days in some small rodents, to as much as three or four years in elephants.

△ FEW OFFSPRING
Larger mammals, such as this baboon, produce just one or two offspring. This allows the parents to put more time into caring for the young, leading to a better chance of survival.

Growing Up

A consequence of increased parental care among mammals is that offspring enjoy a longer "childhood" or period of learning. They can observe their parents and group members, and try out activities such as hunting and testing out different food items in a trial-and-error fashion.

◁ PLAYING
Young mammals such as foxes may seem to have carefree fun through playing sessions. But the play has a serious purpose as it helps to develop keen senses, quick reactions, strength, and agility. Such abilities will be needed in adulthood to catch food, ward off enemies, and compete with rivals at breeding time.

SOCIAL GROUPS

MAMMALS SHOW a whole range of social behaviours. Sociability tends to be more pronounced in plant-eating species than in meat-eaters. However, as part of the mammalian trait of adaptability, social groupings also vary with habitat, season, breeding and life cycle, and food availability. Most mammals live with others of their kind for a time, when young are being reared.

SOLITARY MAMMALS

Carnivorous mammals such as cats, some bears, mustelids (the weasel group), and viverrids (civets and mongooses) tend to be more solitary than group-dwelling. This is partly to reduce competition for food among hunters in one area. However, members of the dog family show the opposite trend and form well-organized packs.

△ LONE STALKER
Cats are solitary, nocturnal stalkers. Two cats together are either a mating pair, or a mother with offspring. Lions are the only exception and form groups or prides.

◁ HOME TERRITORY
Bears, like other large carnivores, each set up a home range. But the ranges of male Brown Bears overlap those of females, so the sexes meet occasionally for mating.

FORMING PAIRS

Some mammals, especially primates, form male–female pairs the whole year round or even for several years. Certain gibbon species pair for life. This system saves the energy and risks of competing anew for partners at each breeding season. However, recent studies show that such partnerships are not always monogamous; one partner may mate outside the pair while the other partner is absent.

△ IN HARMONY
Some gibbons reinforce their pair bond by making duet-like calls, especially at dawn. The calls also warn other gibbons to avoid the pair's territory.

EXTENDED GROUPS

Several species of monkeys and foxes live in family groups consisting of the female, often her male partner, and their offspring. In some species such as lions, wolves, and gorillas, the group is extended by other close relatives and perhaps a further generation or two of offspring. Mature males may live alone and only associate with the group for mating. Many hoofed animals form large herds of one species, or even of mixed species such as zebras, gazelles, and antelopes – this provides safety in numbers.

△ RABBIT WARREN
Rabbits form stable groups of up to about 20 individuals, with approximately equal numbers of females and males. Senior-ranking females occupy the bigger and safer nesting chambers nearest the centre of the warren.

◁ ELEPHANT HERD
The elephant herd is led by a matriarch, and usually contains related females and their offspring. As in many large herbivores, young males form bachelor groups while older males are solitary.

▷ KILLER WHALE POD
The Killer Whale pod, of up to 30, is led by a senior female, with subordinate males and females, and their offspring. It may develop into a multi-generational unit. Mature males leave briefly to mate with females in other pods.

WORKING AS A COMMUNITY

The Naked Mole-rat of East Africa has a social system unique among mammals, and which is more akin to insect societies. It lives in a colony of up to 80, dominated by a "queen". Only she gives birth to young and suckles them. Other members or "courtiers" of the colony care for the offspring. Workers dig tunnels and gather food which they bring back to share in the colony's central chamber.

Senses and Communication

Mammals not only use their senses to find the way, locate food, and identify danger, but also to interact with others of their own kind. Their intricate array of communications, such as calling and scent-marking, is used for courtship, competition for dominance, and repelling intruders from a territory.

Senses

Mammals have the five main senses of most animals: sight, hearing, smell, touch, and taste. Habitat influences which senses are best developed. Underground, some moles and mole-rats are virtually blind but extremely sensitive to touch and vibration. In thick forest, where sight is limited, monkeys use sounds to communicate. Zebras and antelopes on open plains tend to have acute sight.

△ Underground

The Star-nosed Mole lives underground and has little need of vision. The tentacles around its nose are extremely sensitive to touch and, combined with acute smell, allow it to feel and sniff for prey and a mate.

large, sensitive ears

big front-facing eyes

prehensile tail

△ Night Senses

The cycle of daily activity is another factor affecting sense development. The nocturnal Bushbaby, or Galago, which lives in the forests of Africa, has very large, front-facing eyes to pick up the slightest trace of light, as well as big ears to locate faint sounds in the darkness.

Echolocation

Cetaceans and bats use echolocation, or sonar, to navigate and find prey. The mammal makes sounds that bounce off nearby objects and return as echoes, which give information about the size, location, and distance of the object.

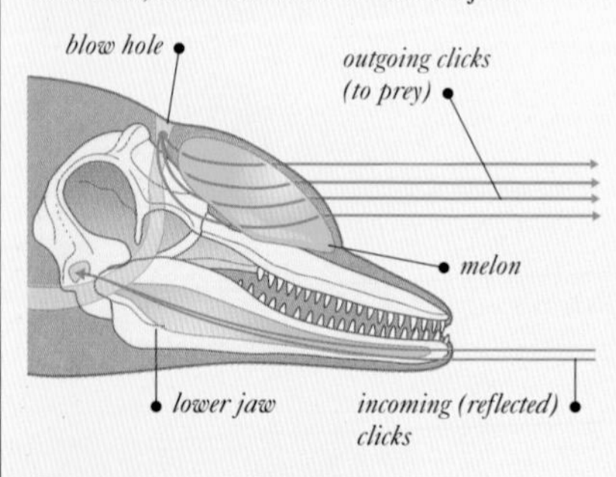

Reflected Sound

Sounds from a dolphin's blowhole are focused by the melon on its forehead. Returning echoes pass through the lower jaw to the ear.

COMMUNICATION

Dawn in a tropical forest is filled with whoops, screeches, howls, and other calls, mainly from monkeys and other primates. These sounds are produced as a form of intra-specific (between members of the same species) communication and have several purposes: they may reinforce bonds between a male–female pair; they allow each member of a troop to know where the other members are; they tell members of a group which individual is dominant, or warn them of danger; they proclaim that a particular patch of forest is occupied as a territory and warn rivals of the same species to keep out. Such communications are found throughout the mammal group, and involve the senses of sight and smell as well as vision.

△ LOUD HOWLS
Howler monkeys of South America produce some of the loudest sounds in the entire animal kingdom. The male howler's dawn call, proclaiming the group's presence, can be heard more than 2 km (1¼ miles) away.

△ SCENT-MARKING
Many mammals leave droppings or spray urine around their home range as a sign to others to keep out. Some rub scents from glands on to rocks or other surfaces, for the same reason.

△ COURTING CALLS
The male humpback whale produces a series of varied moans, screams, clicks, and wails, partly as a courtship "song" to attract females. Each male's unique song lasts up to 30 minutes.

◁ MUTUAL GROOMING
Mammals groom their fur to get rid of dirt and tangles. However, they also groom others of their kind to strengthen bonds within a social group.

LOCOMOTION

A TYPICAL MAMMAL moves around by using its four limbs. This is known as quadrupedal locomotion. However, there are many variations, as well as exceptions, to this basic action. Some mammals are bipedal, such as the kangaroo and wallaby. Bats fly, moles tunnel, gibbons swing with their front limbs, gerbils hop, and seals swim using their rear limbs. Whales and dolphins do not use their limbs at all – their tail flukes contain no limb bones.

LIMB LENGTHS

The length of a mammal's limbs compared to the size of its body gives an indication of its speed. Long legs, as in horses and deer, generally mean faster locomotion. Insectivores such as moles and shrews shuffle slowly since their prey (worms and slugs) does not move rapidly. The short limbs and long bodies of mustelids, such as weasels, slow down surface speed but allow them to enter prey burrows.

▷ PLANTIGRADE

In this gait, the mammal walks on its heel (calcaneum), foot (metapodial), and toe (digit) bones. Bears, badgers, and humans have this type of posture.

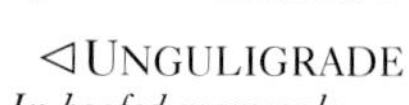

◁ UNGULIGRADE

In hoofed mammals (ungulates), only the tip of each digit touches the ground, and it is capped by the hard hoof (ungus), as in this Elk. The number of digits varies: Artiodactyls have two; Perissodactyls have one.

▷ DIGITIGRADE

In this gait the weight is borne by four or five digits (no hoof). The central part of the foot (metapodial) does not touch the ground. In some cases the foot is very long, making the heel look like a backward-facing knee. Carnivores, such as cats and dogs, walk in this way.

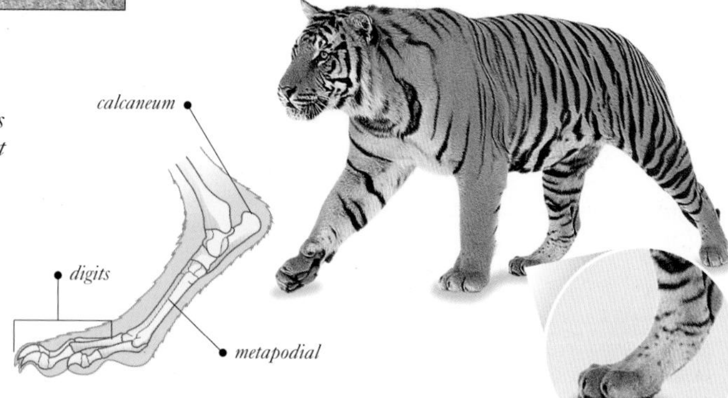

HOPPING AND JUMPING

Two-footed hops and leaps are used by several types of marsupials, such as kangaroos and wallabies, and also rodents such as jerboas and springhares. The foot portion of the rear limb is relatively huge and provides a cushion for the animal to "bounce" along smoothly.

▽ ENERGY EFFICIENT
As a mammal using bipedal hops moves faster and faster, the gait becomes more energy-efficient than running. Energy is stored in the large tendons of the hindlimbs, and the tail swings up and down for momentum.

THROUGH THE AIR

Mammals from various groups, such as squirrels and possums, can glide. But only bats are capable of truly sustained, controlled flight. However, they comprise more than one-fifth of all mammal species, making flight among mammals a relatively common feature. Some species of free-tailed bats can fly at more than 55 kph (35 mph).

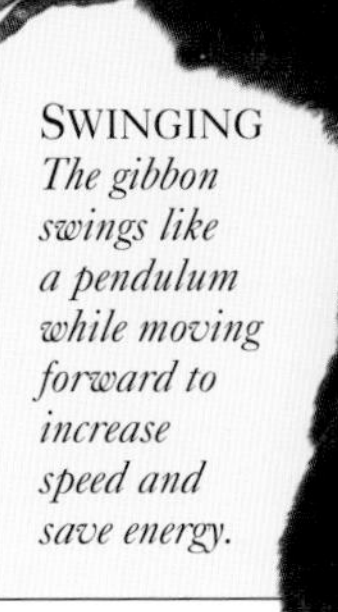

▷ GLIDER
The Flying Lemur has the largest area of gliding membrane, compared to body size, of any mammal.

◁ FLIER
To save weight, a bat has a tiny rear body and hind limbs, compared to its head and chest.

BRACHIATION

The group of apes called gibbons moves by swinging hand-over-hand, suspended from branches. This specialized method of locomotion is known as brachiation. The fingers of the hands have evolved into hooks, and the thumb is not able to grasp as in other apes or humans.

SWINGING
The gibbon swings like a pendulum while moving forward to increase speed and save energy.

IN WATER

Aquatic mammals have evolved fins and flippers to move through water (see p. 42). The whale flexes its backbone up and down, by alternately contracting muscles along it, to move the tail flukes.

◁ WHALE TAIL
The tapering flukes with rear-curved tips are designed to allow water to roll off efficiently, without swirls that would act as brakes.

FEEDING

MAMMALS, BEING warm-blooded, need far more food to burn for body heat than cold-blooded creatures of equivalent size. Almost anything that is organic, from flesh, to eggs, fungus, vegetation, fruits, nuts, bark, sap, honey, droppings, or blood, is food for a mammal of some kind.

MEAT EATERS

The major group of mammals known as the order Carnivora includes mostly animals that eat meat and little else. Families in it include the felids (cats), canids (dogs, foxes, and wolves), mustelids (stoats and otters) and viverrids (mongooses, linsangs, and genets). Exceptions are certain canids and ursids (bears) such as the Spectacled Bear and Giant Panda, which eat very little meat. All seals, sea lions, whales, and dolphins devour flesh, varying from krill to fish, as well as each other. On a smaller scale, the order Insectivora also contains prey-eating species, such as shrews and moles.

▷ CATCHING FISH
The mustelid family has a number of species that are chiefly fish-eaters. These include the aquatic otters and semi-aquatic minks, which have sharp teeth to grasp and tear up their slippery prey.

◁ FAST HUNTER
The Cheetah is a typical cat, hunting fast-moving prey such as hares and gazelles, which it knocks down by the sheer speed of its charge.

△ SCAVENGER
Jackals will eat any scraps left behind by other predators. However, they are also efficient hunters in their own right.

LIQUID DIET

A Vampire Bat consumes only blood, obtaining all the essential nutrients from it. It makes an incision with its teeth and laps the blood that oozes out, rather than sucking it out.

▷ BIG DRINKER
The Vampire Bat licks up half its own body weight in blood in 10 minutes.

△ BIG EATER
A Pygmy Shrew loses body heat fast as it has a large surface area for its size. Its energy needs are so great that it eats its own weight in food daily.

△ TINY PREY
The Aardvark is a specialized consumer of ants and termites. It licks up thousands of these prey daily and chews them with its peg-like cheek teeth.

FILTER FEEDERS

The biggest mammals – the baleen whales – eat some of the smallest prey. The size discrepancy between consumer and consumed, and the passive nature of filter-feeding (the whale simply gulps in vast mouthfuls of water and combs out its food), makes it difficult to think of these mammals as predators or carnivores. A hungry Blue Whale is known to gulp down more than 4 tonnes (4 tons) of its exclusive diet – shrimp-like krill – in the span of just one day.

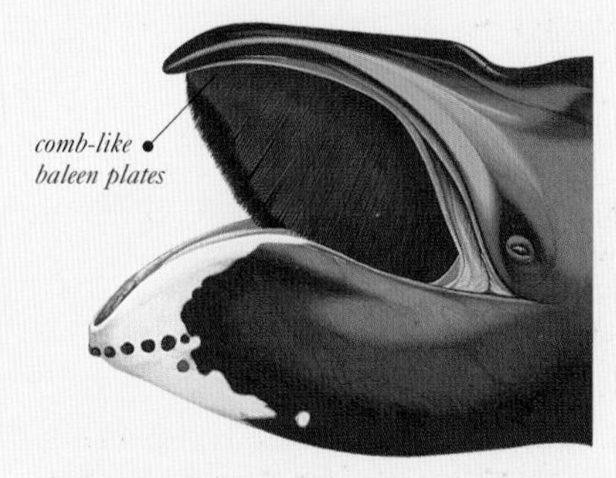

△ GIANT SIEVE
Baleen is a cartilage-like substance that forms bristly filter plates, which hang from the whale's upper jaw. The Bowhead Whale has the longest baleen plates at 4 m (13 ft).

OMNIVORES

Many mammals are omnivores, taking a huge range of foods, from vegetation, nuts and berries to eggs, flesh, insects, crops, and carrion. The largest mammalian omnivores are bears. In this family, only the Polar Bear eats almost nothing but meat; other species are much more adaptable. The American Black Bear's diet is almost nine-tenths vegetarian. However, like all other bears, it takes advantage of seasonal gluts, such as spawning salmon in early autumn. Raccoons are another group of opportunistic feeders, which eat almost anything that comes their way. Certain mammals that are chiefly carnivorous, including wolves, dogs, and foxes, and a few cats, such as the Margay, can also turn to omnivory in times of need.

△ VEGETARIAN BEAR
Despite its bulk, sometimes up to 300 kg (660 lb), the American Black Bear climbs trees and bushes to pluck fruits and berries with its flexible lips.

HERBIVORES

Plant food [illegible] has far less nutrients [illegible] than meat. It is also more [illegible] to break down in the digestive system. Herbivores spend much longer each day, eating huge quantities of food, compared to carnivores. One benefit over meat-eating mammals is that plants are incapable of movement and do not need to be hunted down. However, some plants have forms of self-defence, such as thorns and poison. Most herbivores have well-developed premolars and molars to thoroughly grind their meal.

△ BIGGEST EATERS

The largest land mammals, elephants, take in up to 150 kg (330 lb) of food daily. But due to poor digestion, almost half passes through virtually unchanged.

◁ GRAZERS AND BROWSERS

Grazers feed on grasses and low plants. They include gazelles and zebras. Browsers, such as deer, eat a wider range of vegetation from bushes and trees, as well.

△ NUTS AND SEEDS

Autumn, in temperate regions, sees a surplus of nuts and berries. Rodents, such as chipmunks, hide food or bury it, to eat later.

DIGESTING VEGETATION

Hoofed herbivores are of two types. The hindgut fermenters (wild horses and zebras) chew their food and swallow it into the stomach. It then passes into the caecum, to be digested by a process known as fermentation. The foregut fermenters (or ruminants) comprise most other artiodactyls. The food they swallow is fermented in the first stomach chamber or rumen. It is then regurgitated for further chewing, and swallowed again into the rest of the digestive tract.

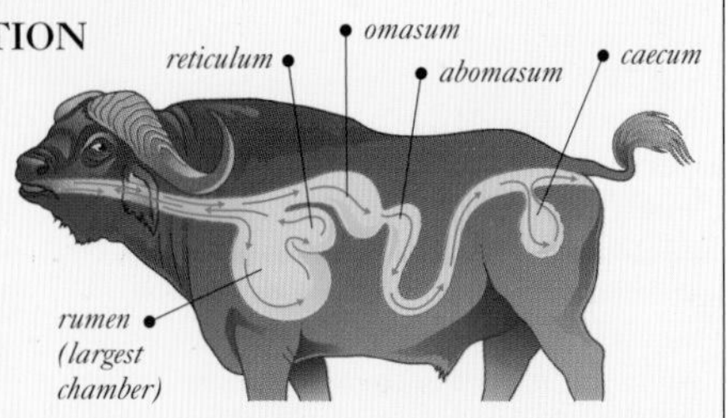

△ RUMINANTS

Buffaloes chew food twice, once as it is gathered, and again after it has part-fermented in the rumen. This is known as chewing the cud.

DIET SYMBOLS

In this book, the wide variety of foods consumed by the different mammals is represented by symbols, which are explained as follows. Wherever relevant, or of particular interest, more details on the diet are given in the text for each species entry.

LARGE MAMMALS
Usually herbivorous ungulates like deer, antelopes, zebras, cattle, sheep, and goats.

AQUATIC MAMMALS
Seals, sea lions, Walrus, whales, dolphins, porpoises, and sea cows.

SMALL MAMMALS
Up to about 30 cm (12 in), mostly herbivorous: mice, rats, voles, rabbits, and squirrels.

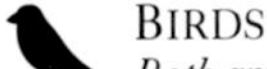

BIRDS
Both ground-dwelling birds (such as quail), as well as water birds (such as ducks); also chicks in nests.

REPTILES
Lizards and snakes, sometimes tortoises, turtles, crocodiles, or alligators.

AMPHIBIANS
Frogs, toads, salamanders, and newts, although many are poisonous.

CEPHALOPODS
Larger molluscs with long tentacles: squid, octopus, and cuttlefish.

FISH
Solitary or shoaling, fast or slow, surface or bottom-dwelling fish.

KRILL
Shrimp-like marine crustaceans, which form swarms of billions.

MOLLUSCS
Snails and slugs, sea-snails, and shellfish: clams, oysters, mussels, and abalones.

WORMS
Soft-bodied dwellers on land such as insect grubs or aquatic crustacean larvae.

ARTHROPODS
Joint-legged animals: insects, centipedes, spiders, crabs, and crayfish.

OTHER INVERTEBRATES
Starfish, sea-cucumbers, and jellyfish.

EGGS
Mainly eggs of birds, often in nests, or of reptiles, which may be buried.

HONEY
Specialized product of bees; also similar substances from other insects.

VEGETATION
Above-ground plant material: leaves; also stems, flowers, and soft twigs.

GRASS
Leafy blades, but also nutritious creeping stems and roots.

FRUITS
Usually soft, fleshy fruits and berries, rather than hard fruits (see Nuts).

NUTS
Very hard-cased fruits or seeds that are split open for the softer kernels within.

SEEDS
Seed-matter of ripe flowers that are not too fleshy or too hard-cased.

GRAIN
Starchy seeds of wild and domesticated grass (such as wheat or rice).

ROOTS
Includes other underground plants like bulbs, corms, and tubers.

FUNGI
Mushrooms, toadstools, bracket fungi, also yeasts and lichens.

BARK
The dead outer layer of a tree (often nibbled through to reach the sap beneath).

AQUATIC PLANTS
Freshwater and marine; aquatic herbivores such as Manatees consume both.

DESERT MAMMALS

DESERTS MAY BE hot or cold, windy or calm, rocky or sandy, high-altitude or low, but the one common factor is that they are always dry. Compared to other major animal groups, such as insects and reptiles, mammals need plenty of water. Mammals that have become adapted to desert life have undergone major changes in the way their bodies take in, handle, and release water.

GETTING WATER

Carnivores living in deserts gain enough moisture from freshly-killed prey, in the form of blood and body fluids, to satisfy their liquid intake. This transfers the challenge of obtaining water to obtaining food. Herbivores get water from plants, especially succulents. Smaller plant-eating mammals store seeds in burrows to retain their moisture, and lick dew from pebbles and stones. Small mammals may also block up the entrances to their tunnels, to retain moist, cool air within.

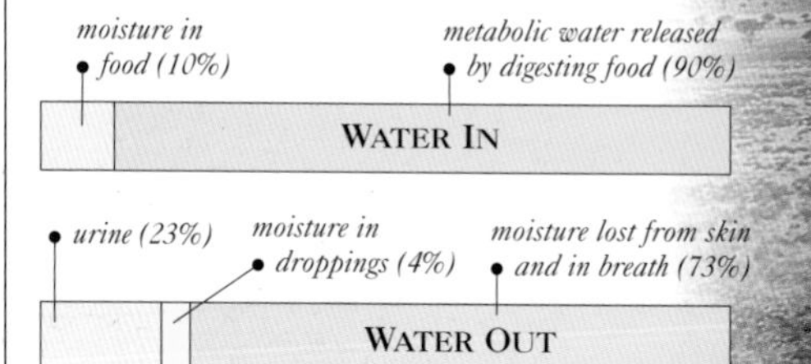

△ WATER BALANCE
The above diagram illustrates how a kangaroo rat survives arid desert conditions. It is entirely dependent on its food, for water. The water taken in has to balance that which is lost in order to prevent the animal from becoming dehydrated.

▽ EFFICIENT ORYX
Oryx have very efficient kidneys that excrete little urine, and they also produce very dry droppings.

◁ FOOD AND WATER STORE
A camel's fatty hump is both stored food as well as on-board moisture, as the fat is broken down to yield energy, the breakdown producing metabolic water. A thirsty camel has been known to drink more than 50 litres (13 gallons) of water in a few minutes.

Finding Food

Large desert herbivores can detect rain using their acute sense of smell and so move towards areas of future plant growth. Even smaller herbivores can travel long distances to temporary food surplus; the Desert Jerboa can cover 10 km (6 miles) in a single night. Some smaller herbivores even collect and store this bounty. The kangaroo rat may cache up to 5 kg (11lb) of spare seeds in its burrow.

▷ On Watch

In open habitats, sight is the main sense for detecting danger. In a colony of meerkats, a few stand upright to scan over the ground for predators. They also peer up to spot birds of prey.

◁ Storing Food

The Fat-tailed Dunnart is a voracious hunter of insects and worms, across Australia's desert and scrubby outback. It stores surplus food as body fat in the base of its tail.

Coping with the Heat

Large desert mammals such as Oryx, Dama Gazelles, Gerenuk, and camels, are active in the cooler dawn and dusk. By day they rest in any shade they find, near bushes or hillocks. They often move to new feeding grounds at night, when it is less hot and they are less visible to predators. Oryx may trek 30 km (20 miles) through the darkness, chewing the cud as they walk. They also allow their body temperature to rise by 5°C (9°F) above normal, before they start to lose water in cooling sweat.

▷ Losing Heat

In warm-blooded mammals, large body surfaces act as radiators to lose heat. The Fennec Fox of the Sahara Desert has huge ears that act as body coolers and also detect the faint scrabbles of small prey on the sand.

◁ Night Foraging

Small desert mammals like the Four-toed Jerboa usually hide away in their tunnels or burrows during daytime heat. They emerge at dusk to forage, using their large eyes, ears, and whiskers both to locate food and to stay alert for danger.

GRASSLAND MAMMALS

GRASSLANDS COVER about one fourth of the world's land surface and support a great diversity of animal communities. Although essentially open habitats dominated by grasses, savanna grassland in tropical and subtropical areas has scattered trees and thickets, often merging into open woodland. Grasslands have long dry periods (with the risk of wildfires) interrupted by occasional rains.

HERDS

On the open plains, there are few places for large mammals to hide. One method of self-protection used by grassland herbivores is to gather in herds for "safety in numbers". While some herd members eat or rest, others are alert with several eyes, ears, and noses ready to sense danger. If one member detects a predator, it can warn the others by calls and actions. Some herd members cooperate to drive off an enemy. The largest herds of mammals are found on the African savanna where zebras, antelopes, and gazelles gather in vast aggregations.

▽ MIGRATION
In the savanna, rainfall is patchy and some regions have fresh plant growth while others are dry. Large herbivores, such as wildebeest, migrate regularly in search of new pasture or water.

△ MIXED HERDS
Different mammal species form mixed herds to pool their resources. Giraffes can see far across the landscape, while zebras have an excellent sense of smell.

◁ HUNTING PACKS
Many grassland predators, such as hyenas, jackals, and lions, hunt in groups or packs. By doing so they can kill large prey such as a wildebeest, which provides food for several days.

SPEED AND MOVEMENT

With relatively little cover available in grassland habitats, speed has become an important feature of many large mammals found here. Evolution favours predators that are fast enough to run down victims, such as the Cheetah in Africa; and prey that are speedy enough to escape hunters, such as the Pronghorn in North America – these are the two fastest runners in the animal kingdom. The wild horse evolved both speed and stamina for travelling long distances over grasslands. The fastest wild equid is the Onager, or Asiatic Wild Ass, which can reach speeds of 70kph (45mph).

LIMB DESIGN
The Maned Wolf has stilt-like legs, yet it is not a fast sprinter. Its legs help it move among the tall grass in which it lives, and also raise its head so that it can look out for prey or danger.

△ SELECTION IN ACTION
A lioness sprints after her prey. To improve their chances of success, these top grassland predators single out the young and weak.

BURROWING

One way of creating hiding places in grassland is by digging. Many small mammals, especially rodents such as prairie dogs, voles, rats and mole-rats, find shelter from predators, extreme climatic conditions, and fire in this way. Often, the burrows have mounds around the entrances, to prevent flooding. The largest burrows, big enough to accommodate a human, are made by the Aardvark of Africa, a specialized ant- and termite-eater. In the Australian grasslands, wombats make shelters almost the same size as those of the Aardvark.

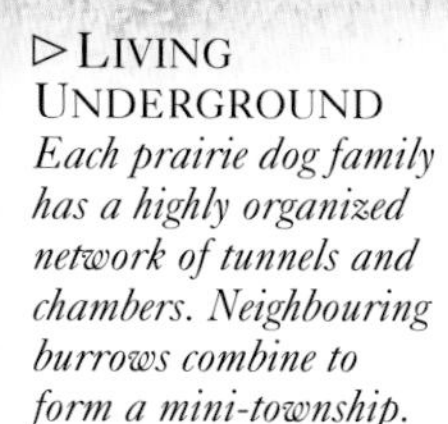

▷ LIVING UNDERGROUND
Each prairie dog family has a highly organized network of tunnels and chambers. Neighbouring burrows combine to form a mini-township.

FOREST MAMMALS

TREES ARE HOSTS TO different species of mammals. They provide food, shelter, safety from ground predators, and nesting sites. Mammals, such as the woolly and spider monkeys, are wholly arboreal ("tree-living") and hardly ever touch the ground. Others, such as martens and the ocelots, are semi-arboreal, and equally at home on the ground or in trees.

MOVEMENT IN TREES

Many arboreal mammals move through the branches by jumps and leaps. Most acrobatic of all, are the Old World gibbons, with their hook-like hands, and the New World Monkeys, which grip with their hands and feet as well as the muscular prehensile tail. Squirrels have sharp claws to dig into bark, muscular paws, a long tail for balance, and large eyes to judge distances before leaping. Tree porcupines, raccoons, and anteaters tend to shuffle along slowly and securely.

▷ LEAPING
Primates tend to keep their bodies upright while jumping. Lemurs use their long, powerful rear legs for propulsion, the long tail for balance, and all four limbs to cushion the landing.

△ GLIDING
Mammals such as possums, gliders, flying lemurs, and flying squirrels parachute through the forest on flaps of skin, which stretch out between their limbs. They are actually gliders rather than true fliers.

LIVING ON THE GROUND

Forests provide abundant food in the form of leaves, blossoms, seeds, fruits, and nuts. Ground-based mammals, from mice and voles to wild pigs, tapirs, deer, and Okapi, take advantage of this rich supply. The decaying carpet of leaves and twigs also encourages a host of worms, slugs, insects, and other small creatures, which become prey for small mammal hunters such as elephant-shrews and solenodons.

◁ CAMOUFLAGE
Most forest mammals are shades of brown to blend in with their surroundings. Young deer and wild boar have dappled coats to merge with the undergrowth. Arboreal cats have spots or patches to blend in with the shadows among the branches.

△ NEST BUILDING
Small mammals use holes in tree trunks to build homes, like the squirrel's nest (drey).

LIFE IN TREES

The tree canopy in a tropical forest holds most of the forest's food, and here leaves, fruits, and flowers grow all year. Consequently, animals thrive that are well adapted to move easily through the branches of trees. In temperate forests, food resources are seasonal and the canopy is generally more open, allowing greater plant growth on the ground. Here, it is often more profitable to venture down from the trees to exploit the available resources.

△ SLOWEST MAMMAL
Among the slowest-moving mammals are sloths. They eat mainly leaves, hanging upside down from branches with their long, sharp, hooked claws. Sloths only come down from trees to defecate or to move to another tree as food runs out. On the ground, sloths are often at the mercy of predators since they are clumsy walkers.

△ FIVE LIMBS
Spider monkeys are named after their long, slim limbs. Their "fifth limb" is the prehensile tail, which can support the whole weight of the body.

▷ ROOSTING
Bats, the true forest fliers, roost during the day on branches or in tree holes. At night, they fly out in search of food.

POLAR AND MOUNTAIN MAMMALS

THE LAND AND SEAS near the Poles, and the tops of tall mountains are some of the harshest habitats. They are cold, with bitter winds and a covering of snow and ice. Food is scarce and hard to access. However, several mammals are adapted to survive in these conditions. They have little competition, except for birds, as temperatures are too low for reptiles and amphibians to survive.

KEEPING WARM

A high priority for mammals in cold climates is to retain body heat. This is done by thick fur, usually an outer coat of longer guard hairs that give protection and shrug off rain and snow, and a dense undercoat to keep in heat. Under the skin is a thick layer of fat that also acts as insulation. Parts projecting from the body such as the nose and ears are small, since they lose heat fastest and are at risk from frostbite. In the worst weather, most mammals simply rest in any sheltered place, curled up to minimize loss of warmth.

△ SHAGGY COATS
The Muskox's outer coat has very long hairs reaching one metre in length around the main body. The Yak of Central Asia has a similar thick overcoat of fur.

COVERED WITH FUR △
The Polar Bear is completely covered with fur, apart from its eyes and nose-tip. Even the soles of its paws are furred, both to keep in heat and to obtain a grip on slippery ice.

◁ UNDER THE FROZEN OCEAN
Seals have an excellent memory for the location of breathing holes in the ice. They even re-cut a hole with their teeth when the ice is several centimetres thick. Seals and sea lions spend hours hauled out of the water, grooming their fur coats with their teeth and flipper claws. This is vital for animals in cold places, since their fur is truly a survival blanket.

CHANGING COATS

Like many polar animals, the Arctic Fox moults its fur twice each year and grows two distinctive coats. Each provides camouflage as the surroundings change with the seasons. Stoats in the far north grow white coats in winter, when they are called Ermines, and moult these for brown fur in the summer. Further south, where snow and ice covers the ground much more briefly, the same species is brownish all year.

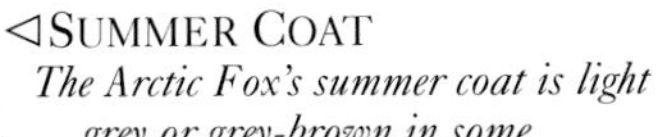

◁ SUMMER COAT
The Arctic Fox's summer coat is light grey or grey-brown in some individuals, and browner and darker in others. This blends in with rocks, earth, and shrubby plants.

▷ WINTER COAT
The winter coat is white and twice as thick as the summer coat. It gives camouflage for stalking in the snow.

ON THE HIGH SLOPES

Mammals living in the mountains have strong limbs and feet, which give good grip on slippery wet rocks or icy slopes. Like many ungulates of the peaks, the Mountain Goat has very large, wide hooves for a foothold on every kind of surface. In common with Chamois, Ibex, and other mountain-dwellers, these goats ascend to the high pastures in spring, to feed on summer vegetation. In autumn they return to the lower slopes and the shelter of bushes and trees.

△ SURE FOOTEDNESS
Each hoof of the wild goat has a hard, sharp outer rim and a soft inner pad, providing excellent grip. Few predators can catch these herbivores as they escape to almost inaccessible vertical rocky cliffs and shelter on tiny, narrow ledges.

MANAGING ALTITUDES

A particular problem facing mountain mammals is the thin air, which contains reduced amounts of oxygen. Many species have adapted to this problem by evolving blood with extra numbers of red cells.

▷ HIGH UP
Vicuñas are found in the Andean tundra at heights of almost 5,000 m (16,500 ft).

AQUATIC MAMMALS

MAMMALS FIRST EVOLVED on land as four-legged animals that walked and ran. Aquatic mammals changed greatly from the original mammalian design to meet the different demands of life in water, and they became the most modified of all mammal species. Some of these modifications in shape and movement are clearly visible, while others are less obvious.

MOVEMENT

Moving through water requires much more energy, as it is a "thicker" medium than air. The Cheetah can reach a speed of about 100 kph (62 mph), but the Killer Whale, weighing about 400 times as much, reaches only half this speed. Aquatic mammals have broad, flat surfaces to push through water. Dolphins, whales, Dugongs, and Manatees propel themselves with their tail flukes and use their flippers for steering. Semi-aquatic species such as otters use their highly webbed feet.

◁PROPULSION
The Killer Whale can gain enough speed to leap clear of the water, despite its massive bulk of up to 5 tonnes (5 tons). This leap, followed by a huge splash as it crashes back in, is known as breaching. *(See also p. 29).*

△BODY SHAPE
Water mammals such as the dolphin have smooth, streamlined bodies to slip through water easily. The dorsal fin stops the body from spiralling and the flippers are used for steering and manoeuvring.

AGILITY△
Agile swimmers such as the Common Seal can catch fast-moving aquatic prey. Sea lions swim with front flippers, while seals use hind flippers.

TEMPERATURE CONTROL

Many aquatic mammals live in the Poles, where seas are cold year round and may freeze in winter. Here, maintaining body temperature is a vital function. This is helped by having smaller extremities, since these lose heat fastest. Mammals such as seals and sea lions have dense coats and hairs coated with water-repelling oils. Size is also a factor, since heat is lost through a mammal's body surface and bigger mammals lose heat more slowly. This is why most truly aquatic mammals are large, with the great whales largest of all.

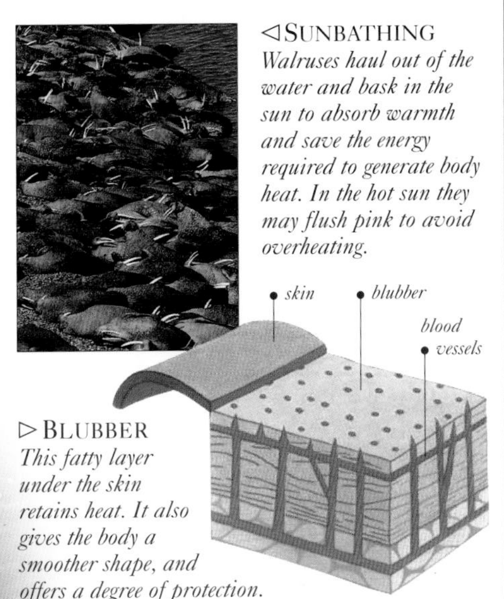

◁ SUNBATHING
Walruses haul out of the water and bask in the sun to absorb warmth and save the energy required to generate body heat. In the hot sun they may flush pink to avoid overheating.

▷ BLUBBER
This fatty layer under the skin retains heat. It also gives the body a smoother shape, and offers a degree of protection.

△ CHAMPION DIVER
The Sperm Whale probably makes the deepest and longest dives of any mammal, staying underwater for almost two hours and reaching depths of 300 m (1,000 ft). The rare beaked whales may make dives as lengthy and deep.

MARINE LIFESTYLE

The Sea Otter is a truly marine mammal. It has the thickest hair of any mammal, to protect it in water, and rarely comes ashore. It feeds while floating on its back and even sleeps afloat, secured in beds of seaweed.

▷ MARINE MEALS
The Sea Otter places a shellfish on its belly and smashes it open with a stone, to eat.

SURVIVING UNDERWATER

Aquatic mammals must come to the surface regularly to obtain fresh supplies of oxygen, since all mammals do so by breathing in air. Pinnipeds (seals and sea lions) and cetaceans (whales, porpoises, and dolphins) are able to retain oxygen in their muscles. The oxygen is stored in a special protein, myoglobin, and can be released gradually during a dive. Some seals have ten times more myoglobin in their muscles than land mammals.

WATCHING MAMMALS

MAMMALS ARE EASY to watch since they are all around us. Depending on where we live, we can see pet cats and dogs in homes, horses in fields, sheep and cows on farms, and squirrels and monkeys in woodland. Studying these mammals can reveal many details of behaviour. However, watching truly wild mammals in their natural habitats needs more planning and preparation.

ACCESSIBLE MAMMALS

Wild mammals are very alert for potential danger. Most people tend to move about noisily, so wild mammals usually hide before humans arrive on the scene. Many mammals are nocturnal, and are able to see people clearly in light levels where humans can hardly see at all. Successful mammal-watching needs planning, preparation, and patience. Plan by finding out about the habits and behaviours of the target species (throughout the day and over the different seasons) from detailed field guides and similar sources. Prepare by surveying sites very observantly, for clues such as burrows, nests, footprints in mud or sand, tufts of fur on thorns, droppings, and signs of eating, such as empty nut shells, gnawed carcasses, claw scratches on tree trunks, trampled grass, or broken branches.

△ CLOSE TO HOME
A host of wild mammals, many active by night, live in gardens and parks. Suitable food may lure them out for study. An infra red lamp or torch helps human vision without frightening the animals.

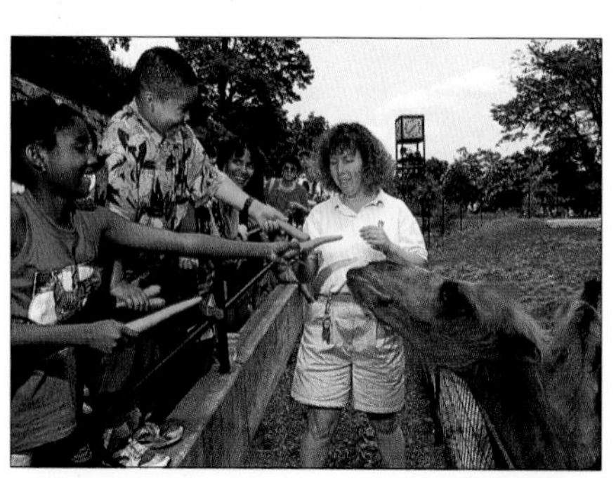

◁ PARKS AND ZOOS
Wildlife parks, zoos, and nature reserves are places to observe mammal species. However, the behaviour of captive animals differs from that of wild ones.

WATCHING GUIDELINES

Much of the advice for watching mammals safely is based on common sense. Avoid going out alone in case of accidents or other mishaps. Do not move about in a place during darkness, unless familiarized with it by daylight. Wear suitable warm, waterproof clothing; weather can change rapidly and nights can get very cold. Avoid steep slopes, slippery rocks, and banks near water. Carry equipment in a backpack, to leave both hands free for balance when clambering about. Never take chances where large and potentially dangerous species may be found.

In the Wild

Getting close to exciting, large mammals in the wild is a complex and potentially hazardous process. There is also a chance of experiencing frustratingly "empty" days. This type of mammal-watching is best undertaken as part of an organized group – there is a large choice of outings from luxury tours to survival-type treks. Guides and rangers know where different species are likely to be at a certain time of the day or season. They can give advice on routes and on using hides, where one can watch wildlife concealed, allowing the animals to behave more naturally.

△ Underwater Action
Snorkelling and scuba-diving reveal an incredible undersea world where dolphins, seals, otters, and other aquatic mammals move with grace and agility.

△ On Safari
Mammals become used to vehicles that act as mobile bases for humans. The viewers are kept safe and the animals relatively undisturbed.

Equipment

Modern, lightweight electronic equipment makes recording mammals – by photograph, video, film, or audio tape – much easier. However, traditional items such as a notebook and pencil, field guide, and binoculars are still essential.

◁ Camera
Telephoto (long) lenses get you close to the subject. Fast-exposure film catches rapid movement or copes with low light.

◁ Binoculars
Practise using the binoculars with one hand only. Keep them ready around your neck to view the subject at once.

◁ Notebook
Keep a notebook to make lists of species observed. A sketch is also better than trying to recall details later.

◁ Camcorder
The video camera should be ready for instant use. Pre-focus on a likely spot and check light settings.

Threatened Mammals

Mammals face many hazards in the modern world. The IUCN Red List (see below) names over 1,000 mammal species, almost one in four, as facing a threat, with 180 species being "critically endangered". The dangers vary for different mammals, but they are nearly all due to human activity.

On the Brink

Many of the critically endangered mammals are relatively large species such as big cats, rhinoceroses, cetaceans, and bovids (cattle and relatives). Information regarding population numbers and breeding rates is easier to collect for these species, and the degree of threat can be assessed more accurately, compared to that of small species such as rodents. Total numbers of a species are important, but so are the populations of its subspecies, and especially how they are distributed. Small, isolated groups are at risk from epidemics, or from genetic problems due to inbreeding, as the total "gene pool" for the species shrinks.

△ Critically Endangered
With a population of probably less than 100, the Javan Rhinoceros is on the verge of extinction, due chiefly to clearance of its lowland forest habitat.

△ Endangered
Tigers number several thousands, as a result of conservation. Their situation, however, is still very serious.

△ Specific Threats
The Amazon River Dolphin faces specific threats such as water pollution by heavy-metal chemicals, and disruption of its echolocation.

The Red List

The Red Lists of Threatened Species are published by the International Union for the Conservation of Nature and Natural Resources (IUCN). Drawn up every few years with the help of data gathered by more than 10,000 scientists from all over the world, the Lists are a global directory to the status of animals, plants, and other living things on our planet. The 2000 List places mammal species in one of eight categories such as lower risk, vulnerable, critically endangered, and so on, with "threatened" used as a general term underlying the entire list.

MAJOR THREAT

The largest overall threat to most mammals is habitat destruction. Expanding human population takes over land to raise crops and livestock, build houses and roads, and to extract natural resources. Forests are felled for the short-term gain of timber and farmed for a few years. They suffer soil erosion, and the displaced species, especially arboreal types such as monkeys and tree kangaroos, never return.

▷MAN-MADE FIRES
The world's richest habitats are being burnt down in tropical forests of the Amazon region, W. Africa, and S.E. Asia.

POLLUTION

Mammals and their habitats are polluted in a variety of ways – including chemical pollution of rivers and oil spills – and also suffer from the effects of climate change. Aquatic species eat polluted prey and suffer from low-grade poisoning known as bio-amplification.

◁OIL SPILL
A mammal's coat is its survival blanket. Polluting oils ruin the fur's protective properties. The mammal swallows the oil while grooming itself and suffers further harm.

HUNTING

Since the 1980s, an agreement between a majority of nations has banned the slaughter of many big whale species. However, a fast-growing threat is the hunting of animals for the trade in bushmeat. Protected species such as monkeys, gorillas, and dolphins are hunted for their flesh, which is often sold openly at market.

△POACHING
Hunting for skins and body parts is now banned for many species. But people are still tempted into poaching to make a living.

◁WHALING
Species not included in the ban on whaling have now become threatened. Cetaceans breed slowly and may take decades to recover.

CONSERVATION

MANY MAMMALS suffer from general threats to wildlife, such as habitat destruction and pollution. Helping these species is part of the overall process of wildlife conservation. Some mammals face more specific problems and need more specialized conservation measures. (See also Introduced Species, p. 51).

HABITAT PROTECTION

Since habitat destruction is the biggest overall threat faced by wild animals (see p. 47), habitat conservation is the most effective way of protecting them. This is the rationale behind establishing national parks and nature reserves. Such preserved habitats provide every resource that a species needs to thrive. However, these areas need to be large enough for mammals such as big cats and ungulate herds to find enough food, establish territories, and reproduce without the problems of inbreeding.

△ REFORESTATION
If soil erosion has not been too severe, trees may be planted in an area which was once forest, as here in a reforestation project in Guatemala.

◁ PROTECTED PARKS
In a protected area, rangers patrol the borders for poachers, while the park animals are prevented from raiding surrounding crops or livestock.

STAR SPECIES

Some animals such as Giant Pandas, dolphins, and tamarins, naturally arouse attention and sympathy as they look cuddly and appealing. Others such as bears and big cats command respect and awe due to their power. The "star status" of these mammals is often used by wildlife organizations to grab headlines and photo-opportunities. However, all threatened mammal species, including those that evoke shudders of horror in the lay person, such as obscure rats and bats, deserve to live securely in their natural habitats.

▷ SYMBOL OF CONSERVATION
The Giant Panda is the globally recognized symbol of the World Wide Fund for Nature (WWF) – a leading organization for nature conservation.

Captive Breeding

Some mammals face such a severe plight that captive breeding is the last resort before extinction. Zoos and parks involved in breeding programmes swap individuals to maintain genetic diversity for a healthy population. They also prepare captive-bred individuals for life in the wild. Arabian Oryx and Golden Lion Tamarins are among many species that have benefitted.

▷ Golden Lion Tamarin
The critically endangered Golden Lion Tamarin has been a target of conservation efforts since the 1960s. It has bred well in primate centres and has been re-introduced in S. E. Brazil since the mid-80s.

Conservation Monitoring

How does one assess which mammals are in greatest danger, and which are the best ways to save them? An army of scientists, wildlife rangers, and conservation volunteers is constantly gathering data, which is studied to reveal the best solutions. Detailed information on blood samples and gut contents, to broad patterns such as migratory routes viewed by aerial survey, is analyzed.

△ Satellite Tracking
Wide-ranging mammals such as this Beluga can be tracked around the world by a radio device attached to them. Its signals are received by a local receiver or a satellite.

△ Dehorning
Conservationists sometimes cut off the rhinoceros's horn (which is insensitive to pain) to counter the threat of poaching.

▷ Back Home
Some mammals, especially primates such as baby Orangutans, are captured for the pet trade. When rescued, they need to be introduced back to the ways of the forest, helped by trained staff at rehabilitation centres.

MAMMALS AND HUMANS

ATTITUDES TO MAMMALS around the world vary in different cultures, even towards the same species. A horse may be a thoroughbred race champion, a beast of burden, a lifelong friend, or a meaty meal. Many species have been domesticated or selectively bred over generations for human use.

DOMESTICATION

One of the chief features in domestication is reduced aggression and increased tameness or response to training. Perhaps the first domesticated species was the dog, bred from the Grey Wolf more than 10,000 years ago. Mammals utilized for their strength include horses, oxen, buffalo, and elephants. Pigs, sheep, and goats also have a long history of domestication. In recent times deer and antelopes have been farmed. There is also huge interest in "fancy" breeds such as rabbits and guinea pigs, while rats and mice are used for medical research.

△ CASH COWS
At least 180 breeds of cattle are selected to cope with varying climates around the world, and to produce milk and meat. The hides are tanned for leather, and the carcass rendered for other uses.

▽ USEFUL CATS
Cats were probably first kept by the Ancient Egyptians more than 5,000 years ago, to hunt mice. They have been worshipped, as well as burned at the stake as devils.

HORSE DOMESTICATION

The domestic horse (*Equus caballus*) may have originated over 4,000 years ago in Asia. Its ancestor possibly resembled Przewalski's Wild Horse. The mule, a popular beast of burden, is the offspring of a male donkey and female horse.

PRZEWALSKI'S WILD HORSE

MULE

THOROUGHBRED

FERAL MAMMALS

Feral species are those that have been domesticated, and have then "escaped" human control to live partly or wholly in the wild again. They include horses such as North American Mustang ponies and English New Forest ponies, various types of dogs, cats, and pigs, and also camels. Dromedaries (one-humped camels) were taken to Australia in the 1840s by explorers and have since established feral herds there.

◁FERAL DOGS

The Dingo may have been a type of domestic dog that has reverted to living wild again over the past few thousand years in Australia and S.E. Asia. It interbreeds readily with domestic dogs – in some parts of Australia half of all dingoes are interbred dingo-dogs.

INTRODUCED SPECIES

A major threat facing some mammals is posed by other mammal species introduced from other regions by humans. In Australia, a wealth of native marsupial mammals has suffered greatly from introduced species such as rabbits and Brown Rats. The newcomers compete with native species for food, prey on them, or carry diseases against which they have little natural resistance.

▽GREY SQUIRREL

The North American Grey Squirrel was brought to Europe around the 1870s. Larger and more aggressive than the native Red Squirrel, it quickly spread as the latter declined.

REINTRODUCED SPECIES

Captive breeding (see p. 49) can help to save a mammal species, which may then be reintroduced into its native area. However, once a species is lost, its habitat may be taken over by other species. The reintroduced individuals may then face unusually severe competition to re-establish a new population in that area.

◁PÈRE DAVID'S DEER

Père David's Deer, extinct in its native China, has existed only in captivity in English parks from about 1900. In the 1980s it was reintroduced to China.

CLASSIFICATION

THE ANIMAL KINGDOM is divided into groups called phyla. One major phylum is Chordata (animals that have a backbone), and mammals are grouped under this in a class called Mammalia. This class is further divided into 21 orders (listed below), each made up of one or more families, in turn made up of one or more genera (singular genus), which is a group of closely related species.

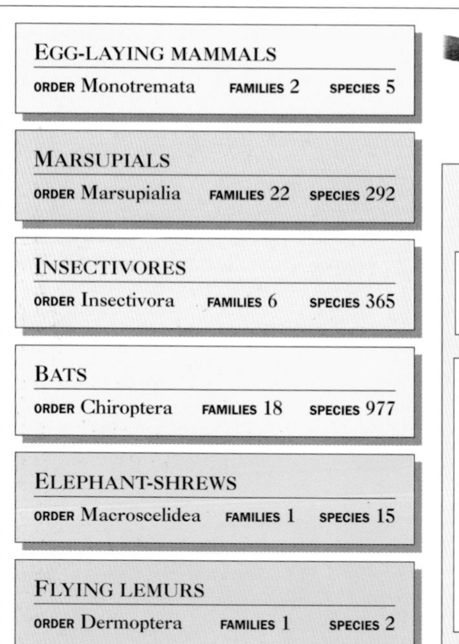

EGG-LAYING MAMMALS
ORDER Monotremata FAMILIES 2 SPECIES 5

MARSUPIALS
ORDER Marsupialia FAMILIES 22 SPECIES 292

INSECTIVORES
ORDER Insectivora FAMILIES 6 SPECIES 365

BATS
ORDER Chiroptera FAMILIES 18 SPECIES 977

ELEPHANT-SHREWS
ORDER Macroscelidea FAMILIES 1 SPECIES 15

FLYING LEMURS
ORDER Dermoptera FAMILIES 1 SPECIES 2

TREE SHREWS
ORDER Scandentia FAMILIES 1 SPECIES 19

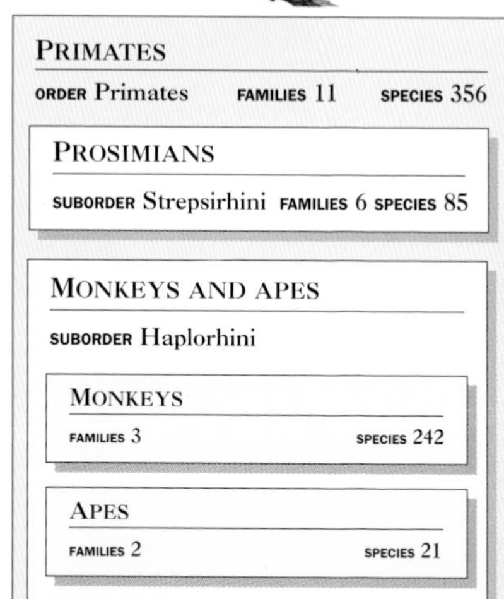

PRIMATES
ORDER Primates FAMILIES 11 SPECIES 356

PROSIMIANS
SUBORDER Strepsirhini FAMILIES 6 SPECIES 85

MONKEYS AND APES
SUBORDER Haplorhini

MONKEYS
FAMILIES 3 SPECIES 242

APES
FAMILIES 2 SPECIES 21

ANTEATERS AND RELATIVES
ORDER Xenarthra FAMILIES 4 SPECIES 29

PANGOLINS
ORDER Pholidota FAMILIES 1 SPECIES 7

RABBITS, HARES, AND PIKAS
ORDER Lagomorpha FAMILIES 2 SPECIES 80

RODENTS
ORDER Rodentia FAMILIES 30 SPECIES 1,702

Cetaceans
ORDER Cetacea FAMILIES 13 SPECIES 83

Baleen whales
SUBORDER Mysticeti FAMILIES 4 SPECIES 12

Toothed whales
SUBORDER Odontoceti FAMILIES 9 SPECIES 71

Carnivores
ORDER Carnivora FAMILIES 7 SPECIES 249

Dogs and relatives
FAMILY Canidae SPECIES 36

Bears
FAMILY Ursidae SPECIES 8

Raccoons and relatives
FAMILY Procyonidae SPECIES 20

Mustelids
FAMILY Mustelidae SPECIES 67

Civets and relatives
FAMILY Viverridae SPECIES 76

Hyenas and aardwolf
FAMILY Hyaenidae SPECIES 4

Cats
FAMILY Felidae SPECIES 38

Seals and sea lions
ORDER Pinnipedia FAMILIES 3 SPECIES 34

Elephants
ORDER Proboscidea FAMILIES 1 SPECIES 3

Hyraxes
ORDER Hyracoidea FAMILIES 1 SPECIES 8

Aardvark
ORDER Tubulidentata FAMILIES 1 SPECIES 1

Dugong and manatees
ORDER Sirenia FAMILIES 2 SPECIES 4

Hoofed mammals

Odd-toed hoofed mammals
ORDER Perissodactyla FAMILIES 3 SPECIES 19

Horses and relatives
FAMILY Equidae SPECIES 10

Rhinoceroses
FAMILY Rhinocerotidae SPECIES 5

Tapirs
FAMILY Tapiridae SPECIES 4

Even-toed hoofed mammals
ORDER Artiodactyla FAMILIES 10 SPECIES 225

Pigs and peccaries
FAMILY Suidae and Tayassuidae SPECIES 17

Hippopotamuses
FAMILY Hippopotamidae SPECIES 2

Camels and relatives
FAMILY Camelidae SPECIES 7

Deer, musk deer, and chevrotains
FAMILY Cervidae, Moschidae, and Tragulidae SPECIES 56

Pronghorn
FAMILY Antilocapridae SPECIES 1

Giraffe and okapi
FAMILY Giraffidae SPECIES 2

Cattle and relatives
FAMILY Bovidae SPECIES 140

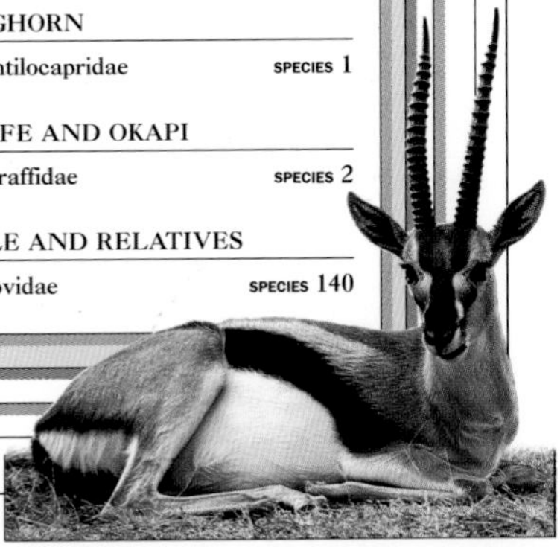

EGG-LAYING MAMMALS

ALSO KNOWN as monotremes, the five species of egg-laying mammals are zoological curiosities of two distinct types. The single species of Duck-billed Platypus dwells in freshwater habitats in eastern Australia. It is a strange mix of duck-like beak, otter-shaped body, webbed feet, and flattened, beaver-like tail. It is also one of the very few poisonous mammals, with a venomous spur in the heel region of juveniles and adult males.

The four species of echidnas (spiny anteaters) are generalist consumers of ants, termites, worms, and grubs. Three long-nosed (long-beaked) species occur on the island of New Guinea. The Short-nosed (Short-beaked) Echidna is found there too, and also across most of Australia, in a variety of habitats.

True to their group name, all five species lay eggs. The female then feeds her hatched young on milk, like all mammals. Adult monotremes lack teeth and grind their food between plates or spines in the mouth.

Family TACHYGLOSSIDAE	Species *Tachyglossus aculeatus*	Status Lower risk

SHORT-NOSED ECHIDNA

AUSTRALIA, TASMANIA, NEW GUINEA

Also known as the Spiny Anteater, the Short-nosed Echidna is covered with long, thick spines, intermixed with short hairs. It has a small head, no external neck, and a protruding snout, with electro-receptors that help it to detect its insect prey. Its long tongue of around 17 cm (6¾ in) is covered with fine barbs for trapping insects. The Short-nosed Echidna is active during the day and night. It becomes torpid in extremely hot or cold weather, its body temperature falling to 4° C (39° F).

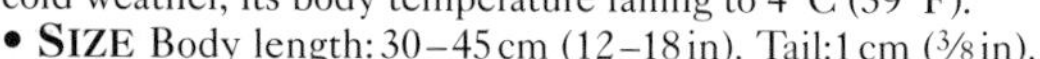

- **SIZE** Body length: 30–45 cm (12–18 in). Tail: 1 cm (⅜ in).
- **OCCURRENCE** Australia, Tasmania, and New Guinea. In all habitats, except tropical and montane rainforest.
- **REMARK** Although it belongs to the most primitive order of mammals, it is the most widespread mammal in Australia.

Social unit Solitary	Gestation 23 days	Young 1	Diet

Family TACHYGLOSSIDAE	Species *Zaglossus bartoni*	Status Endangered

LONG-NOSED ECHIDNA

A long, downward-curving snout, which may exceed 20 cm (8 in) in length, with a tiny mouth at its tip, gives this species its name. The white tips of its defensive spines are only just visible through the sleek black coat of this slow-moving animal, which curls itself into a ball during times of danger. This echidna is active both during the day and at night. Probing the earth for worms with its bill, it impales them with hooked spines at the tip of its long tongue and then swallows them. The female digs a burrow for her egg, but carries and suckles the hatched young in her pouch.

- **SIZE** Body length: 60–100 cm (23 ½–39 in). Tail: None.
- **OCCURRENCE** New Guinea. In montane forest and alpine grassland.
- **REMARK** This is the largest monotreme, or egg-laying mammal.

NEW GUINEA

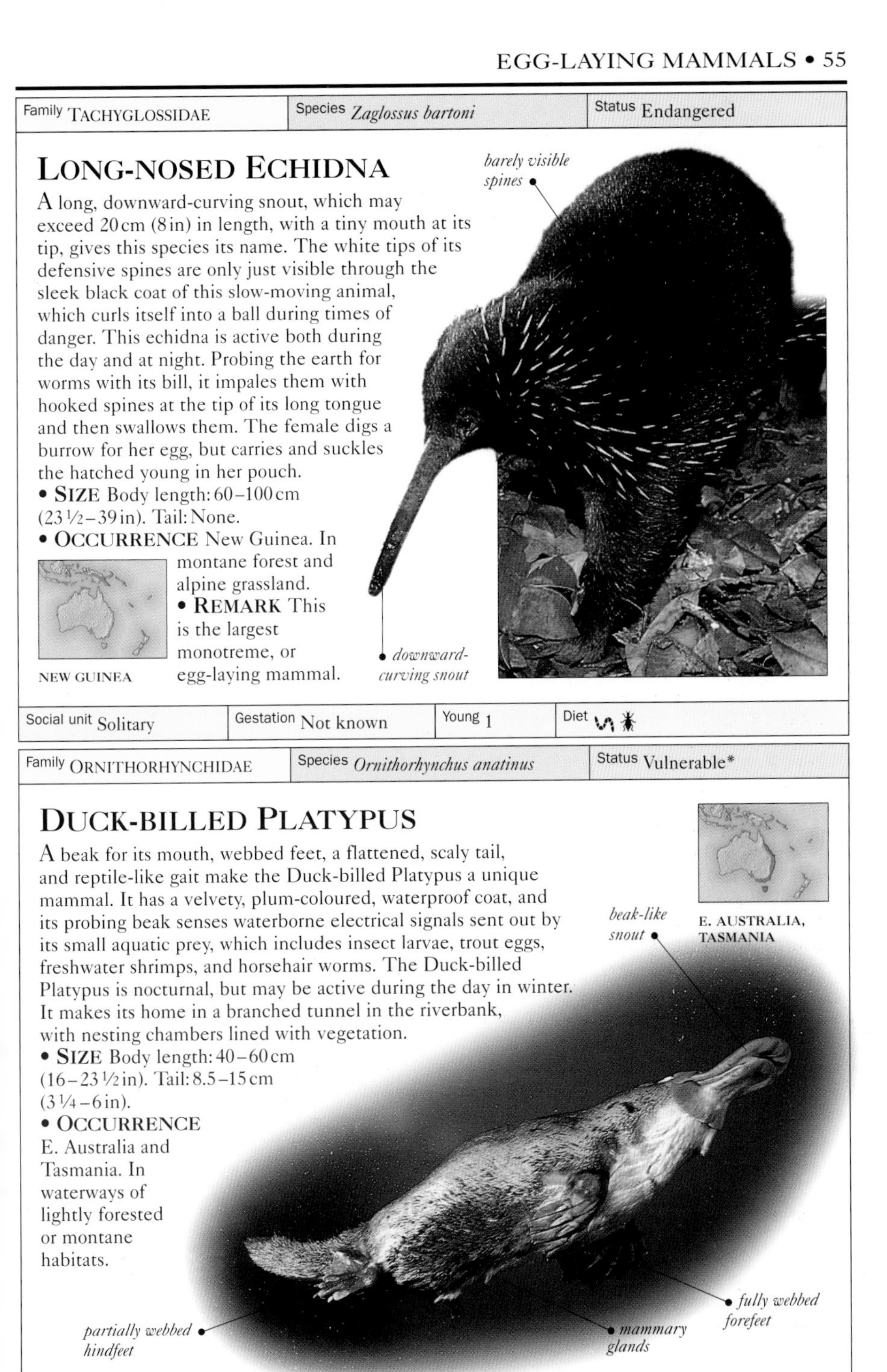

Social unit Solitary	Gestation Not known	Young 1	Diet

Family ORNITHORHYNCHIDAE	Species *Ornithorhynchus anatinus*	Status Vulnerable*

DUCK-BILLED PLATYPUS

A beak for its mouth, webbed feet, a flattened, scaly tail, and reptile-like gait make the Duck-billed Platypus a unique mammal. It has a velvety, plum-coloured, waterproof coat, and its probing beak senses waterborne electrical signals sent out by its small aquatic prey, which includes insect larvae, trout eggs, freshwater shrimps, and horsehair worms. The Duck-billed Platypus is nocturnal, but may be active during the day in winter. It makes its home in a branched tunnel in the riverbank, with nesting chambers lined with vegetation.

- **SIZE** Body length: 40–60 cm (16–23 ½ in). Tail: 8.5–15 cm (3 ¼–6 in).
- **OCCURRENCE** E. Australia and Tasmania. In waterways of lightly forested or montane habitats.

E. AUSTRALIA, TASMANIA

Social unit Solitary	Gestation 1 month	Young 1	Diet

MARSUPIALS

NAMED AFTER the marsupium, or pouch, of the adult female, the 292 species of marsupials are distinguished from other mammals by their method of reproduction. The young are born at a very early stage of development, after a short time in the womb. They have rudimentary limbs, closed eyes, and no fur. They can do little except crawl to the pouch or belly region of the mother. Here, they attach firmly to the nipples and feed on her milk, as they continue development for weeks or months. Gradually, they leave the pouch for longer periods to feed and fend for themselves.

Most marsupials live in Australia, having evolved without competition from other mammals, to take up many lifestyles and habitats. The 22 families have diversified into large grazing kangaroos, smaller wallabies, tiny leaping rat-kangaroos, arboreal and gliding possums, bear-shaped wombats, tunnelling moles, fierce shrew- and cat-like carnivores, and many other types.

Certain marsupial species are also found in New Guinea, including the tree-kangaroos. Around eighty species, mainly small tree-dwellers, are found in South America. One, the Virginia Opossum, has spread to North America.

Family DIDELPHIDAE	Species *Didelphis virginiana*	Status Lower risk*

VIRGINIA OPOSSUM

The largest of the American marsupials, the Virginia Opossum has succeeded in expanding its range rapidly in North America. It has benefited from human habitation both in terms of shelter – it nests in piles of debris or in outbuildings – and of food, since it scavenges on scraps. This nocturnal omnivore has a wide-ranging diet that includes grubs and eggs, flowers, fruits, and carrion, and it sometimes raids poultry or damages garden plants. When threatened, it may feign death or "play possum", sometimes for several hours. Varying from grey to red, brown, and black, it has white-tipped guard hair and thick underfur.

- **SIZE** Body length: 33–50 cm (13–20 in). Tail: 25–54 cm (10–12½ in).
- **OCCURRENCE** W., C., and E. USA, Mexico, and Central America. In grassland, and tropical and temperate forest.

N. & C. AMERICA

unkempt appearance

pale, grey-white face

five long-clawed toes on each foot

hairless, partly prehensile tail

Social unit Solitary	Gestation 12–13 days	Young 5–13	Diet

Family DIDELPHIDAE	Species *Chironectes minimus*	Status Lower risk

WATER OPOSSUM

The only aquatic marsupial, this opossum has fine, dense, water-repellent fur, and long, webbed toes on its rear feet. Both the male and the female have a pouch with a muscular opening that can be closed tightly underwater. Also called the Yapok, this nocturnal animal feeds on fish, frogs, and other freshwater prey, which it grasps using its dextrous, clawless front paws. During the day it rests in a leaf-lined den near water.

• **SIZE** Body length: 26–40 cm (10–16 in). Tail: 31–43 cm (12–17 in).

• **OCCURRENCE** S. Mexico to C. South America. In tropical and temperate forest.

Social unit Solitary	Gestation 2 weeks	Young 2–5	Diet

Family DIDELPHIDAE	Species *Marmosa murina*	Status Lower risk*

COMMON MOUSE OPOSSUM

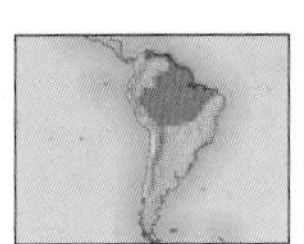

Pale buff to grey on its upperparts, this species has short, velvety fur which is creamy white below. It has a black face mask, prominent eyes, and erect ears. Its strongly prehensile tail is much longer than its head and body, and the female uses it to carry leaves from one place to another. The Common Mouse Opossum is found near forest streams and human habitation, and feeds at night on insects, spiders, lizards, birds' eggs and chicks, and some fruits. A fast, lithe climber, it uses an old bird's nest, tree-hole, or tangle of twigs among tree branches in which to shelter during the day.

• **SIZE** Body length: 11–14.5 cm (4 1/4–5 3/4 in). Tail: 13.5–21 cm (4 3/4–5 1/4 in).

• **OCCURRENCE** N. and C. South America. In tropical forest and rainforest.

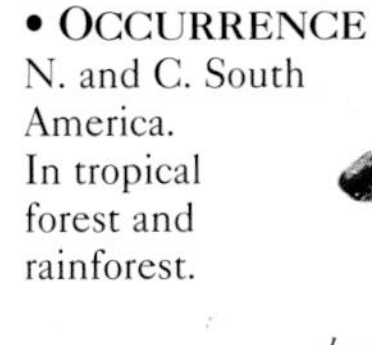

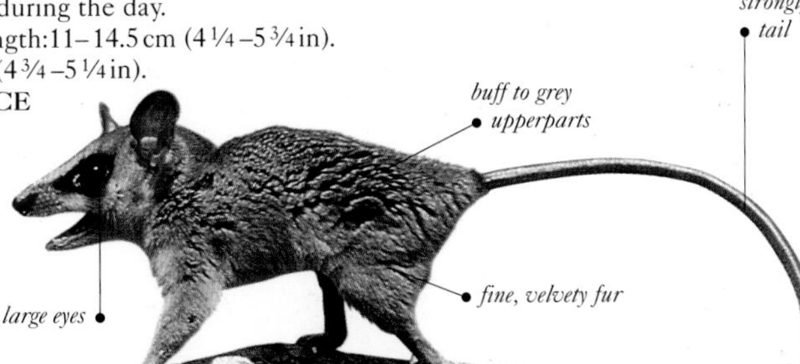

Social unit Solitary	Gestation 13 days	Young 5–10	Diet

Family DASYURIDAE	Species *Ningaui ridei*	Status Lower risk*

INLAND NINGAUI

Also known as the Wongai Ningaui, this shrew-like marsupial has a sharply conical head with small eyes and ears, and a coat that is uniformly brown, with an orange tinge on the cheeks and belly. It is a fierce predator of invertebrates less than 1 cm (3/8 in) in size, such as beetles, crickets, and spiders, and hunts at night among clumps of spinifex (hummock grass), using its sharp senses of smell and hearing. By day it rests in thick undergrowth or the discarded holes of lizards and rodents.

• **SIZE** Body length: 5–7.5 cm (2–3 in). Tail: 5–7cm (2–2¾ in).

• **OCCURRENCE** C. Australia. In spinifex desert, with or without woodland cover.

• **REMARK** This species was identified by zoologists as late as 1975.

Social unit Solitary	Gestation 13–21 days	Young 5–7	Diet

Family DASYURIDAE	Species *Sminthopsis crassicaudata*	Status Lower risk*

FAT-TAILED DUNNART

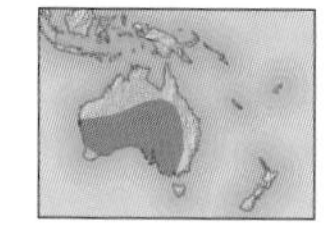

This small mammal has fawn or brown upper parts and white underparts, large eyes, upright ears, and a sharp, pointed nose. Its tail serves as a store for fat and becomes carrot-shaped as food intake increases. Once its food reserves are used up, the Fat-tailed Dunnart goes into torpor, sometimes for as long as 12 hours. It hunts at night on bare soil, or in leaf litter, pouncing upon prey and killing it with a bite on the neck, or subduing larger prey with a "death shake". Solitary during the summer breeding season, it may huddle in small groups in winter.

• **SIZE** Body length: 6–9 cm (2¼–3½ in). Tail: 4–7 cm (1½–2¾ in).

• **OCCURRENCE** S. Australia, except the southwestern tip and eastern coast. In grassland, desert, farmland, and woodland.

• **REMARK** Since it prefers low, open vegetation, this is one native Australian mammal that has benefited from land clearing by European agricultural settlers.

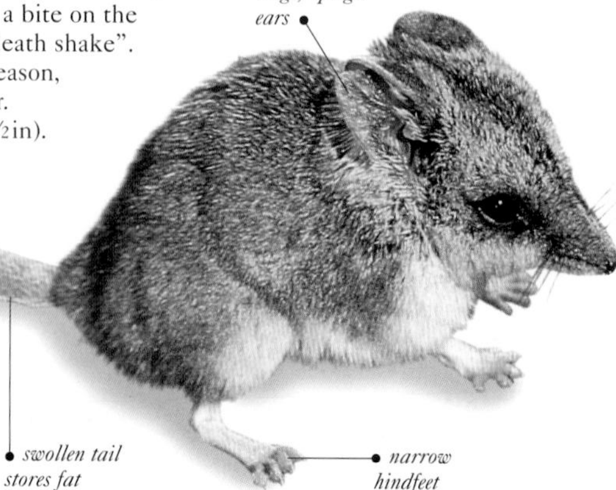

Social unit Solitary	Gestation 13 days	Young 8–10	Diet

Family DASYURIDAE	Species *Antechinomys laniger*	Status Vulnerable

KULTARR

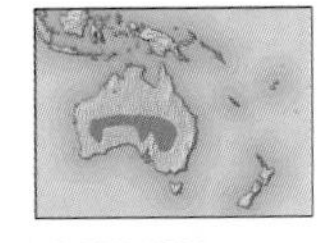

The Kultarr moves with great speed and agility across woodland and semi-desert scrub, leaping on its elongated hindfeet, but landing on its front feet. It is a small, fawn or brown marsupial with white underparts, very large eyes and ears, and a long, thin, tufted tail. Nocturnal by nature, it may dig shallow burrows or occupy crevices, or the dens of other animals. Like the Fat-tailed Dunnart (see opposite), it hunts by pouncing on its prey, giving a fatal bite on the neck.

- **SIZE** Body length: 7–10 cm (2¾–4 in). Tail: 10–15 cm (4–6 in).
- **OCCURRENCE** S. and C. Australia. In varied habitats from woodland and grassland, to desert.
- **REMARK** This is an elusive species: populations may appear in a spot and suddenly "disappear". The reason is unknown, but flooding of the animals' burrows during heavy rainfall may be a cause.

Social unit Solitary	Gestation 12 days	Young 6–8	Diet

Family DASYURIDAE	Species *Parantechinus apicalis*	Status Endangered

DIBBLER

Rediscovered in 1967 after a gap of 80 years, this marsupial has a unique grizzled appearance, due to its white-flecked greyish brown fur. Clear white rings circle the eyes, and it has a white belly and a stout, hairy, tapering tail. This solitary forager hunts mainly invertebrates by day and night. However, its sharp canines also equip it to kill and eat small vertebrates, and it may sip nectar from *Banksia* flowers. Although mostly terrestrial, it can climb 2–3 m (6½–10 ft) up trees. It may dig its own burrow or occupy that of seabirds.

- **SIZE** Body length: 10–16 cm (4–6½ in). Tail: 7.5–12 cm (3–4¾ in).
- **OCCURRENCE** Southwestern tip of Australia, and Whitlock and Boullanger islands. In dense, coastal heath and shrubland, and *Banksia* forest.
- **REMARK** Among the island populations, in most years, all males die immediately after mating.

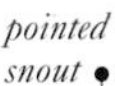

Social unit Solitary	Gestation 7 weeks	Young 8–10	Diet

Family DASYURIDAE	Species *Pseudantechinus macdonnellensis*	Status Lower risk*

RED-EARED ANTECHINUS

Also known as the Fat-tailed Pseudantechinus, this species has a carrot-shaped tail that swells with fat as a food store when it is wellfed. Greyish brown all over, it has red patches behind the ears, and a white-grey belly. The Red-eared Antechinus is terrestrial and prefers rocky habitats with hummock grass and bushes. It is sometimes found in large termite mounds. Nocturnal by nature, it occasionally sunbathes in the morning near rocky shelters. It speedily pursues its prey and kills it with a bite on the neck, holding larger prey with its forepaws for an effective bite.

• **SIZE** Body length: 9.5–10.5 cm (3¾–4¼in). Tail: 7.5–8.5 cm (3–3¼in).

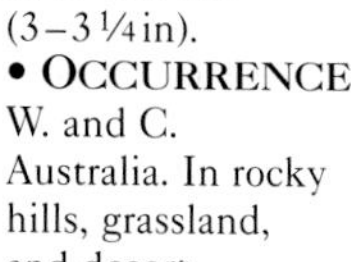
AUSTRALIA

• **OCCURRENCE** W. and C. Australia. In rocky hills, grassland, and desert.

Social unit Solitary	Gestation 45–55 days	Young 6	Diet

Family DASYURIDAE	Species *Dasycercus byrnei*	Status Vulnerable

KOWARI

A burrow-dwelling, nocturnal carnivore with a squirrel-like appearance, the Kowari is light fawn to grey on the back, lighter below, with a bushy black tail. Compact and powerfully built, it has a broad head with large eyes and erect ears. It expands its burrow after rain softens the hard, stony soil, and marks its home range with urine, faeces, and chest gland scents.

• **SIZE** Body length: 13.5–18 cm (5¼–7 in). Tail: 11–14 cm (4¼–5½in).

• **OCCURRENCE** C. Australia, especially "channel country" of Queensland and South Australia. In red clay desert.

AUSTRALIA

Social unit Solitary	Gestation 30–35 days	Young 6–7	Diet

Family DASYURIDAE	Species *Neophascogale lorentzii*	Status Lower risk*

LONG-CLAWED MARSUPIAL MOUSE

NEW GUINEA

Dark grey fur sprinkled with long white hair gives this species the alternative name Speckled Dasyure. It has short, powerful limbs with very long claws on all the digits, to enable it to excavate prey from soft soil, leaf litter, and rotting wood.

• **SIZE** Body length: 17–22 cm (6½–9 in). Tail: 17–22 cm (6½–9 in).

• **OCCURRENCE** New Guinea. In mixed, montane forest at high altitudes.

• **REMARK** This species is concentrated in pockets only; the reasons behind this uneven distribution are not known.

long tail, tipped white

speckled grey and white fur

Social unit Solitary	Gestation Not known	Young 4	Diet

Family DASYURIDAE	Species *Dasyurus viverrinus*	Status Lower risk

EASTERN QUOLL

TASMANIA

Sometimes called the Spotted Native Cat, this cat-like marsupial is slim and agile, with a scattering of distinctive white spots on its brown or black body. It has a deep, narrow face, with large, erect ears. The tail is the same colour as the body, but not spotted. The Eastern Quoll inhabits woody, shrubby, and grassy habitats, common in mixed agricultural areas. It hunts on the ground at night for large insects, small mammals, birds, and lizards, and also eats grasses, fruits, and carrion. The female is longer and 50 per cent heavier than the male. She might bear as many as 20 young at a time, but since she has six teats in her pouch, only six offspring survive.

• **SIZE** Body length: 28–45 cm (11–18 in). Tail: 17–28 cm (6½–11 in).

• **OCCURRENCE** Tasmania. In forest, woodland, scrub, heath, and farmland.

• **REMARK** Previously found in S.E. Australia, the quoll was last sighted on the mainland near Sydney suburbs in the 1960s.

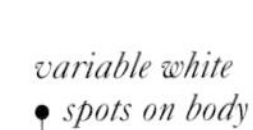

long, erect ears

tail not spotted

Social unit Solitary	Gestation 3 weeks	Young 6	Diet 

Family DASYURIDAE	Species *Sarcophilus harrisii*	Status Lower risk

TASMANIAN DEVIL

TASMANIA

Found across Tasmania, particularly in the forests of the northeast, the "devil" has a compact black body like that of a small bear, with a white band on its chest and sometimes on its rump. This nocturnal hunter and scavenger feeds on animals of various sizes, and has sharp teeth and massive jaws for masticating the bones of its prey.

• **SIZE** Body length: 52–80 cm (20½–32 in). Tail: 23–30 cm (9–12 in).

• **OCCURRENCE** Tasmania. In all major habitats, including urban areas.

• **REMARK** This is the largest marsupial carnivore.

Social unit Solitary	Gestation 30–31 days	Young 4	Diet

Family MYRMECOBIIDAE	Species *Myrmecobius fasciatus*	Status Vulnerable

NUMBAT

Now found only in the extreme southwest tip of Australia, the Numbat has orange-brown body fur with six or seven white transverse bars from behind the shoulder to its rump, which has darker fur. Its narrow head has a dark stripe running along each large eye. Also known as the Banded Anteater, it feeds almost exclusively on termites, using its strong claws to dig for them, and licking them up with its 10cm (4 in) long tongue.

• **SIZE** Body length: 20–28 cm (8–11 in). Tail: 16–21 cm (6½–8½ in).

• **OCCURRENCE** W. Australia. In eucalypt forest and woodland.

AUSTRALIA

• **REMARK** The Numbat has 52 teeth, more than any other land mammal.

Social unit Solitary	Gestation 14 days	Young 4	Diet

Family PERAMELIDAE	Species *Perameles gunnii*	Status Critically endangered

EASTERN BARRED BANDICOOT

Virtually extinct in the wild in Western Australia, this species is now found mainly in Tasmania. It has rabbit-like ears and broad whitish bands across its hindquarters. Resting by day, it forages alone at night.

AUSTRALIA, TASMANIA

• **SIZE** Body length: 27–35 cm (10½–14 in). Tail: 7–11 cm (2¾–4¼ in).

• **OCCURRENCE** Australia and Tasmania. In open grassland and wooded urban areas.

Social unit Solitary	Gestation 12.5 days	Young 1–5	Diet

Family PERAMELIDAE	Species *Macrotis lagotis*	Status Vulnerable

GREATER BILBY

Also known as the Rabbit-eared Bandicoot, this species has huge ears, long hindfeet, and a tricoloured tail: grey, black, and white. A powerful digger, it has an excellent sense of hearing and of smell, and hunts only after dark.

• **SIZE** Body length: 30–55 cm (12–22 in). Tail: 20–29 cm (6–11½ in).

AUSTRALIA

• **OCCURRENCE** N.W. Australia. In open grassland and desert.

• **REMARK** The males are more bulky than females and may be twice as heavy.

Social unit Solitary	Gestation 13–16 days	Young 2	Diet

Family PERORYCTIDAE	Species *Echymipera kalubu*	Status Lower risk

NEW GUINEAN SPINY BANDICOOT

Extremely variable in colour and size, this bandicoot may be brown, copper, yellow, or black, with buff underparts. It has a conical snout, hairless tail, and spiny hairs. This nocturnal species is mostly insectivorous, but also eats fruits.

• **SIZE** Length: 20–50cm (8–20in). Tail: 5–12.5cm (2–5in).

NEW GUINEA

• **OCCURRENCE** New Guinea. In grassland and forest.

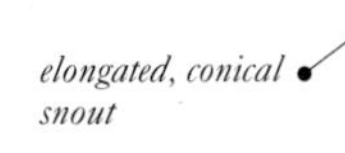

Social unit Solitary	Gestation 14 days	Young 1–3	Diet 

Family NOTORYCTIDAE	Species *Notoryctes typhlops*	Status Endangered

MARSUPIAL MOLE

This small, elusive mammal is well-adapted to its burrowing way of life. Probing with its horny nose, it scoops aside sand with its spadelike foreclaws and kicks it back with its hind feet. It makes tunnels down to 8 ft (2.5 m) in the desert soil, or simply "swims" through loose sand, grubbing for something to eat. Its silky coat is off-white to cinnamon, stained a dark red in iron-rich areas, and its eyes are reduced to vestigial lenses. The female Marsupial Mole's pouch opens backward, so as not to fill up with sand.

- **SIZE** Length: 4¾–7 in (12–18 cm). Weight: 1½–2½ oz (40–70 g).
- **OCCURRENCE** S.W. and S. Australia. In sandy desert.

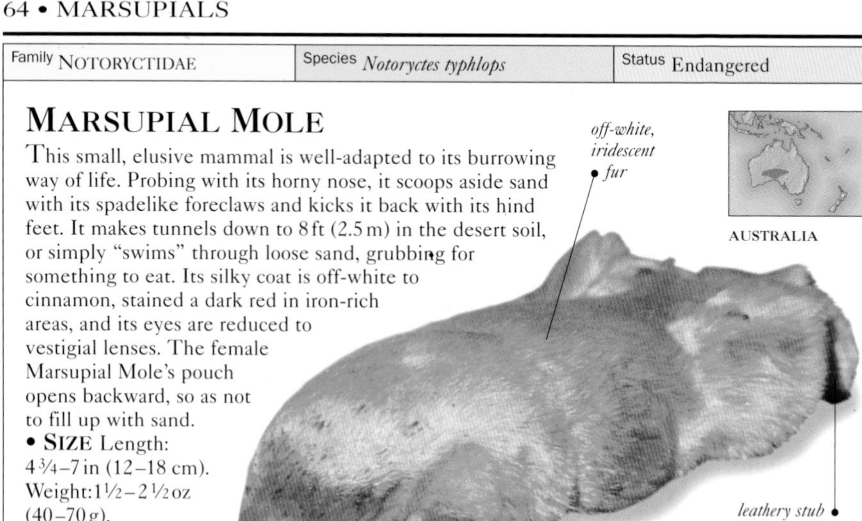

Social unit Solitary	Gestation Not known	Young 1–2	Diet

Family VOMBATIDAE	Species *Vombatus ursinus*	Status Lower risk*

COARSE-HAIRED WOMBAT

Like a small bear in appearance, the Coarse-haired Wombat is a prolific digger and makes a single-exit tunnel up to 655 ft (200 m) long. It emerges at night to graze on slopes above creeks and valleys, and in winter it may sunbathe at dawn and dusk. A powerful body, broad head, stocky limbs, and flattened claws equip it for burrowing.

- **SIZE** Length: 28–47 in (70–120 cm). Weight: 55–88 lb (25–40 kg).
- **OCCURRENCE** E. Australia and Tasmania. In forest, coastal scrub, and Australian alpine heathland.

Social unit Solitary	Gestation 33 days	Young 1	Diet

Family PHASCOLARCTIDAE	Species *Phascolarctos cinereus*	Status Lower risk

KOALA

AUSTRALIA

Living and feeding almost totally in eucalyptus trees, the Koala occasionally descends to change perches or aid digestion by eating gravel. It has a uniquely adapted liver that can break down the toxic substances in eucalypt leaves – its only food. With a compact body and large head, it has a woolly, grayish brown coat that is whitish on the neck and chest, and mottled at the rump. Koalas bellow at each other during the mating season, and dominant males mate more often than the junior ones. The single young of this marsupial "bear" is suckled in the pouch for six months and then clings to the mother's back until it is a year old. This species has few natural enemies, apart from raptors, but forest clearance poses a manmade threat.

- **SIZE** Length: 26–32 in (65–82 cm). Weight: 8¾–33 lb (4–15 kg).
- **OCCURRENCE** E. Australia. In eucalypt forest and woodland below 3,300 ft (1,000 m).
- **REMARK** The Koala feeds for four hours each night, but is inactive for the rest of the day.

SEDENTARY LIFESTYLE
The Koala wedges itself in the fork of a tree and dozes for long periods. The first two toes on its forefeet are opposing, allowing a pincerlike grip of thin branches.

Social unit Solitary	Gestation 35 days	Young 1	Diet

Family PHALANGERIDAE	Species *Phalanger orientalis*	Status Lower risk

COMMON CUSCUS

The colour of this animal varies from white to black across its range of many small islands. Generally, however, it has a dark stripe along its back, and the lower part of the female's tail is hairless and tipped white. Resembling a combination of a sloth and a monkey, this nocturnal species is a deliberate, but agile climber with strong, grasping digits. Its tail is prehensile with a rough underside.

• **SIZE** Body length: 38–48 cm (15–19 in). Tail: 28–43 cm (11–17 in).

• **OCCURRENCE** New Guinea, Solomon Islands, and surrounding smaller islands. In forest.

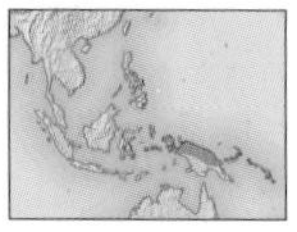

NEW GUINEA, SOLOMON ISLANDS

• **REMARK** The docile Cuscus is a popular pet among the local population.

dark stripe along back

white-tipped tail in female

pointed face with large eyes

Social unit Solitary	Gestation 2–3 weeks	Young 1–2	Diet

Family PHALANGERIDAE	Species *Trichosurus vulpecula*	Status Lower risk

COMMON BRUSH-TAILED POSSUM

A familiar animal found in many habitats, this possum is generally silver-grey, with shorter fur tinged coppery-red in the north of its range, and longer, dark grey to black fur in the south. It bounds and climbs with ease, foraging at night in trees, using its forepaws to manipulate food. A vocal animal, it hisses, chitters grunts, and growls, and chatters its teeth when startled. The Brush-tailed Possum nests in tree hollows, fallen logs, rock piles, holes in creek banks, or roofs. Both sexes maintain exclusive areas around their nest sites, although the male may live with the female for a short while during the mating season.

• **SIZE** Body length: 35–58 cm (14–23 in). Tail: 25–40 cm (10–16 in).

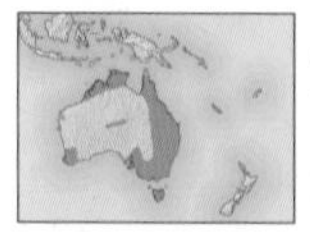

AUSTRALIA, TASMANIA

• **OCCURRENCE** Australia and Tasmania. In all habitats with tree cover, and suburban parks and gardens.

• **REMARK** Once found all over Australia, this possum now has a much reduced range.

Social unit Solitary	Gestation 16–18 days	Young 1	Diet

Family PETAURIDAE	Species *Gymnobelideus leadbeateri*	Status Endangered

LEADBEATER'S POSSUM

Grey with a dark stripe along its back, the Leadbeater's Possum is a speedy and elusive nocturnal species that forages in trees for small insects, resin, sap, and nectar. Colonies of up to eight comprise a breeding pair and its offspring. The females defend the group's territory.

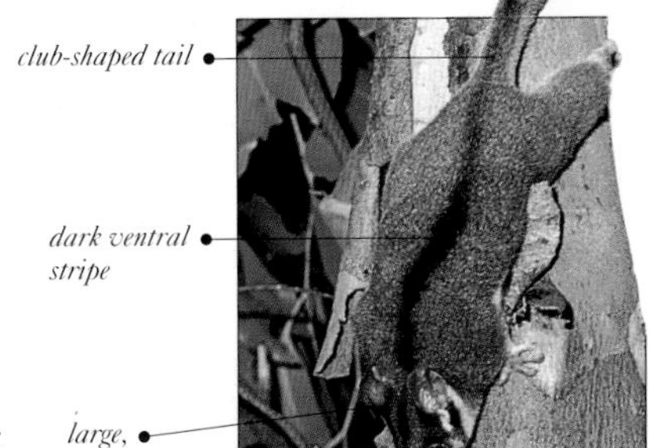

AUSTRALIA

- **SIZE** Body length: 15–17 cm (6–7 in). Tail: 14.5–20 cm (6–8 in).
- **OCCURRENCE** Australia (Victoria). In montane ash forest.
- **REMARK** It was rediscovered in 1961 after 52 years.

Social unit Social	Gestation 20 days	Young 1–2	Diet

Family PETAURIDAE	Species *Dactylopsila trivirgata*	Status Lower risk

STRIPED POSSUM

Conspicuous, skunk-like stripes, and a bushy, white-tipped tail are the chief features of this small nocturnal possum. Like the skunk (see pp. 256–257), it can also emit a foul odour from its anal glands. Slender-bodied and agile, it has long fingers and strong claws for gripping, and an elongated fourth digit that it uses to probe branches for insects, ants, and termites.

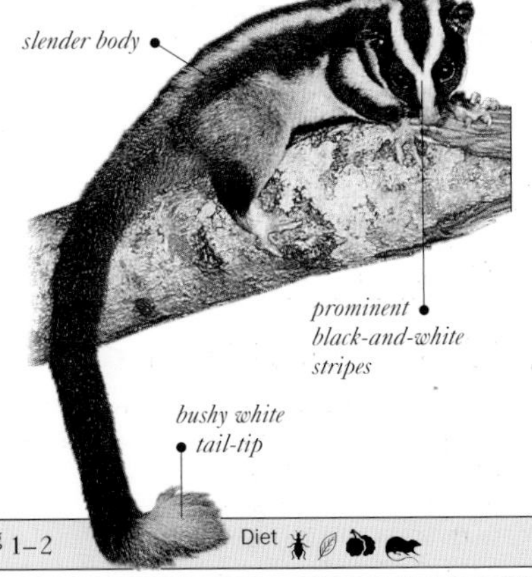

AUSTRALIA, NEW GUINEA

- **SIZE** Body length: 24–28 cm (9½–11 in). Tail: 31–39 cm (12–15½ in).
- **OCCURRENCE** New Guinea and N.E. Australia. In rainforest.

Social unit Variable	Gestation Not known	Young 1–2	Diet

Family PETAURIDAE	Species *Petaurus norfolcensis*	Status Lower risk

SQUIRREL GLIDER

A furry gliding membrane, extending from the fifth front toe to the back foot, and a bushy tail used as a rudder allow this squirrel to parachute over distances of 50 m (165 ft). Blue-grey to brown-grey on the back, and creamy white below, it has a dark stripe on its head, and erect, hairless ears. Highly active, agile, and speedy, it has strong, sharply clawed digits for climbing. The nocturnal Squirrel Glider lives in a leaf-lined nest in a tree hollow. It feeds on nectar and pollen, and strips bark with its incisors to eat resin and larvae.

AUSTRALIA

- **SIZE** Body length: 18–23 cm (7–9 in). Tail: 22–30 cm (9–12 in).
- **OCCURRENCE** Australia (N.E. Queensland to C. Victoria). In eucalypt and coastal forest.

Social unit Social	Gestation 20 days	Young 1–2	Diet

Family PSEUDOCHEIRIDAE	Species *Pseudocheirus peregrinus*	Status Lower risk

COMMON RINGTAIL

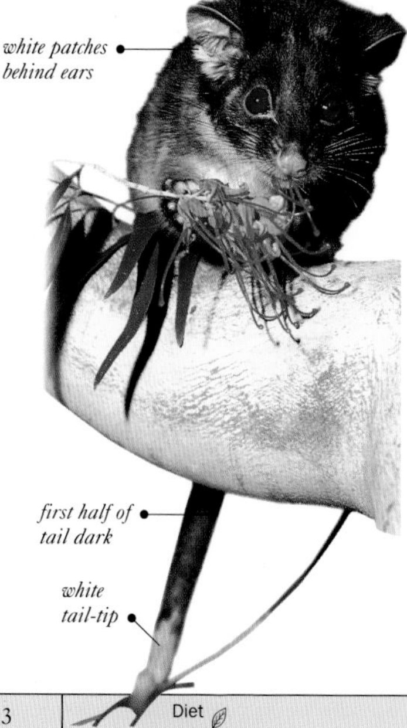

Adapted to lead a life in the trees, the Common Ringtail has sharp molars to chew its leafy diet, a large intestine that efficiently breaks down plant fibre, a strongly prehensile tail that is naked and rough on the last third of its underside, and forefeet with two opposing first digits to allow a strong grip of branches. Its coat is rufous, grey, or dark brown-grey, and its small ears have white patches behind them. Feeding at night on tender leaves, it savours eucalyptus, acacia, tea-tree, and paperbark. In the north of its range it nests in a tree-hole, while in the south it makes a squirrel-like drey or spherical nest of twigs, bark, and grass.

• **SIZE** Body length: 30–35 cm (12–14 in). Tail: 30–35 cm (12–14 in).

• **OCCURRENCE** E. Australia and Tasmania. In forest, thickets, coastal scrub, and suburban parks and gardens.

AUSTRALIA, TASMANIA

• **REMARK** The male helps care for the young by carrying them on its back, watching over them in the nest, and feeding them – a rare occurrence among marsupials.

Social unit Social	Gestation Up to 30 days	Young 1–3	Diet

Family PSEUDOCHEIRIDAE	Species *Petauroides volans*	Status Lower risk

GREATER GLIDER

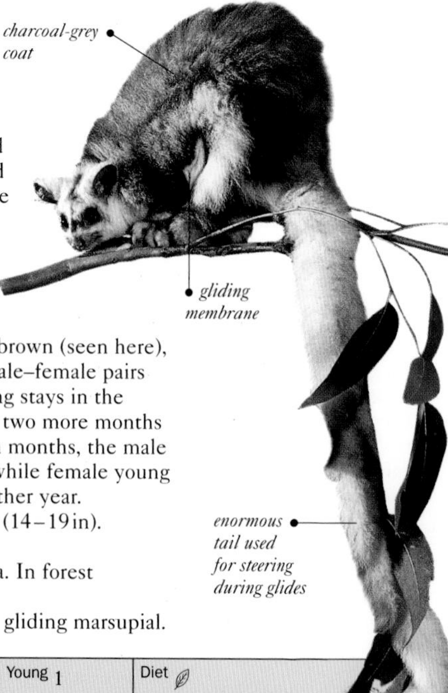

This highly arboreal marsupial can glide over distances of 100 m (330 ft) by spreading out the furred gliding membrane between its elbows and rear feet. Trees used as a landing pad by this animal have regular scratch marks created by its sharp claws, which help it to grip bark. The large ears and eyes of this nocturnal species face the front, so that it can judge distances accurately in the dark, with the help of stereophonic hearing and stereoscopic vision. The Greater Glider occurs in two colour forms: one is charcoal-black to grey, tinged with brown (seen here), while the other is pale grey or mottled cream. Male–female pairs share the same den in a tree hollow. The offspring stays in the mother's pouch for five months and then spends two more months in the den or riding on the mother's back. By ten months, the male offspring is driven out of the den by the father, while female young remain with the mother for another year.

• **SIZE** Body length: 35–48 cm (14–19 in). Tail: 45–60 cm (18–23½ in).

• **OCCURRENCE** E. Australia. In forest dominated by eucalypt species.

AUSTRALIA

• **REMARK** This is the largest gliding marsupial.

Social unit Pair	Gestation Not known	Young 1	Diet

Family TARSIPEDIDAE	Species *Tarsipes rostratus*	Status Lower risk

HONEY POSSUM

Living wholly on a diet of nectar and pollen, the Honey Possum is a tiny, nocturnal animal with a bristle-tipped tongue, a pointed snout, and long, prehensile tail. Its toes have padded tips and sharp claws to grip bark and glossy leaves.

• **SIZE** Body length: 6.5–9 cm (2 1/2–3 1/2 in). Tail: 7–10.5 cm (2 3/4–4 1/4 in).

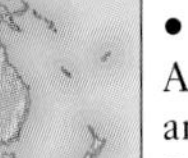

AUSTRALIA

• **OCCURRENCE** S.W. Australia. In sandy heathland and tropical woodland.

• **REMARK** This is one of the smallest possums.

Social unit Social	Gestation 21–28 days	Young 1–4	Diet

Family ACROBATIDAE	Species *Distoechurus pennatus*	Status Lower risk

FEATHER-TAILED POSSUM

Identified by its white face with four black stripes, and a quill-like, prehensile tail, this nocturnal possum darts across branches, gripping them with its sharp claws, trapping cicadas. It has a sixth pad on its hindfeet, a feature shared only with the Pygmy Glider (see below).

• **SIZE** Body length: 10.5–13.5 cm (4 1/2–5 1/2 in). Tail: 12.5–15.5 cm (5–6 1/2 in).

NEW GUINEA

• **OCCURRENCE** New Guinea. In rainforest, lower mossy forest, and regenerating forest near villages.

light brown to grey fur

slender, quill-like tail

black eye-stripes

Social unit Social	Gestation Not known	Young 1–2	Diet

Family ACROBATIDAE	Species *Acrobates pygmaeus*	Status Lower risk

PYGMY GLIDER

Also called the Feather-tailed Glider, this tiny, nocturnal marsupial has a bristled tail that it uses as a rudder. Its gliding membrane extends between its front and rear limbs, and its toes have sharp claws to dig into bark with, and padded tips that allow it to grip smooth and shiny surfaces.

• **SIZE** Body length: 6.5–8 cm (2 3/4–3 in). Tail: 7–8 cm (2 3/4–3 in).

AUSTRALIA

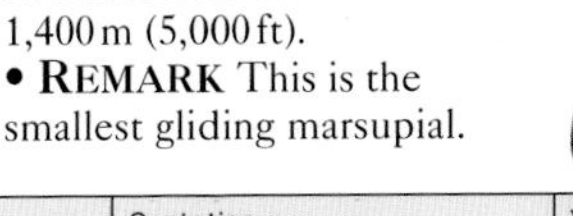

• **OCCURRENCE** E. Australia. In forest from sea level to 1,400 m (5,000 ft).

• **REMARK** This is the smallest gliding marsupial.

Social unit Social	Gestation Not known	Young 1–4	Diet

Family BURRAMYIDAE	Species *Cercartetus lepidus*	Status Lower risk*

LITTLE PYGMY-POSSUM

This is the only pygmy-possum with a grey underside; the upperparts are fawn or brown. It has a blunt face and large, erect ears, and can hang from its prehensile tail, which stores excess food as fat at its base. It feeds at night in low bushes or on the ground.

AUSTRALIA, TASMANIA

- **SIZE** Body length: 5 – 6.5 cm (2–2¾ in). Tail: 6–7.5 cm (2½–3 in).
- **OCCURRENCE** S.E. Australia and Tasmania. In all wooded habitats, except rainforest.
- **REMARK** This is the smallest possum species.

Social unit Solitary	Gestation 30 – 51 days	Young 3 – 4	Diet

Family POTOROIDAE	Species *Hypsiprymnodon moschatus*	Status Lower risk

MUSKY RAT-KANGAROO

Bounding along on all fours, this potoroid marsupial (see below) is medium chocolate-brown above and lighter below, with five-toed hindfeet. Both sexes emit a musky odour during breeding.

- **SIZE** Body length: 16 – 28 cm (6½–11 in). Tail: 12–17 cm (4¾–7 in).

AUSTRALIA

- **OCCURRENCE** N.E. Queensland. In dense rainforest.
- **REMARK** Unusually for a marsupial, it hoards excess food.

Social unit Solitary	Gestation Not known	Young 2	Diet

Family POTOROIDAE	Species *Potorous longipes*	Status Endangered

LONG-FOOTED POTOROO

Resembling a combination of a rat and a kangaroo, this potoroo bounds on its back feet and scrabbles for food using its short forelimbs with strong claws. A nocturnal fungus-eater, it digs conical pits for food and helps disperse fungi spores.

- **SIZE** Body length: 38 – 42 cm (15–16½ in). Tail: 31 – 33 cm (12–13 in).
- **OCCURRENCE** S.E. Australia. In temperate rainforest.

AUSTRALIA

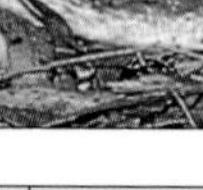

Social unit Solitary	Gestation 38 days	Young 1	Diet 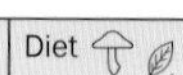

Family POTOROIDAE	Species *Bettongia penicillata*	Status Lower risk

BRUSH-TAILED BETTONG

A fungus-eater like the potoroos (see opposite), the Brush-tailed Bettong is pale to orange-grey above and pale grey below. Its long tail has a black crest on the upperside and is used for carrying nesting material. When disturbed, this bettong bounds at high speed with its head down, and tail held out straight, with the crest raised. A nocturnal forager, it scrapes up woodland soil to find fungi, and by day it shelters in a domed nest made of bark, leaves, and grass.

- **SIZE** Body length: 30–38 cm (12–15 in). Tail: 29–36 cm (11½–14 in).
- **OCCURRENCE** S.W. Australia. In open forest and woodland.

Social unit Solitary	Gestation 21 days	Young 1	Diet

Family MACROPODIDAE	Species *Petrogale penicillata*	Status Vulnerable

BRUSH-TAILED ROCK WALLABY

Adapted for leaping and scrambling over rocky surfaces, the Brush-tailed Rock Wallaby can leap as high as 4 m (13 ft), and has specially padded, rough soles on its hindfeet for excellent grip. Its coat is medium to dark brown on the back, greyer on the shoulders, and reddish on the flanks and rump; its feet and bushy tail-tip are black or dark brown. Feeding by night, it rests in cool rock crevices in the day.

- **SIZE** Body length: 50–60 cm (20–23½ in). Tail: 50–70 cm (20–28 in).
- **OCCURRENCE** S.E. Australia. Introduced in New Zealand and Hawaii. In rocky forest.
- **REMARK** Once abundant, this species saw a rapid decline in the early twentieth century as it was killed for its fur and was also perceived as an agricultural pest. Many populations now comprise only 10–12 animals, indicating future declines.

Social unit Social	Gestation 30–32 days	Young 1	Diet

Family MACROPODIDAE	Species *Lagorchestes conspicillatus*	Status Lower risk

SPECTACLED HARE WALLABY

Adapted to its arid habitat in various ways, this white-grizzled grey-brown wallaby uses the least amount of water for any mammal its size. It hardly drinks, does not sweat or pant unless the temperature is above 30°C (86°F), and produces concentrated urine. The orange eye-patches give it the name Spectacled Hare Wallaby. It nibbles grass or leaves at night, burrowing under large clumps of spinifex.

• **SIZE** Body length: 40–48 cm (16–19 in). Tail: 37–50 cm (14½–20 in).

• **OCCURRENCE** N. Australia. In tropical grassland, open forest, and woodland.

AUSTRALIA

Social unit Solitary	Gestation 29–31 days	Young 1	Diet

Family MACROPODIDAE	Species *Thylogale stigmatica*	Status Lower risk

RED-LEGGED PADEMELON

With a compact body, slender head, and long, erect ears, the Red-legged Pademelon (a type of small wallaby) tends to be brown-grey in rainforest, but paler fawn in open woodland. Active both by day and night, it is usually solitary, but may gather in groups to feed. Often living on the forest edge, it depends on leaf cover for protection from predators, such as Dingoes, Tiger Quolls, and large pythons. When threatened, it gives a loud alarm thump with its hindfeet and flees rapidly for shelter under dense cover.

• **SIZE** Body length: 38–58 cm (15–23 in). Tail: 30–47 cm (12–18½ in).

• **OCCURRENCE** N. and E. Australia, and New Guinea. In rainforest, mixed savanna and woodland, and open forest.

AUSTRALIA, NEW GUINEA

Social unit Solitary	Gestation 20–30 days	Young 1	Diet

Family MACROPODIDAE	Species *Setonix brachyurus*	Status Vulnerable

QUOKKA

This marsupial has a compact, rounded body and coarse brown fur tinged red around the face and neck. Very rare in mainland Australia, the Quokka survives on the Rottnest and Bald Islands off the southwest coast, where introduced predators such as foxes are absent. These wallabies are often harmed when tourists feed them inappropriate food.

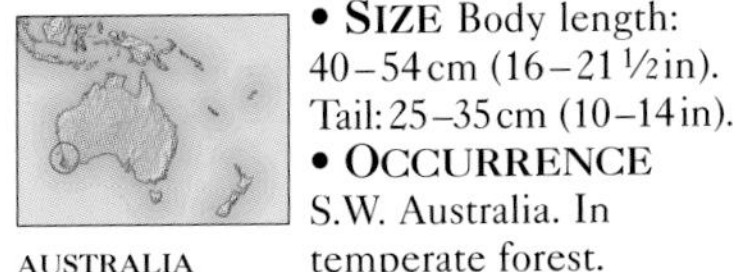

• **SIZE** Body length: 40–54 cm (16–21 ½ in). Tail: 25–35 cm (10–14 in).
• **OCCURRENCE** S.W. Australia. In temperate forest.

reddish facial fur
dense brown coat

Social unit Social	Gestation 27 days	Young 1	Diet

Family MACROPODIDAE	Species *Wallabia bicolor*	Status Lower risk

SWAMP WALLABY

Also referred to as the Stinker or Black Wallaby, this brown-black species has a much darker face, snout, and feet, and the blackish tail is often white-tipped. It has an unusual loping gait with its head held low and tail straight behind. It feeds at night on a wide-ranging diet of plant matter, including toxic plants such as hemlock. The Dingo and Red Fox are its enemies.

• **SIZE** Body length: 66–85 cm (26–34 in). Tail: 65–86 cm (26–34 in).
• **OCCURRENCE** E. Australia. In tropical and temperate forest.

blackish face
brown to black fur

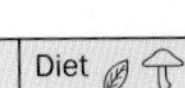

Social unit Solitary	Gestation 33–38 days	Young 1	Diet

Family MACROPODIDAE	Species *Macropus robustus*	Status Lower risk

WALLAROO

This dark grey to reddish brown wallaby is found in a range of habitats, but prefers rocky outcrops, cliffs, and boulder piles. It shelters there by day, moving by late afternoon to forage on grasses and other kinds of leaves. Also called the Euro or Hill Kangaroo, it resembles other brown wallabies, but its posture, with shoulders, back, and elbows together, and wrists raised, is distinctive.

• **SIZE** Body length: 0.8–1.4 m (2 ½–4 ½ ft). Tail: 60–90 cm (23 ½–35 in).
• **OCCURRENCE** Almost all of Australia. In grassland, desert, and tropical and temperate forest.

lightly furred ears
shaggier fur than other kangaroos
nostrils bare and black

Social unit Solitary	Gestation 32–34 days	Young 1	Diet

Family MACROPODIDAE	Species *Macropus fuliginosus*	Status Lower risk

WESTERN GREY KANGAROO

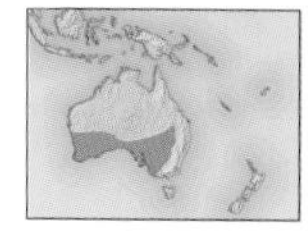

AUSTRALIA

This species is one of the largest and most abundant of kangaroos, and has thick, coarse fur which varies from light grey-brown to chocolate brown, with a dark brown to black muzzle, and a paler chest and belly. Powerfully built, it has a strong and tapering tail. The Western Grey Kangaroo lives in stable groups of 2–15, whose members recognize one another through smell. The male kangaroo has an especially distinctive body odour and is about twice the size of the female. Aggressive during the mating season, males fight each other for females, also battling for food and resting sites. Antagonists lock arms and attempt to push each other over, and may also lean back on their tails to kick with their rear feet. This species grazes at night, usually feeding on grasses, it also browses on shrubs and low trees.

- **SIZE** Body length: 0.9–1.4 m (3–4½ ft). Tail: 75–100 cm (30–39 in).
- **OCCURRENCE** S. Australia. In open forest, woodland, shrubland, wet and dry heathland, and grassy savanna.
- **REMARK** This species is culled since it is regarded as an agricultural pest. It is also hunted for its meat and skin.

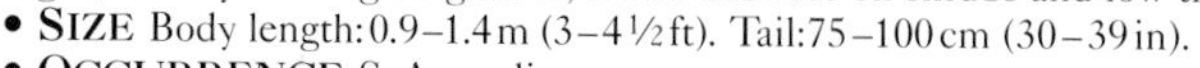

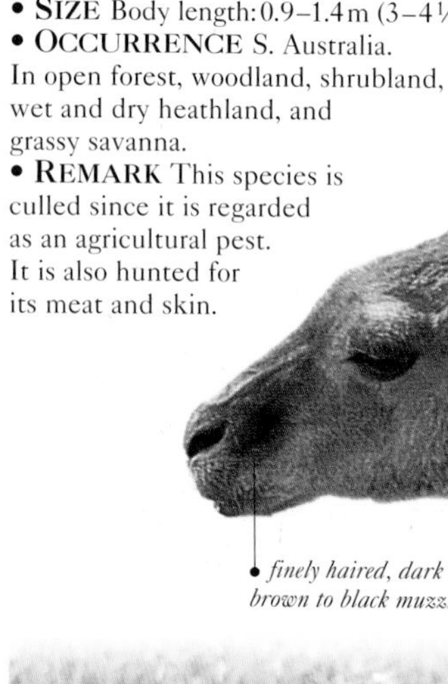

light grey-brown to chocolate brown body

large ears

finely haired, dark brown to black muzzle

Social unit Social	Gestation 30–31 days	Young 1	Diet

POWERFUL MOVER This kangaroo lopes rabbit-like when moving slowly, using all four limbs, with the tail as a brace. However, it bounds swiftly on its hindfeet over longer distances, and the male covers as much as 10 m (33 ft) in one leap.

CARRYING THE YOUNG The young, or joey, of the Western Grey Kangaroo remains attached to the teat in its mother's pouch for 130–150 days. It leaves the safety of the pouch for short periods once it is about 250 days old, but returns quickly when threatened by predators – usually dingoes or large birds such as wedge-tailed eagles.

Family MACROPODIDAE	Species *Macropus parma*	Status Lower risk

PARMA WALLABY

The smallest member of its genus, this red- or grey-brown wallaby is characterized by a black stripe in the centre of its back, and a white stripe on each cheek. It has erect, rounded ears, shorter than those of most wallabies, and a white throat and belly. The male is slightly heavier than the female, with a broader chest and stronger forelimbs. This solitary and reclusive animal remains camouflaged in dense undergrowth by day and feeds at night, grazing and browsing on a wide range of plants, under cover of vegetation or with its head held low to the ground. Its existence is adversely affected by forest clearance and the burning of forest floors.

- **SIZE** Body length: 45–53 cm (18–21 in). Tail: 41–54 cm (16–21 ½ in).
- **OCCURRENCE** E. New South Wales; introduced in Kawan Island, New Zealand. In eucalyptus forest and rainforest, with dense undergrowth.
- **REMARK** Believed to be extinct for about a century, the Parma Wallaby was "rediscovered" on Kawan Island, New Zealand, in 1965 and on the Australian mainland in 1967.

RESTING ON ITS HAUNCHES
The Parma Wallaby seldom lies down and usually rests by sitting with its tail between its legs. When hopping, its body is held horizontally close to the ground, with forepaws folded against its chest.

Social unit Solitary	Gestation 34–35 days	Young 1	Diet

Family MACROPODIDAE	Species *Macropus rufus*	Status Common

RED KANGAROO

The largest living marsupial, the Red Kangaroo is found over much of Australia. Its population varies greatly from year to year. When rainfall is high, numbers may reach 12 million, falling to less than half this figure during drought. Females will not conceive if food is scarce due to lack of rain, and in prolonged drought males do not produce sperm. Red Kangaroos live in groups of 2–10, with a single dominant male, although up to 1,500 kangaroos may gather at a water hole in dry conditions. A kangaroo warns its group members of danger by stamping its foot or thumping its tail. The group then flees, in short bursts of speed reaching 50 kph (30 mph). This kangaroo is nocturnal but it may sometimes forage in the early morning or late afternoon.

• **SIZE** Body length: 1–1.6 m (3 1/4 – 5 1/4 ft). Tail: 75–120 cm (30–47 in).
• **OCCURRENCE** Australia. In savanna grassland, desert, and temperate forest.
• **REMARK** Regarded as a pest in Australia, the Red Kangaroo is hunted widely for its meat and skin.

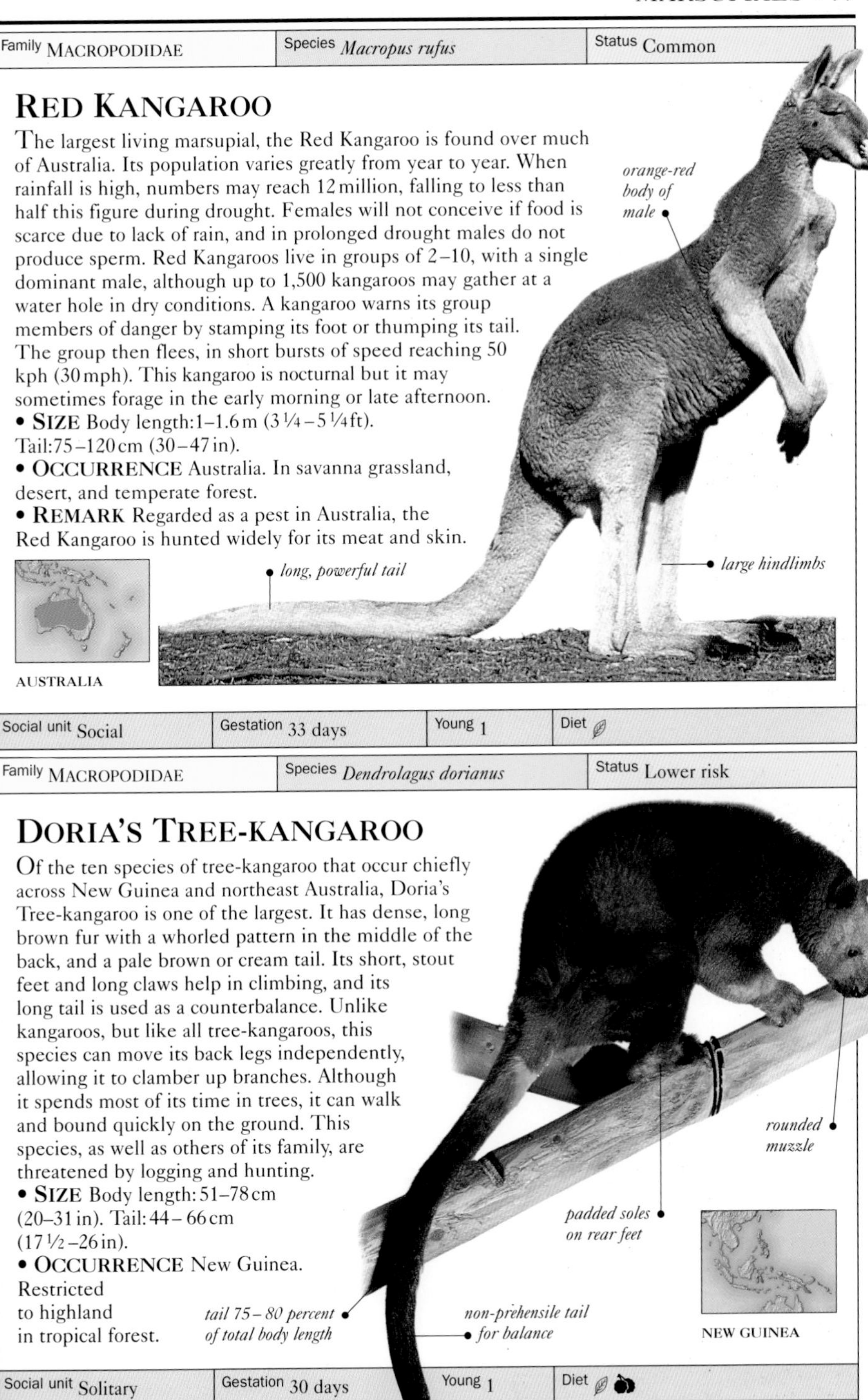

Social unit Social	Gestation 33 days	Young 1	Diet

Family MACROPODIDAE	Species *Dendrolagus dorianus*	Status Lower risk

DORIA'S TREE-KANGAROO

Of the ten species of tree-kangaroo that occur chiefly across New Guinea and northeast Australia, Doria's Tree-kangaroo is one of the largest. It has dense, long brown fur with a whorled pattern in the middle of the back, and a pale brown or cream tail. Its short, stout feet and long claws help in climbing, and its long tail is used as a counterbalance. Unlike kangaroos, but like all tree-kangaroos, this species can move its back legs independently, allowing it to clamber up branches. Although it spends most of its time in trees, it can walk and bound quickly on the ground. This species, as well as others of its family, are threatened by logging and hunting.

• **SIZE** Body length: 51–78 cm (20–31 in). Tail: 44–66 cm (17 1/2–26 in).
• **OCCURRENCE** New Guinea. Restricted to highland in tropical forest.

Social unit Solitary	Gestation 30 days	Young 1	Diet

INSECTIVORES

DESPITE THEIR NAME, insectivores do not restrict their diet to insects, but also eat other small prey such as worms, spiders, and slugs. Nor are they the only animals to eat insects – many other mammal species do too.

There are 365 insectivores, in six families, and they have much in common. They are mostly small, nocturnal, and they retain anatomical features shown by their ancient ancestors, the early mammals. These include up to 48 simple teeth; a small brain compared to body size, with few surface folds; and in the male, testes that stay in the abdomen. An insectivore's eyes and ears are small, but the snout is elongated, flexible, and very sensitive. There are five clawed toes on each foot and, usually, a mid-length tail.

Insectivores follow three main lifestyles. Shrews, hedgehogs, moonrats (gymnures), tenrecs, and solenodons are terrestrial, busily foraging for small creatures at night and perhaps by day too. Moles and golden moles are burrowers, spending most of their lives tunnelling in search of soil insects, worms, and grubs. Web-footed tenrecs, water shrews, otter-shrews, and desmans are semi-aquatic and catch a variety of water creatures including fish and frogs.

Various types of hedgehogs, moonrats, shrew-moles, and especially shrews are widespread in many parts of the world. However, three insectivore families are more localized. These are solenodons on the Caribbean islands of Cuba and Hispaniola, tenrecs in Madagascar and equatorial Africa, and golden moles in Africa, south of the Sahara.

Family ERINACEIDAE	Species *Erinaceus europaeus*	Status Common

WEST EUROPEAN HEDGEHOG

Roaming urban parks, gardens, hedgerows, fields, and woods at night, the West European Hedgehog snuffles, hog-like, for worms, insects, birds' eggs, and carrion. By day, it shelters in a nest of grass and leaves under a bush, log, or outbuilding, or in an old burrow. Its rotund body is covered with short, pointed spines, except for the head, undersides, and limbs. Uniformly grey-brown in colour, it has small ears and a short tail. In defence, the West European Hedgehog tucks its head and legs into its belly and curls itself into a spiny ball. It can also run or climb with surprising agility.

- **SIZE** Body length: 22–27 cm (9–10½ in). Tail: None.
- **OCCURRENCE** W. Europe to C. Scandinavia, N. Russia, and Siberia.
- **REMARK** The spines appear on a newborn hedgehog within hours of its birth.

uniform grey-brown coat

short spines on back

EUROPE

Social unit Solitary	Gestation 31–35 days	Young 4–6	Diet

Family ERINACEIDAE	Species *Hemiechinus auritus*	Status Locally common

LONG-EARED DESERT HEDGEHOG

This desert hedgehog has black, brown, yellow, or white bands on its spines, and coarse fur on its whitish face, limbs, and belly. Its long ears help radiate heat, and it can survive in dry habitats for long periods without food or water.

• **SIZE** Body length: 15–27 cm (6–10½ in). Tail: 1–5 cm (⅜–2 in).

• **OCCURRENCE** Ukraine to Mongolia; Libya to Pakistan. In steppe and arid habitats.

• **REMARK** It may hibernate in summer if food is scarce.

AFRICA, ASIA

Social unit Solitary	Gestation 35–42 days	Young 1–6	Diet

Family ERINACEIDAE	Species *Echinosorex gymnura*	Status Lower risk*

MOONRAT

black, grey, and white streaks on coat

rough, coarse outer fur

Looking like a mix between a hedgehog and a small pig, the Moonrat has rough, spiky fur and a long scaly tail. A nocturnal hunter, it forages for prey and swims after fish and crustaceans. Its territory-marking scent smells of rotten onions.

• **SIZE** Body length: 26–46 cm (10–18 in). Tail: 16–30 cm (6½–12 in).

• **OCCURRENCE** S.E. Asia. In lowland tropical forest, often near water.

ASIA

Social unit Solitary	Gestation 35–40 days	Young 2	Diet

Family SOLENODONTIDAE	Species *Solenodon paradoxus*	Status Endangered

HISPANIOLAN SOLENODON

Fast and agile, the Hispaniolan Solenodon scrabbles on the forest floor for prey, sniffing it out with its long, mobile snout and stunning it with a poisonous bite, which it also uses for defence. This shrew-like, nocturnal insectivore varies from black to reddish brown, and its feet, long tail, and ears-tips are almost hairless.

CARIBBEAN

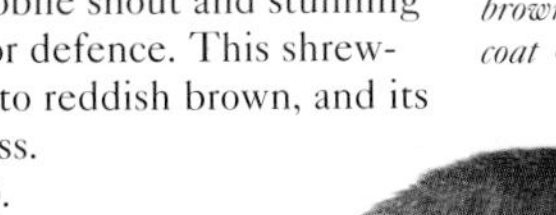

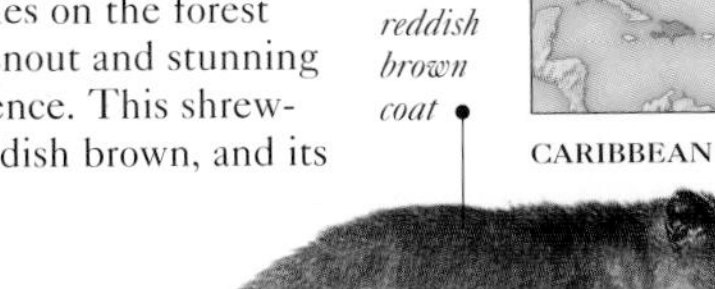

• **SIZE** Body length: 28–32 cm (11–12½ in). Tail: 17–26 cm (6½–10 in).

• **OCCURRENCE** Dominican Republic as well as Haiti. In forest and shrubland.

• **REMARK** There are two species of solenodon: Hispaniolan and Cuban. Both are endangered.

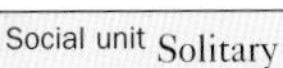

Social unit Solitary	Gestation Not known	Young 1–3	Diet

Family TENRECIDAE	Species *Tenrec ecaudatus*	Status Common

COMMON TENREC

One of the largest insectivores, the Common Tenrec resembles both shrew and hedgehog, as do the 25 other tenrec species found in C. Africa and Madagascar. It has sharp spines and coarse, grey to reddish fur, and uses its pointed snout to root among leaves and litter for prey at night. If threatened, the tenrec squeals, erects its spiny neck hair, jumps, and bucks, biting readily.

• **SIZE** Body length: 26–39 cm (10–15 ½ in). Tail: 1–1.5 cm (⅜–½ in).

• **OCCURRENCE** Madagascar. In tropical forest and savanna.

MADAGASCAR

coarse hair and spines

grey to reddish fur

long, mobile snout

Social unit Solitary	Gestation 50–60 days	Young 10–12	Diet

Family CHRYSOCHLORIDAE	Species *Eremitalpa granti*	Status Vulnerable

GRANT'S GOLDEN MOLE

With a small, compact body covered in soft, silky hair, Grant's Golden Mole may vary from steel-grey to buff or almost white. Adapted to burrowing in various ways, this mole has tiny, almost invisible eyes, a hard, naked nose-pad, and three long, broad claws on each foot. Hardly ever seen above ground, it pushes through loose sand as if swimming, making burrows in deeper soil or near the surface in compacted sand. Active during the day and at night, it detects the movement of its prey while digging in the soil.

• **SIZE** Body length: 7–8 cm (2 ¾–3 ¼ in). Tail: None.

• **OCCURRENCE** S.W. Africa. In coastal sand dunes.

• **REMARK** A very specialized species, this mole is particularly vulnerable to habitat destruction or isolation due to parts of its range being separated by mining and other industrial activities.

AFRICA

grey or buff coat

blunt head for burrowing

Social unit Solitary	Gestation Not known	Young Not known	Diet

Family SORICIDAE	Species *Scutisorex somereni*	Status Unconfirmed

ARMOURED SHREW

Also known as the Hero Shrew – the only species in its genus – this insectivore has a grey coat and a characteristically arched and immensely strong back. This is because its vertebrae have interlocking flanges, or spines, not only along their sides, as in other mammals, but also at the top and the bottom. Stories of people actually standing on this shrew with it escaping unhurt, have been recorded. It is active in the day as well as the night.

- **SIZE** Body length: 10–15 cm (4–6 in). Tail: 6.5–9.5 cm (2¾–3¾ in).
- **OCCURRENCE** C. to E. Africa. In tropical forest.
- **REMARK** Local people believe that it has magical properties.

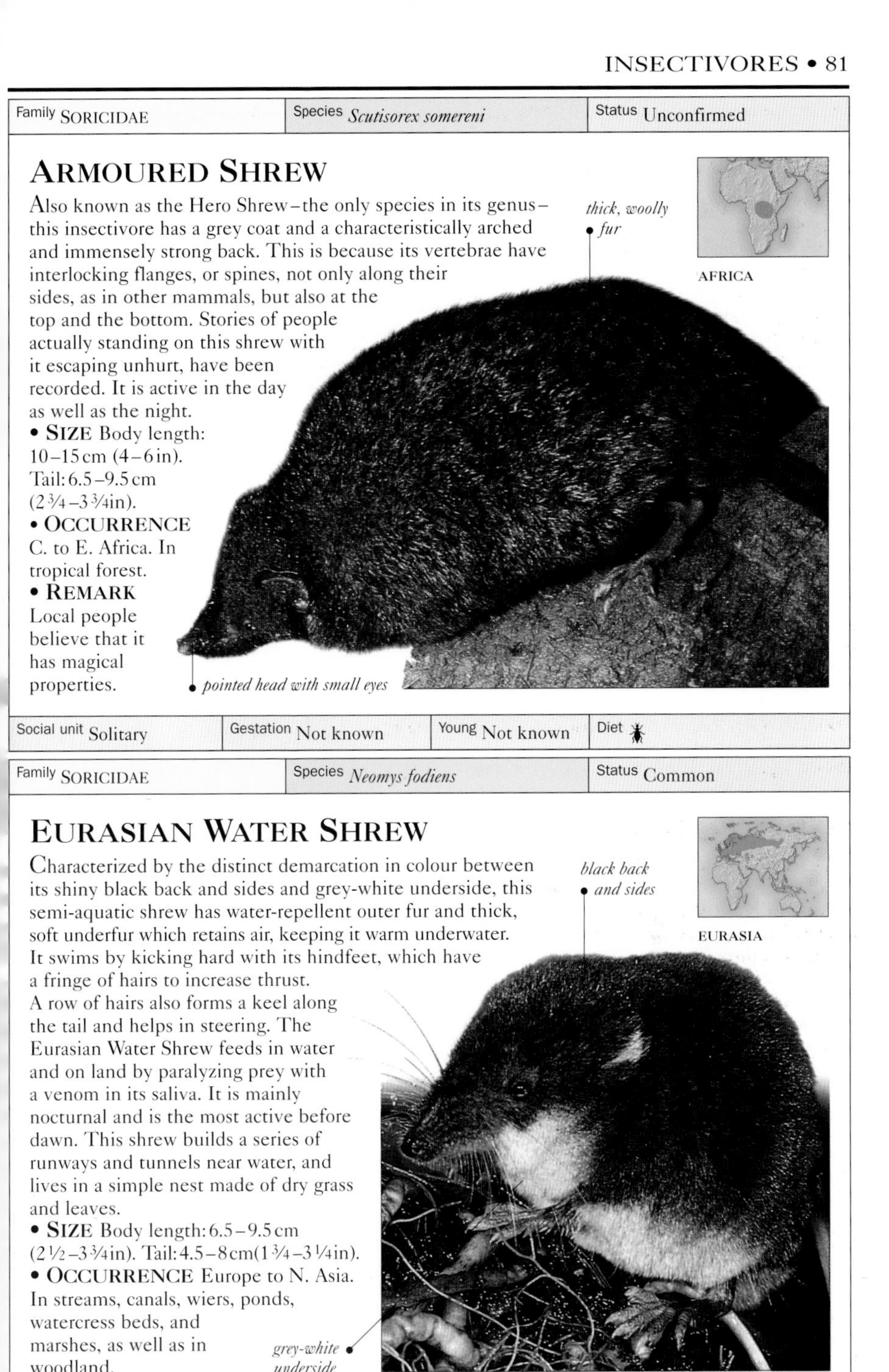

Social unit Solitary	Gestation Not known	Young Not known	Diet

Family SORICIDAE	Species *Neomys fodiens*	Status Common

EURASIAN WATER SHREW

Characterized by the distinct demarcation in colour between its shiny black back and sides and grey-white underside, this semi-aquatic shrew has water-repellent outer fur and thick, soft underfur which retains air, keeping it warm underwater. It swims by kicking hard with its hindfeet, which have a fringe of hairs to increase thrust. A row of hairs also forms a keel along the tail and helps in steering. The Eurasian Water Shrew feeds in water and on land by paralyzing prey with a venom in its saliva. It is mainly nocturnal and is the most active before dawn. This shrew builds a series of runways and tunnels near water, and lives in a simple nest made of dry grass and leaves.

- **SIZE** Body length: 6.5–9.5 cm (2½–3¾ in). Tail: 4.5–8 cm (1¾–3¼ in).
- **OCCURRENCE** Europe to N. Asia. In streams, canals, wiers, ponds, watercress beds, and marshes, as well as in woodland.

Social unit Solitary	Gestation 14–21 days	Young 6	Diet

Family SORICIDAE	Species *Blarina brevicauda*	Status Common

NORTH AMERICAN SHORT-TAILED SHREW

This robust shrew has short, greyish black fur, tiny eyes, concealed ears, and a hairy tail. Mostly active at night, it uses its senses of smell and touch to hunt out its prey from the soil, and disables it with a toxic bite. Unusually for a shrew it eats voles, mice, and vegetation.

• **SIZE** Body length: 12–14 cm (4¾–5½ in). Tail: 3 cm (1¼ in).

N. AMERICA

• **OCCURRENCE** S. Canada to N. and E. USA. In deciduous and coniferous forest, bogs, and grassland.

Social unit Variable	Gestation 17–22 days	Young 3–7	Diet

Family SORICIDAE	Species *Suncus etruscus*	Status Unconfirmed

PYGMY WHITE-TOOTHED SHREW

Living on a diet of small insects, this species hunts actively for a few hours, and then rests, throughout the day and night. It relies on its speed to surprise and attack prey that is slower and larger than itself, and also scavenges on recently dead insects. This minute mammal is a typical example of a shrew with its pointed snout; it has large ears and fur that is mainly greyish brown.

• **SIZE** Body length: 4–5 cm (1½–2 in). Tail: 2–3 cm (¾–1¼ in).

• **OCCURRENCE** S. Europe, S. and S.E. Asia, N., E., and W. Africa.

EUROPE, ASIA, AFRICA

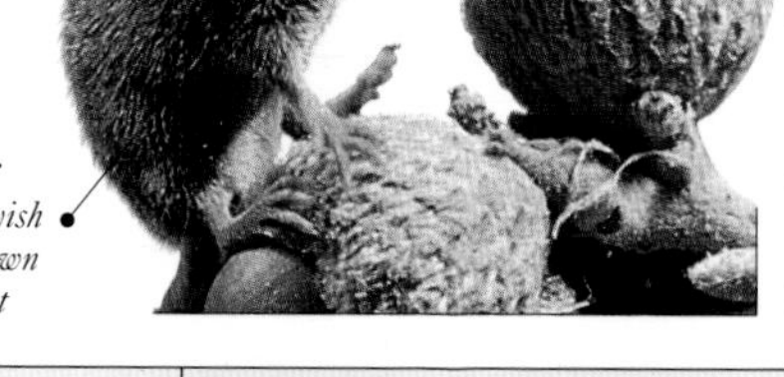

Social unit Solitary	Gestation 27–28 days	Young 2–5	Diet

Family TALPIDAE	Species *Condylura cristata*	Status Locally common

STAR-NOSED MOLE

An unmistakable star-shaped nose, with 22 radiating tentacles to sniff out its prey in water, gives this mole its name. It forages during the day and night and is an expert swimmer.

• **SIZE** Body length: 18–19 cm (7–7½ in). Tail: 6–8 cm (2½–3 in).

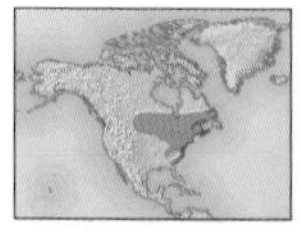

N. AMERICA

• **OCCURRENCE** E. Canada to N.E. USA. In marshes, and along streams and lakes.

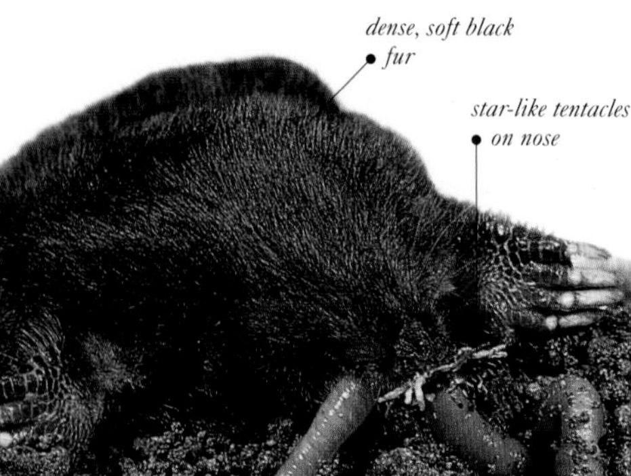

Social unit Variable	Gestation Not known	Young 2–7	Diet

Family TALPIDAE	Species *Talpa euoropaea*	Status Common

EUROPEAN MOLE

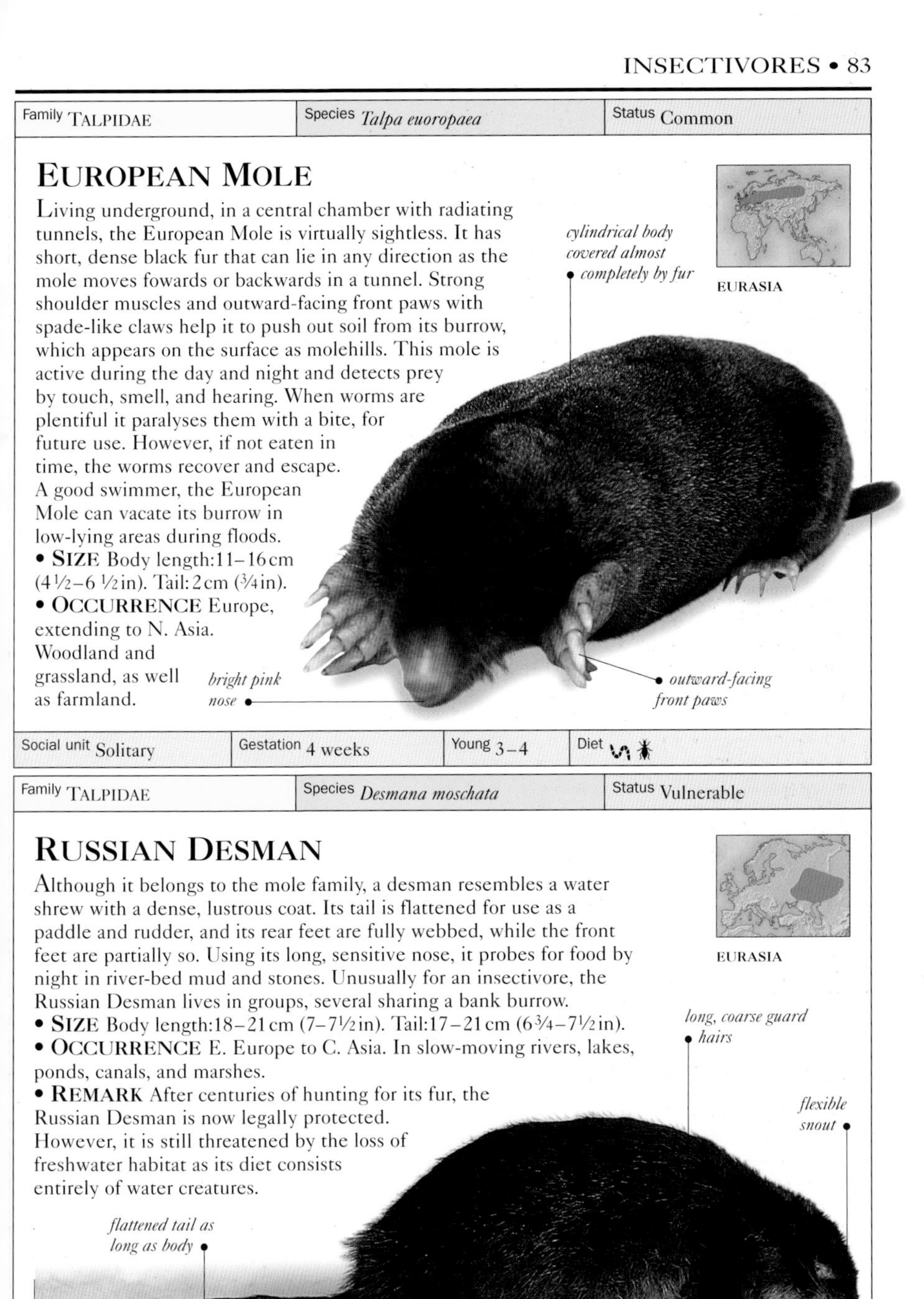

Living underground, in a central chamber with radiating tunnels, the European Mole is virtually sightless. It has short, dense black fur that can lie in any direction as the mole moves fowards or backwards in a tunnel. Strong shoulder muscles and outward-facing front paws with spade-like claws help it to push out soil from its burrow, which appears on the surface as molehills. This mole is active during the day and night and detects prey by touch, smell, and hearing. When worms are plentiful it paralyses them with a bite, for future use. However, if not eaten in time, the worms recover and escape. A good swimmer, the European Mole can vacate its burrow in low-lying areas during floods.

- **SIZE** Body length: 11–16 cm (4½–6 ½in). Tail: 2 cm (¾in).
- **OCCURRENCE** Europe, extending to N. Asia. Woodland and grassland, as well as farmland.

Social unit Solitary	Gestation 4 weeks	Young 3–4	Diet

Family TALPIDAE	Species *Desmana moschata*	Status Vulnerable

RUSSIAN DESMAN

Although it belongs to the mole family, a desman resembles a water shrew with a dense, lustrous coat. Its tail is flattened for use as a paddle and rudder, and its rear feet are fully webbed, while the front feet are partially so. Using its long, sensitive nose, it probes for food by night in river-bed mud and stones. Unusually for an insectivore, the Russian Desman lives in groups, several sharing a bank burrow.

- **SIZE** Body length: 18–21 cm (7–7½in). Tail: 17–21 cm (6¾–7½in).
- **OCCURRENCE** E. Europe to C. Asia. In slow-moving rivers, lakes, ponds, canals, and marshes.
- **REMARK** After centuries of hunting for its fur, the Russian Desman is now legally protected. However, it is still threatened by the loss of freshwater habitat as its diet consists entirely of water creatures.

Social unit Social	Gestation Not known	Young 3–5	Diet

BATS

MORE THAN one-fifth of all mammals, almost 1,000 species, are bats. These are the only mammals capable of sustained flight. A bat's front limbs are wings consisting of a thin, extensible, leathery membrane, the patagium, held out by long, slim finger and arm bones.

Bats live mainly in tropical and temperate forests worldwide. Some species are found in more open habitats, and in recent times a few species have adapted to live in human habitations.

Most bats belong to the suborder Microchiroptera and fly at night, using a sound-radar system to catch small aerial insects. Certain species have specialized diets such as fish, snails, or blood. In temperate regions they survive winter's lack of food by hibernating.

The suborder Megachiroptera consists of the larger flying foxes and fruit bats. As their name suggests, they consume fruit and other plant material.

Family MEGADERMATIDAE	Species *Macroderma gigas*	Status Vulnerable

AUSTRALIAN FALSE VAMPIRE BAT

Also known as the Ghost Bat, this species derives its name from the misconception that it feeds on blood. Large ears, a forked tragus (ear projection), and prominent noseleaf, make this bat unmistakable. It has pale grey or light brown fur and pale cream to brown wings. It roosts in large numbers in rocky crevices, but is affected by increased mining activity.

• **SIZE** Body length: 10–12 cm (4–4 3/4 in). Forearm: 9.5–11 cm (3 3/4–4 1/4 in).

• **OCCURRENCE** W. and N. Australia. In tropical rainforest, grassland, and rocky hillsides.

AUSTRALIA

Social unit Social	Gestation 11–12 weeks	Young 1	Diet

Family EMBALLONURIDAE	Species *Taphozous mauritianus*	Status Common

MAURITIAN TOMB BAT

A medium-sized "sheath-tailed" bat, the Mauritian Tomb Bat is distinguished from other members of its genus by its grizzled brown-black coat with white wings and underparts. Watchful as it roosts by day on walls and tree trunks, it uses echolocation to detect flying prey. It also produces a range of vocalizations that are audible to humans.

• **SIZE** Body length: 7.5–9.5 cm (2 3/4–3 1/2 in). Forearm: 6 cm (2 1/4 in).

• **OCCURRENCE** W., C., E., and S. Africa, Madagascar, and Mauritius. In forest, dry habitats, and urban settlements.

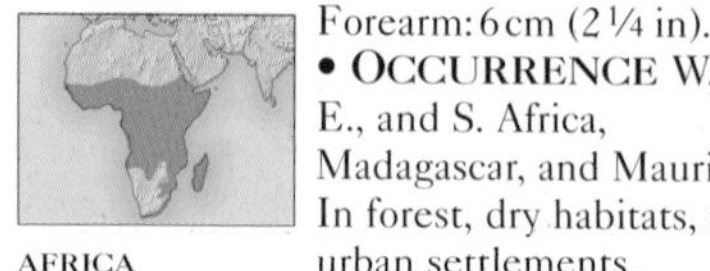

AFRICA

Social unit Social	Gestation 90 days	Young 1	Diet

Family EMBALLONURIDAE	Species *Rhynchonycteris naso*	Status Common

PROBOSCIS BAT

With its long, pointed nose that extends over its mouth, this species is also known as the Sharp-nosed Bat. Grizzled grey-brown to yellowish fur, with two faint, wavy, cream stripes from shoulder to rump, make this streamlined bat distinctive. A fairly typical, small, insectivorous bat, it does, however, show unique roosting behaviour. Groups of 5–10 (rarely more than 40) roost by day in a line, nose-to-tail on a branch or wooden beam. Each line includes a single dominant male that defends the feeding area, which is often near a patch of water. A horizontal line of Proboscis Bats on a beam in a boathouse, or a branch hanging over water, is a common, but striking sight.

• **SIZE** Body length: 3.5–5 cm (1 ½–2 in). Forearm: 3.5–4 cm (1 ½–1 ¾ in).

• **OCCURRENCE** Mexico, Peru, Bolivia, Brazil, French Guiana, Surinam, Guyana, and Trinidad. In lowland tropical forest along waterways.

DISTINCTIVE LOOKS
The most outstanding feature of this bat is its proboscis-like nose. Its fur is grizzled, with protruding tufts on its forearms, and its head is small in proportion to its body.

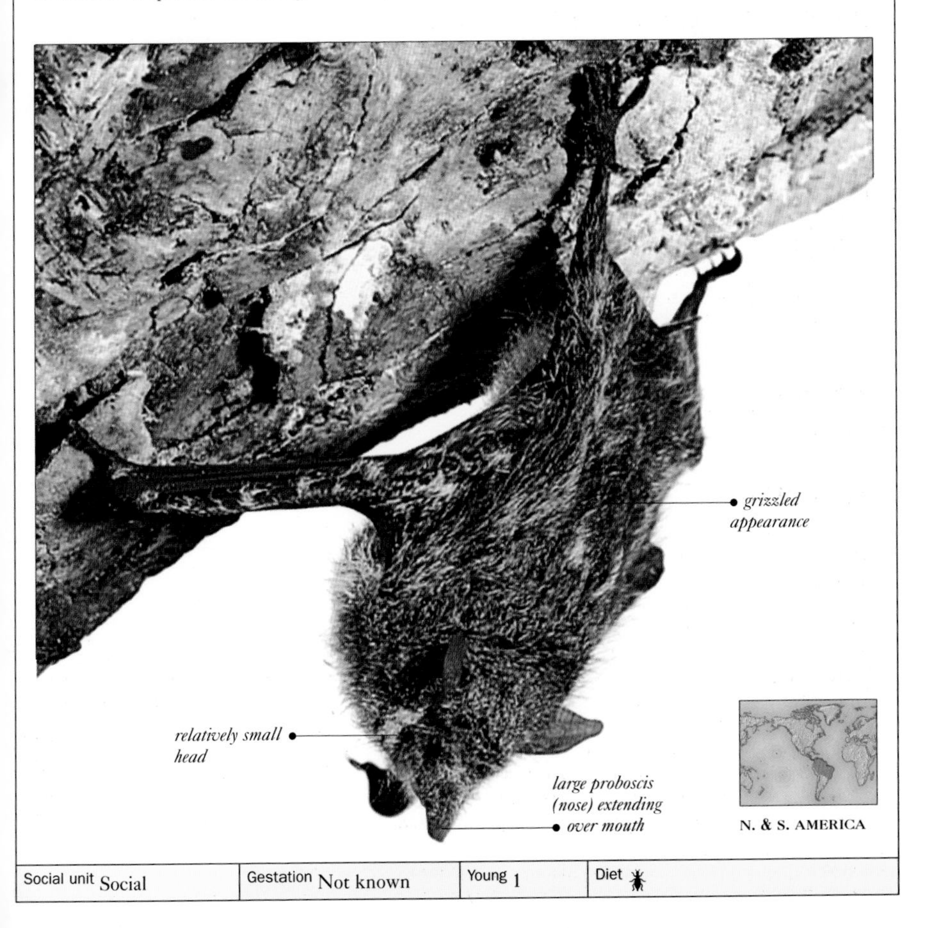

Social unit Social	Gestation Not known	Young 1	Diet

Family PTEROPODIDAE	Species *Epomops franqueti*	Status Unconfirmed

FRANQUET'S EPAULETTED BAT

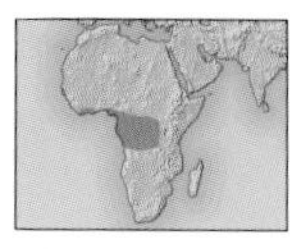

This species is a medium-sized fruit bat, the male of which is slightly heavier than the female, and has shoulder patches of long white hair resembling epaulettes. These may be drawn into the fur and hidden. Males make high-pitched whistling calls from their perches during the night, to attract mates, and their collective din is a common sound in the African night. While males defend their calling perches from other males, these bats do not display territorial behaviour while feeding or roosting. The Franquet's Epauletted Bat may breed at any time of the year, females bearing two litters in areas where food – consisting chiefly of fruits such as figs, guavas, and bananas, as well as soft, young leaves – is available in plenty. Usually found roosting in groups in the daytime, these bats prefer the foliage of trees and vines. Individuals rarely make physical contact, mothers alone taking care of their young until they are weaned. The pattern of palatal ridges on the roof of the mouth, specialized for drawing out fruit pulp and juices, distinguishes one species from another in this genus.

• **SIZE** Body length: 11–15 cm (4 1/4–6 in).
Forearm: 8–9 cm (3 1/4–3 1/2 in).

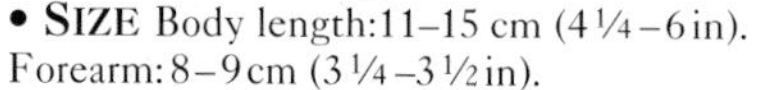

• **OCCURRENCE** W. and C. Africa.
In rainforest.

pale patch of hair at base of ear

wide wingspan

brownish wing membrane

thumb

extremely long forearms

broad molars for crushing fruits to pulp

broad, rounded ears

SPECIALIZED FEEDERS
Like other fruit bats, this bat uses its broad molars to crush fruits, rubbing its tongue against the palatal ridges of its mouth and sucking to draw out juices and pulp before spitting out the fruit fibres.

Social unit Social	Gestation Not known	Young 1	Diet 

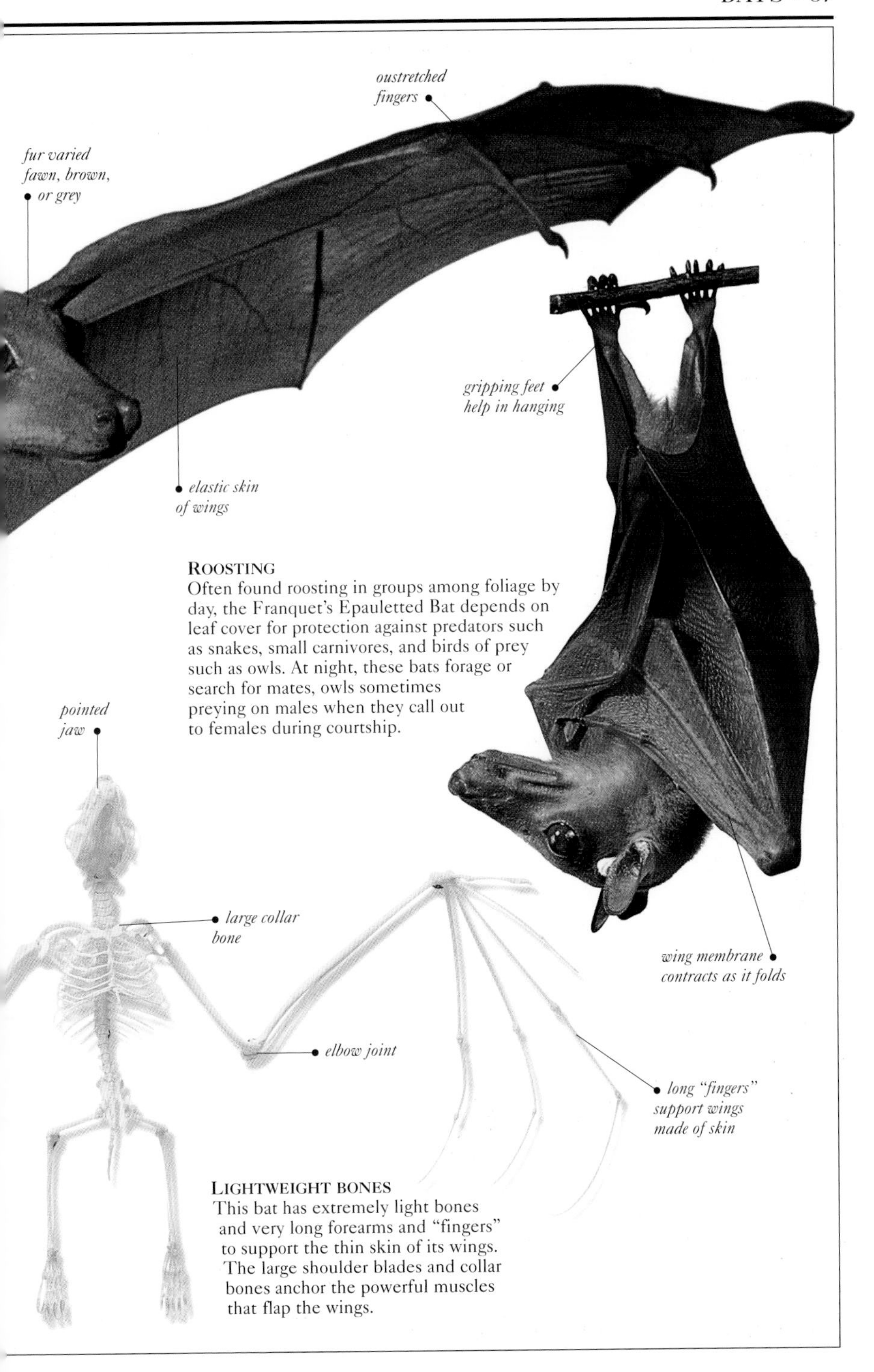

ROOSTING

Often found roosting in groups among foliage by day, the Franquet's Epauletted Bat depends on leaf cover for protection against predators such as snakes, small carnivores, and birds of prey such as owls. At night, these bats forage or search for mates, owls sometimes preying on males when they call out to females during courtship.

LIGHTWEIGHT BONES

This bat has extremely light bones and very long forearms and "fingers" to support the thin skin of its wings. The large shoulder blades and collar bones anchor the powerful muscles that flap the wings.

Family PTEROPODIDAE	Species *Rousettus egyptiacus*	Status Common

EGYPTIAN FRUIT BAT

Also known as the Egyptian Rousette Bat, this species may vary from dark brown to slate-grey on its upperparts, with lighter, smoky grey underparts. It is unusual among bats in having fur that extends about halfway along each forearm. Unlike most fruit bats, this species uses high-pitched echolocation clicks which allow it to find its way and roost in cool, dark caves.

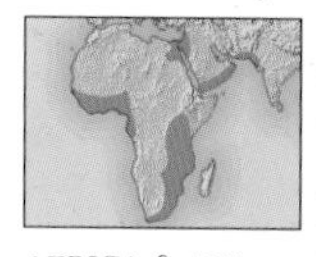
AFRICA & ASIA

• **SIZE** Body length: 14–16cm (5½–6½in). Forearm: 8.5–10cm (3¼–4in).
• **OCCURRENCE** N., W., E., and S. Africa, W. Asia. In desert, tropical forest, and urban areas.

dark brown to grey body
yellow or buff collar

Social unit Social	Gestation 4 months	Young 1	Diet

Family RHINOPOMATIDAE	Species *Rhinopoma hardwickei*	Status Common

LESSER MOUSE-TAILED BAT

This bat is one of four *Rhinopoma* species, sometimes called Long-tailed Bats, all of which have long, thin, trailing, mouse-like tails, that may be as long as the head and body. An inhabitant of arid, open areas, the Lesser Mouse-tailed Bat is able to withstand drier conditions than many other bats. When food is available in abundance, it may store fat for the long dry season, during which it remains inactive.

AFRICA & ASIA

• **SIZE** Body length: 5.5–7cm (2½–2¾in). Forearm: 4.5–6cm (1¾–2¼in).
• **OCCURRENCE** N. and E. Africa, W. to S. Asia. In desert and tropical forest – both covered and open areas.

very long tail
leaf-like structure on nose

Social unit Social	Gestation 123 days	Young 1	Diet

Family NOCTILIONIDAE	Species *Noctilio leporinus*	Status Lower risk*

GREATER BULLDOG BAT

Velvety orange, brown, or grey fur with a distinctive pale stripe along its back characterize the Greater Bulldog Bat. By night, it hunts for fish, fiddler crabs, and other prey such as arthropods, snatching them from the ground or surface of water in rivers and estuaries, using its large, powerful, sharp-clawed hindfeet.

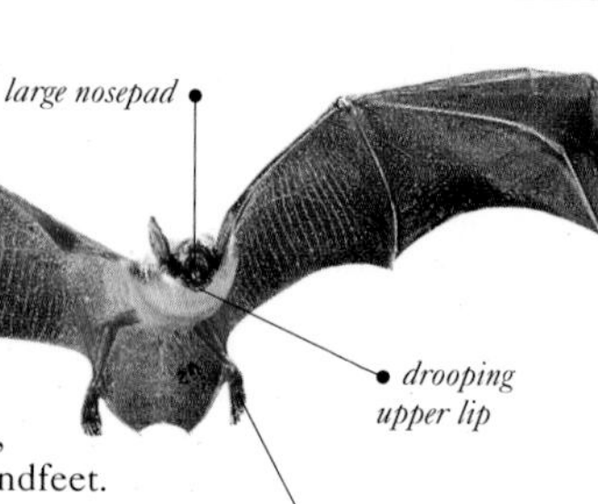

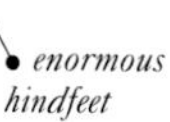

C. & S. AMERICA

• **SIZE** Body length: 9–10cm (3½–4in). Forearm: 8–9cm (3¼–3½in).
• **OCCURRENCE** Central America, N., E., and C. South America. In forest and near rivers.

Social unit Social	Gestation 60–70 days	Young 1	Diet

Family RHINOLOPHIDAE	Species *Rhinolophus hipposideros*	Status Vulnerable

LESSER HORSESHOE BAT

This bat has a head and body smaller than a human thumb. Widespread in woods and scrub, it faces the threat of habitat destruction – both its winter hibernation sites, such as deep caverns, and summer daytime roosts in tree-holes, caves, chimneys, and mine shafts, have been disturbed.

EUROPE, AFRICA, ASIA

- **SIZE** Body length: 4 cm (1 ½ in). Forearm: 3.5–4.5 cm (1 ½–1 ¾ in).
- **OCCURRENCE** Europe, N. Africa to W. Asia. In desert and temperate forest, in covered places.

Social unit Social	Gestation 2 months	Young 1	Diet

Family NATALIDAE	Species *Natalus stramineus*	Status Unconfirmed

MEXICAN FUNNEL-EARED BAT

Tiny and delicate, with rounded ears, soft, woolly fur, and a tail joined by flight membranes to the legs, this is one of five species of American Funnel-eared Bats, all found in the tropics. This species may be distinguished by its rapid, almost butterfly-like flight. Often encountered roosting by day in caves, which may house hundreds of individuals hanging separately, it feeds by night on small, flying insects.

N., C., & S. AMERICA

- **SIZE** Body length: 4–4.5 cm (1 ½–1 ¾ in). Forearm: 3.5–4 cm (1 ½ in).
- **OCCURRENCE** W. USA to N. South America. In tropical forest and dry, semi-deciduous forest, mostly deep within caves.

tail longer than head and body

orange to yellowish brown fur

Social unit Social	Gestation Not known	Young 1	Diet

Family MORMOOPIDAE	Species *Pteronotus davyi*	Status Common

DAVY'S NAKED-BACKED BAT

A common sight at night in urban areas, where it feeds on moths, flies, and other flying insects attracted to street lights, the Davy's Naked-backed Bat roosts by day in large colonies in caves and mines, usually some distance away from its main feeding areas. It is the smallest among moustached bats (mormoopids), and its wings meet along the centre of its back, obscuring the fur beneath, and giving it a "naked-backed" appearance.

N. & S. AMERICA

- **SIZE** Body length: 4–5.5 cm (1 ½–2 ½ in). Forearm: 4–5 cm (1 ½–2 in).
- **OCCURRENCE** Mexico to N. and E. South America. In tropical forest.

Social unit Social	Gestation Not known	Young 1	Diet

Family PHYLLOSTOMIDAE	Species *Anoura geoffroyi*	Status Common

GEOFFROY'S TAIL-LESS BAT

Also known by the alternative name, Geoffroy's Hairy-legged Bat, this bat's tail is reduced to a furred membrane. Its triangular, upright nose also gives it a third name: Geoffroy's Long-nosed Bat. It hovers in front of night-blooming flowers, feeding on nectar and pollen with its long, brush-tipped tongue – playing an important role in pollinating several plant species. Large colonies of bats roost in caves and tunnels.

N., C., & S. AMERICA

- **SIZE** Body length: 6–7.5 cm (2¼–3 in). Forearm: 4–4.5 cm (1½–1¾ in).
- **OCCURRENCE** Mexico, Caribbean, N. South America. In tropical and evergreen forest.

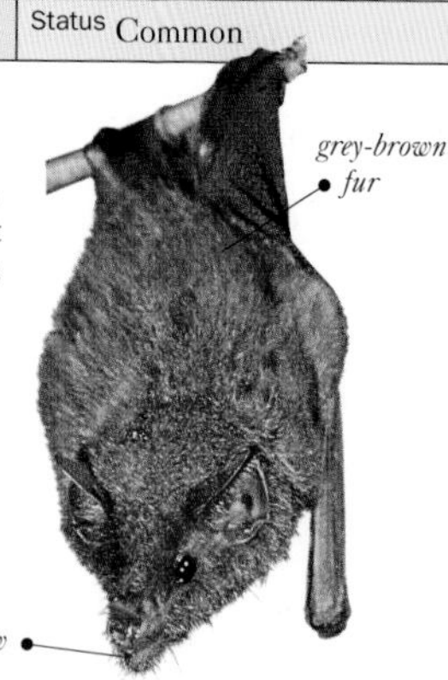

Social unit Social	Gestation Not known	Young 1	Diet

Family PHYLLOSTOMIDAE	Species *Uroderma bilobatum*	Status Common

TENT-BUILDING BAT

A grey-brown species with white stripes on its face and back, this is one of about 15 species of leaf-nosed bats that roost in "tents" – shelters made by biting palm or banana leaves so that they fold over. Around 2–50 individuals live in these tents, which protect them from sunlight, rain, and predators. This bat chews on fruits, sucking out their juices.

N., C., & S. AMERICA

- **SIZE** Body length: 6–6.5 cm (2¼–2½ in). Forearm: 4–4.4 cm (1½–1¾ in).
- **OCCURRENCE** Mexico to C. South America. In deciduous and evergreen forest.

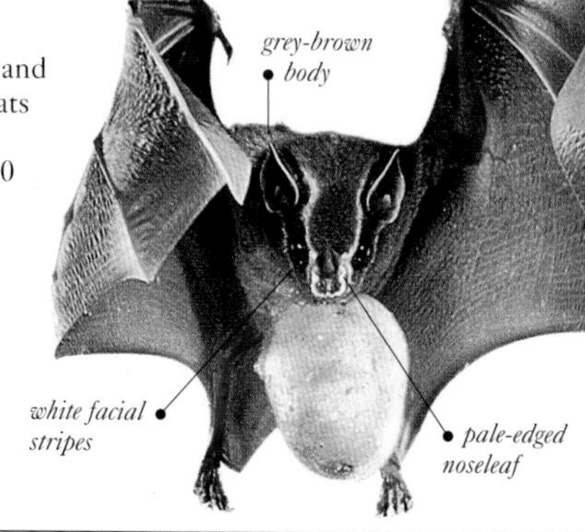

Social unit Social	Gestation 4–5 months	Young 1	Diet

Family PHYLLOSTOMIDAE	Species *Vampyrum spectrum*	Status Lower risk

FALSE VAMPIRE BAT

A wingspan of 1 m (3¼ ft) makes this bat (also known as Linnaeus' False Vampire Bat), the largest in the Americas. It is not a blood-sucker, but a powerful predator that hunts other bats, rodents such as mice and rats, and birds such as wrens, orioles, and parakeets. It grabs its prey directly in its mouth and kills it with a powerful bite. By day it roosts in groups of up to five, inside hollow trees. This species is at risk from habitat loss.

N., C., & S. AMERICA

- **SIZE** Body length: 13.5–15 cm (5¼–6 in). Forearm: 10–11 cm (4–4¼ in).
- **OCCURRENCE** Mexico to N. South America, and Trinidad. In evergreen forest.

Social unit Variable	Gestation Not known	Young 1	Diet

Family PHYLLOSTOMIDAE	Species *Desmodus rotundus*	Status Common

VAMPIRE BAT

The Vampire Bat is distinguished by its pointed, blade-like canines and upper incisors. From dusk this strong flier searches for a warm-blooded victim – a bird, domestic animal, or even a human being. It bites away the fur or feathers without the victim noticing, and laps up to 25 ml (1 fl oz) of blood over 30 minutes.

- **SIZE** Body length: 7–9.5 cm (2 3/4 – 3 3/4 in). Forearm: 5.3 cm (2 1/4 in).
- **OCCURRENCE** Mexico, Central and South America. In forest, woodland, grassland, and cities.
- **REMARK** Among the most social of mammals, it roosts in groups and often shares blood by regurgitation.

Social unit Social	Gestation 7 months	Young 1	Diet Blood

Family PHYLLOSTOMIDAE	Species *Pipistrellus pipistrellus*	Status Common

COMMON PIPISTRELLE

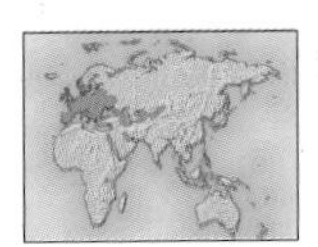

The 70 or more species of pipistrelles are distinguished from each other by their size, colour, and dental features. The Common Pipistrelle is the most widespread of these. Emerging from its roost in the early evening, it preys on small insects such as common flies, caddis flies, mayflies, and moths. By day, the Common Pipistrelle roosts in cracks, crevices, buildings, and bat boxes. It hibernates through the winter, like other bats from temperate regions. This bat communicates by an array of vocal signals, males calling out to females during courtship. Nursery colonies may contain 1,000 mothers, each recognizing and nursing her single young.

- **SIZE** Body length: 3.5–4.5 cm (1 1/2 – 2 in). Forearm: 2.8 – 3.5 cm (1 1/4 – 1 1/2 in).
- **OCCURRENCE** Europe to N. Africa, W. and C. Asia. In temperate and coniferous forest, parkland, riverbanks, suburban gardens, and urban areas.
- **REMARK** Pipistrelle bats often emerge from their roosts in enormous swarms to prevent being ambushed by predators.

Social unit Social	Gestation 44 days	Young 1	Diet (insect symbol)

Family VESPERTILIONIDAE	Species *Nyctalus noctula*	Status Lower risk*

COMMON NOCTULE BAT

This bat is one of the largest and the most widespread of the six noctule species. It has sleek, golden, ginger, or reddish fur, and short, broad ears. The female is slightly larger than the male. A powerful flier, the Common Noctule Bat dives steeply to catch large flying insects such as crickets and chafers, using echolocation to track them. It also forages on insects that are attracted to light or are found over rubbish dumps. During the breeding season in late summer and early autumn, the male bat becomes territorial, calling out to females from tree hollows. In spring, the female may produce up to three young, whereas most small bats have only one offspring. In summer, the females form nursery colonies of over 100 individuals in trees and buildings. Later in the year, they gather in groups around the usually solitary males.

• **SIZE** Body length: 7–8 cm (2 3/4–3 in). Forearm: 4.7–5.5 cm (1 3/4–2 1/4 in).

• **OCCURRENCE** Europe to W., E., and S. Asia. In temperate forest and urban areas.

• **REMARK** The Common Noctule Bat hibernates for part of the winter and then migrates up to 2,000 km (1,200 miles) to newer summer feeding grounds.

ROOSTING

By day, the Common Noctule Bat roosts in the hollows of trees, often competing with starlings for tree-holes, as well as in buildings, or among rocks. It emerges early in the evening from its roost and forages high above the ground or in the tree canopy.

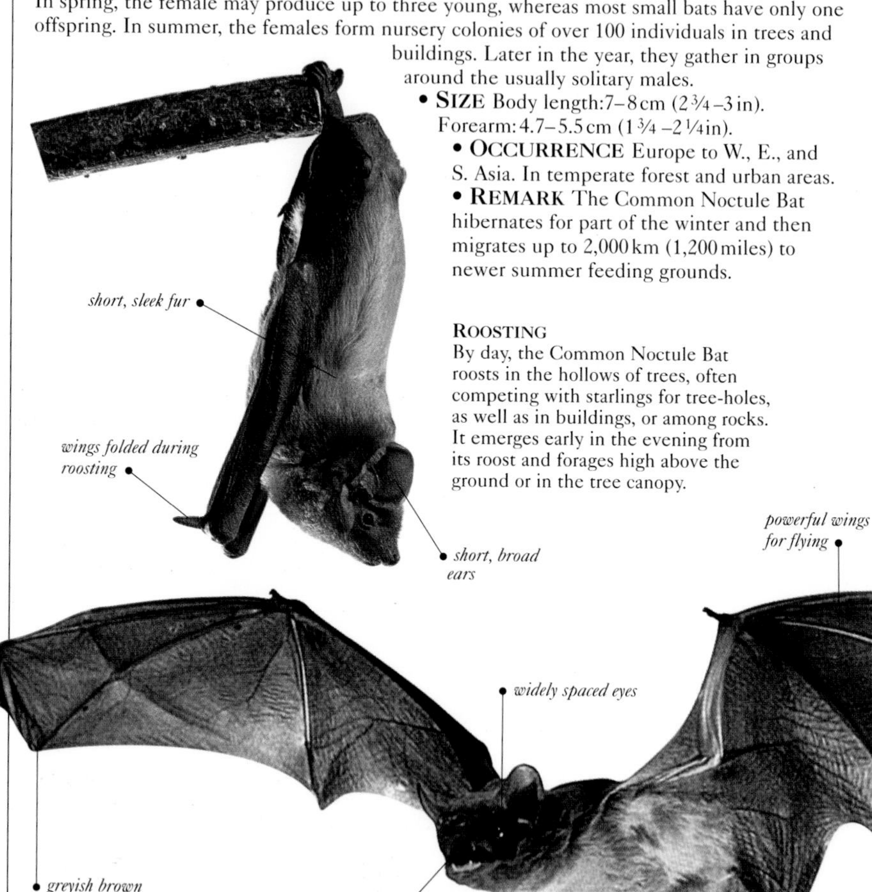

Social unit Social	Gestation 70–73 days	Young 1–3	Diet

Family VESPERTILIONIDAE	Species *Plecotus auritus*	Status Lower risk*

BROWN LONG-EARED BAT

The large ears, about three-quarters the length of its head and body combined, distinguish this bat. The ears help it detect prey, such as moths, midges, mosquitoes, and flies, which it picks off the vegetation in swooping flights. The Brown Long-eared Bat roosts in buildings and trees, females and young forming nursery colonies. In winter it hibernates in caves, mines, and cellars.

• **SIZE** Body length: 4–5 cm (1 ½–2 in). Forearm: 23–28 cm (9–11 in).

• **OCCURRENCE** Europe to C. Asia. In sheltered, lightly wooded areas.

EURASIA

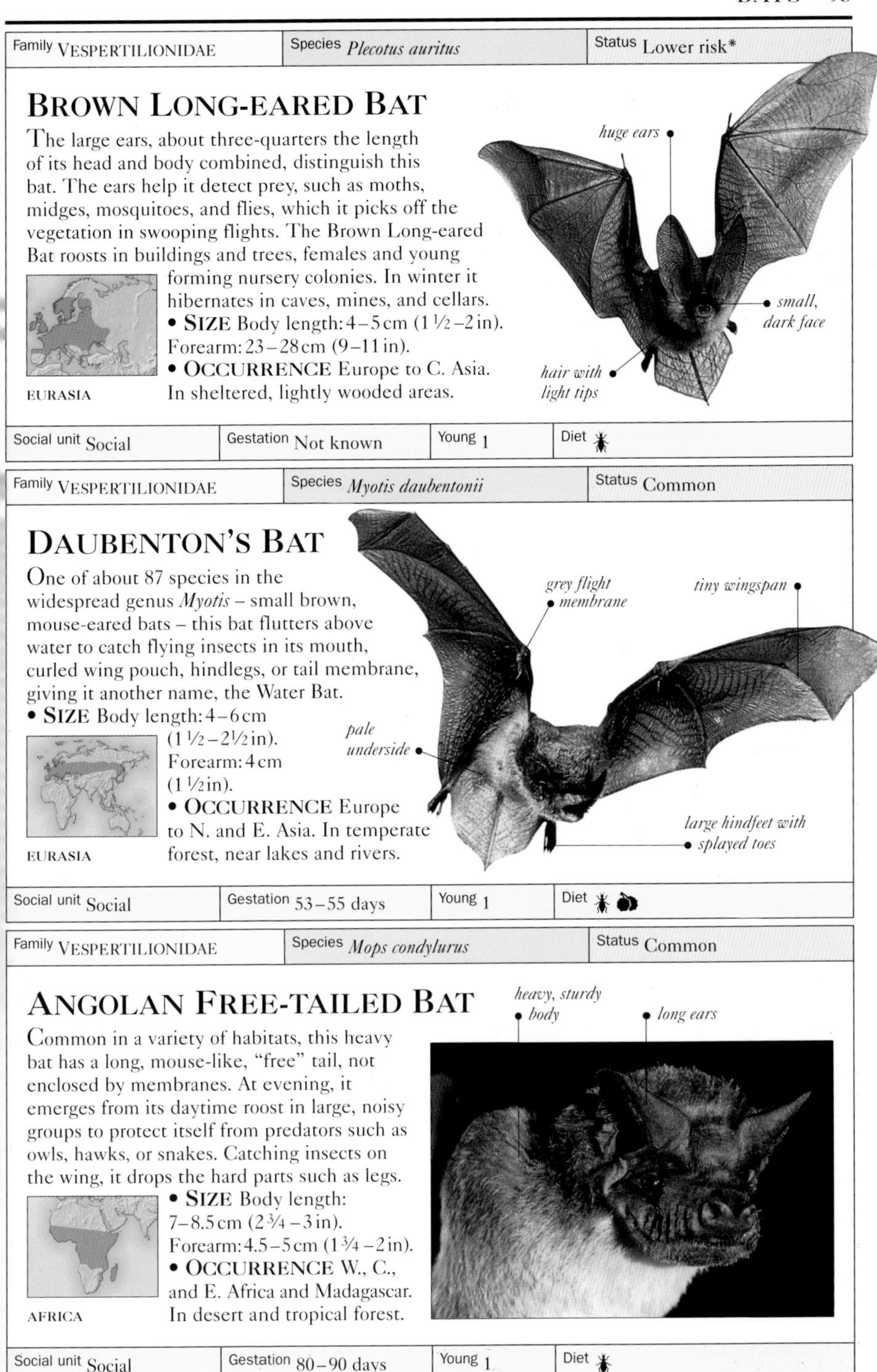

Social unit Social	Gestation Not known	Young 1	Diet

Family VESPERTILIONIDAE	Species *Myotis daubentonii*	Status Common

DAUBENTON'S BAT

One of about 87 species in the widespread genus *Myotis* – small brown, mouse-eared bats – this bat flutters above water to catch flying insects in its mouth, curled wing pouch, hindlegs, or tail membrane, giving it another name, the Water Bat.

• **SIZE** Body length: 4–6 cm (1 ½–2½ in). Forearm: 4 cm (1 ½ in).

• **OCCURRENCE** Europe to N. and E. Asia. In temperate forest, near lakes and rivers.

EURASIA

Social unit Social	Gestation 53–55 days	Young 1	Diet

Family VESPERTILIONIDAE	Species *Mops condylurus*	Status Common

ANGOLAN FREE-TAILED BAT

Common in a variety of habitats, this heavy bat has a long, mouse-like, "free" tail, not enclosed by membranes. At evening, it emerges from its daytime roost in large, noisy groups to protect itself from predators such as owls, hawks, or snakes. Catching insects on the wing, it drops the hard parts such as legs.

• **SIZE** Body length: 7–8.5 cm (2 ¾–3 in). Forearm: 4.5–5 cm (1 ¾–2 in).

• **OCCURRENCE** W., C., and E. Africa and Madagascar. In desert and tropical forest.

AFRICA

Social unit Social	Gestation 80–90 days	Young 1	Diet

ELEPHANT-SHREWS

FOUND ONLY IN Africa, the 15 species of elephant-shrews (order Macroscelidea) occupy a variety of habitats, from rocky hillsides to grassland and the forest floor. The group's common name is derived from the elongated, trunk-like nose, which is extremely sensitive to touch and scent. Hearing and sight are also keen.

Elephant-shrews bound around their territories by day, using their powerful back legs and long tails in the manner of a tiny kangaroo. They chase away intruders of their own kind, and follow well-maintained trails from their burrows to forage for small prey such as insects; some species also eat buds, berries, and other plant material.

Elephant-shrews resemble true shrews in appearance and habits, and have been classified as insectivores. However, certain aspects of their anatomy suggest that they may be more closely related to rabbits and hares (lagomorphs).

Family MACROSCELIDIDAE	Species *Rhynchocyon chrysopygus*	Status Endangered

GOLDEN-RUMPED ELEPHANT-SHREW

Distinctive and colourful in appearance, this elephant-shrew has a russet head and body, golden patches on its forehead and rump, and hairless, black legs, feet, and ears. When alarmed, it slaps its white-tipped tail on leaf litter as a warning and bounds away.

• **SIZE** Length: 27–29 cm (10 1/2–11 1/2 in). Weight: 525–550 g (1 1/4 lb).

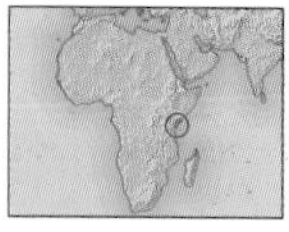

• **OCCURRENCE** Kenyan coast to the border of Somalia. In coastal, dry forest and scrub, including coral rag scrub.

distinctive golden rump patch

proboscis-like long beak

Social unit Solitary/Pair	Gestation 42 days	Young 1	Diet

Family MACROSCELIDIDAE	Species *Elephantulus rufescens*	Status Lower risk

RUFOUS ELEPHANT-SHREW

Grey or brown in colour, with whitish underparts, this fleet-footed animal probes leaf litter with its long nose and licks up small invertebrates, especially termites, with its sticky tongue; it also feeds on fruits and other plant matter. It does not nest or burrow, but creates foraging trails, along which it flees from predators.

• **SIZE** Length: 12–12.5 cm (4 3/4–5 in). Weight: 50–60 g (1 3/4–2 1/8 oz).

• **OCCURRENCE** E. Africa. In arid woodland and grassland.

• **REMARK** This shrew is monogamous, each sex chasing away its rivals.

Social unit Solitary/Pair	Gestation 60 days	Young 1–2	Diet

FLYING LEMURS

ALSO KNOWN AS colugos, the two species of flying lemur make up the order Dermoptera. Both species live in the forests of Southeast Asia. Similar in size and form, they have unusual, ridged lower teeth to sieve sap and juices from fruits and other plant matter.

They are not lemurs, but they do have lemur-like faces, with large, front-facing eyes enabling them to judge distances precisely. They are not true fliers, but they are the most accomplished gliding mammals, able to swoop more than 100 m (330 ft) horizontally with minimal height loss, and swerve on the way with excellent aerial control. The extensive gliding membrane of tough skin-like tissue is known as the patagium.

Family CYNOCEPHALIDAE	Species *Cynocephalus variegatus*	Status Common

MALAYAN FLYING LEMUR

This arboreal species has a small head and ears, blunt snout, and large eyes. The fine, short coat is brownish grey with red or grey on the back, sometimes with lighter flecks to mimic lichen-covered branches of trees. Active at twilight and night, this flying lemur eats soft plant parts and scrapes up sap and nectar with its comb-like incisors.

• **SIZE** Length: 33–42 cm (13–16½ in). Front "wingspan": 65–75 cm (26–30 in).

• **OCCURRENCE** S.E. Asia. In forest, from coastal lowland to mountainous areas up to 1,000 m (3,300 ft).

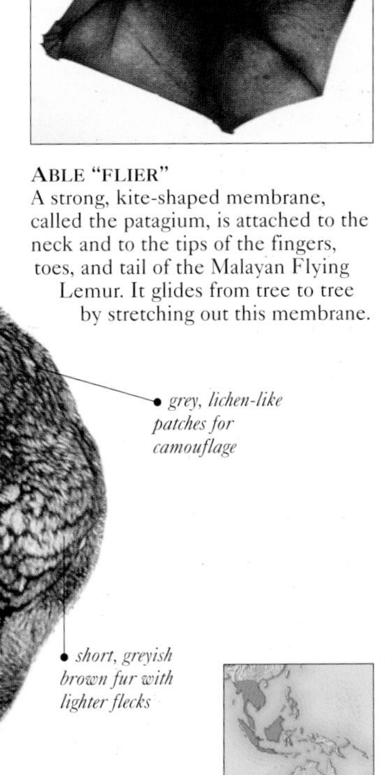

ABLE "FLIER"
A strong, kite-shaped membrane, called the patagium, is attached to the neck and to the tips of the fingers, toes, and tail of the Malayan Flying Lemur. It glides from tree to tree by stretching out this membrane.

Social unit Social	Gestation 60 days	Young 1	Diet

TREE SHREWS

NEITHER WHOLLY arboreal, nor really shrews, tree shrews are squirrel-like inhabitants of tropical forests in South and Southeast Asia, belonging to the order Scandentia. Most are adept climbers, but a few of the 19 species hardly ever venture into trees.

Also called tupaids, from their family name, most tree shrews are long-bodied, agile, solitary, and diurnal. They forage for insects, worms, other small creatures, and occasionally fruits and berries. Their senses of sight, hearing, and smell are well-developed; however, unlike true shrews, they lack whiskers. The long, furred tail aids balancing, and the clawed fingers and toes grip well.

Previously, tree shrews were classified with insectivores, due to their diet of invertebrates; or with primates, due to common anatomical features such as a relatively large brain for the body size and, in the male, testes that descend into a bag or scrotum.

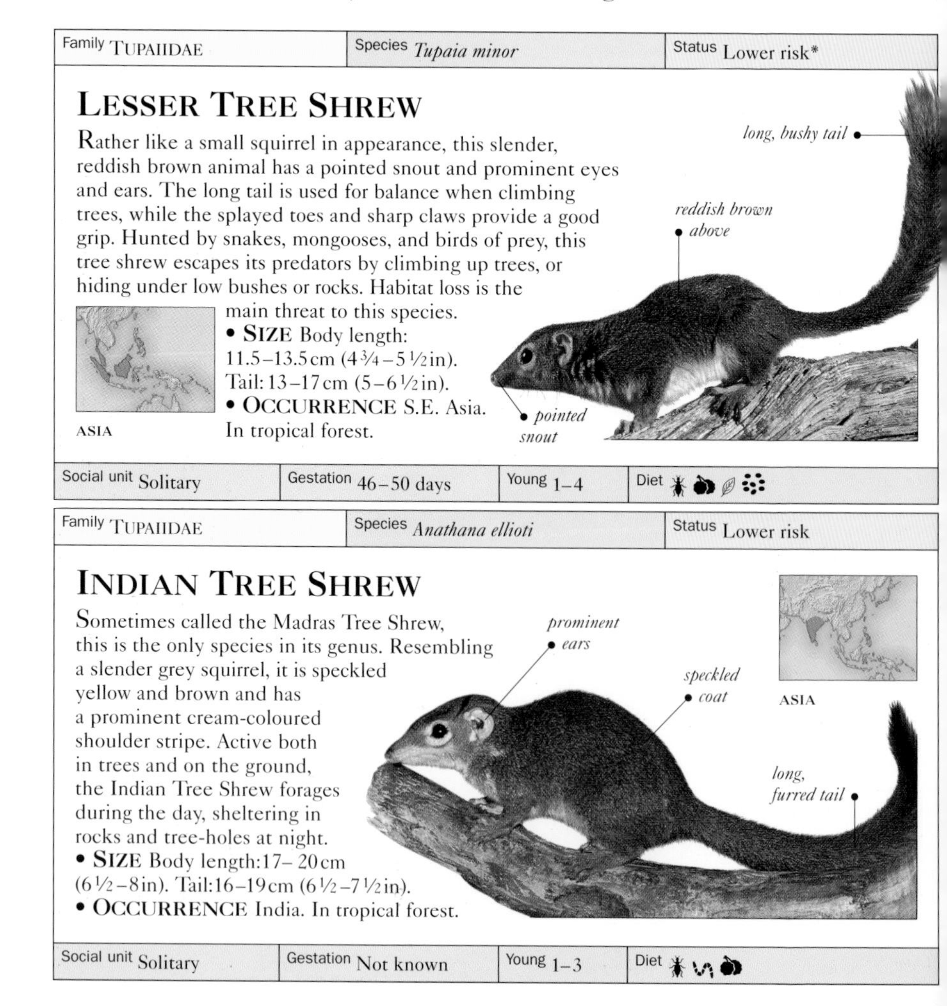

Family TUPAIIDAE	Species *Tupaia minor*	Status Lower risk*

LESSER TREE SHREW

Rather like a small squirrel in appearance, this slender, reddish brown animal has a pointed snout and prominent eyes and ears. The long tail is used for balance when climbing trees, while the splayed toes and sharp claws provide a good grip. Hunted by snakes, mongooses, and birds of prey, this tree shrew escapes its predators by climbing up trees, or hiding under low bushes or rocks. Habitat loss is the main threat to this species.

- **SIZE** Body length: 11.5–13.5cm (4¾–5½in). Tail: 13–17cm (5–6½in).
- **OCCURRENCE** S.E. Asia. In tropical forest.

Social unit Solitary	Gestation 46–50 days	Young 1–4	Diet

Family TUPAIIDAE	Species *Anathana ellioti*	Status Lower risk

INDIAN TREE SHREW

Sometimes called the Madras Tree Shrew, this is the only species in its genus. Resembling a slender grey squirrel, it is speckled yellow and brown and has a prominent cream-coloured shoulder stripe. Active both in trees and on the ground, the Indian Tree Shrew forages during the day, sheltering in rocks and tree-holes at night.

- **SIZE** Body length:17– 20cm (6½–8in). Tail:16–19cm (6½–7½in).
- **OCCURRENCE** India. In tropical forest.

Social unit Solitary	Gestation Not known	Young 1–3	Diet

PRIMATES
PROSIMIANS

BUSHBABIES AND POTTOS in Africa, lemurs in Madagascar, and lorises in Asia, together forming the suborder Strepsirhini, are a step below monkeys and apes (simians) in the evolutionary ladder.

Most prosimians are arboreal, nocturnal, forest-dwellers, with large front-facing eyes, long limbs, and gripping digits equipped with nails rather than claws. Many have long, balancing tails, and communicate by calls, scents, and visual displays to attract mates or group members, or to repel intruders. Lemurs, in particular, are social, and several kinds form troops and are diurnal.

Numbering 77 species, prosimians are regarded as less specialized in form than simians. In this book, the tarsiers, from Southeast Asia, have been included in this group, although they appear to be in between prosimians and simians.

Family TARSIIDAE	Species *Tarsius bancanus*	Status Unconfirmed

WESTERN TARSIER

A nocturnal prosimian, the Western Tarsier is adapted to climbing and grasping the branches of trees. It has a small, compact body, slender fingers, and padded toes with sharp claws. These help it to grip branches firmly. An opportunistic hunter that feeds chiefly on insects, it can rotate its head through 360 degrees, using its huge eyes and keen ears to detect prey or predators. It ambushes its victim by creeping up and leaping to seize it with its front paws. Its predators include nocturnal birds of prey, such as owls. This species usually sleeps on branches, rarely using nests for shelter. The young are carried by the mother at first but soon learn to cling to her fur.

• **SIZE** Body length: 12–15 cm (4¾–6 in). Tail: 18–23 cm (7–9 in).
• **OCCURRENCE** S.E. Asia. In primary and secondary tropical forest, and mangrove swamps.
• **REMARK** Habitat degradation has adversely affected the population of the little-known Western Tarsier.

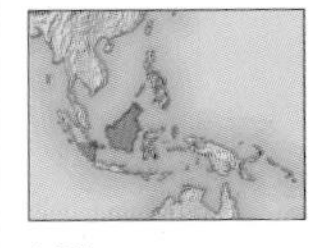
ASIA

round head
large eyes for spotting prey
pale olive to reddish brown fur

Social unit Solitary	Gestation 180 days	Young 1	Diet

Family LORISIDAE	Species *Loris tardigradus*	Status Vulnerable

SLENDER LORIS

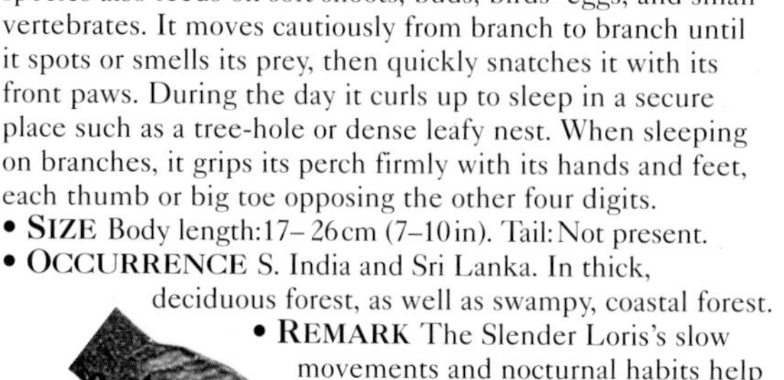

Like all of its family, this small, slender primate is nocturnal, and has large, round eyes equipped with binocular and night vision. It has a dark face mask with a pale central stripe, and its soft, thick fur ranges from yellow-grey to dark brown above, with silvery grey underparts. Mainly an insect-eater, this species also feeds on soft shoots, buds, birds' eggs, and small vertebrates. It moves cautiously from branch to branch until it spots or smells its prey, then quickly snatches it with its front paws. During the day it curls up to sleep in a secure place such as a tree-hole or dense leafy nest. When sleeping on branches, it grips its perch firmly with its hands and feet, each thumb or big toe opposing the other four digits.

- **SIZE** Body length: 17–26 cm (7–10 in). Tail: Not present.
- **OCCURRENCE** S. India and Sri Lanka. In thick, deciduous forest, as well as swampy, coastal forest.
- **REMARK** The Slender Loris's slow movements and nocturnal habits help it to escape the notice of predators, but serious habitat degradation is threatening its existence.

VICE-LIKE GRIP
The Slender Loris is able to sleep on tree branches, holding onto them with firmly locked thumbs and toes.

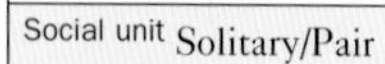

Social unit Solitary/Pair	Gestation 165–170 days	Young 1–2	Diet

Family LORISIDAE	Species *Nycticebus coucang*	Status Unconfirmed

SLOW LORIS

This species is named after its slow and deliberate movement, a characteristic that singles out lorises from all other primates, which are generally known for their leaps and bounds. The Slow Loris is pale grey-brown to red-brown, with a brown stripe from the top of the head to the middle of the back or base of the tail, and dark rings around its eyes and ears. It shares common features with the Slender Loris (see opposite), such as digits adapted for a strong grip, binocular vision, and an arboreal and nocturnal lifestyle.

- **SIZE** Body length: 26–38 cm (10–15 in). Tail: 1–2 cm (3/8 – 3/4 in).
- **OCCURRENCE** S. Asia and S.E. Asia. In tropical forest, gardens, plantations, and bamboo groves.

ASIA

- **REMARK** This loris continues to be hunted in some regions as its body parts are used to make traditional medicines.

Social unit Variable	Gestation 190 days	Young 1–2	Diet

Family LORISIDAE	Species *Perodicticus potto*	Status Locally common

POTTO

A secretive and solitary nocturnal hunter, the Potto has very mobile limbs and can reach out at any angle to bridge gaps between branches, although it cannot jump. Its arms and legs are of similar length, with extremely powerful grasping hands and feet. It can remain immobile in trees for several hours to escape attention; if attacked, it lowers its head to batter its enemy with a spiny shield of bony protrusions on its neck. The Potto may be grey, brown, or red in colour. Compared to a Galago (see p.100), its eyes and ears are relatively small. This species feeds on fruits, leaves, sap, fungi, and small animals.

AFRICA

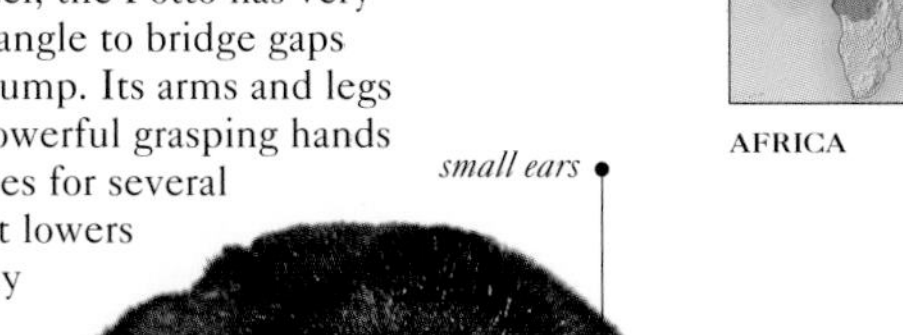

- **SIZE** Body length: 30–40 cm (12–16 in). Tail: 3.5–15 cm (1 1/2–6 in).
- **OCCURRENCE** W. and C. Africa. In tropical forest (especially the margins), lowland, and swamps.

Social unit Variable	Gestation 194–205 days	Young 1–2	Diet

Family LORIDAE	Species *Arctocebus calabarensis*	Status Lower risk

ANGWANTIBO

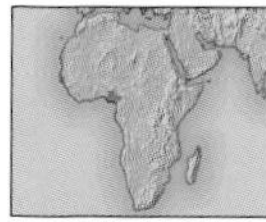

This primate is one of the two *Arctocebus* species (the other is the Golden Angwantibo found further south), and is golden brown on its upperparts but buff underneath. It has four equal-length limbs which it uses to climb slowly up branches. Its tiny second toe and the first toe, widely separated from the other three, provide a clamp-like grip. Nocturnal by nature, the Angwantibo prefers newly grown secondary vegetation in treefall zones, clearings, and roadsides. It forages on insects, chiefly caterpillars, using sight and smell, rubbing hairy caterpillars to remove the irritating hairs. When threatened it rolls itself into a ball.

- **SIZE** Body length: 22–26 cm (9–10 in). Tail: 1 cm (3/8 in).
- **OCCURRENCE** W. Africa. In woodland.

Social unit Social	Gestation 133 days	Young 1	Diet

Family GALAGONIDAE	Species *Galago crassicaudatus*	Status Locally common

THICK-TAILED GALAGO

The largest of all galagos, this species has huge eyes and ears to locate insects at night, catching them rapidly with its powerful hands. Its comb-like teeth are used to scrape gum from trees. Also known as the Thick-tailed Greater Bushbaby, it ranges from the size of a large squirrel to that of a pet cat, and varies from silver to grey, brown, or black. It tends to move on all fours and is unable to land on its hindlimbs like some of the specialized, leaping galagos.

- **SIZE** Body length: 25–40 cm (10–16 in). Tail: 34–49 cm (13 1/2–19 1/2 in).
- **OCCURRENCE** C., E., and S. Africa. In tropical forest, woodland, and plantations.
- **REMARK** This animal's child-like wail gives its group the common name, Bushbaby.

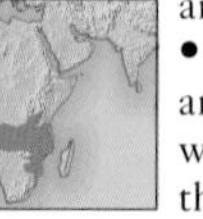

Social unit Social	Gestation 126–135 days	Young 1–3	Diet

Family GALAGONIDAE	Species *Galago moholi*	Status Common

South African Galago

Sometimes called the Lesser Bushbaby, this species makes kangaroo-like vertical leaps up branches to a height of 5 m (16 ft), using its hands and feet, regularly moistened with urine, to maintain its grip. It catches insects in mid-air by hand, or scrapes gum from trees with its comb-like, lower front teeth. A solitary forager, the South African Galago sleeps huddled in groups during the day and may congregate again at night to groom and socialize. It is a shy animal, and is especially cautious before descending to the ground. Checking all directions, it may make several trial runs, calling out loudly when it spots a predator. Lithe and agile, it easily leaps out of trouble.

prominent ears
diamond-shaped black eye-rings
large rear feet
furred tail

AFRICA

- **Size** Body length: 15–17 cm (6–6½ in). Tail: 12–27 cm (4¾–10½ in).
- **Occurrence** E., C., and S. Africa. In acacia savanna and woodland.

Social unit Variable	Gestation 121–124 days	Young 1–2	Diet

Family CHEIROGALEIDAE	Species *Cheirogaleus medius*	Status Lower risk

Fat-tailed Dwarf Lemur

With soft, woolly fur, buff or grey-red on its upperparts and whitish yellow underneath, the Fat-tailed Dwarf Lemur has prominent eyes, ringed with dark circles, and the large ears are mainly naked. It stores food as fat in its body and tail during the rains, to survive the 6–8 month dry season. At this time, it may become torpid in order to overcome food shortage, remaining huddled with others of its kind. When it revives, it becomes solitary again and clambers around trees and bushes at night, picking up and holding food in its forepaws while feeding. The Fat-tailed Dwarf Lemur usually eats fruits and other vegetation at the beginning of the year, while insects become a more important part of its diet as the year passes. By day, it rests in a nest made of leaves and twigs, in a tree-hole or at the top of a tree.

soft, woolly fur
prominent, naked ears
dark circles around eyes
yellow whitish underparts
"fat" tail to store food

MADAGASCAR

- **Size** Body length: 17–26 cm (7–10 in). Tail: 19–30 cm (7½–12 in).
- **Occurrence** W. and S. Madagascar. In primary and secondary dry forest.
- **Remark** Like most lemurs, this species is threatened by the destruction of its forest habitat.

Social unit Variable	Gestation 61–64 days	Young 1–4	Diet 

Family LEMURIDAE	Species *Lemur catta*	Status Vulnerable

RING-TAILED LEMUR

MADAGASCAR

Distinctively feline in its graceful appearance and movements, the Ring-tailed Lemur has brownish grey to rosy brown upperparts, with whitish grey underparts. It has distinctive dark, triangular eye patches, a black nose, and a striking black-and-white ringed tail that it uses to send visual signals. Unlike other lemurs, this skilled climber is also often found on the ground, running to take refuge in the forest canopy as soon as it is threatened. It feeds on fruits, vegetation, bark, and sap at all levels of the forest, using its hands to put food into its mouth. Very sociable by nature, these lemurs are often found in groups of 5–25 with a hierarchical core of adult females, who dominate the males and defend their territory with loud calls. Young females remain with their mothers and sisters, whereas juvenile males move to other groups. The young first clings to its mother's underside and later rides on her back.

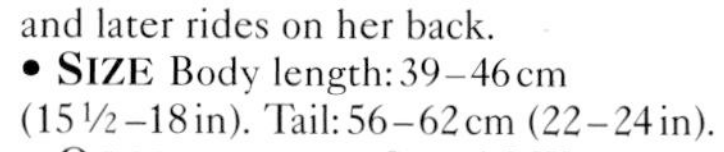

• **SIZE** Body length: 39–46 cm (15½–18 in). Tail: 56–62 cm (22–24 in).

• **OCCURRENCE** S. and S.W. Madagascar. In dry forest and bush of closed canopy deciduous forest, adjoining rocky mountain outcrops, and gallery forest along rivers.

• **REMARK** The main threat to the Ring-tailed Lemur is from loss of habitat through deforestation and forest fires. It is also hunted by humans for food and frequently kept as a pet.

SCENT GLANDS
This lemur communicates by means of scent secreted from glands in its body. One of these glands is visible here on the inside of the right arm.

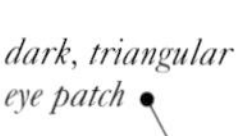

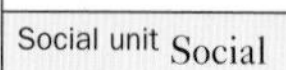

Social unit Social	Gestation 134–138 days	Young 1	Diet

Family LEMURIDAE	Species *Lemur fulvus*	Status Lower risk*

BROWN LEMUR

Despite its name, this lemur is extremely variable in colour and ranges from brown to yellow or grey, according to the subspecies. However, it usually has a dark face with lighter patches above the eyes and a thickly furred tail, carried arch-like over its back when moving on the ground or along branches. Found in fluid groups, Brown Lemurs are highly adaptable to forest habitats and forage in trees and on the ground, feeding on fruits, vegetation, and sap. Each smears urine on itself for scent recognition. Home ranges overlap, but neighbouring groups avoid contact.

- **SIZE** Body length: 38–50 cm (15–20 in). Tail: 46–60 cm (18–23½ in).
- **OCCURRENCE** N. and W. Madagascar. In tropical forest.

Social unit Social	Gestation 120 days	Young 1–2	Diet

Family LEMURIDAE	Species *Lemur macaco*	Status Lower risk*

BLACK LEMUR

All Black Lemurs have soft and relatively long fur, but only the male has the black coat after which the species is named. Females vary considerably in colour from reddish brown to grey. The small head, with a pointed snout and large eyes, has a distinctive ruff around the neck and shoulders. Groups of 5–15 Black Lemurs, led by a single female, forage in trees, using their forelimbs to pick and tear food. Unlike most lemurs, this species is active for part of the night–human disturbance may be the cause of such behaviour. Experts recognize two subspecies of this poorly studied lemur: *Lemur macaco macaco* and *Lemur macaco flavifrons*.

- **SIZE** Body length: 30–45 cm (12–18 in). Tail: 40–60 cm (16–23½ in).
- **OCCURRENCE** N. Madagascar. In tropical, evergreen forest.
- **REMARK** The Black Lemur is under threat from a variety of activities including forest fires, land clearing for agriculture, as well as hunting by humans.

MADAGASCAR

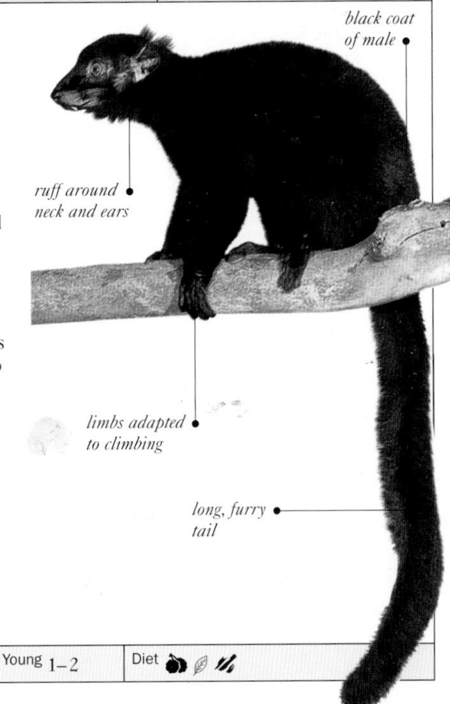

Social unit Social	Gestation 125 days	Young 1–2	Diet

Family LEMURIDAE	Species *Varecia variegata*	Status Endangered

RUFFED LEMUR

The largest lemur, this species is white or reddish white in colour, with long, soft fur and a black face. It usually has black shoulders, chest, flanks, feet, and tail, although there is considerable colour variation within its range. The Ruffed Lemur is unusual in that it builds a simple leafy nest, in a tree-hole or fork, for its young. The mother carries the young about by the scruff of the neck from a few days after birth. Once they are several weeks old, they learn to cling to their mother's body. This species lives in groups of 2–20, with several dominant females defending a common territory. Eating the highest proportion of fruit of all lemurs, it forages at dawn and in the late afternoon.

- **SIZE** Body length: 55 cm (22 in). Tail: 1.1–1.2 m (3 ½–4 ft).
- **OCCURRENCE** E. Madagascar. Also introduced in the island of Nosy Mangabe. In rainforest.
- **REMARK** A dwindling habitat poses the major threat to this species; it also continues to be hunted for food.

MADAGASCAR

black face

ears tufted with white hair

soft, dense coat

black tail twice as long as body

Social unit Social	Gestation 90–102 days	Young 2–3	Diet

Family LEMURIDAE	Species *Hapalemur griseus*	Status Unconfirmed

GREY GENTLE LEMUR

Grey all over, with a distinctive blunt snout, this highly specialized lemur is the only primate that is adapted to living among reeds and rushes fringing lakes. Jumping from one reed stalk to another, it clings to them, chewing on the bark and pith, and plucking leaves, shoots, and buds to stuff into its mouth; feeding in the early morning and late afternoon. Unlike any others in its family, this species may be able to swim, but this has not been confirmed as its habits have yet to be properly studied. Groups of Grey Gentle Lemurs, led by one dominant male, normally number 3–5, but may be as large as 40. The single young is born around January or February and is carried by the mother on her back.

- **SIZE** Body length: 40 cm (16 in). Tail: 40 cm (16 in).
- **OCCURRENCE** N. and E. Madagascar, around Lake Alaotra. In reed beds and papyrus beds.
- **REMARK** Habitat loss and degradation is affecting the existence of this extremely localized species – already thought to be extinct in parts of its range. Burning reed beds along the lakeshore to make way for agriculture is a particular problem.

MADAGASCAR

Social unit Social	Gestation Not known	Young 1	Diet

Family LEMURIDAE	Species *Lepilemur mustelinus*	Status Lower risk

WEASEL LEMUR

This nocturnal prosimian has long, soft, brown fur, a grey head, and dark tail-tip. It has prominent naked ears, and large eyes for good night vision. As with other leaping primates, both eyes face forwards and so provide stereoscopic vision, allowing this lemur to judge distances precisely. Adapted to life in the trees, the Weasel Lemur leaps from one branch to another, feeding on fruits and leaves, and is rarely seen on the ground. It has grasping feet, but its tail lacks the ability to grip.

- **SIZE** Body length: 30–35 cm (12–14 in). Tail: 25–35 cm (10–14 in).
- **OCCURRENCE** N.E. Madagascar. In rainforest.
- **REMARK** The habits of this species are poorly studied, and like all lemurs, its population is threatened by habitat loss, especially due to slash-and-burn cultivation.

Social unit Social	Gestation Not known	Young Not known	Diet

Family INDRIIDAE	Species *Propithecus verreauxi*	Status Critically endangered*

VERREAUX'S SIFAKA

Generally white all over, Verreaux's Sifaka has brown-black areas on its face, crown, and the undersides of its limbs. Its hindlegs and tail are extremely long, and its palms and soles are black. While walking on the ground it uses two legs, awkwardly hopping sideways, with its arms held aloft. In desert areas, it moves among the cactus-like vegetation without injuring itself, using high, springing leaps. When resting, one sifaka crouches behind another, in a line. Verreaux's Sifaka lives in varying social groups, calling out to dispute territorial boundaries between two groups. The name "Sifaka" is derived from the sound of its call.

- **SIZE** Body length: 43–45 cm (17–18 in). Tail: 56–60 cm (22–23 ½ in).
- **OCCURRENCE** S. and W. Madagascar. In evergreen, gallery, and dry, deciduous forest, as well as spiny desert vegetation.
- **REMARK** Rapid habitat degradation is severely threatening this animal.

Social unit Social	Gestation 150–162 days	Young 1	Diet

Family INDRIIDAE	Species *Indri indri*	Status Endangered

INDRI

The largest member of the lemur group, the Indri has a coat that is predominantly black with variable white patches on the back of the head, neck, and limbs; the ears have conspicuous black tufts. It has long hindlimbs for taking enormous leaps, and virtually no tail. A vertical clinger and leaper, it remains inactive for long periods in the day, even though it is diurnal. Male–female pairs live with their offspring, with the male defending its territory, and the female having first access to food. Once common in Madagascar, this species is threatened by the loss of its rainforest habitat. Known locally as the Babakoto or "little father", its English name was actually derived from the local exclamation "indri indri", meaning "there it is", when first shown to explorers.

• **SIZE** Body length: 60 cm (23 ½ in). Tail: 5 cm (2 in).

• **OCCURRENCE** E. Madagascar. At all levels of montane rainforest.

• **REMARK** The Indri is the only species in its genus.

MADAGASCAR

Social unit Pair	Gestation 172 days	Young 1	Diet

Family DAUBENTONIIDAE	Species *Daubentonia madagascariensis*	Status Endangered

AYE-AYE

Also dubbed the "primate woodpecker", the Aye-aye listens intently for woodboring grubs inside trees, exposes them by gnawing the bark with its enormous front teeth, and finally picks them out with its long middle finger, to eat. It has a coarse black coat with a mantle of white guard hairs. Shy and reclusive, it hides in its nest, made of sticks, during the day and emerges at night to feed.

• **SIZE** Body length: 40 cm (16 in). Tail: 40 cm (16 in).

• **OCCURRENCE** N.W. and E. Madagascar. In forest, spiny desert, and plantations.

• **REMARK** Thought to be extinct, it was rediscovered in 1957.

MADAGASCAR

Social unit Variable	Gestation 120–150 days	Young 1	Diet

PRIMATES

MONKEYS

THE 242 species of monkeys, along with the apes, make up the primate suborder Haplorhini.

Two monkey families are found in South and Central America: the small, soft-haired marmosets and tamarins in the family Callitrichidae; and the sakis, uakaris, and titis, which together with the spider, woolly, and howler monkeys are known as the New World Monkeys and belong to the family Cebidae.

A third family, from Africa and Asia, is the Cercopithecidae, or the Old World monkeys: guenons, colobus, mangabeys, macaques, baboons, and langurs.

Typical monkeys are forest dwellers, with five grasping digits on each of the four flexible limbs, a long tail, and a large brain for the body size, giving considerable intelligence. Most species live in large social groups and eat a mixed diet of plants and small creatures.

Family CEBIDAE	Species *Lagothrix cana*	Status Vulnerable

GREY WOOLLY MONKEY

Woolly monkeys are characterized by their thick, closely curled fur, which is a darker shade on the head, hands, feet, and tail-tip. The Grey Woolly Monkey is grey with black flecks, and is stout and pot-bellied, with powerful shoulders, hips, and tail for swinging on branches. It has a large forehead and braincase and is known to be highly intelligent. These monkeys live in mixed troops, which may break up into subgroups for foraging, with a dominance hierarchy based on age.

- **SIZE** Body length: 50–65 cm (20–26 in). Tail: 55–77 cm (22–30 in).
- **OCCURRENCE** Brazil, Peru, and Bolivia. In primary forest, especially flooded forest.
- **REMARK** Forest fragmentation has adversely affected this monkey, which needs large tracts of uninterrupted forest to thrive.

GENTLE-NATURED This monkey displays little aggression, and allows members of another troop to encroach on its territory.

Social unit Social	Gestation 233 days	Young 1	Diet

Family CEBIDAE	Species *Ateles geoffroyi*	Status Vulnerable

BLACK-HANDED SPIDER MONKEY

Sometimes called Geoffroy's Spider Monkey, this species is distinguished by its black head, hands, and feet, and the "cowl" that surrounds its face. Like other spider monkeys, it uses its thumbless hands as hooks to help it swing easily through trees, or to pull branches towards its mouth in order to eat fruits, leaves, and flowers. The prehensile tail is often used as a fifth limb.

N. & C. AMERICA

• **SIZE** Body length: 50–63 cm (20–25 in). Tail: 63–84 cm (25–33 in).
• **OCCURRENCE** S. Mexico and Central America. In tropical forest and mangrove swamps.

Social unit Social	Gestation 226–232 days	Young 1	Diet

Family CEBIDAE	Species *Ateles chamek*	Status Lower risk

BLACK SPIDER MONKEY

This monkey has long black fur and black facial skin. A sociable species, it lives in troops, each with a large territory occupying 150–230 hectares (370–570 acres). The troop splits into variable subgroups to forage, greeting other groups with whoops and wails. Females use a quarter to a third of the troop's territory, leaving when mature to join other groups. The young of dominant females survive better into adulthood.

black face
long, black fur

S. AMERICA

• **SIZE** Body length: 40–52 cm (16–20½ in). Tail: 80–88 cm (32–35 in).
• **OCCURRENCE** Peru, Brazil, and Bolivia, in the upper Amazon tributaries. In tropical forest.

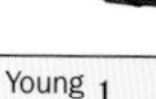

Social unit Social	Gestation 225 days	Young 1	Diet

Family CEBIDAE	Species *Brachyteles arachnoides*	Status Critically endangered

MURIQUI

One of the largest monkeys in the Americas, the Muriqui or Woolly Spider Monkey has dense fawn fur and a black face. It is a slow forager, feeding on fruits, seeds, and leaves by pulling off branches with its hands.
• **SIZE** Body length: 55–61 cm (22–24 in). Tail: 67–84 cm (26–33 in).

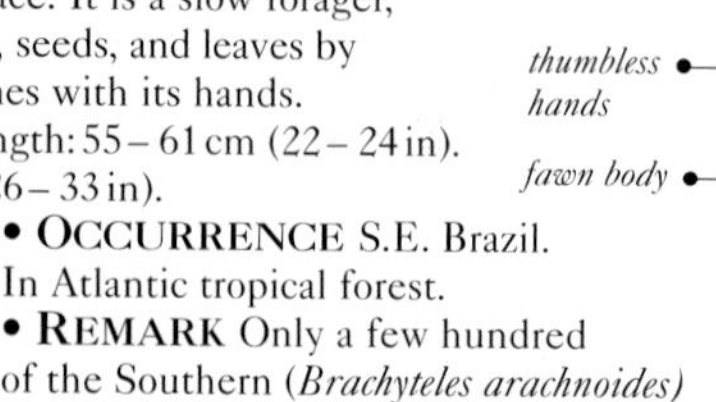

S. AMERICA

• **OCCURRENCE** S.E. Brazil. In Atlantic tropical forest.
• **REMARK** Only a few hundred of the Southern (*Brachyteles arachnoides*) and Northern (*B. hypoxanthus*) Muriqui survive today.

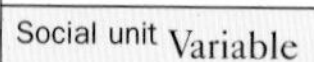

Social unit Variable	Gestation 210–255 days	Young 1	Diet

Family CEBIDAE	Species *Alouatta pigra*	Status Vulnerable

MEXICAN HOWLER MONKEY

Also known as the Guatemalan Black Howler Monkey, this species is uniformly black, except for the male's white scrotal sac. Found in territories sometimes up to 25 hectares (62 acres) in area, each group of around seven members comprises females, juveniles, and a single adult male, which may be twice as heavy as the female. The Mexican Howler Monkey calls loudly at dusk and dawn to establish the troop's territory. It feeds on fruits, flowers, and leaves, subsisting on a single species of tree during poor seasons. It pulls branches towards its mouth and bites off fruits, after sniffing to see if they are ripe.

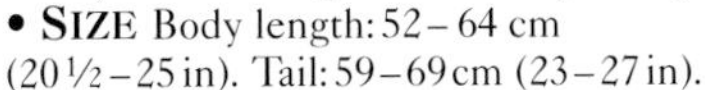

• **SIZE** Body length: 52–64 cm (20½–25 in). Tail: 59–69 cm (23–27 in).
• **OCCURRENCE** Mexico and Central America. In tropical forest.
• **REMARK** This species was formerly thought to be a subspecies of the Mantled Howler (*Alouatta palliata*) of Central America. The two species overlap without interbreeding in Tabasco.

Social unit Social	Gestation 190 days	Young 1	Diet

Family CEBIDAE	Species *Alouatta seniculus*	Status Lower risk

RED HOWLER MONKEY

S. AMERICA

One of the nine howler species, this monkey's loud calls carry more than 2.5 km (1½ miles) to inform others in the area of its territorial presence. It is usually found in small groups of an adult male, females, and young. When a new male supersedes the existing male, he may kill any offspring, so that the females mate with him and raise his young. Howlers lead a sluggish lifestyle due to a low nutrient diet.
• **SIZE** Body length: 51–63 cm (20–25 in). Tail: 55–68 cm (22–27 in).
• **OCCURRENCE** Colombia, Venezuela, Brazil, and Peru. In rainforest, mangrove swamps, and savanna woodland.

Social unit Social	Gestation 191 days	Young 1	Diet

Family PITHECIIDAE	Species *Pithecia pithecia*	Status Lower risk

WHITE-FACED SAKI

No other New World monkey has such a striking difference between the sexes as the White-Faced Saki. The male is black with a white or pale gold face, and a black nose. The female is grey-brown with long white tips to her hair, and has a blackish face with a white stripe on either side of the nose. The fur on both is lank, falling to the sides from midback and nape to form a cowl on the crown. Both have a bushy tail.

- **SIZE** Body length: 34–35 cm (13½–14 in). Tail: 34–44 cm (13½–17½ in).
- **OCCURRENCE** North of the Amazon. In tropical forest: gallery, palm, and savanna.
- **REMARK** One of the five species of true sakis, it is closely related to the Bearded Saki and uakaris (see below).

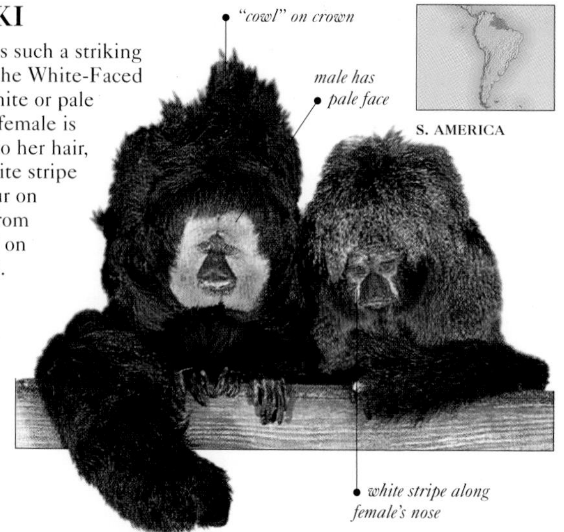

Social unit Social	Gestation 170 days	Young 1	Diet

Family PITHECIIDAE	Species *Cacajao calvus*	Status Endangered

BALD UAKARI

A bright red, skull-like, hairless face, and in adults a bald crown, characterize Bald Uakaris. They are classified into subspecies according to colour such as white, golden, and red. Shown here is the Red Bald Uakari which has shaggy red hair over its thin body. When this monkey is excited, it wags its short tail. Found in groups of up to 100, it inhabits the black-water forests of Brazil: part-flooded forest along small rivers, lakes, and swamps.

- **SIZE** Body length: 38–57 cm (15–22½ in). Tail: 14–18.5 cm (5½–7½ in).
- **OCCURRENCE** Brazil. In black-water forest.
- **REMARK** The Bald Uakari is the only short-tailed New World monkey.

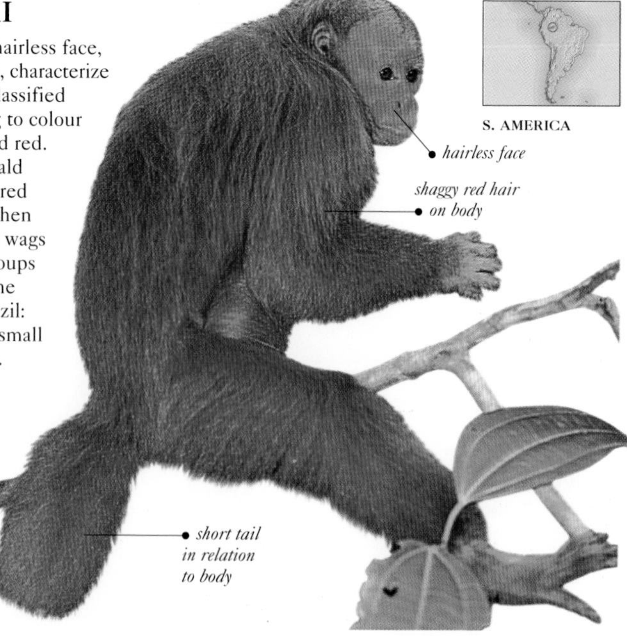

Social unit Social	Gestation Not known	Young 1	Diet

Family PITHECIIDAE	Species *Callicebus moloch*	Status Lower risk

DUSKY TITI MONKEY

The 20 or more titi species typically have thick, soft fur, a stocky body, short limbs, and ears almost hidden beneath fur. The Dusky Titi has a speckled brown back, while its underparts are orange. Its drab coloration and slow movements provide effective camouflage in the lower level of trees where it is found. The male and female form a monogamous pair, grooming frequently and defending a home range that extends over 6–12 hectares (15 – 30 acres). Just before dawn, they entwine tails and "sing" a duet to maintain their family and pair-bonds, and to declare their territory. The young stay with the pair for up to three years, and the male carries the infant and shares food with it for around a year.

• **SIZE** Body length: 27–43 cm (10 1/2–17 in). Tail: 35–55 cm (14–22 in).

S. AMERICA

• **OCCURRENCE** Brazil. In swamps and flooded tropical forest, notably forest edges.

• **REMARK** The Dusky Titi belongs to a group of 6–8 titi species that are found in the southern Amazonian forest.

Social unit Pair	Gestation 155 days	Young 1	Diet

Family AOTIDAE	Species *Aotus lemurinus*	Status Vulnerable

NIGHT MONKEY

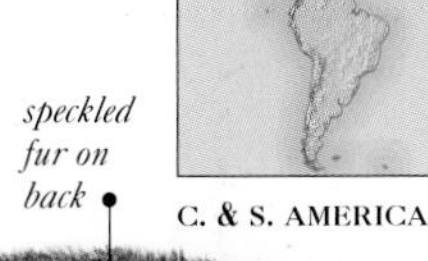

C. & S. AMERICA

Genetic studies show that there are probably ten species of night monkeys, rather than one as previously believed. Also called Dourocouli, or owl monkeys, from their hooting calls in the darkness, they clamber cautiously through the branches at night, searching for food. The Grey-bellied Night Monkey (shown here) is yellow or grey beneath, while the monkeys found further south have red bellies. However, all night monkeys are speckled grey above, with white cheeks and chin, large white spots above the eyes, and three dark stripes from the crown to the face.

speckled fur on back

dark, bushy tail-tip

grasping feet

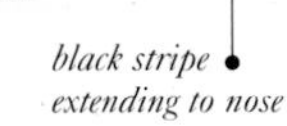

• **SIZE** Body length: 30–42 cm (12–16 1/2 in). Tail: 29–44 cm (11 1/2–17 1/2 in).

• **OCCURRENCE** Ecuador and Colombia (west of the Andes) to Panama. In tropical forest.

• **REMARK** Night monkeys are the only nocturnal monkeys and are equipped with night vision.

Social unit Social	Gestation 120 days	Young 1	Diet 

Family CEBIDAE	Species *Cebus apella*	Status Lower risk

BROWN CAPUCHIN

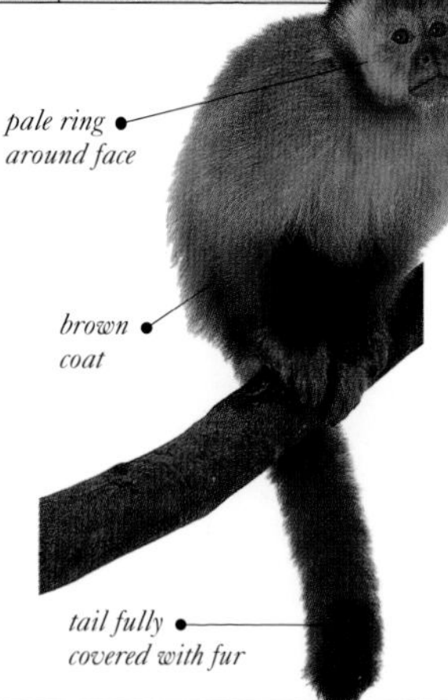

This New World monkey is found foraging with other species of monkeys, both to exploit their ability to find food and for protection from predators. It uses tools such as sticks and stones to crack hard nuts or to flush prey out of tree trunks. Also called the Tufted Capuchin for the short, upright crown hairs that form "horns" above its ears, this monkey is unusual in using the reverse-mount to mate. It has the widest range of all the monkeys found in the Americas.

- **SIZE** Body length: 33–42 cm (13–17 in). Tail: 41–49 cm (16–19½ in).
- **OCCURRENCE** N., C., and E. South America. In tropical forest.
- **REMARK** The Golden-bellied Capuchin (*Cebus xanthosternos*), a closely related species, is critically endangered.

S. AMERICA

Social unit Social	Gestation 5 months	Young 1	Diet

Family CEBIDAE	Species *Saimiri boliviensis*	Status Lower risk*

BOLIVIAN SQUIRREL MONKEY

This monkey has speckled orange and black fur, turning orange at the limbs and underside, and a white face and black muzzle. During courtship, the male becomes "fatted" around the shoulders and several males compete to win the largest number of females. Moving in groups of 40–200, Squirrel Monkeys, of which there are five species, form the largest and most active troops of all the monkeys found in South America. Extremely vocal, they disturb insects to feed on them, or follow other monkeys, catching insects in their wake.

- **SIZE** Body length: 27–32 cm (10½–12½ in). Tail: 38–42 cm (15–16½ in).
- **OCCURRENCE** C. South America. In primary and secondary tropical forest, and swamp forest.
- **REMARK** The different species of squirrel monkeys are known as the "Gothic" or "Roman" types, depending on the shape of their arched white eyebrows. The Bolivian Squirrel Monkey belongs to the "Roman arch" group.

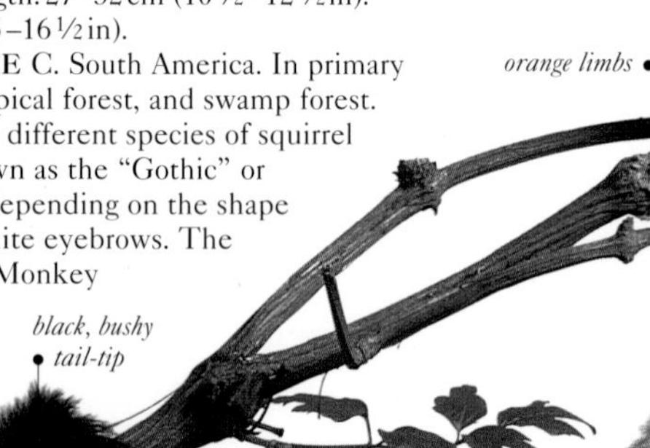

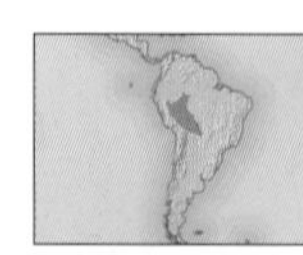

S. AMERICA

Social unit Social	Gestation 170 days	Young 1	Diet 

Family CALLITRICHIDAE	Species *Callimico goeldii*	Status Vulnerable

GOELDI'S MARMOSET

Larger than most marmosets and tamarins, Goeldi's Marmoset has long black fur and a "cape" of longer hair around its head and neck. Using its incisors, it makes gashes in trees and feeds on sap and gum. It also eats fruits, insects, and small vertebrates. Moving in stable, close-knit groups of up to ten, these monkeys are usually found in dense vegetation.

- **SIZE** Body length: 22–23 cm (9 in). Tail: 26–32 cm (10–12 ½ in).
- **OCCURRENCE** N.W. South America. In tropical forest and bamboo groves.
- **REMARK** Unlike other marmosets this species has wisdom teeth.

S. AMERICA

Social unit Social	Gestation 154 days	Young 1	Diet

Family CALLITRICHIDAE	Species *Leontopithecus rosalia*	Status Critically endangered

GOLDEN LION TAMARIN

Weighing twice as much as most marmosets and tamarins, the Golden Lion Tamarin has a mane of long, silky, red-gold fur and a grey face. The slender, long-clawed hands are used to hold fruits while eating and to probe into tree-holes and bark for grubs. Usually found in groups of 4–11 individuals, these monkeys differ from other marmosets in that the sexual activity of subordinate monkeys is not suppressed by the dominant pair. However, it is only the dominant pair that produces the young. The juvenile monkeys of the group often assist in rearing new infants.

- **SIZE** Body length: 20–25 cm (8–10 in). Tail: 32–37 cm (12 ½–14 ½ in).
- **OCCURRENCE** E. South America. In Atlantic tropical forest.
- **REMARK** Deeply affected by deforestation, this species has been the focus of conservation efforts since the 1960s. Having bred well in captivity, it has been reintroduced in the wild in S.E. Brazil. Its situation, however, is still precarious.

S. AMERICA

Social unit Social/Pair	Gestation 129 days	Young 2	Diet

Family CALLITRICHIDAE	Species *Saguinus imperator*	Status Vulnerable

EMPEROR TAMARIN

Known for its flamboyant white moustache, the Emperor Tamarin has speckled grey- or red-brown fur, a black head, and a fiery red-orange tail that is white underneath. It is a small monkey, often forming a mixed group with a related species, such as the Saddleback Tamarin (S. fuscicollis). Each species responds to the other's alarm calls. Feasting on fruits in the wet season, the Emperor Tamarin feeds on nectar and sap in the dry season, and on insects, particularly crickets, round the year. It pulls a plant to its mouth and scans it for insects before pouncing on them. Its two offspring are carried by the father, except when being suckled by the mother.

black face
curly, white moustache
speckled fur on body
claws on digits
tail reddish orange above and white underneath
tail longer than body

• **SIZE** Body length: 23–26 cm (9–10 in). Tail: 39–42 cm (15½–16½ in).

• **OCCURRENCE** W. South America. In tropical forest as well as mountain areas.

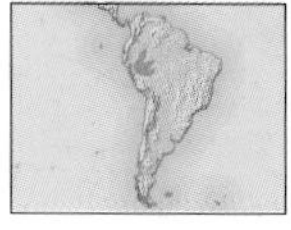

S. AMERICA

• **REMARK** Marmosets and tamarins form a distinct group of about 35 American primate species. They differ biologically from other New World monkeys, with claws instead of nails, and bear two offspring rather than a single young.

Social unit Social	Gestation 140–145 days	Young 2	Diet

Family CALLITRICHIDAE	Species *Callithrix pygmaea*	Status Lower risk*

PYGMY MARMOSET

Found only in the wetlands and tropical forests of the Upper Amazon Basin, this tiny monkey can fit into a human palm. Its fur is speckled tawny with a long cape of hair on the head that hangs over its ears, while its face has a patch of bare skin in a distinctive three-lobed shape. A specialized gum-eater, the Pygmy Marmoset differs from other marmosets in the way it feeds: it gouges out ten or more holes in bark each day and scent marks them, returning to these and older holes at intervals to scrape up the sticky, oozing fluid with its long lower incisors. Each group of 5–10 monkeys has one breeding pair, with "helpers" to care for the young. However, the father takes care of the offspring for the first few weeks.

S. AMERICA

• **SIZE** Body length: 12–15 cm (4¾–6 in). Tail: 17–23 cm (7–9 in).

• **OCCURRENCE** Upper Amazon Basin. In tropical forest and wetland.

• **REMARK** The Pygmy Marmoset is the smallest monkey in the world.

Social unit Social/Pair	Gestation 137–140 days	Young 2	Diet Sap

Family CALLITRICHIDAE	Species *Callithrix argentata*	Status Lower risk*

SILVERY MARMOSET

One of 10–15 similar species found in the Amazon basin, the Silvery Marmoset has pale silvery-grey fur on its back, creamy underparts, and a black tail. Its face and ears have pink skin but the huge ears have concave outer edges. Like other marmosets, each troop has one breeding pair, with helpers (usually siblings) who help bring up the young.

• **SIZE** Body length: 20–23 cm (8–9 in). Tail: 30–34 cm (12–13 ½ in).
• **OCCURRENCE** South of the Amazon basin. In tropical, especially inundated, forest.
• **REMARK** There is a great taxonomic diversity among the Amazonian marmosets, their distributions separated by the major and minor Amazon tributaries

S. AMERICA

pink skin on face
huge ears with concave edges
creamy white underparts

Social unit Social	Gestation 140 days	Young 2	Diet

Family CERCOPITHECIDAE	Species *Papio papio*	Status Lower risk

GUINEA BABOON

The male of this species is larger than the female, with a black face and a mane that almost reaches its dark red rump. The Guinea Baboon usually forages in groups of about 40, but troops of up to 200 are not uncommon. Some males may have "harems" within the group whom they herd with a neck-bite. However, females may sometimes try to mate surreptitiously with males from another group. The infant first clings to the mother's belly and is transferred to her back after a few weeks. The Guinea Baboon has a wide diet that varies from tough roots to juicy grubs and eggs, and sometimes farm crops. It plucks fruits with one hand, vigorously digs the earth with its hands for roots, and rips its food with its hands and teeth.

• **SIZE** Body length: 69 cm (27in). Tail: 56 cm (22 in).
• **OCCURRENCE** W. Africa. In woodland savanna, gallery forest, and scrub.
• **REMARK** The smallest baboon species in the world, it is also the least known and has the smallest range of all baboons.

AFRICA

Social unit Social	Gestation 184 days	Young 1	Diet

Family CERCOPITHECIDAE	Species *Papio anubis*	Status Lower risk

OLIVE BABOON

Typically powerful and dog-like, the Olive Baboon is speckled olive-green with a black face and rump. Both sexes have a grey ruff around the cheeks, although the male may be twice as large as the female. With plenty of stamina, this baboon runs fast on its long legs. It eats vegetation, insects, lizards, and even prey as large as gazelle fawns and lambs. Clad in their dark, "baby" fur, the young are tolerated in the troop; as they take on adult coloration, females move to the bottom of the social hierarchy, while males are driven away and must battle their way into a new troop.

• **SIZE** Body length: 60–86 cm (20–34 in). Tail: 41–58 cm (16–23 in).

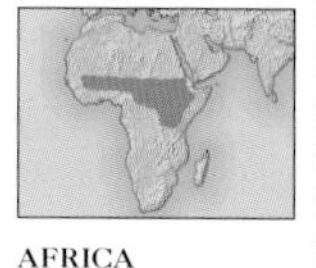

AFRICA

• **OCCURRENCE** W. and E. Africa. In tree savanna and thornbush, up to the forest edge.

• **REMARK** This is one of the largest baboons in Africa.

Social unit Social	Gestation 180 days	Young 1	Diet

Family CERCOPITHECIDAE	Species *Mandrillus sphinx*	Status Vulnerable

MANDRILL

A scarlet nose, with bright blue flanges on either side, a yellow beard, and a mauve-blue rump make the male Mandrill an outstanding primate. The female's coloration is more subdued and she is about a third the size of the male. A forest-floor dweller, the Mandrill spends the day in troops, looking for fruits, seeds, eggs, and insects and small animals. Groups as large as 250 may split into smaller units, with a single male having a harem of 20 females.

• **SIZE** Body length: 63–81 cm (25–32 in). Tail: 7–9 cm (2¾–3½ in).

• **OCCURRENCE** W. C. Africa. In primary and secondary rainforest.

• **REMARK** The largest Old World monkey, the Mandrill is hunted for its meat.

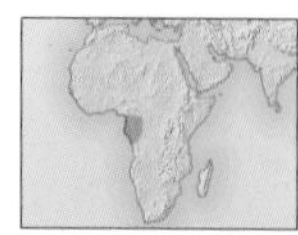

AFRICA

Social unit Social	Gestation 152–182 days	Young 1	Diet

Family CERCOPITHECIDAE	Species *Theropithecus gelada*	Status Lower risk

GELADA

The most outstanding feature of the Gelada, a close cousin of the baboons, is the bare pink patch on its chest. It has brown fur, and adult males have a furry mane on the head and shoulders. Sitting on its buttock pads, the Gelada shuffles over grassland, rapidly plucking grass blades and seeds and stuffing them into its mouth with its dextrous hands. It lives in huge, loose troops, made up of smaller units consisting of a dominant male and his harem of inter-related females. In time the male is driven away by a younger contender, who kills his predecessor's offspring.

- **SIZE** Body length: 70–74 cm (28–29 in). Tail: 46–50 cm (18–20 in).
- **OCCURRENCE** Ethiopia. In montane plateaux and grassland.
- **REMARK** The only non-human primate in its habitat, the Gelada is threatened by the rapid expansion of human activity such as agriculture.

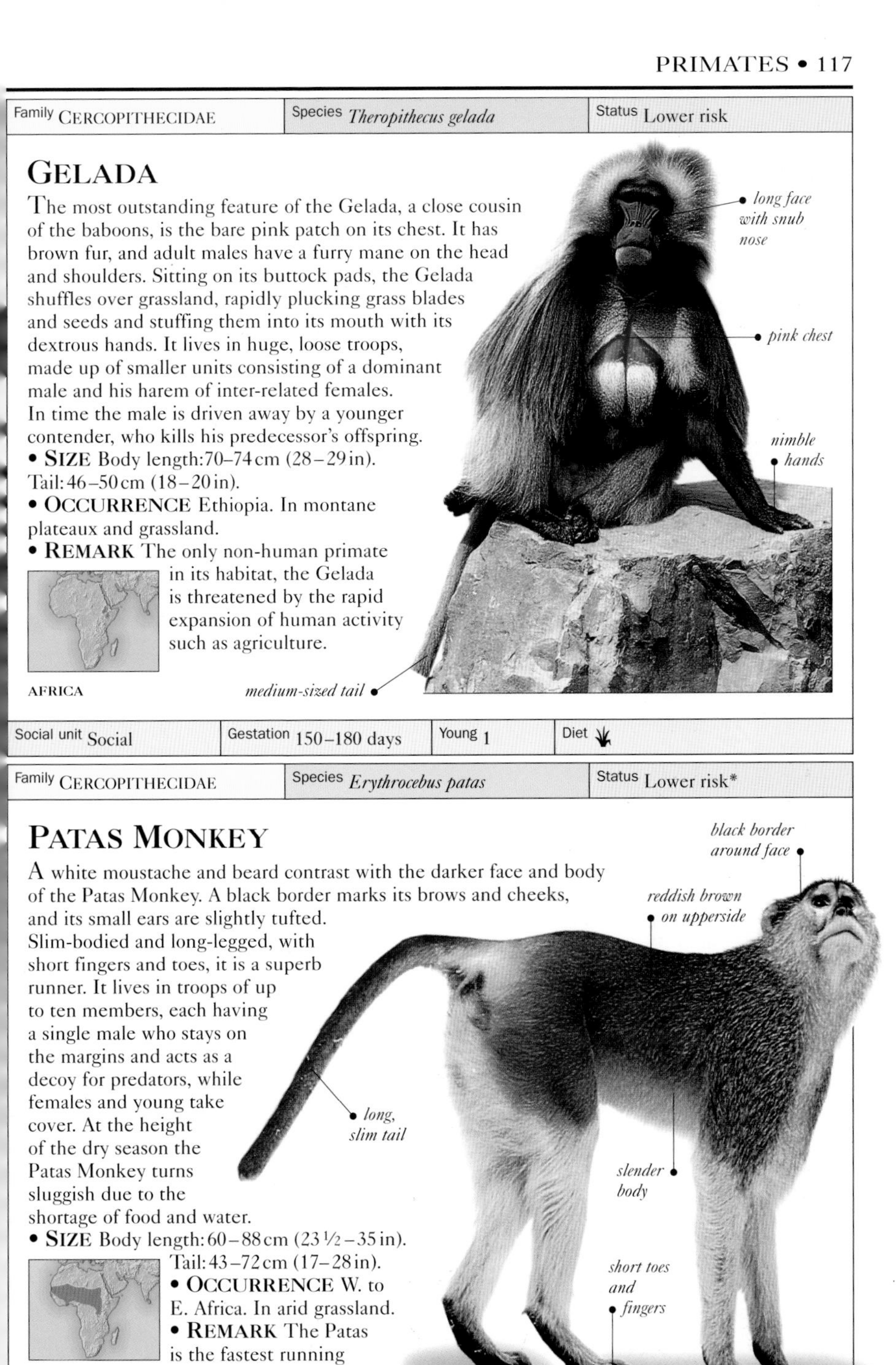

Social unit Social	Gestation 150–180 days	Young 1	Diet

Family CERCOPITHECIDAE	Species *Erythrocebus patas*	Status Lower risk*

PATAS MONKEY

A white moustache and beard contrast with the darker face and body of the Patas Monkey. A black border marks its brows and cheeks, and its small ears are slightly tufted. Slim-bodied and long-legged, with short fingers and toes, it is a superb runner. It lives in troops of up to ten members, each having a single male who stays on the margins and acts as a decoy for predators, while females and young take cover. At the height of the dry season the Patas Monkey turns sluggish due to the shortage of food and water.

- **SIZE** Body length: 60–88 cm (23 ½–35 in). Tail: 43–72 cm (17–28 in).
- **OCCURRENCE** W. to E. Africa. In arid grassland.
- **REMARK** The Patas is the fastest running monkey species.

Social unit Social	Gestation 167 days	Young 1	Diet

Family CERCOPITHECIDAE	Species *Cercocebus torquatus*	Status Critically endangered

WHITE-COLLARED MANGABEY

This monkey has a sooty body and a pink-grey face with a long muzzle, deep pits below its cheeks, known as "fossae", and pale or white eyelids. It stuffs its cheek pouches with nuts, to bring out later in its hand and crack with its powerful teeth and jaws. Mainly terrestrial, the White-collared Mangabey forms large troops of up to 90 monkeys, consisting of males, females, and young. There is a hierarchy among males, but subordinate males are allowed to mate, sometimes more often than the seniors. Troops occupy overlapping home ranges, foraging for food usually along rivers.

• **SIZE** Body length: 50–60 cm (20–23 ½ in). Tail: 60–75 cm (23 ½–30 in).

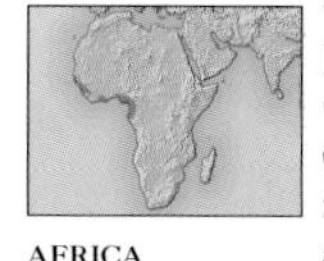
AFRICA

• **OCCURRENCE** W. Africa. In rainforest.

• **REMARK** Six species of *Cercocebus* ("white eyelid" mangabeys) are found across C. Africa.

Social unit Social	Gestation 167 days	Young 1	Diet

Family CERCOPITHECIDAE	Species *Cercopithecus neglectus*	Status Lower risk*

DE BRAZZA'S MONKEY

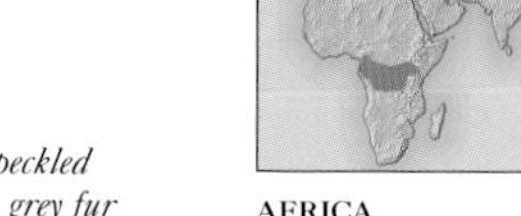
AFRICA

One of the most terrestrial of the 20 or so members of its family, De Brazza's Monkey has speckled grey fur with a black crown, and a white-bordered orange strip on its forehead. Its upper lip and chin are covered with bluish white fur. A thin white stripe runs across its thigh, and its tail and limbs are darker. The male monkey is considerably larger than the female and has a bright blue scrotum. This species eats mainly seeds and fruits, and plucks and holds its food with one hand. Widespread, yet inconspicuous over its large range, it marks its territory with saliva and scent, and communicates by deep booming calls. Although territorial in nature, it tries to avoid, rather than confront, intruders.

• **SIZE** Body length: 50–59 cm (20–23 in). Tail: 59–78 cm (23–31 in).

• **OCCURRENCE** C. to E. Africa. In rainforest, swamps, and submontane forest.

• **REMARK** This is the only monogamous monkey in its family.

Social unit Pair	Gestation 168 days	Young 1	Diet

Family CERCOPITHECIDAE	Species *Macaca nigra*	Status Endangered

CELEBES MACAQUE

Also known as the Celebes Black Ape, this monkey is completely black in colour, and has a very short tail. It has a crest that runs from its forehead to the back of its crown, which rises when the animal is aroused, and high, bony ridges on either side of its nose. This inconspicuous forest-dweller forms large, mixed-sex troops of more than 100. Gentle by nature, males display little aggression towards each other.

- **SIZE** Body length: 52–57 cm (20 ½ – 22 ½ in). Tail: 2.5 cm (1 ¼ in).
- **OCCURRENCE** S.E. Asia (Sulawesi, formerly Celebes). In lowland rainforest, including some secondary forest.

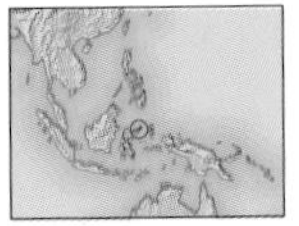

ASIA

- **REMARK** The Celebes Macaque is one of the six or seven macaque species, that inhabit this region, each with its own particular range.

Social unit Social	Gestation 174–196 days	Young 1	Diet

Family CERCOPITHECIDAE	Species *Macaca fascicularis*	Status Lower risk

CRAB-EATING MACAQUE

A grey-white moustache on a pink face, and often a small, pointed crest on the crown, distinguish this macaque, probably the most common primate in Southeast Asia after humans. A good climber and swimmer, it also spends time on the ground, often around human habitations, especially temples in Bali, where it is revered. The Crab-eating Macaque moves around in noisy, quarrelsome groups that may sometimes number up to 100. However, the hierarchy is less defined than among other macaques. When threatened, it flees through trees or drops into the water and swims away.

- **SIZE** Body length: 37–63 cm (14 ½ – 23 in). Tail: 36–72 cm (14–28 in).
- **OCCURRENCE** S.E. Asia. In forest and mangrove swamps, along rivers, coasts, and offshore islands.
- **REMARK** The Crab-eating Macaque is sometimes trapped for the biomedical research trade.

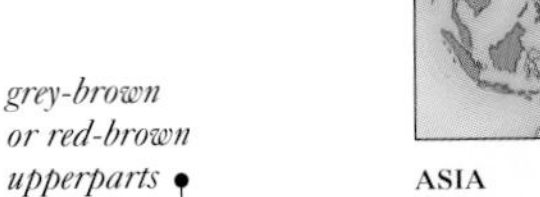

ASIA

Social unit Social	Gestation 160–170 days	Young 1	Diet 

Family CERCOPITHECIDAE	Species *Colobus guereza*	Status Lower risk*

EASTERN BLACK-AND-WHITE COLOBUS

With a white border to its face and white "cloak" down its flank and rump, this monkey is also known as the Mantled Guereza. Its exceptionally long tail is also tipped white. The newborn monkey is fully white and later takes on the black-and-white coloration. A single male monkey leads a group of four or five females and young, defending his territory with roars and spectacular jumping displays. Although diurnal, it is known to wake up and roar at night. This thumbless monkey pulls branches towards its mouth and bites off leaves and fruits.It has a monotonous diet, over 70 percent of which may consist of a single tree species. Like other leaf monkeys, it has a complex, three-part stomach, which houses gut microbes that help break down cellulose, enabling it to gain optimum nutrition from its leafy diet.

• **SIZE** Body length: 52–57 cm (20 ½–22 ½ in). Tail: 53–83 cm (21–33 in).

• **OCCURRENCE** S. Cameroon, east to Ethiopia, Kenya, and N. Tanzania. In light forest.

AFRICA

white border around face

large nose reaching lip

thumbless hands

white "cloak" over back

long black tail, tipped white

Social unit Social	Gestation 170 days	Young 1	Diet

Family CERCOPITHECIDAE	Species *Semnopithecus entellus*	Status Lower risk

HANUMAN LANGUR

A striking black face set in a hairy head, black extremities, and a grey to brown or golden-fawn body, characterize this monkey, which is identified with the demigod Hanuman and revered by Hindus. It has long limbs, and a long tail, held up in a curl. The social units of the Hanuman Langur are flexible, from peaceful troops with a number of males to those with only one male that are invaded by bachelors. Communication is made through whoops. An adaptable species, troops of Hanuman Langurs live near villages, feeding on leftovers and "offerings" from local people.

• **SIZE** Body length: 51–78 cm (20–31 in). Tail: 69–102 cm (27–40 in).

• **OCCURRENCE** Pakistan, India, Bhutan, Nepal, and Sri Lanka. All habitats, except rainforest.

• **REMARK** There are several species of langurs in the Indian subcontinent, ranging from the brown, white-headed Himalayan Langur to the small, pale fawn southern species of Sri Lanka that is half its size.

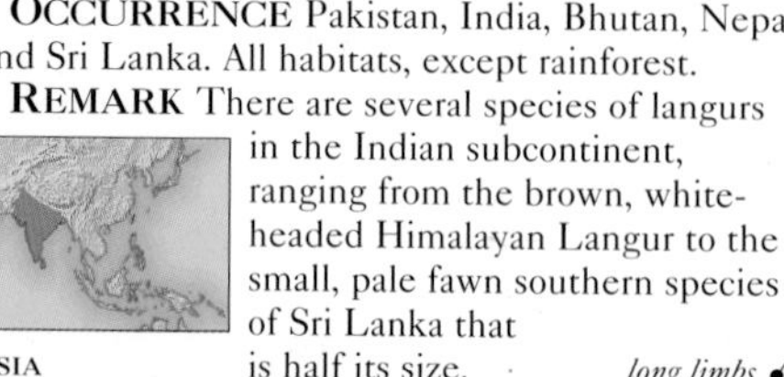

ASIA

Social unit Social	Gestation 200 days	Young 1	Diet

Family CERCOPITHECIDAE	Species *Nasalis larvatus*	Status Endangered

PROBOSCIS MONKEY

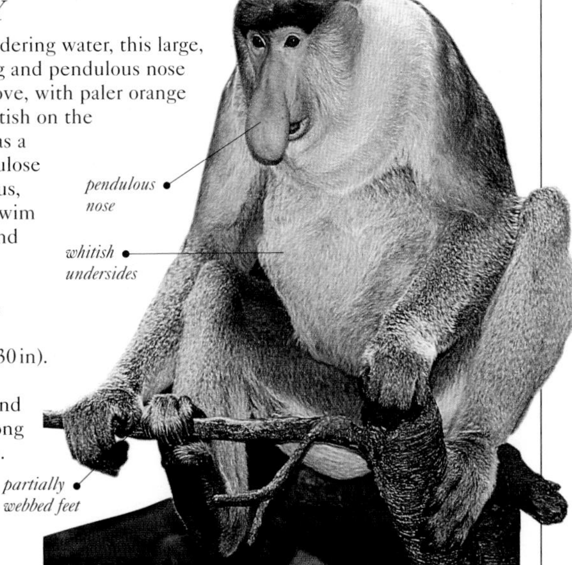

Living in very restricted habitats bordering water, this large, arboreal monkey is known for its long and pendulous nose (more so in males). It is brick-red above, with paler orange flanks, throat, and cheeks, and is whitish on the underside. The Proboscis Monkey has a complex stomach to break down cellulose (see Eastern Black-and-White Colobus, opposite). Troops of 6–10 monkeys swim expertly across creeks and streams, and forage among trees, led by a single dominant male, which bares its teeth and uses loud nasal honks and penile erection to ward away invaders.

• **SIZE** Body length: 73–76 cm (29–30 in). Tail: 66 cm (26 in).

• **OCCURRENCE** Borneo. In lowland rainforest, mangrove swamps, and along rivers and coastlines.

ASIA

• **REMARK** This rare and localized monkey is difficult to keep in captivity.

Social unit Social	Gestation 166 days	Young 1	Diet

Family CERCOPITHECIDAE	Species *Rhinopithecus roxellana*	Status Vulnerable

GOLDEN SNUB-NOSED MONKEY

This monkey endures winter temperatures of -5°C (23°F). Its long fur and bushy tail provide insulation as it moves in trees or on the ground on its strong, stout limbs. Heavily built, it has a pale blue, triangular face, with an upturned nose, and prominent jaws. Large troops of several hundred monkeys split into small bands of one male and several females, to forage and breed. Males are up to twice the size of the females, with black hair on the back. This monkey is preyed upon by eagles and, perhaps, Leopards. It is also threatened by deforestation and hunting for its pelt.

• **SIZE** Body length: 54–71 cm (21½–28 in). Tail: 52–76 cm (20½–30 in).

• **OCCURRENCE** W. China. In montane forest.

• **REMARK** One of the four species of mountain monkeys, three found in China and one in Vietnam, the Golden Snub-nosed Monkey is the only species that is not critically endangered.

ASIA

Social unit Social	Gestation 195 days	Young 1	Diet 

PRIMATES

APES

THE 21 SPECIES of apes, comprising two families, make up the primate suborder, Haplorhini, along with monkeys.

Apes resemble Old World monkeys in many respects, being forest-dwellers with a flat face, forward-pointing eyes, flexible limbs, and grasping hands and feet. However, they are larger than monkeys, with a more upright posture, and have no tail.

The lesser apes, or gibbons, swing through Southeast Asian forests. The great apes are our closest relatives, and the most intelligent of animals. They include the Gorilla, Chimpanzee, and Pygmy Chimpanzee (Bonobo) of Africa, and Orangutan of Southeast Asia.

Gibbons form monogamous pairs, adult male Orangutans tend to be solitary, while African apes live in highly social groups.

Family HYLOBATIDAE	Species *Hylobates lar*	Status Endangered

LAR GIBBON

This gibbon has black skin with a white fur fringe around its face, and on its hands and feet. The rest of its body may vary from cream to red, brown, or almost black. The feet, like the palms of the hands, have bare, leathery soles providing effective grip. The big toe can grasp in opposition to the other toes, enabling this ape to walk upright along branches. The gibbon's arm-to-arm swinging movement from one tree to another, known as brachiation, saves energy by maintaining momentum, using the body as a pendulum. Active shortly after dawn, the male and the female "duet" to reinforce their pair-bond: the female begins with long, loud hoots that rise to a crescendo, and the male responds to these with simpler, quivering hoots.

- **SIZE** Length: 42–59 cm (16½–23 in). Weight: 4.5–7.5 kg (10–17 lb).
- **OCCURRENCE** S. China, Burma, Laos, Thailand, Malaysia, and N. Sumatra. In dry deciduous and moist evergreen rainforest, from lowlands to mountains.
- **REMARK** The Lar Gibbon is hunted in China.

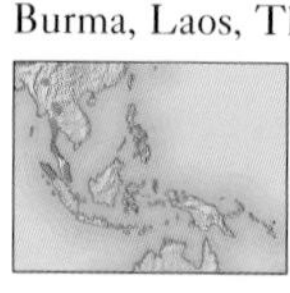

ASIA

arms around 40 percent longer than legs

young clinging to mother's chest

Social unit Pair	Gestation 7–8 months	Young 1	Diet 

Family HYLOBATIDAE	Species *Hylobates syndactylus*	Status Lower risk

SIAMANG

The largest of all gibbons, the robust, muscular Siamang stands at a height of 1.5 m (5 ft). Both sexes have uniform, black, shaggy fur; the male is slightly larger than the female with a tuft of hair on his genital region, which at first glance may be mistaken for a tail. This species lives in very close-knit family groups consisting of the female (who is dominant), male, and one or two young. They rarely stray more than 30 m (100 ft) from each other, and are usually found less than 10 m (33 ft) apart. The family occupies a home range of about 47 hectares (116 acres), but defends only 60 percent of this territory, using powerful calls. The Siamang has dark grey, elastic skin on its throat, which inflates to the size of a grapefruit and amplifies its calls – the loudest among gibbons.

- **SIZE** Length: 90 cm (35 in). Weight: 10–15 kg (22–33 lb).
- **OCCURRENCE** Central part of the Malay Peninsula and Sumatra. In both primary and secondary rainforest.
- **REMARK** The Siamang can inhabit montane forest at higher altitudes than most other gibbons, because of its ability to retain body heat.

LONG ARMS
The exceptionally long arms of the Siamang enable it to swing from tree to tree using its hands. Sometimes it walks on its feet with arms held aloft or sideways.

Social unit Social/Pair	Gestation 6½–7½ months	Young 1	Diet

Family PONGIDAE	Species *Pongo pygmaeus*	Status Endangered

BORNEAN ORANGUTAN

With arms that are twice the length of its body, and feet that can grasp branches like hands, the Bornean Orangutan is the largest arboreal animal. It has extremely flexible limbs, with wrist, hip, and shoulder joints allowing a wider range of movement than any other great ape. The Bornean Orangutan spends its life in the forest canopy. The male is more likely to descend to the ground, and even then, only rarely. He is bigger than the female with a long beard, large throat pouch, and long neck and arm hair which hangs like a cape. The female gives birth to a tiny infant once every seven or eight years (the longest inter-birth interval of any animal), in a nest atop a tree, and the male–female pair remain together until the young is about eight years old. The male defends his home range by emitting long calls, and if necessary by fighting. Genetic research has led to two different species of orangutan being recognized: the Bornean Orangutan and the Sumatran Orangutan (*Pongo abelii*).

- **SIZE** Height:1.1–1.4 m (3 ½–4 ½ ft). Weight: 40–80 kg (88–175 lb).
- **OCCURRENCE** Borneo. In primary rainforest.
- **REMARK** Although protected by law, young orangutans are still illegally captured and sold as pets. Rehabilitation projects have a good success rate, but some animals find it difficult to readjust to their natural habitat, which is also under threat.

ASIA

very long arms compared to body

cape-like arm hair in male

coat varies from orange-red to grey-brown

slow, cautious movement on the ground

digits grip like thumb and fingers

FIST-WALKING
On the rare occasion when an Orangutan leaves the trees, it walks on the soles of its feet and clenched fists (not solely the knuckles). The arms of this ape are so long that it can almost stand upright, yet still touch the ground with its hands.

Social unit Variable	Gestation 8 ½ months	Young 1	Diet

EATING WITH HANDS
Orangutans use their fingers and teeth to prepare their food, stripping plants and scraping off the peel of fruits to expose the juicy flesh within. Their diet also includes honey, small animals such as lizards, termites, nestling birds, and eggs.

MEMORY AND EXPERIENCE
The orangutan appears to build up a four-dimensional "mental map" of its forest area. It knows where fruit trees are located, and which ones are likely to be fruiting for the time of year.

Family HOMINIDAE	Species *Pan troglodytes*	Status Critically endangered

CHIMPANZEE

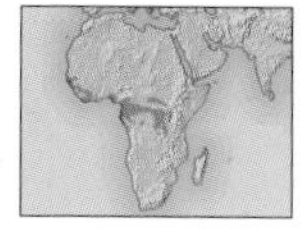

AFRICA

One of the apes that most resemble humans, the Chimpanzee is highly expressive, often using its flexible, protrusible lips to make grimace-like "smiles" that actually indicate fear. It has arms that are much longer than its legs and it walks on its knuckles and flat feet, with the big thumb-like toe of the foot opposing the other toes and providing a good grip while climbing. Chimpanzees live in groups of 15–120, and parties of adult males are known to attack and kill intruding males. Although chiefly herbivorous, Chimpanzees may cooperate to kill and eat animal prey such as monkeys, small antelopes, and birds. The Chimpanzee not only uses tools but has also learnt to make them. It uses stripped branches to scoop out termites from their nests and then licks them up.

• **SIZE** Length: 63–90 cm (23–35 in). Weight: 30–60 kg (66–130 lb).

• **OCCURRENCE** W. to C. Africa. In montane, primary, and secondary rainforest to woodland savanna.

• **REMARK** One of the world's most endangered animals, the Chimpanzee is close to humans in intelligence, emotions, and learning skills.

BUILDING A HOME

The mother chimpanzee builds a new nest in a tree for her young almost every night, by bending over and intertwining many branches to make a firm, leafy platform, away from predators.

Social unit Social	Gestation 8 months	Young 1	Diet

Family HOMINIDAE	Species *Pan paniscus*	Status Endangered

PYGMY CHIMPANZEE

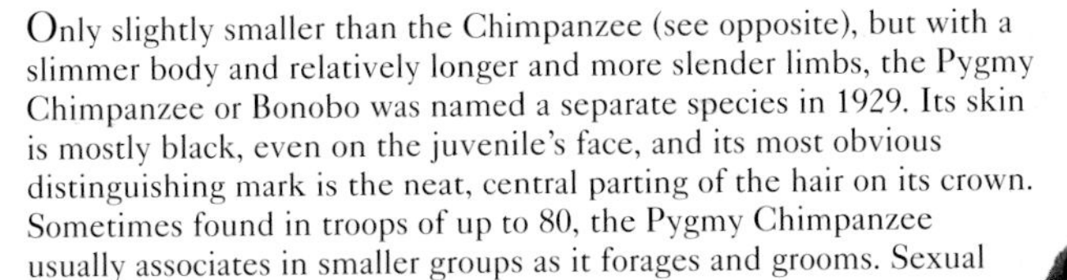

Only slightly smaller than the Chimpanzee (see opposite), but with a slimmer body and relatively longer and more slender limbs, the Pygmy Chimpanzee or Bonobo was named a separate species in 1929. Its skin is mostly black, even on the juvenile's face, and its most obvious distinguishing mark is the neat, central parting of the hair on its crown. Sometimes found in troops of up to 80, the Pygmy Chimpanzee usually associates in smaller groups as it forages and grooms. Sexual relations are common between males, females, and young at various levels, and may be used to ease tensions within the group. Female chimpanzees, who are usually dominant, leave their family unit on reaching maturity, although the males tend to stay on.

- **SIZE** Length: 70–83 cm (28–33 in). Weight: Up to 39 kg (86 lb).
- **OCCURRENCE** C. Africa. In tropical forest and at the southern limit of their range, in savanna.
- **REMARK** Seriously threatened by hunting, the Pygmy Chimpanzee may soon become the first great ape to be extinct in the wild.

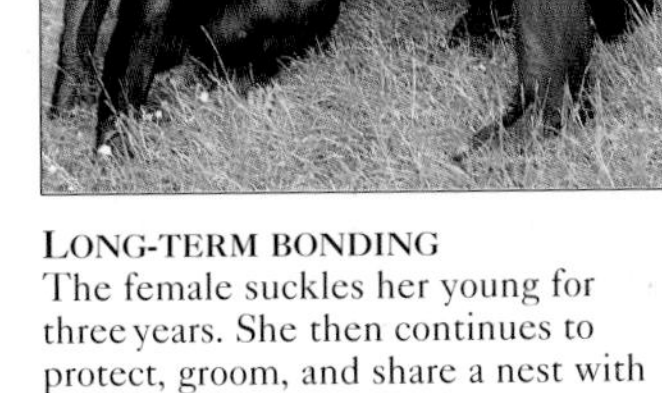

LONG-TERM BONDING
The female suckles her young for three years. She then continues to protect, groom, and share a nest with it for another year or two.

Social unit Social	Gestation 8 months	Young 1	Diet

Family HOMINIDAE	Species *Gorilla beringei*	Status Critically endangered

EASTERN GORILLA

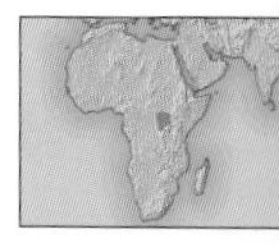

The largest living primates, gorillas are classified as the Western Gorilla (*Gorilla gorilla*) and the Eastern Gorilla (*Gorilla beringei*). The latter includes both the Eastern Lowland Gorilla (*G. beringei graueri*), as well as one or more subspecies of mountain gorillas (*G. beringei beringei*). The Eastern Gorilla is distinguished by its dark, shaggy coat, extremely long arms, and chestnut brown eyes; the mountain subspecies having longer fur to retain body warmth at higher altitudes. The mature male Eastern Gorilla has a saddle of white fur on his back and is often referred to as the silverback. These gorillas are found in groups, each of which roam a home range of 400–800 hectares (1,000–2,000 acres). Apart from a core area, the ranges may overlap with territories of neighbouring groups. The dominant silverback male fathers most or all of the young in the group. He gains the attention of receptive females in various ways, including mock feeding, thumping plants, hooting loudly, chest-beating, or jump kicking. When threatened by intruders, the male begins to hoot. He then stands upright, beats his chest with cupped hands, and throws vegetation. If all else fails, he charges forward with a huge roar, and knocks down the aggressor with a massive hand-swipe.

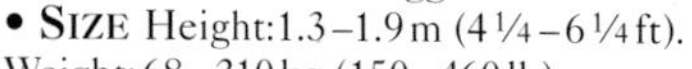

- **SIZE** Height: 1.3–1.9 m (4 1/4–6 1/4 ft). Weight: 68–210 kg (150–460 lb).
- **OCCURRENCE** C. and E. Africa. In montane rainforest, bamboo forest, swampy glades, and Afro-Alpine zone.
- **REMARK** The Western Gorilla, found in C. Africa, is more commonly seen in zoos and parks. However, it is endangered and faces the risks of poaching and forest clearance.

FUSSY EATER
Gorillas select their food carefully and then prepare each mouthful before they eat it. Their diet includes leaves, shoots, and stems, especially bamboo, as well as wild celery, nettles, thistles, fruits, roots, soft bark, and fungi.

Social unit Social	Gestation 8 1/2 months	Young 1	Diet

NESTING

Eastern Gorillas sleep in a new nest every evening. Adult males usually rest on the ground, while females sleep with their current offspring either on the ground or in forks of trees. An expectant female may build several nests within a few paces of each other, until she feels comfortable and gives birth. The infant is never left behind in the nest and accompanies the mother wherever she goes by clinging on to her fur.

SLOTHS, ANTEATERS, ARMADILLOS

FORMERLY PART OF the order Xenarthra (or Edentata), these 29 species, are now classified into two separate orders: Cingulata (armadillos), and Pilosa (anteaters and sloths).

Found in a variety of habitats from S. USA to South America, members of these orders have unique strengthening joints called xenarthrales, in the lower spine. They also have a relatively small brain for the body size.

Apart from these features, howeve the three groups are markedly differen Armadillos have protective armou burrow extensively, and eat sma animals as well as occasional plan material. Anteaters have a tubular snou and a long tongue for licking up ant and termites. Sloths, among the mos arboreal of all mammals, have long fu and a small, rounded head, and ea mainly leaves and fruits.

Family BRADYPODIDAE	Species *Choloepus didactylus*	Status Unconfirmed

LIME'S TWO-TOED SLOTH

Characterized by its slender limbs and slow movements, this sloth has two hooked claws on its front feet and three on its rear feet. Its coarse, grey-brown fur may be tinged with green algae, and helps in camouflage as it hangs upside down from forest branches. Active at night, it moves so slowly through the canopy that its presence can barely be detected. It hardly ever descends to the ground – perhaps once a week to defecate – and at such times becomes vulnerable to predators such as Jaguars, Ocelots, and large eagles. Occasionally, it is hunted by humans.

- **SIZE** Length: 46–86 cm (18–34 in). Weight: 4–8.5 kg (8¾–19 lb).
- **OCCURRENCE** E. Venezuela, Guianas south to Ecuador, Peru, and the Amazon Basin of Brazil. In mature, disturbed, and secondary forest.
- **REMARK** The body temperature of this sloth is thought to be the lowest of any mammal.

S. AMERICA

long forelimbs with two hooked claws

three hooked claws on hindfeet

grey-brown, coarse fur

Social unit Solitary	Gestation 11 months	Young 1	Diet

Family BRADYPODIDAE	Species *Bradypus torquatus*	Status Endangered

MANED THREE-TOED SLOTH

Active both during the day and night, this sloth blends remarkably well with its forest habitat, its greyish brown coat tinged green with algae and infested with ticks, beetles, and moths. The long, coarse outer hairs form a mane around its head and shoulders. Like Lime's Two-toed Sloth (see opposite), it only rarely descends to the ground, dragging itself along on its strong front legs. Its metabolic rate and body temperature are very low. While its main defence is to stay unnoticed, it will lash out with its claws when confronted.

• **SIZE** Length: 45–50 cm (18–20 in). Weight: 3.5–4 kg (7 ¾–8 ¾ lb).

• **OCCURRENCE** Brazil (Bahia, Espirito Santo, and Rio de Janeiro). In coastal tropical forest.

S. AMERICA

• **REMARK** In some areas, conservationists are trying to capture and relocate these sloths before tree felling takes place.

Social unit Solitary	Gestation 5–6 months	Young 1	Diet

Family MYRMECOPHAGIDAE	Species *Tamandua tetradactyla*	Status Vulnerable*

SOUTHERN TAMANDUA

S. AMERICA

Sometimes called the Lesser Anteater, this species is pale yellow, with a black "vest" around its middle, a small pointed head, and a sparsely haired prehensile tail. Arboreal as well as terrestrial, it uses its long claws and powerful limbs to break up rotting logs and insect nests. It is active for eight-hour stretches, even at night.

• **SIZE** Length: 53–88 cm (21–35 in). Weight: 3.5–8.5 kg (7 ¾–19 lb).

• **OCCURRENCE** S. Venezuela to N. Argentina and Uruguay. In diverse habitats, from rainforest to gallery forest, savanna, and plantations.

Social unit Solitary	Gestation 4–5 months	Young 1	Diet

Family MYRMECOPHAGIDAE	Species *Myrmecophaga tridactyla*	Status Vulnerable

GIANT ANTEATER

This species has a long, tubular snout widening to its small face, with tiny eyes and ears, massive forelegs, smaller rear legs, and a huge, bushy brown tail. Ambling along on its knuckles, it rips open ant hills and termite nests with its large front claws, lapping up its prey with its sticky, spine-covered tongue, which may be up to 60 cm (2 ft) long. The Giant Anteater defends itself from Pumas and Jaguars by rearing up, roaring, and slashing with its claws. Habitat destruction and hunting by humans for meat have threatened this species.

• **SIZE** Length: 1–2 m (3 ¼–6 ½ ft). Weight: 22–39 kg (49–80 lb).

• **OCCURRENCE** Central to South America. In grassland and forest.

Social unit Solitary	Gestation 190 days	Young 1	Diet

Family MYRMECOPHAGIDAE	Species *Cyclopes didactylus*	Status Unconfirmed

SILKY ANTEATER

As the name suggests, this anteater has long, dense, silky fur that is usually smoky grey with a silvery sheen, often with a brown stripe running from shoulder to rump. Adapted to life in trees, the Silky Anteater grasps branches with its feet and hook-like claws, and long, thickly furred tail. This opportunistic feeder breaks open hollow stems inhabited by tree-ants and licks them out with its long, sticky tongue. Hanging onto branches, it rests curled up in thick foliage or lianas; it is not known to build a nest. The male's territory appears to include those of several females.

• **SIZE** Length: 16–21 cm (6 ½–8 ½ in). Weight: 150–275 g (5–10 oz).

• **OCCURRENCE** Mexico, and Central to N. South America. In moist, lowland rainforest.

• **REMARK** The Silky Anteater rarely comes to the ground and is especially vulnerable to habitat loss by deforestation.

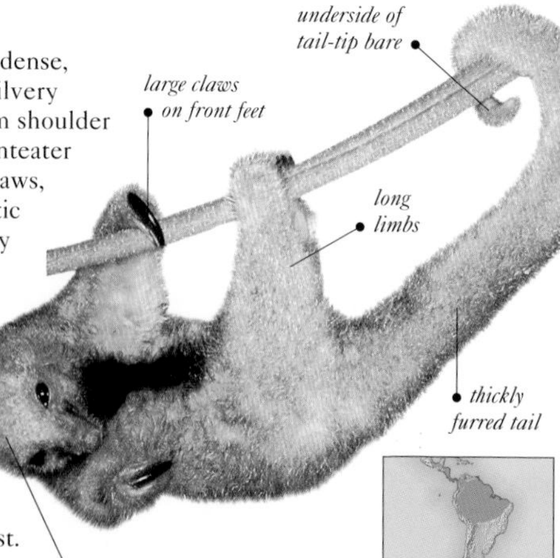

Social unit Solitary	Gestation Not known	Young 1–2	Diet

Family DASYPODIDAE	Species *Chaetophractus villosus*	Status Unconfirmed

HAIRY ARMADILLO

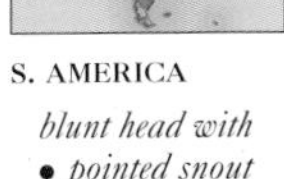

A creature of arid habitats, this species has coarse hair protruding from its thick armour, which comprises 18 or so bony, skin-covered bands. Around seven or eight bands are movable, allowing the armadillo to roll into a ball to protect its vulnerable furry underside. A largely solitary species, it growls when threatened and defends itself by running away or digging burrows with its sharp claws. In summer, the Hairy Armadillo is chiefly nocturnal, feeding on a range of small prey, from insects and rodents to reptiles and carrion. In winter, it is more active during the day and feeds on a greater quantity of plant material. The Hairy Armadillo is hunted by humans throughout its range.

- **SIZE** Length: 22–40 cm (9–16 in). Weight: 1–3 kg (2½–6½ lb).
- **OCCURRENCE** S. South America. In sandy, semi-desert areas.

Social unit Solitary	Gestation 60–75 days	Young 1–2	Diet

Family DASYPODIDAE	Species *Priodontes maximus*	Status Endangered

GIANT ARMADILLO

S. AMERICA

The largest of all armadillos, this species has 11–13 slightly movable, hinged plates over its body, and three or four plates over its neck. The long, tapering tail is also armoured. This armadillo is dark brown, except for a pale, yellow-white head, tail, and band along the lower edges of its bony plates. The unusually large third front claw is used to rip through the earth for small prey such as termites, ants, worms, and snakes. The front claws are also used to dig a burrow in which it shelters by day. This armadillo is not social or territorial, and moves onto a new feeding area every 2–3 weeks.

- **SIZE** Length: 75–100 cm (30–39 in). Weight: 30 kg (66 lb).
- **OCCURRENCE** N. and C. South America. In a range of habitats, from grassland to tropical rainforest.
- **REMARK** This armadillo is the only species in its genus.

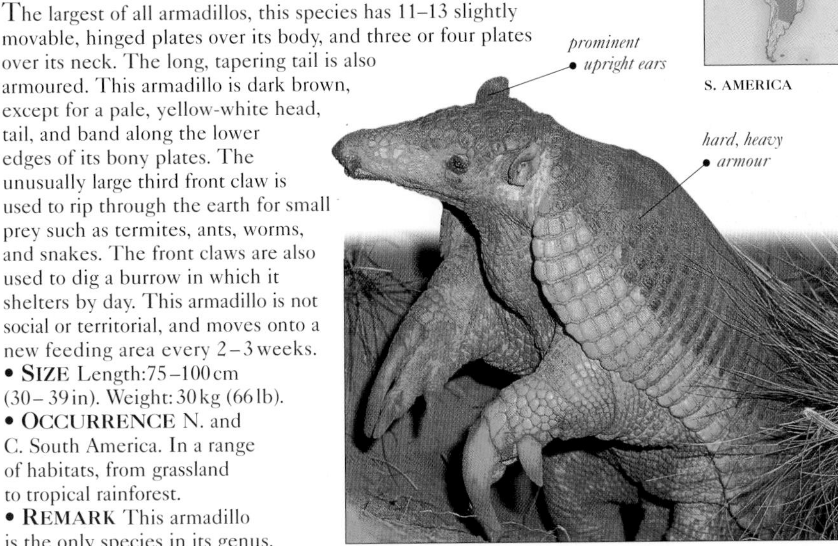

Social unit Solitary	Gestation 4 months	Young 1–2	Diet

Family DASYPODIDAE	Species *Dasypus novemcinctus*	Status Locally common

NINE-BANDED ARMADILLO

Characterized by 8–10 flexible bands across its carapace, this most commonly seen armadillo digs an extensive burrow system, like others in its family. It produces multiple young of the same sex from a single fertilized egg.

N., C., & S. AMERICA

• **SIZE** Length: 35–57 cm (14–22 ½ in). Weight: 2.5–6.5 kg (5 ½–14 lb).
• **OCCURRENCE** Mexico, and Central and South America. In grassland and forest.

Social unit Solitary	Gestation 8–9 months	Young 4	Diet

Family DASYPODIDAE	Species *Zaedyus pichiy*	Status Common

PICHI

When in danger, this small armadillo, the only species in its genus, grips the ground with its sharp-clawed feet, and relies on its bony shell for protection; or wedges itself into a burrow, its armour facing outwards. It digs a tunnel for shelter.
• **SIZE** Length: 26–34 cm (10–13 ½ in). Weight: 1–2 kg (2 ¼–4 ½ lb).

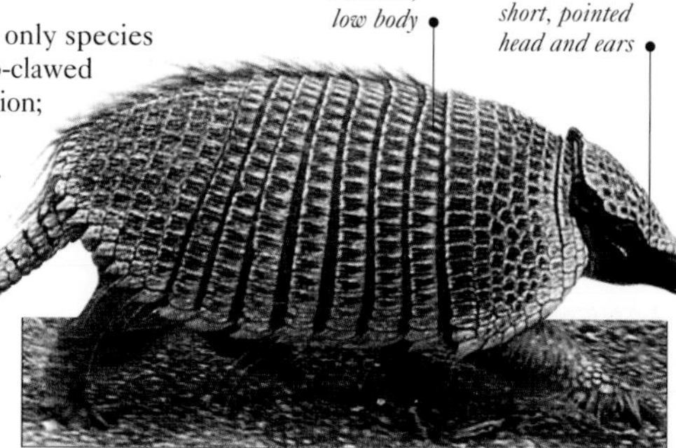

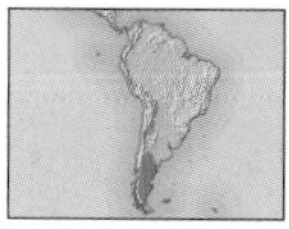
S. AMERICA

• **OCCURRENCE** Argentina and Chile, to Magellan's Strait. In grassland.

Social unit Solitary	Gestation 60 days	Young 1–3	Diet

Family DASYPODIDAE	Species *Cabassous centralis*	Status Unconfirmed

NORTHERN NAKED-TOED ARMADILLO

Found in a wide range of habitats, this large-eared armadillo has a specially enlarged middle claw on each forefoot for digging. This slow-moving species tears open ant and termite nests with its claws and licks up the insects with its long, sticky tongue, like an anteater. When threatened, it burrows into the ground, leaving its armour exposed.

C. & S. AMERICA

• **SIZE** Length: 30–40 cm (12–16 in). Weight: 2–3.5 kg (4 ½–7 ¾ lb).
• **OCCURRENCE** C. and N. South America. In grassland and forest.

Social unit Solitary	Gestation Not known	Young 1	Diet

PANGOLINS

IN BOTH LIFESTYLE and anatomy, pangolins resemble armadillos, anteaters, and sloths. However, they are unrelated and are classed in a separate order known as Philodata. The similarities between them are due to the fact that, in the course of evolution, they have all adapted in a similar way to their environment.

There are seven pangolin species, and they are all found in Africa and South Asia. Some are tree-dwelling, while others stay on the ground. None have any teeth. They gather their food, mainly ants and termites, with their extremely long, flexible tongue, swallowing it whole and grinding it up in their stomach.

The most noticeable features of pangolins are their sharp-edged scales, made of horn, which cover most of the outward-facing parts of the body, and the tapering head and tail. The scales offer protection and camouflage. They are tilted by muscles at their base in the skin and are replaced periodically.

Family MANIDAE	Species *Manis pentadactyla*	Status Lower risk

CHINESE PANGOLIN

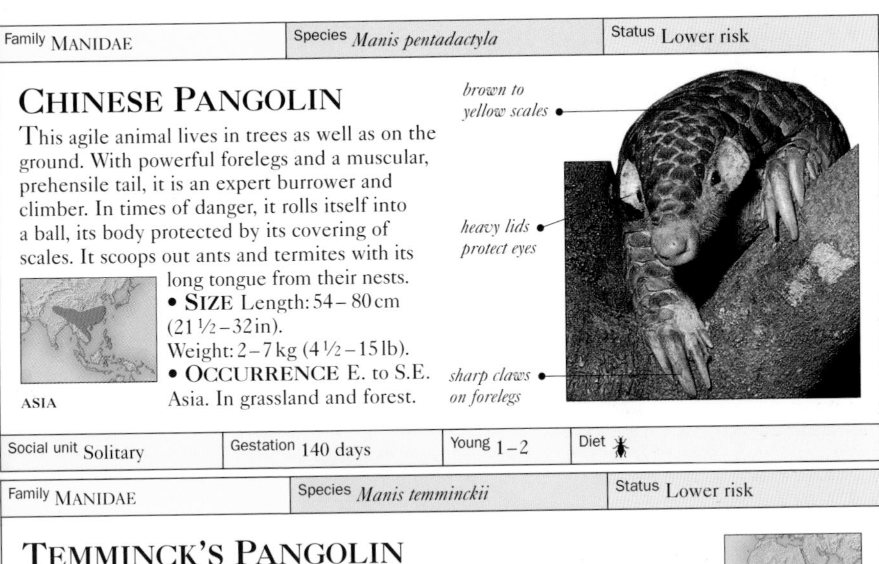

This agile animal lives in trees as well as on the ground. With powerful forelegs and a muscular, prehensile tail, it is an expert burrower and climber. In times of danger, it rolls itself into a ball, its body protected by its covering of scales. It scoops out ants and termites with its long tongue from their nests.

• **SIZE** Length: 54–80 cm (21½–32 in). Weight: 2–7 kg (4½–15 lb).

• **OCCURRENCE** E. to S.E. Asia. In grassland and forest.

Social unit Solitary	Gestation 140 days	Young 1–2	Diet

Family MANIDAE	Species *Manis temminckii*	Status Lower risk

TEMMINCK'S PANGOLIN

The streamlined body of this pangolin is covered with brown or yellow-brown overlapping scales, making it resemble the Chinese Pangolin (see above), although it is smaller in size. It uses its large claws to rip open ant and termite nests, both in trees and on the ground, for food.

• **SIZE** Length: 50–60 cm (20–23½ in). Weight: 15–18 kg (33–40 lb).

• **OCCURRENCE** E. to S. Africa. In grassland, tropical and temperate forest.

overlapping scales

small head

AFRICA

Social unit Solitary	Gestation 120 days	Young 1–2	Diet

RABBITS, HARES, AND PIKAS

THE ORDER LAGOMORPHA, aptly meaning "leaping shape", includes rabbits, hares, and pikas. As targets for many predators, lagomorphs breed fast to maintain numbers, and detect danger with their extremely keen senses. Long ears catch slight sounds, and eyes high on the sides of the head give all-round vision. The powerful legs, especially the elongated hind pair, allow rapid escape.

As gnawing mammals, lagomorphs are sometimes mistakenly thought to be rodents. Important differences include a second set of incisors in the upper jaw, lighter skull structure, slit-like nostrils, and a small, rounded, "bob" tail. The more mouse-like pikas, however, have rounded ears, four limbs of equal length, and no visible tail.

The 80 lagomorph species occur mainly in open habitats, from tundra to desert, and on all continents.

Family OCHOTONIDAE	Species *Ochotona princeps*	Status Vulnerable

NORTH AMERICAN PIKA

The egg-shaped, short-legged body of this pika, with no external tail, is typical of all its family. With a soft, dense coat, its back is greyish to cinnamon-brown, often richly tinged tawny or ochre, while its sides are a lighter buff. The soles of its feet are densely furred. It is usually found near a talus – an area of piled up, broken rocks surrounded by alpine meadows. Each talus is occupied by a solitary pika, which defends its territory with a short whistle. A patchwork of alternate male and female territories, of equal size, is formed across the area. The same short whistle, when used repeatedly, serves as a warning against predators, while the male emits a long, song-like call during courtship. The North American Pika does not hibernate. Instead, it makes a "hay pile" of grasses, herbs, and other plants near its burrow during late summer, as a winter store, often selecting plants that contain the most protein.

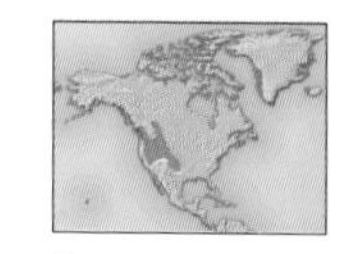

N. AMERICA

- **SIZE** Length: 16–22 cm (6½–9 in). Weight: 121–176 g (4–6 oz).
- **OCCURRENCE** S.W. Canada and W. USA. In mountainous regions – mostly alpine meadows.

Social unit Solitary	Gestation 30 days	Young 3	Diet

Family OCHOTONIDAE	Species *Ochotona curzoniae*	Status Common

BLACK-LIPPED PIKA

With sandy-brown upperparts and a yellow-white underside, the Black-lipped Pika or Plateau Pika has a distinctive black nose and lips. Extremely sociable, it lives in extended families occupying a single burrow system. Although this pika rarely lives beyond a year, in certain areas it is found in high density and is viewed as a pest.

- **SIZE** Length: 14–18.5 cm (5 ½–7 ½ in). Weight: 124–171 g (4–6 oz).
- **OCCURRENCE** Himalayas, Nepal, Tibet, and W. China. On high altitude grassland.
- **REMARK** This pika is a key species in maintaining the biodiversity of the Tibetan Plateau.

ASIA

Social unit Social	Gestation 21 days	Young 2–8	Diet

Family LEPORIDAE	Species *Pentalagus furnessi*	Status Endangered

AMAMI RABBIT

A uniform black coat, small eyes and ears, and a pointed snout characterize this rare species, found only on two islands of Japan. Nocturnal and herbivorous, the Amami Rabbit, also known as the Ryukyu Rabbit, feeds on plants such as pampas grass leaves, sweet potato runners, bamboo sprouts, and bark. It uses its long-nailed paws to dig nest-holes, and communicates by making clicking sounds. Little is known about the social and breeding habits of this rabbit.

- **SIZE** Length: 42–51 cm (16 ½–20 in). Weight: 2–5 kg (4 ½–11 lb).
- **OCCURRENCE** Only on Amami and Tokuno islands of Japan. In tropical forest.
- **REMARK** Habitat destruction and predation are serious threats to this endangered species.

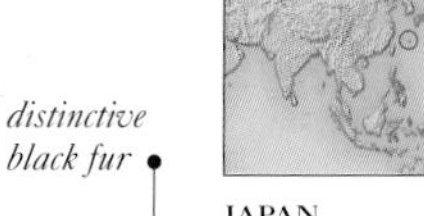

JAPAN

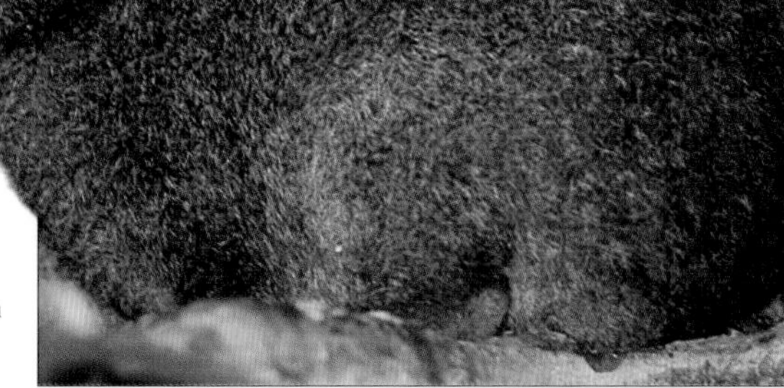

Social unit Variable	Gestation Not known	Young 2–3	Diet

Family LEPORIDAE	Species *Brachylagus aquaticus*	Status Locally common

SWAMP RABBIT

Highly territorial by nature, this able swimmer lives in wetlands and is always associated with water. It goes into water readily, especially when alarmed, and feeds on sedges, rushes, and aquatic plants. Active both day and night, it builds a nest on the ground made of weeds, lined with fur. This rabbit has a rusty brown or blackish coat with a white vent.

- **SIZE** Length: 45–55 cm (18–22 in). Weight: 1.5–2.5 kg (3 1/4–5 1/2 lb).
- **OCCURRENCE** S.E. USA. In swamps, marshes, and forest.

N. AMERICA

cinnamon ring around eyes

short, sleek coat

Social unit Social	Gestation 37 days	Young 1–6	Diet

Family LEPORIDAE	Species *Brachylagus floridanus*	Status Common

EASTERN COTTONTAIL

This rabbit is the most widely distributed species in its genus, occupying diverse habitats. It has long and dense fur, usually brown or grey, with a rusty nape, and a cottony white tail tipped reddish brown. The Eastern Cottontail feeds from late morning to evening, on lush green vegetation in summer, and bark and twigs in winter.

- **SIZE** Length: 38–49 cm (15–19 1/2 in). Weight: 1–1.5 kg (2 1/4–3 1/4 lb).
- **OCCURRENCE** S.E. Canada to Mexico, Central America, N. South America, and Europe. In grassland, desert, and forest.

N., C., & S. AMERICA

Social unit Social	Gestation 26–30 days	Young 3–7	Diet

Family LEPORIDAE	Species *Brachylagus idahoensis*	Status Lower risk

PYGMY RABBIT

The smallest of all North American rabbits, the Pygmy Rabbit is the only leporid species in the region that digs extensive burrow systems, with four or five widely spaced entrances. Feeding on big sagebrush and other related species, this rabbit is found in areas where these are easily available. It mostly forages at dawn and dusk, but in winter, could be active at any time of the day. This rabbit's coat is grey in winter and brown in summer, and nape, chest, legs, and tail are cinnamon-buff.

- **SIZE** Length: 22–29 cm (9–11 1/2 in). Weight: 350–450 g (13–16 oz).
- **OCCURRENCE** Great Basin Desert, USA. In areas of dense big sagebrush.
- **REMARK** This is the world's smallest rabbit.

N. AMERICA

Social unit Solitary	Gestation 26–28 days	Young 4–8	Diet

Family LEPORIDAE	Species *Oryctolagus cuniculus*	Status Common

EUROPEAN RABBIT

EUROPE, AFRICA, AUSTRALIA, S. AMERICA

The ancestor of all breeds of domestic rabbit, this species has been introduced in many countries throughout the world. It has a coat of black to light brown, with a dark collar, and a buff-coloured nape. Buffish white underneath, it has long, black-tipped ears, long hindlegs, and extremely furry feet. Nocturnal in habit, the European Rabbit is the most social species of the order, with a strict hierarchy of dominance. It digs elaborate underground tunnels or warrens, with large breeding burrows, mothers of lower "rank" digging small burrows outside the main warren. Males are known to protect juveniles, irrespective of paternity, from females, who may attack and even kill any strange young.

• **SIZE** Length: 34–50 cm (13 ½–20 in). Weight: 1–2.5 kg (2 ¼–5 ½ lb).

• **OCCURRENCE** Native only to S.W. Europe and probably N.W. Africa. Introduced in South America, Australia, and New Zealand and other parts of Europe. In grassland.

• **REMARK** Responsible for widespread habitat destruction, the rapidly multiplying European Rabbit has created a virtual wildlife management disaster wherever it has been introduced.

CUDDLY PETS
In the Middle Ages, French monks who reared wild rabbits for food began cross-breeding those with different features. Now popular pets, domestic rabbits come in different sizes, ear shapes, fur types, and colour patterns. The rabbit shown above is the Bi-coloured French Lop.

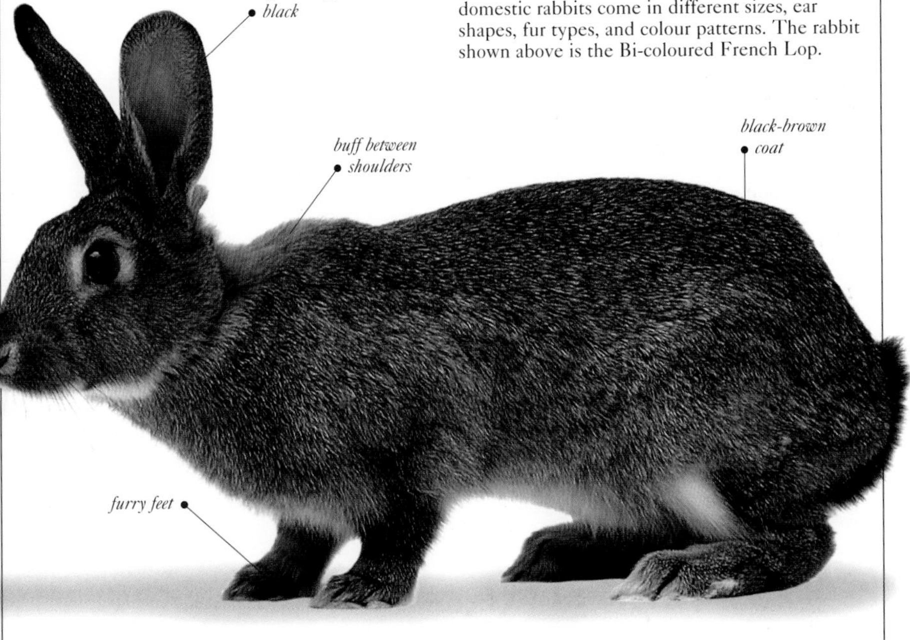

Social unit Social	Gestation 28–33 days	Young 3–12	Diet

Family LEPORIDAE	Species *Romerolagus diazi*	Status Endangered

VOLCANO RABBIT

Endemic to the volcanic region surrounding Mexico City, this rabbit is found in pine forests that have a dense undergrowth of bunch grasses, known as "zacaton". Found in groups of up to five, it is especially active in the early morning and evening. Unusual for its small, round ears, this species also has a small tail that is not visible externally. Its short, dense fur is yellow-black on the back and sides. The tip and base of the guard hairs on the underside are black.

• **SIZE** Length: 23–32 cm (9–12½ in). Weight: 375–600 g (13–21 oz).

• **OCCURRENCE** Around Mexico City. On forested, volcanic slopes with zacaton grasses.

N. AMERICA

• **REMARK** Threats to this species include habitat loss, hunting, and the seasonal burning of the zacaton.

Social unit Social	Gestation 38–40 days	Young 2	Diet

Family LEPORIDAE	Species *Lepus europaeus*	Status Common

BROWN HARE

This long-legged, large-eared hare has a prominent tail that is black on top and white below, and a distinctive, triangular black patch on the back of each ear-tip. Uniform in appearance, it is tawny or rusty on the chest and sides and darker on the back. This solitary hare spends the day hiding in a form – a shallow depression made in open fields or under cover of long grass or shrubs. During courtship it often displays what is called "mad March Hare behaviour", akin to a boxing bout, with unreceptive females chasing away interested males.

• **SIZE** Length: 48–70 cm (19–28 in). Weight: 2.5–7 kg (5½–15 lb).

• **OCCURRENCE** Europe and Asia. Introduced in E. Canada, N.E. USA, South America, S.E. Australia, and New Zealand. In open country, farmland, steppe, and woodland.

• **REMARK** Although the Brown Hare has been introduced to many countries, there is general concern over the decline of its numbers in Europe.

EURASIA

Social unit Solitary	Gestation 42 days	Young 1–10	Diet

Family LEPORIDAE	Species *Lepus arcticus*	Status Common

ARCTIC HARE

ARCTIC

A large, stocky hare with long fur, this species is highly adapted to survive in the severe conditions of the Arctic tundra. Some subspecies remain white throughout the year, but others turn grey in summer. The Arctic Hare's front incisors are modified to eat snow-covered plants, such as mosses and lichens, the buds, berries, leaves, and bark of other low-growing vegetation, and the roots of shrubs such as willow. It stamps on the ground with its hindfeet to expose plants beneath the snow. Unlike other hares, this species also takes meat from hunters' traps. Except during the mating season, it usually lives in groups of up to 300 and displays flocking behaviour, with the whole group suddenly changing direction at once. It uses speed to escape from predators, often running as fast as 64 kph (40 mph).

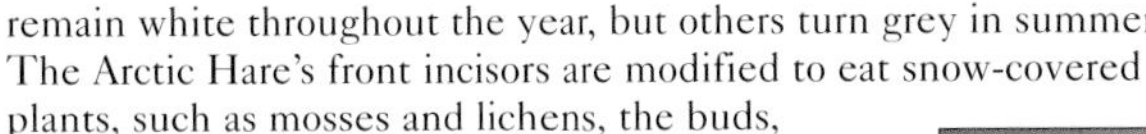

• **SIZE** Length: 43–66 cm (17–26 in). Weight: 3–7 kg (6 ½–15 lb).

• **OCCURRENCE** N. Canada, (Ellesmere Island, Northwest Territories, Newfoundland) and Greenland. On hillsides or rocky plateaux.

WHITE WINTER WEAR
The Arctic Hare's winter coat is completely white, except for black ear-tips, providing camouflage in the snow.

Social unit Variable	Gestation Not known	Young 1–8	Diet

Family LEPORIDAE	Species *Lepus timidus*	Status Locally common

MOUNTAIN HARE

Also known as the Blue Hare, the Mountain Hare is smaller than the Brown Hare (see p.140) and has long, black-tipped ears and large, furry feet. Moulting in late autumn and spring, its winter coat is completely white, while in summer it turns brown. A browser of woody plants such as heather, gorse, and juniper, it prefers grasses when available, and digs up the snow in winter to expose its food.

black ear-tips

long ears

brown summer coat

- **SIZE** Length: 43–61 cm (16½–24½ in). Weight: 2–3.5 kg (4½–7¾ lb).
- **OCCURRENCE** Arctic region of Europe and Asia. In coniferous forest, tundra, and mountains.
- **REMARK** It is also grouped with the Arctic Hare, expanding its range across Canada and USA.

EURASIA

Social unit Solitary	Gestation 50 days	Young 1–3	Diet

Family LEPORIDAE	Species *Lepus californicus*	Status Common

BLACK-TAILED JACKRABBIT

Enormous, black-tipped ears, up to 15 cm (6 in) long, help this hare to release excess body heat in hot, arid habitats, and to detect the faintest sounds of predators. It has a greyish brown to sandy coat, long legs, and a black tail with a black line extending to the rump. Found in different vegetation, such as sagebrush, creosote bush, mesquite, snakeweed, and juniper, it prefers succulent grasses and herbs, but can survive nibbling on woody twigs during drought or winter. When threatened it "freezes" or lies still with its head pressed to the ground, or it runs very fast on a jagged course. It builds a shallow "form" or nest under cover of a bush, and primarily stays within its home range, rather than defends a larger territory. This hare displays complex courtship behaviour, including long chases, pair jumping, and frequent fights between male and female.

- **SIZE** Length: 47–63 cm (18½–25 in). Weight: 1.5–3.5 kg (3¼–7¾ lb).
- **OCCURRENCE** W. USA and N. Mexico. In arid grassland and desert.
- **REMARK** One of the speediest lagomorphs, it can run at 56 kph (35 mph).

N. AMERICA

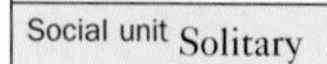

Social unit Solitary	Gestation 41–47 days	Young 1–6	Diet

RODENTS

MORE THAN TWO out of every five mammal species are rodents. Typified by mice and rats, they are found on every continent, except Antarctica, and in almost every habitat.

Despite a variety of lifestyles, from treetop squirrels, to semi-aquatic beavers and Capybaras, lemmings that live under the snow in winter, leaping jerboas, and burrowing mole rats, most rodents share common features. They are relatively small animals, with compact bodies, move on four limbs, and have clawed feet, a long tail, and powerful teeth and jaws specialized for gnawing. In particular, the four incisor teeth are long, deep-rooted, and grow continually. Rodents are equipped with acute senses of smell and hearing which, aided by numerous long and sensitive whiskers, make them extremely aware of their surroundings.

A few species, such as the Woodchuck, are solitary. However, most rodents are social, and some – like the lemmings, rats, mice, and prairie dogs – form enormous, loose-knit communities. Many small rodents breed very fast – some voles produce more than ten litters yearly. This counteracts predation and persecution by humans.

Family APLODONTIDAE	Species *Aplodontia rufa*	Status Lower risk

MOUNTAIN BEAVER

N. AMERICA

Known locally as the Sewellel, the Mountain Beaver has a long-haired coat which is black to red-brown above and yellow-brown below. It has a white spot below each ear and a flattened head. Its tail is comparatively short compared to those of beavers. This nocturnal rodent digs elaborate nests and tunnels under tree logs, with the entrances leading directly to a source of food such as bark, twigs, shoots, and succulents. Unlikely to travel far in its search for food, it carries back supplies to its nest, where they are consumed or stored. A good climber of trees, the Mountain Beaver can reach heights up to 7 m (22 ft) and is known to destroy large numbers of young trees such as fir and spruce by gnawing. Although it is found in mountainous areas, it prefers lower elevations. It is preyed upon by skunks, weasels, foxes, coyotes, and eagle owls.

- **SIZE** Body length: 30–46 cm (12–18 in). Tail: 2–4 cm (¾–1½ in).
- **OCCURRENCE** S.W. British Columbia (Canada) to C. California. In mountains and along coastlines.
- **REMARK** It has benefited from felling since it builds its home under felled logs.

long-haired coat

wide, flattened head

Social unit Solitary	Gestation 28–30 days	Young 2–6	Diet

Family SCIURIDAE	Species *Marmota monax*	Status Common

WOODCHUCK

Also called the Groundhog, this species is one of the largest and strongest squirrels. It has a stout body, short legs, a bushy tail, and brown fur that is grizzled or white-tipped. This able climber and swimmer feeds mainly in the afternoon, sometimes in groups, and eats grasses, clover, seeds, fruits, and small animals such as snails and grasshoppers. It often shows aggression to its own kind, especially when defending its burrow, and in the mating season when males fight for dominance. Defensive behaviour includes back arching, jumping, stiffening the tail, and teeth chattering. In autumn, the Woodchuck digs a deeper burrow for its hibernation.

• **SIZE** Body length: 32–52 cm (12½–20½ in). Tail: 7.5–11.5 cm (3–4½ in).

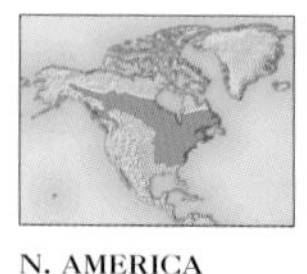

N. AMERICA

• **OCCURRENCE** Canada, Alaska, and E. USA. In woodland and fields.

• **REMARK** February 2 is known as Groundhog Day in North America, when this animal supposedly peers out to assess the weather.

Social unit Variable	Gestation 31–32 days	Young 1–8	Diet

Family SCIURIDAE	Species *Marmota flaviventris*	Status Locally common

YELLOW-BELLIED MARMOT

This marmot has a thickset body with a short, broad head and strong claws. Its ears are small and covered with fur, and it has soft, woolly underfur on the back and flanks. In many, the coarse outer guard hair varies from yellow-brown to tawny, frosted with paler tips, although some individuals may be black. Adapted to a wide range of habitats, the Yellow-bellied Marmot has an extensive diet of grasses, flowers, herbs, and seeds. It feeds mainly in the morning and late afternoon, then grooms and suns itself with other members of its colony, which usually consists of one male and several females. Preyed upon by Coyotes, Bobcats, hawks, owls, and eagles, it takes cover in its burrow when threatened. In autumn, this rodent goes into a long hibernation in its burrow, and may stay there for up to eight months.

• **SIZE** Length: 34–50 cm (13½–20 in). Tail: 13–22 cm (5–9 in).

N. AMERICA

• **OCCURRENCE** W. Canada and W. USA. In varied habitats, from the alpine zone, woodland, and forest clearings, to semi-desert.

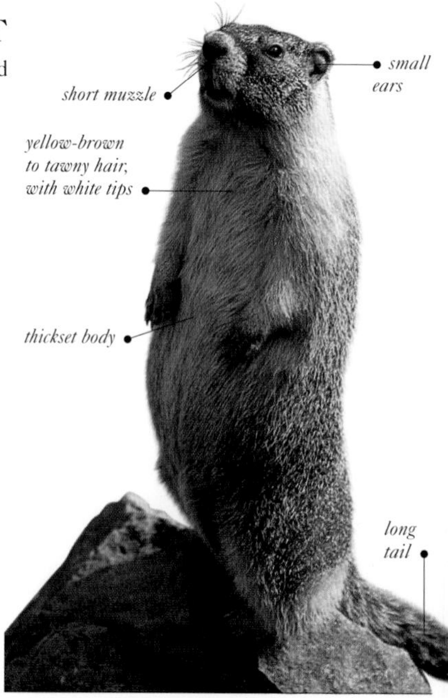

Social unit Social	Gestation 30 days	Young 3–8	Diet

mily SCIURIDAE	Species *Cynomys ludovicianus*	Status Lower risk

BLACK-TAILED PRAIRIE DOG

Also known as the Plains Prairie Dog, this squirrel is one of the five species of prairie dogs that derive their common name from their grassy habitat and characteristic dog-like bark or "yip". It has brown or reddish brown fur, tipped white in summer and black in winter, and its whiskers and tail-tip are black. During territorial disputes, these squirrels aggressively flare their tails and bark, in what are known as "jump-yip displays". They feed on wheat grass, buffalo grass, globemallow, and rabbitbrush in summer, and thistles, cacti, and roots in winter.

- **SIZE** Body length: 28–30 cm (11–12 in). Tail: 7–11.5 cm (2¾–4½ in).
- **OCCURRENCE** Great Plains of North America. In grassland.
- **REMARK** The Black-tailed Prairie Dog causes massive destruction to crops and has been the subject of very successful extermination campaigns in some regions.

cial unit Social	Gestation 33–38 days	Young 1–8	Diet

mily SCIURIDAE	Species *Xerus inauris*	Status Locally common*

CAPE GROUND SQUIRREL

Like the other ground squirrels, this African species has large, strong claws, enabling it to dig the extensive system of tunnels in which it lives. The fur on its back and head is brownish pink, and it has a white flank stripe, white face, feet, and belly, and black bands near the base and tip of its white tail. An opportunistic feeder, its diet ranges from seeds, bulbs, and fleshy roots to insects and birds' eggs. This squirrel lives in colonies of 5–10, occasionally as many as 30. It waits until well after sunrise before it emerges from its burrow, although on cold or overcast days it remains underground.

- **SIZE** Body length: 20–30 cm (8–12 in). Tail: 18–26 cm (7–10 in).
- **OCCURRENCE** S. Angola, Zimbabwe, Botswana, South Africa, Namibia. In open country.

cial unit Social	Gestation 42–49 days	Young 1–3	Diet

Family SCIURIDAE	Species *Tamias striatus*	Status Locally common

EASTERN CHIPMUNK

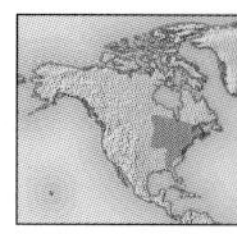

N. AMERICA

This rodent is greyish or reddish brown all over with a central dark brown stripe running down its back, which is bordered on either side by greyish or reddish brown hairs, followed by stripes of dark brown or yellow-orange, leading to a white belly. Light and dark facial stripes border its eyes. Its bushy tail is dark on the upperside with a light grey border, and its rump is yellowish or reddish brown. However, the overall colour and patterns on the coat vary from one region to another. Although sedentary by nature, the Eastern Chipmunk looks for food over a wide range of areas. Foraging activity peaks around mid-morning and mid-afternoon, with males more active in the morning and females more so in the afternoon. This solitary animal occupies a single burrow system and chases away trespassers from its core home territory, although a number of individuals may share a larger range. Both sexes make loud "chip" and "cuk" sounds that act as alarm calls to fellow chipmunks and other small animals living nearby. They are preyed upon by snakes, hawks, foxes, bobcats, and weasels. Chipmunks hibernate from autumn to early spring, but they may come out of their burrows to feed on fine winter days.

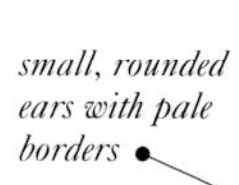

small, rounded ears with pale borders

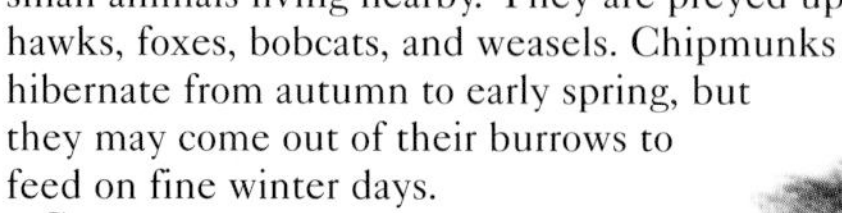

• **SIZE** Body length:15.5–16.5 cm (6–6 1/2 in). Tail:7–10 cm (2 3/4–4 in).

• **OCCURRENCE** Canada (Lake Manitoba to Nova Scotia), south to Louisiana, Alabama, Georgia, and N. Florida. In deciduous woodland and bushy areas with abundant rock crevices.

• **REMARK** Female chipmunks become intolerant of their young soon after birth, prompting the offspring to set up their own homes when only two weeks old.

dark stripes with pale borders along back

Social unit Solitary	Gestation 31–32 days	Young 4–5	Diet  

LONG AND SLIM
The chipmunk stretches its body to move through its burrow system, the tunnels of which are extremely narrow, to keep out larger predators such as stoats. Each burrow system may be 10 m (33 ft) long.

OPPORTUNISTIC FEEDER
The Eastern Chipmunk is familiar in the wild to humans as a bold visitor at picnic sites. Although it can climb well, this chipmunk forages mainly on the ground for seeds, acorns, and nuts, which it manipulates with its forepaws. Its stuffs food into its cheek pouches and hoards in its burrow to eat later.

Family SCIURIDAE	Species *Spermophilus columbianus*	Status Locally common

COLUMBIAN GROUND SQUIRREL

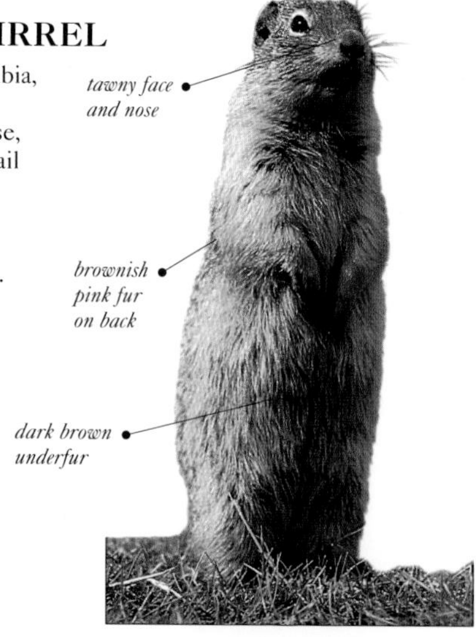

Found in the mountain meadows of British Columbia, Canada, the Columbian Ground Squirrel is mainly brownish pink in colour, with a tawny face and nose, pale grey patches on the sides of the neck, and a tail that is black above and grey below. Its fur is short and dense, and the underfur is darker in colour. Adult squirrels "kiss" each other on meeting, or tilt their heads to sniff the scent glands near the mouth. Feeding on flowers, seeds, bulbs, and fruits, this species customarily takes a few steps, pauses to eat, and takes a few steps again. It also catches insects mid-flight and occasionally becomes cannibalistic. Adult males have overlapping territories, while females have exclusive ones.

N. AMERICA

- **SIZE** Body length: 25–29cm (10–11½in). Tail: 8–11.5cm (3¼–4½in).
- **OCCURRENCE** W. Canada to N.W. USA. In alpine and subalpine meadows.

Social unit Social	Gestation 24 days	Young 2–5	Diet

Family SCIURIDAE	Species *Sciurus vulgaris*	Status Lower risk

EURASIAN RED SQUIRREL

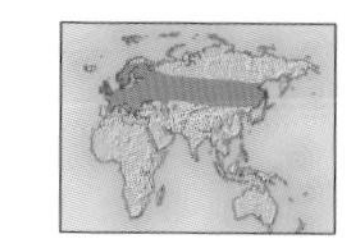
EURASIA

Not uniformly red, this squirrel varies from grey to red, brown, and black on its back, and is white below. In winter its coat may turn bright or dark greyish brown. The bushy tail is the same length as the body, while the ears are tufted, more so in winter. Adept at climbing and leaping, the Red Squirrel feeds on the ground or on branches, eating pine seeds, beechnuts, acorns, mushrooms, shoots, fruits, and bark, manipulating food with its forepaws. The drey (nest) is a ball of twigs in a branch fork or tree-hole. The female makes a larger breeding nest, lined with soft material, for her young.

- **SIZE** Body length: 20–25cm (8–10in). Tail: 15–20cm (6–8in).
- **OCCURRENCE** W. Europe to E. Asia. In forest, parks, and gardens up to the tree line.
- **REMARK** This species is threatened by deforestation and competition with the introduced Grey Squirrel.

Social unit Solitary	Gestation 38 days	Young 2–5	Diet

Family SCIURIDAE	Species *Sciurus carolinensis*	Status Common

EASTERN GREY SQUIRREL

A native of North America that has now been introduced in parts of Europe, the Eastern Grey Squirrel is grey on the back and white to grey or pinkish brown on its underparts. Its face, back, and forelegs are tinged brown and its tail is white or pale grey. Unlike the Red Squirrel (see opposite), its ears are not tufted. This medium-sized tree squirrel is an opportunistic feeder and "scatter-hoards" its food, carrying it in its mouth and burying it more than 2 cm (3/4 in) below the soil surface. It displays a strong homing tendency, although males are known to travel long distances. A social animal, it warns other squirrels of danger through different sounds, and chatters its teeth when confronted.

- **SIZE** Body length: 23–28 cm (9–11 in). Tail: 15–25 cm (6–10 in).
- **OCCURRENCE** S. and S.E. Canada to C. and E. USA. Introduced to the UK and Italy. In temperate forest, especially where winter food is plentiful.
- **REMARK** This introduced species has largely replaced the Red Squirrel in the UK.

TREE-TOP HOME
The Eastern Grey Squirrel makes a "drey", or nest, out of twigs in the branches of trees, and lines it with grass or bark. Although it becomes inactive in winter, it may emerge from its nest to forage.

N. AMERICA, EUROPE

Social unit Variable	Gestation 44 days	Young 1–5	Diet

Family SCIURIDAE	Species *Ratufa indica*	Status Vulnerable

INDIAN GIANT SQUIRREL

The huge, bushy tail of this large squirrel is longer than its head and body combined. Usually black above, its limbs and head are red-brown, while its underside is white. Its hands and feet are broad, and the claws are strong and well developed. Always alert, this squirrel makes giant leaps of 6 m (20 ft) from branch to branch in search of food. It builds a typical squirrel-type drey (nest) for resting and breeding. When alarmed, it flattens itself against a branch or hides behind a thick trunk.

• **SIZE** Body length: 35–40 cm (14–16 in). Tail: 35–60 cm (14–23 in).

• **OCCURRENCE** Peninsular India. In deciduous and moist evergreen forest.

• **REMARK** Unlike other squirrels, this squirrel does not feed upright, but leans forwards on its hindlegs, using its tail for balance.

ASIA

Social unit Solitary/Pair	Gestation 28 days	Young Not known	Diet

Family SCIURIDAE	Species *Heliosciurus gambianus*	Status Lower risk*

SUN SQUIRREL

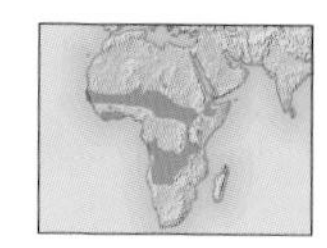

AFRICA

Speckled olive-brown in appearance, the Sun Squirrel has hairs banded in yellow, brown, and grey along its coat. There are 14 black bands on its tail and a white ring around each eye. When alarmed, this typical ground-and-tree squirrel climbs a tree away from the source of danger. It lines its nest, in a tree-hole, with fresh leaves each night.

• **SIZE** Body length: 15.5–21 cm (6–8½ in). Tail: 15.5–31 cm (6–12 in).

• **OCCURRENCE** Senegal to Sudan, south to Angola and N. Zambia. In woodland savanna and secondary forest.

• **REMARK** The Sun Squirrel gets its name from its habit of basking in the sun.

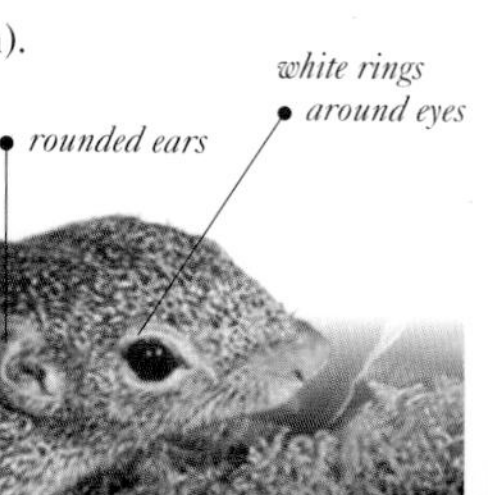

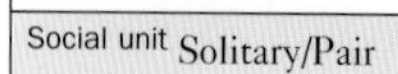

Social unit Solitary/Pair	Gestation Not known	Young 1–5	Diet

Family SCIURIDAE	Species *Callosciurus prevostii*	Status Lower risk*

PREVOST'S SQUIRREL

A striking and largely uniform colour pattern characterizes this arboreal squirrel. Black on its upperparts and chestnut-red below, it has a gleaming white band running between these two colours from the nose to the thigh. Large, protruding eyes equip it with sharp vision. Prevost's Squirrel lives alone or in small family groups, communicating with bird-like calls and tail displays.

- **SIZE** Body length: 13–28 cm (5–11 in). Tail: 8–26 cm (3¼–10 in).
- **OCCURRENCE** S.E. Asia. In lowland and montane forest, farms, and gardens.
- **REMARK** It is one of the most brilliantly coloured mammals.

ASIA

white band along side from thigh to nose

protruding eyes

chestnut limbs

black tail

Social unit Variable	Gestation 46–48 days	Young 2–3	Diet

Family SCIURIDAE	Species *Petaurista elegans*	Status Common

GIANT FLYING SQUIRREL

A glider rather than a flier, this squirrel stretches out a thin, furred membrane between its front and rear limbs to travel over distances of more than 400 m (1,310 ft), from tree to tree. Usually gliding to avoid danger, it covers three times the distance that it loses in height, using its forelegs to navigate. The Giant Flying Squirrel has a tawny to reddish brown coat that is densely furred, with lighter underparts, and a black-tipped tail. An agile climber, it nests in tree-holes and forks of branches, emerging at night in search of food.

- **SIZE** Body length: 30–45 cm (12–18 in). Tail: 32–61 cm (12½–24 in).
- **OCCURRENCE** E. Afghanistan, N. India, W. China, and S.E. Asia. In coniferous forest.

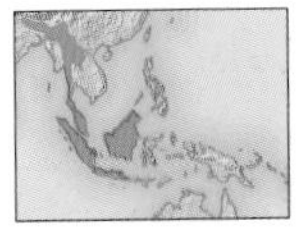
ASIA

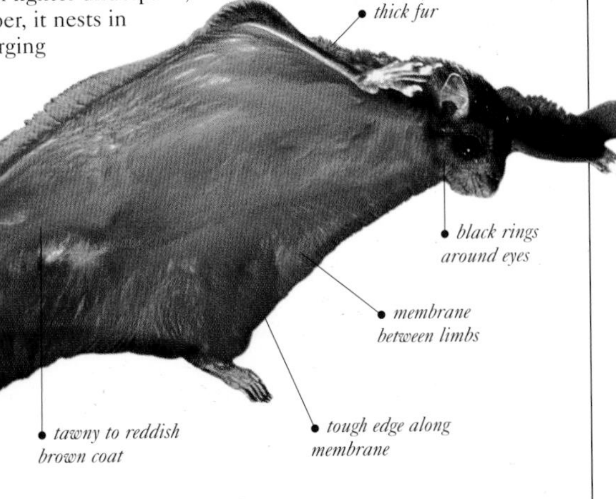

Social unit Pair	Gestation Not known	Young 1–3	Diet

Family GEOMYIDAE	Species *Thomomys bottae*	Status Common

BOTTA'S POCKET GOPHER

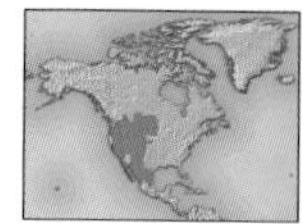

N. AMERICA

This solitary, burrowing rodent has greyish brown upperparts and brownish orange underparts, with dark hair bases. It has external fur-lined storage pouches on its cheeks. Its flat skull, small ears closed by flaps, strong shoulders and forelegs, and rapidly growing, middle three claws, equip it to dig in loose and wet soil. Extremely territorial, this animal mostly stays underground, creating extensive subterranean burrow systems, in which it rests during the hot, dry summer months. Botta's Pocket Gopher communicates using a wide variety of sounds, ranging from scolding shrieks to soft mumbling and squeaking.

- **SIZE** Body length: 11.5–30 cm (4 ½–12 in). Tail: 4–9.5 cm (1 ½–3 ¾ in).
- **OCCURRENCE** W. USA to N. Mexico. In desert to open forest.

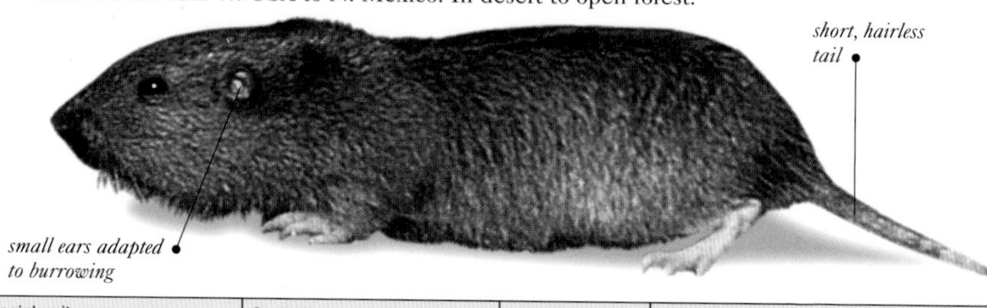

Social unit Solitary/Pair	Gestation 18–19 days	Young 3–7	Diet

Family HETEROMYIDAE	Species *Dipodomys merriami*	Status Common

MERRIAM'S KANGAROO RAT

N. AMERICA

A small rodent with a white stripe from its flank to the base of its tail, and distinctive white spots above its eyes and behind its ears, this nocturnal rodent moves with great speed over sandy desert soil. It hops like a kangaroo on its large hindfeet and uses its long, slender tail for balance. This rat digs, burrows, and searches for food, especially cockle- and sand-burrs in winter, and prickly pear and other cactus seeds in summer. It is often found grooming itself by taking rapid and energetic dust-baths. Hardy and industrious, it also moves on snow-covered ground in winter.

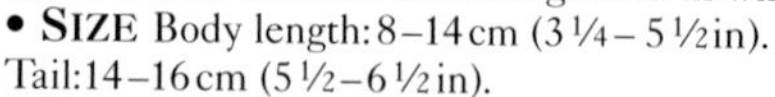

- **SIZE** Body length: 8–14 cm (3 ¼–5 ½ in). Tail: 14–16 cm (5 ½–6 ½ in).
- **OCCURRENCE** S.W. USA to N. Mexico. In desert.

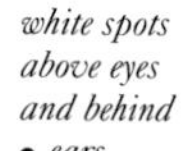

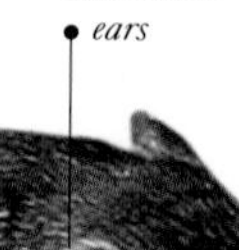

Social unit Social	Gestation Not known	Young 1–5	Diet

Family PEDETIDAE	Species *Pedetes capensis*	Status Vulnerable

SPRINGHARE

The only member in its family, the Springhare or Springhaas has a bushy, black-tipped tail and long, upright ears. It leaps across arid or semi-arid country on its powerful hindlegs, easily covering 2–3 m (6½–9¾ft) per bound. This nocturnal rodent feeds on seeds, bulbs, and stems, as well as small creatures such as locusts and beetles. While eating, it bends forwards, loping rabbit-like on all four limbs, and while resting on its haunches, it tucks its head between its rear legs and wraps its tail around its body. Always alert, it has acute senses of sight, hearing, and smell. Occasionally found in pairs, this animal makes several extensive burrows to use on different days.

- **SIZE** Body length: 27–40 cm (10½–16 in). Tail: 30–47 cm (12–18½ in).
- **OCCURRENCE** C. and E. Africa to S. Africa. In arid or semi-arid country with scanty vegetation or light woodland.
- **REMARK** The Springhare is prized for its meat by African bushmen.

Social unit Solitary/Pair	Gestation 2 months	Young 1	Diet

Family CASTORIDAE	Species *Castor fiber*	Status Lower risk

EURASIAN BEAVER

An adept builder of dams, burrows, lodges, and canals, the Eurasian Beaver is similar to its American cousin (see pp.154–155) in habits, lifestyle, and appearance, but may be more heavily built. As with the American Beaver, it has oil glands at the base of its tail, which produce a lubricating secretion that it spreads through its fur while grooming, to waterproof it. Webbed hindfeet and a rudder-like tail enable this nocturnal swimmer to manoeuvre in water, and it can remain submerged for almost 20 minutes. In areas which have many natural waterways, it does not build a lodge, but digs tunnels in the riverbank.

- **SIZE** Body length: 83–100 cm (33–39 in). Tail: 30–38 cm (12–15 in).
- **OCCURRENCE** Europe and W. Siberia. Isolated populations in China and Mongolia. In lakes and rivers.

Social unit Social	Gestation 105–107 days	Young 1–5	Diet

Family CASTORIDAE	Species *Castor canadensis*	Status Locally common

AMERICAN BEAVER

This mammal is adapted to aquatic life in various ways: its eyes have a nictitating membrane (a transparent third eyelid) that enables it to see underwater, its ears and nose shut with valve-like flaps, and the lips close behind the incisors, allowing it to nibble with its mouth shut. Webbed feet facilitate swimming, as does its long, flattened tail, which it also flaps on the water surface as an alarm signal. The American Beaver's long, coarse, reddish brown guard hair and dense grey underfur retain body heat underwater and its long whiskers help it to feel its way in the dark. It grasps its food with its front paws and gnaws it with its teeth. Hissing, grunting, and tooth-sharpening may sometimes accompany the competition for food. Usually living in colonies of 4–8 related individuals, American Beavers are found grooming each other, touching noses, wrestling, or dancing. Members of a colony rest together in the lodge at daytime, with females being more sedentary than males.

- **SIZE** Length: 74–88 cm (29–35 in). Weight: 11–26 kg (24–57 lb).
- **OCCURRENCE** North America – except the southwestern desert, peninsular Florida, and Arctic tundra. In streams, ponds, and lakes.

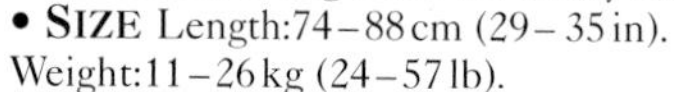

Social unit Social	Gestation 107 days	Young 3–4	Diet

ECOLOGICAL BIND
The American Beaver's extensive tree-felling and dam-building activity often lays waste wide tracts of land and water, considerably altering the local ecology. However, an alternative viewpoint is that beaver dams reduce floods, helping to preserve the habitat.

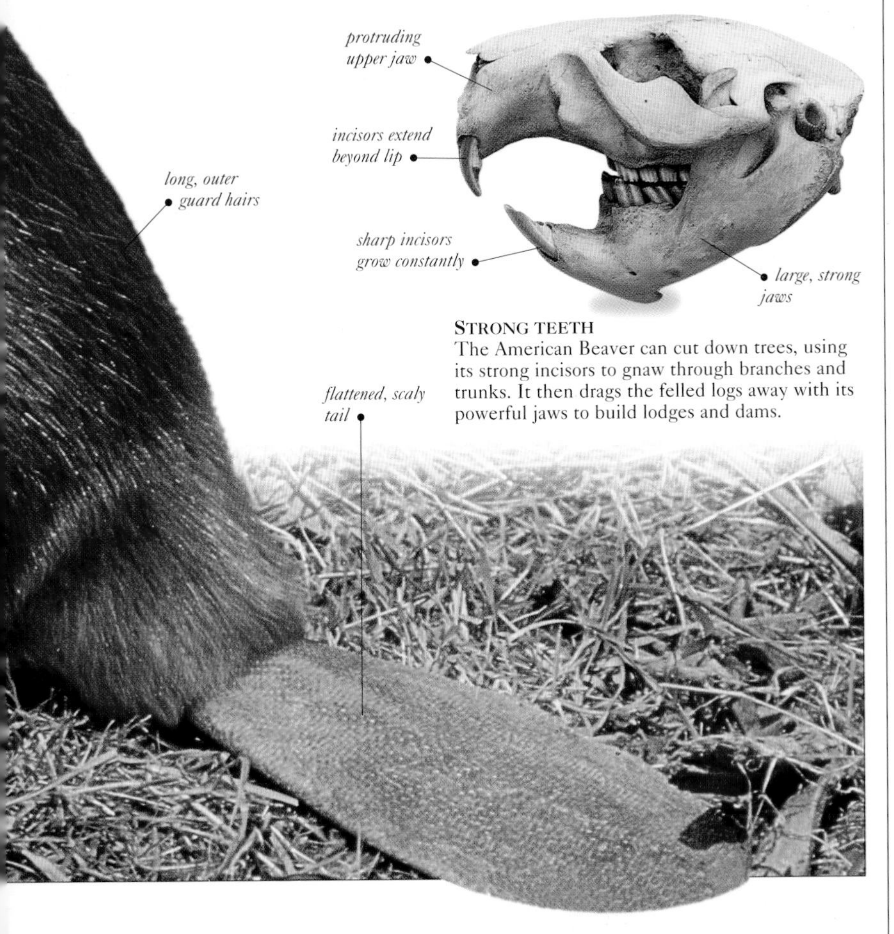

STRONG TEETH
The American Beaver can cut down trees, using its strong incisors to gnaw through branches and trunks. It then drags the felled logs away with its powerful jaws to build lodges and dams.

Family MURIDAE	Species *Reithrodontomys raviventris*	Status Vulnerable

AMERICAN HARVEST MOUSE

Similar to the House Mouse in size and appearance (see p.167), the American Harvest Mouse has more hair on its slender and scaly tail. It is pinkish brown to brownish grey on the upperparts and greyish white underneath, and has large ears and grooved upper incisors. An expert climber, this nocturnal mouse builds sturdy summer nests in tufts of grass or shrubs, but also burrows in tunnel systems of other rodents in winter. During this season, the American Harvest Mouse uses the runways of these rodents to feed on seeds, from the ground or from grass stems, as well as on green shoots and insects.

- **SIZE** Body length: 7–7.5 cm (2¾–3 in). Tail: 4.5–11.5 cm (1¾–4½ in).
- **OCCURRENCE** USA (San Fransisco Bay area of California), and N. Mexico. In grassland.

Social unit Solitary	Gestation 21–24 days	Young 1–7	Diet

Family MURIDAE	Species *Peromyscus leucopus*	Status Common

WHITE-FOOTED MOUSE

This species is characterized by large eyes, prominent ears, white feet and underparts, a brown back, and a long, sparsely haired tail. Active at night, it lives in pairs in a small den, which it may dig, or which may be an abandoned nest, or a hollow under a stone, log, or thicket. This den also houses the breeding nest and the store of food, generally seeds and insects, which is lightly covered with soil.

- **SIZE** Body length: 9–10.5 cm (3½–4¼ in). Tail: 6–10 cm (2¼–4 in).
- **OCCURRENCE** S.E. Canada, C. and E. USA, and N. Mexico. In thickets, deciduous forest, and grassland.

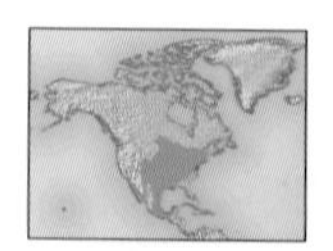

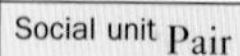

Social unit Pair	Gestation 22–23 days	Young 1–6	Diet

Family MURIDAE	Species *Nyctomys sumichrasti*	Status Locally common

SUMICHRAST'S VESPER RAT

Adapted well to its arboreal existence, this species is found in colonies of nests made of twigs, leaves, and creepers. Its long, thumb-like big toes help it grip branches efficiently and it rarely climbs down to the ground. This brightly coloured rat has tawny or pinkish brown upperparts, with darker hair along the centre of its back. A dark ring encircles each eye, extending to the whiskers, and its ears are finely furred. Its tail is scaly with longer and heavier hairs on the tip. This nocturnal rat feeds on fruits such as wild figs and avocados.

- **SIZE** Body length: 11–13 cm (4 1/4–5 in). Tail: 8.5–15.5 cm (3 1/4–6 in).
- **OCCURRENCE** Mexico and Central America, to C. Panama. In tropical forest.

Social unit Social	Gestation Not known	Young 2–4	Diet

Family MURIDAE	Species *Sigmodon hispidus*	Status Lower risk

HISPID COTTON RAT

There are ten cotton rat species in the Americas, some quite rare, and some so plentiful that they are considered to be pests in the harvest season. The Hispid Cotton Rat, often viewed as a menace, has stiff brown-grey fur on the back and a greyish white coat underneath. It feeds on plants, insects, and crops. A good swimmer, it also catches crayfish, crabs, and frogs, as well as climbs up reeds into birds' nests, to eat eggs and chicks. Active during the day and night, this cotton rat lives in a sheltered depression in the ground, or in a burrow up to 75 cm (30 in) deep. It digs shallow pits as it feeds, and creates a network of well-worn foraging runways to its source of food.

- **SIZE** Body length: 13–20 cm (5–8 in). Tail: 8–16.5 cm (3 1/4–6 1/2 in).
- **OCCURRENCE** S. USA, Central America, and N. South America. In forest, desert, mountains, and grassland.
- **REMARK** This species may cause damage to sweet potato and sugar cane crops.

N., C., & S. AMERICA

Social unit Solitary	Gestation 27 days	Young 12	Diet

Family MURIDAE	Species *Mesocricetus auratus*	Status Endangered

GOLDEN HAMSTER

EUROPE, ASIA

Familiar around the world as a pet, the Golden Hamster is restricted in the wild to a small area from E. Europe to W. Asia. A small, robust rodent with a very short tail, it has a blunt muzzle but a broad face, with small eyes and prominent ears. Its soft, woolly coat is deep orange on the back, a lighter shade on the face, cheeks, and flanks, and grey-white underneath. The forehead may have a black patch between, and in front of, the eyes, while a black stripe may mark each cheek and the nape of the neck. The Golden Hamster is solitary by nature and is known to be aggressive towards other hamsters. It prefers cultivated grain fields and excavates a burrow down to 2m (6½ft), rarely coming up except to feed at night, dawn, and dusk. Its varied diet includes seeds, nuts, and insects such as ants, flies, cockroaches, bugs, and wasps. It stuffs its large cheek pouches with food, to eat later.

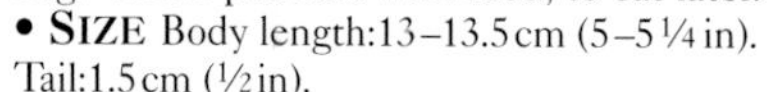

• **SIZE** Body length:13–13.5cm (5–5¼in). Tail:1.5cm (½in).

• **OCCURRENCE** A few, scattered sites in E. Europe and W. Asia. In grassland.

• **REMARK** The Golden Hamster is known to hibernate in captivity.

CAREFUL GROOMING
The Golden Hamster cleans its coat, and that of its young, of dirt, old fur, tangles, and pests, such as fleas, by using both its front teeth and foreclaws.

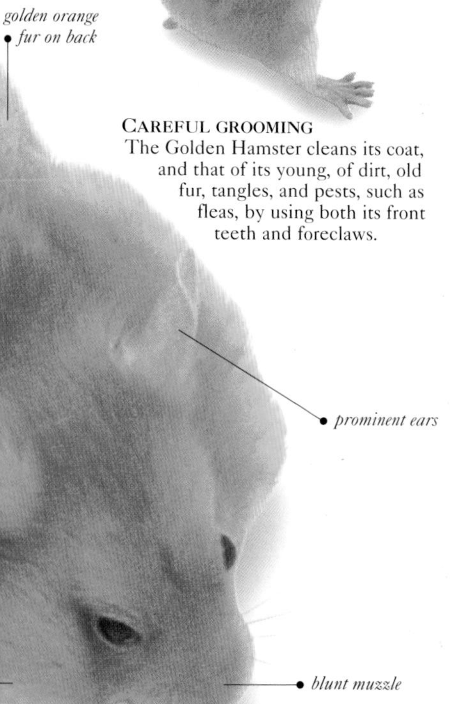

Social unit Solitary	Gestation 16–19 days	Young 6–10	Diet

Family MURIDAE	Species *Phodopus roborovskii*	Status Locally common

DESERT HAMSTER

Also called the Dwarf Hamster, this short-tailed, large-eared rodent is pale brown on its upperparts, with pure white underparts. Its rear feet are short and broad, with dense fur on the underside to protect them when jumping across hot, loose desert sand. The Desert Hamster digs a separate nesting burrow in firm, moist sand and lines it with hair shed by camels and sheep. Like other hamsters, it stuffs its cheeks with food, usually millet and grass seed, to take back into its burrow for storage. It also eats insects such as beetles, locusts, and earwigs.

• **SIZE** Body length: 5.5–10 cm (2¼–4 in). Tail: 7–11 cm (2¾–4¼ in).

• **OCCURRENCE** Russia (Tuva), E. Kazakhstan, Mongolia, and neighbouring parts of China. In desert.

ASIA

• **REMARK** Scrupulously clean, the Desert Hamster performs an elaborate daily grooming ritual.

Social unit Solitary	Gestation 20–22 days	Young 3–9	Diet

Family MURIDAE	Species *Cricetus cricetus*	Status Lower risk*

COMMON HAMSTER

EURASIA

The largest hamster, this species has distinctively thick fur, which is red-brown on the back, mainly black on the underside, with white patches on the nose, cheeks, throat, flanks, and paws. It has large cheek pouches, which it inflates while swimming for extra buoyancy, or when alarmed. The Common Hamster feeds at dawn and dusk. In autumn, it hoards food in its burrow and then hibernates until spring, waking every 5–7 hours to feed. The burrow size varies with the age of the animal, and there are several oblique entrances and compartments for nesting, storage, and excretion.

• **SIZE** Body length: 20–34 cm (8–13½ in). Tail: 4–6 cm (1½–2¼ in).

• **OCCURRENCE** E. Europe to C. Asia. In steppe and farmland, and along riverbanks.

• **REMARK** The darker underside is very unusual in a mammal.

white patches on parts of body

red-brown fur

broad feet with long claws

Social unit Solitary	Gestation 18–20 days	Young 4–12	Diet

Family MURIDAE	Species *Meriones unguiculatus*	Status Locally common

MONGOLIAN GERBIL

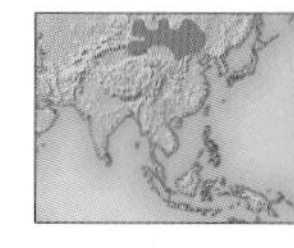

This rodent, familiar as a pet, is among the 13 species of gerbils found in the Middle East and Asia. Its coat has pale brown hair, tipped with black, giving it a speckled appearance, while its underparts are grey or white. Protruding black eyes dominate its short, broad head, and its long tail averages around 90 per cent of its head and body length. Moving on all four feet, it leaps expertly on its long hindlegs and pounds its rear feet in staccato fashion during courtship. The Mongolian Gerbil makes elaborate underground burrows in the dry sand or clay soil of the steppe, and is often found sitting upright in front of the entrance. It shares the burrow with its mate and up to 12 young, which are weaned 20–30 days after birth. Active by day and night, summer and winter, it eats mainly seeds of buckwheat, millet, grasses, and sedges, storing excess food in the burrow.

• **SIZE** Body length: 10–12.5 cm (4–5 in). Tail: 9.5–11 cm (3¾–4¼ in).

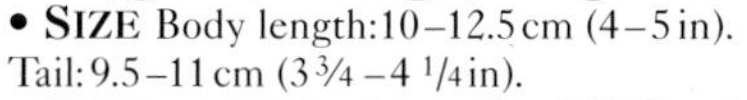

• **OCCURRENCE** Mongolia, S.W. Russia, and N. China. In steppe and plains.

• **REMARK** Compared with other rodents, the Gerbil has a recent history of being kept as a pet, since the 1960s. Its appealing looks and adaptable nature make it a popular choice.

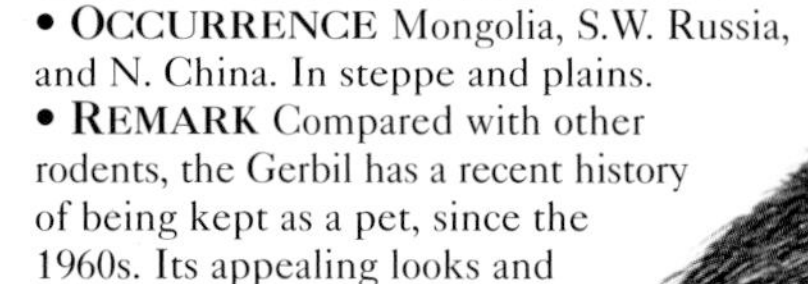

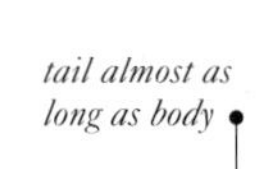

Social unit Social	Gestation 19–21 days	Young 1–12	Diet

STANDING UP

By standing up on its hindlegs, the Gerbil has a better view of its surroundings and can detect the presence of predators. If a male Gerbil is threatened by another male, he stands up on his hindlegs to face the intruder. The two then box each other with their forepaws. The Gerbil's large hindfeet and long hairy tail help to provide it stability in this upright stance.

HIGH JUMPS

The Gerbil is an agile rodent, capable of making spectacular leaps. This is essential for its survival in an arid habitat which provides little cover from predators such as foxes and polecats.

Family MURIDAE	Species *Pachyuromys duprasi*	Status Locally common

FAT-TAILED GERBIL

Found in the Sahara Desert, the Fat-tailed Gerbil has a long, fluffy coat, that is pale cinnamon on the back and sides, with blackish tips on the back hair. It has white underparts and feet, partly haired soles, long hindfeet, and a distinctive club-shaped tail. Large, sensitive ears enable it to receive long-range, low-frequency sounds across the desert sand. This docile burrower becomes active at dusk and emerges from its burrow to search for insects such as crickets, and for leaves, seeds, and other plant material.

- **SIZE** Body length: 9.5–13 cm (3¾–5 in). Tail: 5.5–16.5 cm (2¼–6½ in).
- **OCCURRENCE** Sahara Desert. In sparsely vegetated, semi-desert of sand and gravel.
- **REMARK** The Fat-tailed Gerbil has a store of body fat in its tail, which is used for nourishment and water when food is scarce.

Social unit Variable	Gestation Not known	Young 1–6	Diet

Family MURIDAE	Species *Hypogeomys antimena*	Status Endangered

MALAGASY GIANT RAT

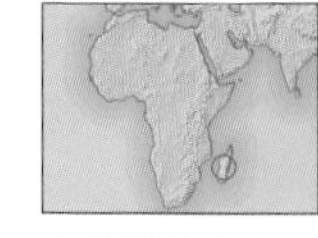

Rotund and rabbit-like, this large rat has long ears and hindfeet, and hops rather than runs. The hair on its upperparts is coarse, dense, and grey to reddish brown, with a darker V-shaped patch over its nose. The limbs and belly are white, the tail is muscular and covered with stiff, short hairs. Locally known as the Votsotsa, the Malagasy Giant Rat digs burrow systems, with up to six entrances, in the sandy soil. It eats mainly fallen fruits and bark at night, manipulating the food with its forefeet in an upright position.

- **SIZE** Body length: 30–35 cm (12–14 in). Tail: 21–25 cm (8½–10 in).
- **OCCURRENCE** W. Madagascar. In coastal forest.
- **REMARK** This rat is threatened by the introduced Black Rat and by habitat loss.

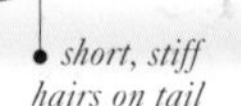

Social unit Social	Gestation Not known	Young 1–2	Diet

Family MURIDAE	Species *Tachyoryctes macrocephalus*	Status Lower risk*

GIANT AFRICAN MOLE RAT

Mole-like in habit, this rat spends a great deal of time in its burrow, which often exceeds 50 m (165 ft) in length. Like a typical burrower, it has a blunt, rounded head, small eyes and ears, a robust body, thick fur, and short limbs. This rat gnaws and digs with its large and protruding orange-yellow incisors.

AFRICA

- **SIZE** Body length: Up to 31 cm (12 in). Tail: 9–10 cm (3½–4 in).
- **OCCURRENCE** Ethiopia. In wet highland, grassland, moorland, and cultivated areas.

Social unit Variable	Gestation Not known	Young Not known	Diet

Family MURIDAE	Species *Clethrionomys glareolus*	Status Common

BANK VOLE

Active mostly at dawn and night, this vole nests in burrows, thickets, and tree stumps. Varying geographically in size and weight, it is reddish to yellowish brown, with grey flanks and hindquarters.

- **SIZE** Body length: 7–13.5 cm (2¾–5¼ in). Tail: 3.5–6.5 cm (1½–2½ in).
- **OCCURRENCE** W. Europe to N. Asia. In temperate and coniferous forest, and wooded banks of rivers.

EURASIA

reddish or yellowish brown coat

large black eyes

Social unit Social	Gestation 18 days	Young 2–8	Diet 

Family MURIDAE	Species *Microtus arvalis*	Status Common

COMMON VOLE

Among the most common rodents in grassy and farmland habitats, this stocky vole has short, sandy to grey-brown fur on the back and grey fur underneath. The Common Vole feeds mainly on plants in summer and bark in winter, digging underground tunnels for larders and nests, and is active during the day as well as the night.

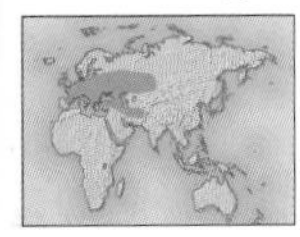

EURASIA

- **SIZE** Body length: 9–12 cm (3½–4¾ in). Tail: 3–4.5 cm (1¼–1¾ in).
- **OCCURRENCE** Europe, Ukraine, and Russia. In cultivated land and meadows.

Social unit Social	Gestation 16–24 days	Young 2–12	Diet

Family MURIDAE	Species *Arvicola terrestris*	Status Locally common

EUROPEAN WATER VOLE

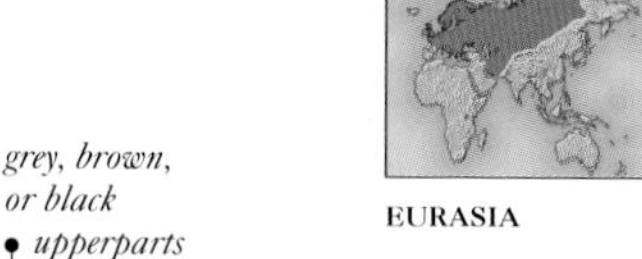

This burrowing rodent displays much variation in colour and size. Its thick coat can be grey, black, or brown on the upperparts, and dark grey to white underneath; those voles that inhabit woods and meadows are half the size of those found near lakes, rivers, and marshes. The tail is generally rounded and about half the length of the body. This adept swimmer and diver is most active at dusk and dawn, feeding on plants, tree roots, rhizomes, bulbs, and tubers. Its main enemies are eagles, owls, wild and domestic cats, and the introduced predator, the American Mink (see p. 250). Water pollution and loss of habitat are also posing a growing threat to this species.

- **SIZE** Body length: 12–23 cm (4¾–9 in). Tail: 7–11 cm (2¾–4¼ in).
- **OCCURRENCE** Europe, Russia, and Iran. In lakes, rivers, streams, marshes, meadows, and woods.

Social unit Solitary/Pair	Gestation 21–22 days	Young 2–10	Diet

Family MURIDAE	Species *Ondatra zibethicus*	Status Common

MUSKRAT

The largest species of burrowing vole, the Muskrat is adapted to swimming – its large back feet have small webs between the toes and are edged by stiff bristles or a "swimming fringe", and its long, furless tail is flattened and can be used as a rudder. Its nostrils and ears are covered by flaps during dives, which may last 20 minutes. The Muskrat lives in groups of up to ten, digging tunnels in banks or building beaver-like lodges with mud and reeds. The female builds her nest in a dry tunnel chamber or lodge platform. This species derives its name from the musky secretions from glands around its genital area.

- **SIZE** Body length: 25–35 cm (10–14 in). Tail: 20–25 cm (8–10 in).
- **OCCURRENCE** North America, W. Europe, and Asia. Along banks of rivers, streams, and ponds.
- **REMARK** The Muskrat is valued commercially for its pelt.

N. AMERICA, EURASIA

Social unit Social	Gestation 25–30 days	Young 1–3	Diet

Family MURIDAE	Species *Lemmus sibericus*	Status Common

BROWN LEMMING

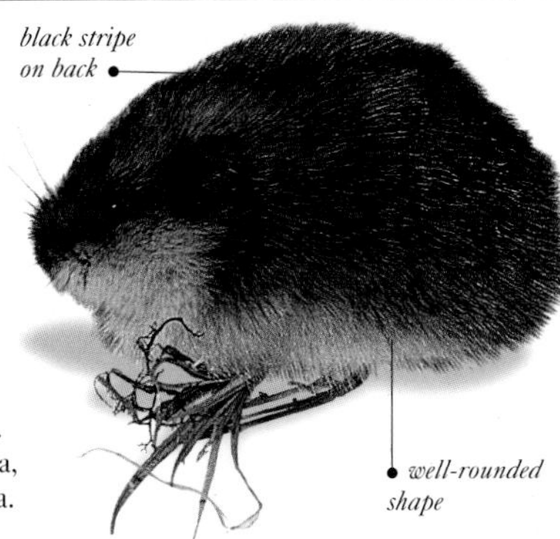

This chiefly nocturnal animal lives in large colonies and breeds prolifically. In autumn it migrates to low tundra areas and lakeshores, where it digs tunnels under peaty mounds or nests in vegetation. As the snows melt in spring, it again ascends to drier areas. With a diet of moss, herbs, and soft twigs, it feeds along worn trails, even under the snow in winter.

N. AMERICA, EURASIA

- **SIZE** Body length: 12–15 cm (4 3/4 – 6 in). Tail: 1–1.5 cm (3/8 – 1/2 in).
- **OCCURRENCE** Alaska to Canada, N.E. Europe to N. Asia. In the tundra.

Social unit Social	Gestation 18 days	Young 4–13	Diet

Family MURIDAE	Species *Lagurus lagurus*	Status Lower risk

STEPPE LEMMING

Small and thickset, this nocturnal rodent is fully covered with long and waterproof fur to protect it from the harsh climate of the steppes. It digs burrows around 1 m (3 ft) deep in which to live, allowing it to survive the cold without hibernating, and it may also dig shallow temporary burrows for defence. The Steppe Lemming is light or cinnamon-grey in colour, with continuously growing molar teeth that enable it to chew abrasive grasses in large quantities.

EURASIA

- **SIZE** Body length: 8 –12 cm (3 1/4 – 4 3/4 in). Tail: 0.7–2 cm (1/3 – 3/4 in).
- **OCCURRENCE** E. Europe to E. Asia. In steppe and semi-desert areas.

Social unit Variable	Gestation 20 days	Young 8–12	Diet

Family MURIDAE	Species *Lemniscomys striatus*	Status Common

STRIPED GRASS MOUSE

This mouse is light buff or reddish orange, with paler forms found in W. Africa and darker ones in E. Africa. Foraging at dawn and dusk, it lives in holes with runways radiating towards its feeding grounds. It sheds the skin of its tail or feigns death when attacked.

- **SIZE** Body length: 10–14 cm (4–5 1/2 in). Tail: 10–15.5 cm (4–6 in).
- **OCCURRENCE** W., E., and S. Africa. In moist, grassy habitats.

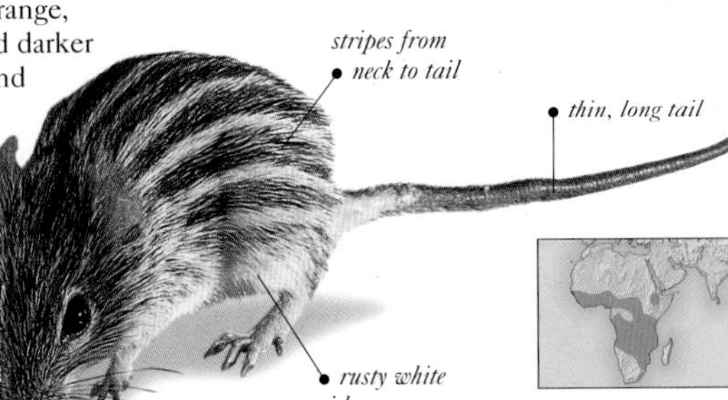

AFRICA

Social unit Solitary	Gestation 28 days	Young 4–5	Diet

Family MURIDAE	Species *Apodemus flavicollis*	Status Common

YELLOW-NECKED FIELD MOUSE

EURASIA

Named for the yellow patch on its throat, this long-tailed rodent has a brown back and yellowish white underparts. Large, prominent eyes and ears are adaptations to its nocturnal habits, and large hindfeet allow it to leap across long distances. The Yellow-necked Field Mouse climbs up to 6 m (20 ft) in trees to forage and establish feeding places in crevices and birds' nests. It is aggressive towards its own kind and similar species such as the Wood Mouse (see below). The Yellow-necked Field Mouse communicates using shrill, high-pitched chirps and squeaks.

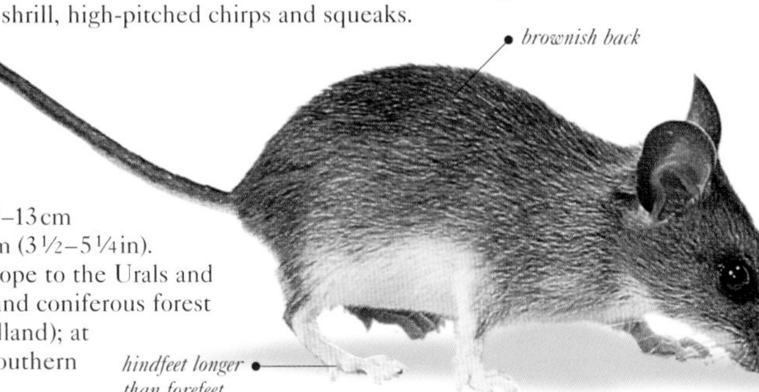

brownish back

tail longer than body

hindfeet longer than forefeet

• **SIZE** Body length: 8.5–13 cm (3 1/4–5 in). Tail: 9–13.5 cm (3 1/2–5 1/4 in).
• **OCCURRENCE** Europe to the Urals and Armenia. In temperate and coniferous forest (prefers deciduous woodland); at higher altitudes in the southern part of its range.

Social unit Solitary	Gestation 21–23 days	Young 3–8	Diet

Family MURIDAE	Species *Apodemus sylvaticus*	Status Common

WOOD MOUSE

EURASIA, AFRICA

Similar to but smaller than the Yellow-necked Field Mouse (see above), the Wood Mouse has a yellow or orange-brown chest patch, a reddish to greyish brown coat, an ochre stripe between its forefeet, and whitish grey underparts. An adept runner and climber, this species digs burrows or inhabits tree-holes, taking food to a safe place before eating it.

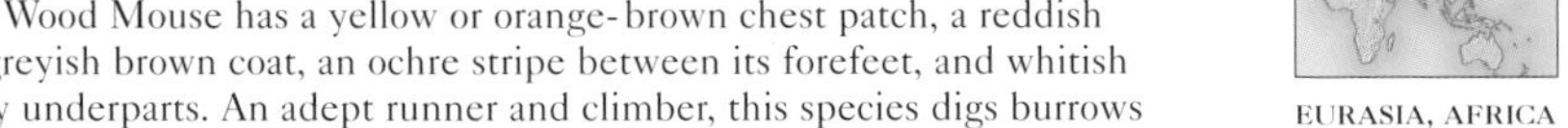

• **SIZE** Body length: 8–11 cm (3–4 1/4 in). Tail: 7–11 cm (2 3/4–4 1/4 in).
• **OCCURRENCE** Europe (including Iceland), C. Asia, and N. Africa. In farmland, riverbanks, moors, plantations, woodland, and urban areas.

greyish to reddish brown coat

tail shorter than body

Social unit Solitary	Gestation 23 days	Young 3–9	Diet

Family MURIDAE	Species *Mus musculus*	Status Common

HOUSE MOUSE

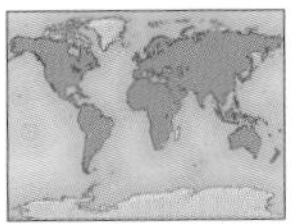

WORLDWIDE

One of the most widely distributed mammals, the House Mouse thrives in human habitations, feeding on a huge range of food at night, dawn, and dusk. A small, slim rodent, it has small eyes, a pointed nose, and large ears. Its upperparts are greyish black to reddish brown, with shorter hair on the rump than on the back, and its underparts are white. The House Mouse lives in a family group of one dominant male and several females, and marks its territory with scent and urine. The female mouse is larger and stronger than the male.

- **SIZE** Body length: 7–10.5 cm (2¾–4¼ in). Tail: 5–10 cm (2–4 in).
- **OCCURRENCE** Worldwide, except polar regions. In houses and other man-made structures – avoids woodland and arid areas.
- **REMARK** This species has been widely bred as a pet and for scientific research.

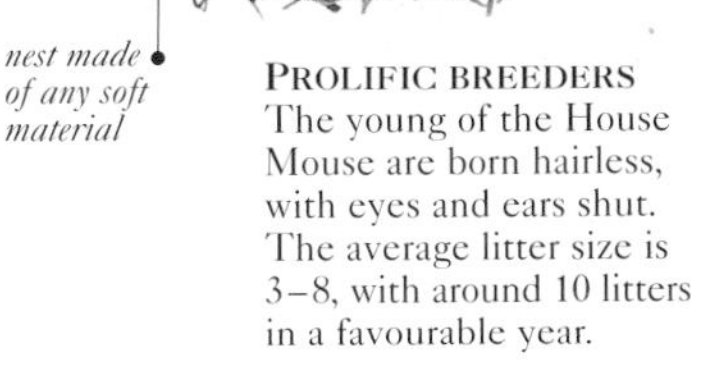

PROLIFIC BREEDERS
The young of the House Mouse are born hairless, with eyes and ears shut. The average litter size is 3–8, with around 10 litters in a favourable year.

pointed nose

FEEDING IN A GROUP
The House Mouse is known to feed on a variety of food ranging from shoots to seeds, such as corn, oats, barley, and millet, and other human foodstuff.

greyish black to reddish brown upperparts

hairless tail

shorter hair on rump

Social unit Social	Gestation 18–24 days	Young 3–8	Diet

Family MURIDAE	Species *Micromys minutus*	Status Lower risk

EURASIAN HARVEST MOUSE

Small and elegant, the Eurasian Harvest Mouse is yellowish or reddish brown on its upperparts, and white below. A good climber, its hindlegs are longer than its forelegs and it has a long, hairless, prehensile tail. It feeds on seeds, farmed cereals, berries, and spiders, as well as insects by day and night. This mouse builds a spherical nest of shredded grass in a thicket or grassy clump, 50–130 cm (20–51 in) above ground level, usually using an old bird's nest as a base. If food becomes scarce, it is known to feed on its own young.

• **SIZE** Body length: 5–8 cm (2–3 in). Tail: 4.5–7.5 cm (1¾–3 in).

EURASIA

• **OCCURRENCE** Europe to Japan. In fields, gardens, wetland, grassland, and fringes of humid tropical forest.

• **REMARK** The only Old World rodent that has a prehensile tail.

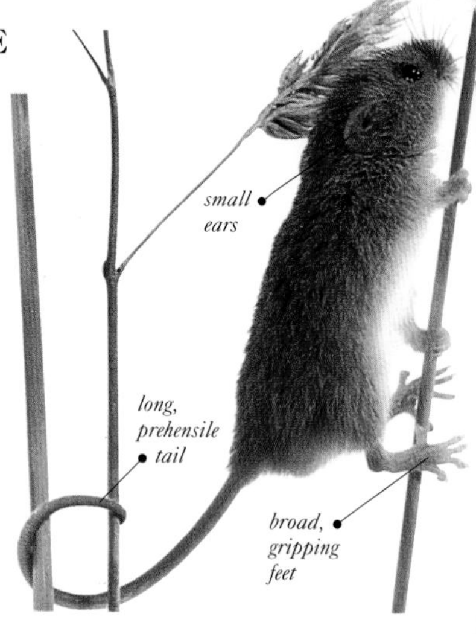

Social unit Solitary	Gestation 21 days	Young 2–6	Diet

Family MURIDAE	Species *Acomys minous*	Status Vulnerable

SPINY MOUSE

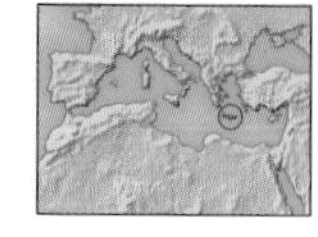

CRETE

Found only on the island of Crete, in Europe, where it was probably carried by ships from Africa, this nocturnal mouse has inflexible spines on its back and tail, and may be yellow, red, brown, or dark grey. It readily sheds its tail to escape capture. Gregarious by nature, it builds only a rudimentary nest, and other females assist the mother at birth, helping with cleaning and biting the cord.

• **SIZE** Body length: 9–12 cm (3½–4¾ in). Tail: 9–12 cm (3½–4¾ in).

• **OCCURRENCE** Crete. In arid areas.

• **REMARK** This rodent has a long gestation period (5–6 weeks); the young are born remarkably well-developed.

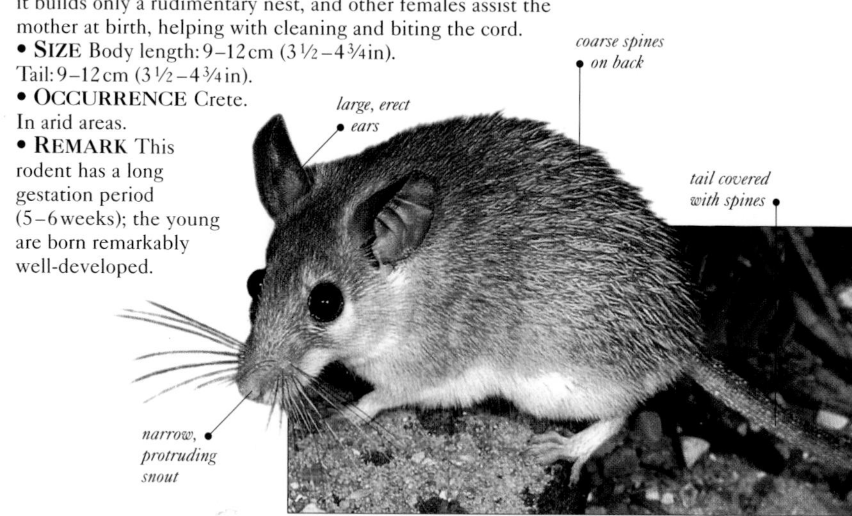

Social unit Variable	Gestation 35–42 days	Young 1–5	Diet

Family MURIDAE	Species *Notomys alexis*	Status Common

SPINIFEX HOPPING MOUSE

Locally called the Dargawarra, this mouse lives in desert habitat covered by bushy, spiky grass known as spinifex. Obtaining all its moisture from its food, it does not need to drink water. However, it avoids dehydration in many ways: it produces highly concentrated urine, lives in deep underground burrows during the day, sleeps in groups (raising humidity levels), and feeds only at night. Sociable by nature, groups of ten mice including males, females, and young, share a common nest.

- **SIZE** Body length: 9–18 cm (3½–7 in). Tail: 12.5–23 cm (5–9 in).
- **OCCURRENCE** W. and C. Australia. In sand dunes, grassland, heath, and wood.
- **REMARK** The most widespread of all hopping mice, it is found across Australia's vast, sandy soil cover.

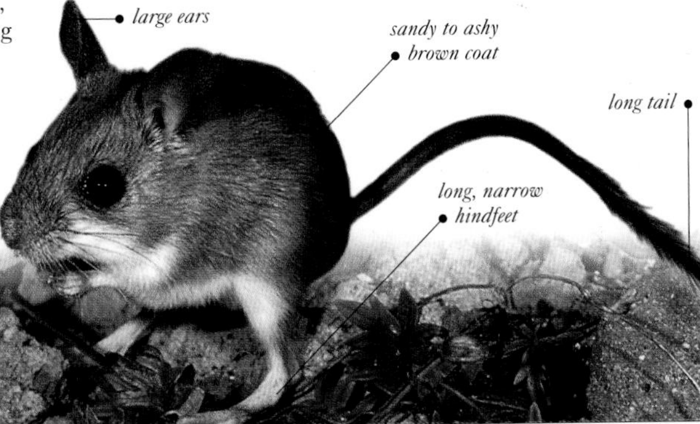

Social unit Social	Gestation 32–34 days	Young 1–9	Diet

Family MURIDAE	Species *Cricetomys gambianus*	Status Common

GIANT POUCHED RAT

This large, nocturnal rat eats a variety of moist or fleshy food, from termites to avocados, and groundnuts to maize. It stores it in its huge cheek pouches, and also carries it back to its burrow. This is extensive, comprising several different chambers for food, resting, breeding, and defecation, with male and female rats occupying separate tunnels. The Giant Pouched Rat has a bristly buff-brown coat, fading to white on the throat, flanks, and underside, with dark brown rings around its eyes, and a brown, thinly haired tail that is often tipped white. Its well-developed hindlegs make it a good jumper.

- **SIZE** Body length: 35–40 cm (14–16 in). Tail: 37–45 cm (14½–18 in).
- **OCCURRENCE** W., C., E., and S. Africa. From savanna to evergreen forest.
- **REMARK** This rat is reared both for its meat and as a pet. The skin is used for tobacco pouches in the Ruwenzori mountains.

AFRICA

Social unit Variable	Gestation 32–42 days	Young 2–4	Diet

Family MURIDAE	Species *Rattus norvegicus*	Status Common

BROWN RAT

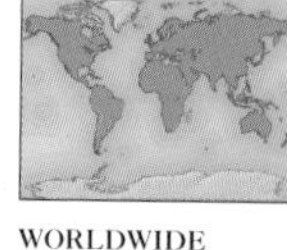

WORLDWIDE

Also known as the Common or Norway Rat, the Brown Rat has established its presence all over the world largely due to its sharp senses, great agility, and varied and opportunistic diet. Varying from brown to grey-brown or black in colour, it has comparatively small eyes and ears and a long, sparsely haired tail. A good swimmer, diver, and climber, it can roam a range of 3 km (1.8 miles) in a single night, hunting out food with its keen sense of smell. Originally a herbivore, this nocturnal or dusk-and-dawn feeder eats mainly seeds, fruits, vegetables, and leaves, but also preys on fish, snails, and water insects. Packs of up to 200 rats, dominated by the larger males, are known to hunt creatures as big as rabbits, poultry, and other large birds. Male rats defend their territory, using scent to mark it and to identify other members of the pack. The female makes a nest of grass, leaves, paper, rags, or any other material, and breeds frequently.

- **SIZE** Body length: 20–28 cm (8–11 in). Tail: 17–23 cm (7–9 in).
- **OCCURRENCE** Worldwide, except at the poles – more common in colder countries. In riverbanks, fields, and human habitations, including sewage systems.

WATER WORLD
The Brown Rat is an excellent swimmer and diver, frequenting riverbanks and sewage networks. It swims with its tail held out for balance. Hunting for fish, crayfish, snails, and aquatic insects, it crunches these up with its sharp teeth.

BRED FOR RESEARCH
This species is the ancestor of rats bred for scientific research and as pets, which may be black-and-white or white.

Social unit Social	Gestation 22–24 days	Young 6–9	Diet

Family MURIDAE	Species *Rattus rattus*	Status Common

BLACK RAT

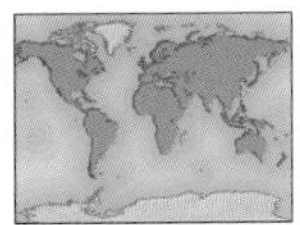
WORLDWIDE

In early Roman times, this rat was spread around the world from Asia, in cargo ships, also giving it the name Ship Rat. Slim-bodied and black, it has whitish pink feet and a grey to white belly. Its tail is unusually long. It is active at night, dawn and dusk, and prefers feeding on plant matter, but also eats insects, faeces, garbage, and carrion. This rat roams in packs of 20–60, and may confront larger animals such as dogs. The female makes a nest of grass or any other material, often in a roof cavity.

• **SIZE** Body length: 16–24 cm (6¼–9½ in). Tail: 18–26 cm (7–10 in).

• **OCCURRENCE** Worldwide – common in Mediterranean countries. In human settlements, ports, and farms.

• **REMARK** This rat is notorious for carrying the fleas that spread plague.

Social unit Variable	Gestation 20–24 days	Young 4–10	Diet

Family MURIDAE	Species *Hydromys chrysogaster*	Status Locally common

AUSTRALIAN WATER RAT

AUSTRALASIA

Australia's heaviest native rodent, this rabbit-sized water rat has broad hindfeet and webbed toes to survive in its permanently aquatic habitat. Brown to grey on its upperparts, and brown to golden yellow or cream underneath, it has abundant whiskers on its long, blunt muzzle, and a thick, white-tipped tail. Active at dusk and dawn, it is a powerful predator of shellfish (prising them open with its incisors), water snails, fish, frogs, turtles, birds, mice, and even bats.

• **SIZE** Body length: 29–39 cm (11½–15½ in). Tail: 23–33 cm (9–13 in).

• **OCCURRENCE** New Guinea, Australia, and Tasmania. In lakes, rivers, and along coastlines; it requires permanent water and prefers inhabited areas.

• **REMARK** It is a protected species in Australia.

Social unit Solitary	Gestation 35 days	Young 1–7	Diet 

Family GLIRIDAE	Species *Glis glis*	Status Lower risk

EDIBLE DORMOUSE

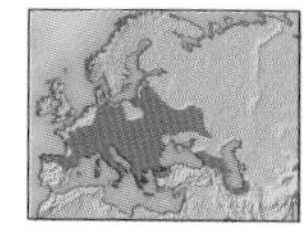

EURASIA

Resembling a squirrel with its bushy tail and semi-upright stance, the Edible or Fat Dormouse is brown to silver-grey, with white underparts and black eye-patches. It inhabits woods and outbuildings, nesting in tree-holes and roof crevices. In autumn it accumulates fat reserves in its body to prepare for its hibernation in a large, deep nest. Like other dormice, it communicates by squeaks and twitters, and lives in loose social groups with no hierarchy; however, the female remains solitary during pregnancy.

- **SIZE** Body length: 13–20 cm (5–8 in). Tail: 10–18 cm (4–7 in).
- **OCCURRENCE** S. and C. Europe, Asia Minor, Caucasus, and N.W. Iran. In deciduous and mixed forest.
- **REMARK** It was bred for food in ancient Rome, hence its name.

Social unit Social	Gestation 30–32 days	Young 2–11	Diet

Family GLIRIDAE	Species *Muscardinus avellanarius*	Status Lower risk

HAZEL DORMOUSE

EUROPE

Also known as the Common Dormouse, this species is as small as a House Mouse and has a yellow- or red-brown coat with white underparts, and a yellowish rump. The skin of its tail can detach if seized by a predator. An excellent climber and jumper, this specialist seasonal feeder eats flowers, grubs, and birds' eggs in spring; seeds and fruits in summer; and nuts in autumn.

- **SIZE** Body length: 6.5–8.5 cm (2½–3¼ in). Tail: 5.5–8 cm (2¼–3¼ in).
- **OCCURRENCE** From Mediterranean (excluding Iberian Peninsula) to S. Sweden, east to Russia.
- **REMARK** It hibernates deeply like all dormice, "dor" meaning "to sleep".

Social unit Solitary	Gestation 22–24 days	Young 2–7	Diet

Family DIPODIDAE	Species *Allactaga tetradactyla*	Status Endangered

FOUR-TOED JERBOA

A vestigial fourth toe on each hindfoot gives this species its name. In other respects it is like a typical jerboa, with tall, rabbit-like ears and long, hopping hindfeet. Its coat is speckled black and orange, with grey sides, an orange rump, and white underparts. Its long tail, used for balancing, has a black band near the feathery white tip. A keen burrower, this nocturnal animal shelters in tunnels from the heat of the midday sun.

- **SIZE** Body length:10–12 cm (4–4¾ in). Tail:15.5–18 cm (6–7 in).
- **OCCURRENCE** N. Africa. In salt marshes and on coasts, and in clay desert, near barley fields.
- **REMARK** The existence of this species is threatened by desert reclamation.

Social unit Solitary	Gestation 25–42 days	Young 2–6	Diet

Family DIPODIDAE	Species *Jaculus jaculus*	Status Common

DESERT JERBOA

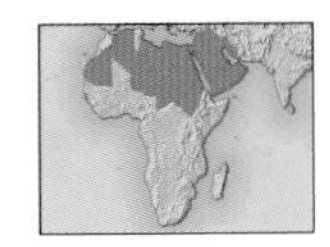

Adapted for hopping at high speed on sandy soil, the Desert Jerboa has long hindfeet, each with three toes on a pad of hairs, and a long tail that it uses for balance. Its fur is brownish orange on the back, greyish orange along the sides, and white on the underparts. Digging burrows with many emergency exits, it plugs these holes to keep out the heat and predators. It travels long distances at night, and takes sand baths to keep its coat clean.

- **SIZE** Body length:10–12 cm (4–4¾ in). Tail:16–20 cm (6½–8 in).
- **OCCURRENCE** N. Africa to W. Asia. In hills of loose sand, desert, and gravel plains.

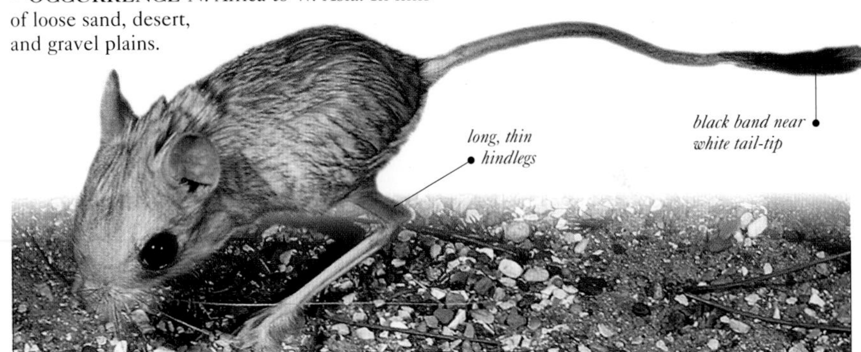

Social unit Solitary	Gestation Not known	Young 4–10	Diet

Family HYSTRICIDAE	Species *Hystrix africaeaustralis*	Status Common

CAPE PORCUPINE

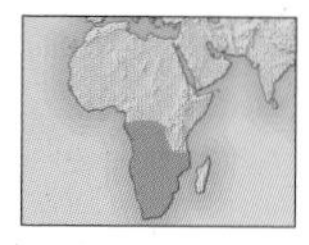

The largest rodent in southern Africa, this stout-bodied porcupine has a dark brown to black coat with thick, cylindrical spines (quills) that are interspersed with ordinary hair. The white-tipped, brown-and-white banded quills are denser towards the back. Despite popular belief, the Cape (or Crested) Porcupine cannot shoot out its quills. Instead, it raises them when annoyed or threatened, and charges backwards at the enemy. The quills detach easily and pierce the enemy's flesh, causing much pain. This porcupine uses grunts, piping calls, or quill rattles to communicate. The male helps to take care of the young.

• **SIZE** Body length: 63–80 cm (25–32 in). Tail: 10.5–13 cm (4½–5 in).

• **OCCURRENCE** C. to S. Africa. In a variety of habitats – especially rocky, hilly country with scrub cover.

• **REMARK** Cape Porcupines effectively defend themselves from larger animals such as lions, leopards, and hyenas, but are hunted by humans in large numbers because of the damage they cause to crops, and for their meat.

BURROWING FOR FOOD
An excellent burrower, the Cape Porcupine forages at night, aided by its keen sense of smell, and travels up to 15 km (9 miles) in search of food. It lives singly, in pairs, or in small groups, resting in caves or crevices with narrow openings during the day.

Social unit Variable	Gestation 6–8 weeks	Young 1–4	Diet

Family ERETHIZONTIDAE	Species *Erethizon dorsatum*	Status Common

NORTH AMERICAN PORCUPINE

More vocal than other porcupines, especially during its mating season in early winter, this North American species moans, grunts, sniffs, snorts, squeaks, sobs, hoots, and chatters. Its quills are yellowish white with black or brown tips and form a distinctive crest, which may be up to 8 cm (3 in) long, on the head. Short and stocky, it has small limbs and its heavy feet have naked soles with claws designed to aid gripping. Active at night, the North American Porcupine climbs trees extensively, feeding on soft bark and conifer needles in winter. In summer it eats roots, stems, leaves, seeds, flowers, and water plants. Its vision is mediocre, but its senses of smell and hearing are excellent. Mostly solitary, several individuals may share a den in winter, or shelter together in trees.

yellowish spines and brown fur on body

crest of long quills

N. AMERICA

• **SIZE** Body length: 65–80 cm (26–32 in). Tail: 15–30 cm (6–12 in).

• **OCCURRENCE** Canada, USA, and N. Mexico. In temperate forest and tundra – mainly riverbanks.

Social unit Solitary	Gestation 205–217 days	Young 1	Diet

Family ERETHIZONTIDAE	Species *Coendou prehensilis*	Status Locally common

PREHENSILE-TAILED PORCUPINE

This large, muscular porcupine is almost entirely arboreal and uses its flesh-coloured prehensile tail and naked-soled feet with long, curved claws to climb slowly up branches. Much of its body is covered with yellow and white, black-tipped spines, but its flanks are virtually naked. The Prehensile-tailed Porcupine shelters in thick foliage, in cavities in tree trunks, or in the ground, during the day. It forages at dusk, travelling over several hundred metres at night to a new tree. This porcupine communicates through a variety of sounds, and isolated animals use a moaning call to contact each other.

• **SIZE** Body length: 52 cm (20 ½ in). Tail: 52 cm (20 ½ in).

• **OCCURRENCE** N. and E. South America. In tropical forest.

• **REMARK** Its tail is as long as its body.

S. AMERICA

Social unit Solitary	Gestation 195–210 days	Young 1	Diet

Family CAVIIDAE	Species *Cavia aperea*	Status Common

GUINEA PIG

The smallest of all cavy-like rodents, the Guinea Pig or Cavy is the ancestor of the Domestic Guinea Pig (*Cavia porcellus*). The five species of Guinea Pig all have a large head with a blunt snout, a tail-less body, and short legs, with four toes on the forefeet and five on the hindfeet. Although usually dark greyish brown, some animals may be almost black. Guinea Pigs are nocturnal and live in grassland, sharing communal feeding runways, but inhabit separate nests.

- **SIZE** Body length: 20–30 cm (8–12 in). Tail: None.
- **OCCURRENCE** N.W. to E. South America. In dry savanna, brushland, and mountains.
- **REMARK** Newborn Guinea Pigs feed on solid food and walk almost immediately after birth.

Social unit Social	Gestation 60 days	Young 1–4	Diet

Family CAVIIDAE	Species *Dolichotis patagonum*	Status Lower risk

MARA

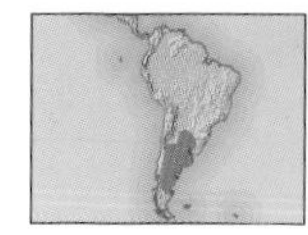

Also known as the Patagonian Cavy, this unusually large and long-legged rodent resembles a deer in appearance. Its behaviour, too, varies from rodent-like tendencies when young and when raising its young, to that of hoofed animals in the adult stage. The Mara has brownish orange upperparts, a white collar-like neck patch, and a whitish fringe on its short tail. Its long muzzle has dark bristles, and it has large eyes and ears. Adult Maras can run, jump, and dig expertly. Male and female pairs graze together on grass and low shrubs, and may be found in groups when raising their young. The female digs a large burrow for the offspring.

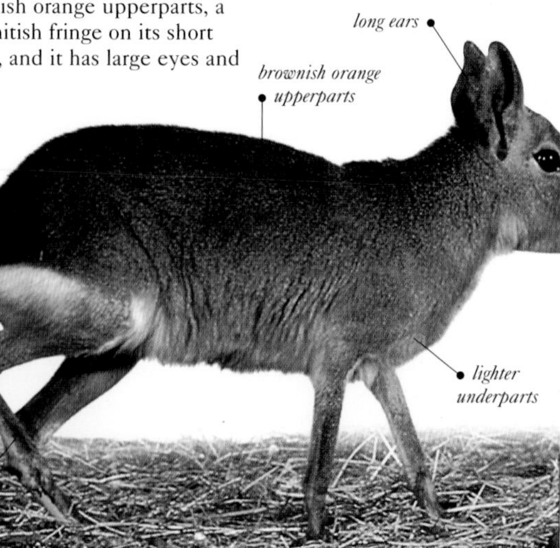

- **SIZE** Length: 43–78 cm (17–31 in). Tail: 2.5 cm (1 in).
- **OCCURRENCE** C. and S. Argentina. In pampas grassland.
- **REMARK** Unusually among rodents, male and female Maras are known to pair for life.

Social unit Pair	Gestation 70–80 days	Young 1–3	Diet

Family HYDROCHAERIDAE	Species *Hydrochaerus hydrochaeris*	Status Common

CAPYBARA

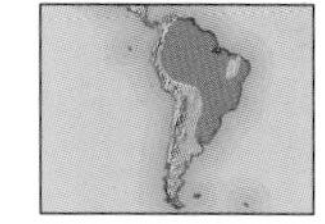
S. AMERICA

This species is the world's largest rodent and an excellent swimmer and diver, with eyes, nostrils, and ears set on the top of its head, and partially webbed toes. The young are born fully furred, and run, swim, and dive within hours of birth. Mixed groups including male–female pairs, and larger herds dominated by a single male, defend their home territory, moving on to find fresh grazing. Capybaras feed at dusk and dawn, sometimes raiding crops, and wallow during the midday heat.

- **SIZE** Length: 1.1–1.3 m (3 ½–4 ¼ ft). Tail: Vestigial.
- **OCCURRENCE** N. and E. South America. In lowland habitats near water: forested riverbanks, wetland, as well as mangrove swamps.

Social unit Variable	Gestation 150 days	Young 1–8	Diet

Family DASYPROCTIDAE	Species *Dasyprocta azarae*	Status Vulnerable

AZARA'S AGOUTI

Large and robust-bodied, this rodent is speckled light to medium brown, and may sometimes have a yellowish underside. Its short legs are distinctive as the forefeet have five toes, and the hindfeet have three. Azara's Agouti has prominent eyes, nostrils, and lips. It barks when alarmed, erecting its rump hairs to appear larger.

- **SIZE** Length: 50 cm (20 in). Tail: 2.5 cm (1 in).
- **OCCURRENCE** C. and S. Brazil, E. Paraguay, and N.E. Argentina. In tropical forest, riverbanks, and mangrove swamps.
- **REMARK** It is often hunted for meat.

S. AMERICA

Social unit Variable	Gestation 120 days	Young 1–2	Diet

Family AGOUTIDAE	Species *Agouti paca*	Status Common

PACA

One of the largest living rodents, this expert swimmer rests in its shelter during the day and emerges to eat at night. Brown, red, or light grey on its upperparts, with four lines of pale dots along each side, it has a white or buff underside and a tiny tail. This species is sometimes hunted for meat and sport.

N. & S. AMERICA

- **SIZE** Body length: 60–80 cm (23½–32 in). Tail: 1.5–3.5 cm (½–1½ in).
- **OCCURRENCE** S. Mexico to E. South America. In tropical forest, preferably areas near water.

Social unit Solitary	Gestation 114–119 days	Young 1	Diet

Family CAPROMYIDAE	Species *Capromys pilorides*	Status Endangered

DESMAREST'S HUTIA

This vole-like, nocturnal rodent has a large head, blunt nose, stocky body, and short limbs. Its strong, tapering tail and sharp, curved claws allow it to grip branches when foraging in trees for food.

- **SIZE** Body length: 55–60 cm (22–23½ in). Tail: 15–26 cm (6–10 in).

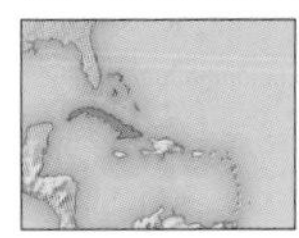

CARRIBEAN

- **OCCURRENCE** Cuban Archipelago. In tropical forest.
- **REMARK** Most species of hutia are severely threatened or extinct.

Social unit Solitary/Pair	Gestation 120–126 days	Young 1–4	Diet

Family CHINCHILLIDAE	Species *Lagostomus maximus*	Status Endangered*

PLAINS VISCACHA

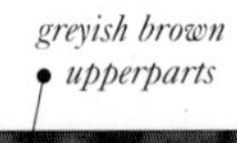

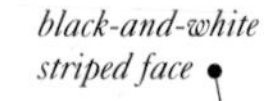

The largest member of the chinchilla family, this rodent has a large, thick head and strong legs. Found in noisy colonies of 20–50, this night-feeder damages pastures by digging tunnel systems, piling up the entrances with sticks, stones, and bones.

- **SIZE** Body length: 47–66 cm (18½–26 in). Tail: 15–20 cm (6–8 in).

S. AMERICA

- **OCCURRENCE** S. South America. In pampas and brushland.

Social unit Social	Gestation 153 days	Young 1–4	Diet

Family CHINCHILLIDAE	Species *Chinchilla lanigera*	Status Vulnerable

CHINCHILLA

S. AMERICA

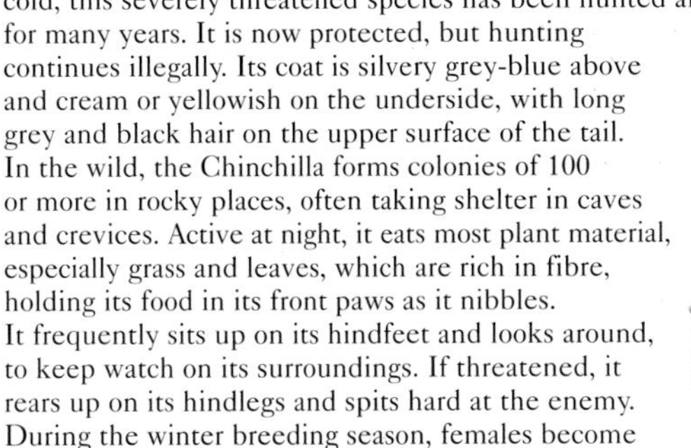

Prized for its soft, silky fur, which protects it from the bitter mountain cold, this severely threatened species has been hunted and farmed for many years. It is now protected, but hunting continues illegally. Its coat is silvery grey-blue above and cream or yellowish on the underside, with long grey and black hair on the upper surface of the tail. In the wild, the Chinchilla forms colonies of 100 or more in rocky places, often taking shelter in caves and crevices. Active at night, it eats most plant material, especially grass and leaves, which are rich in fibre, holding its food in its front paws as it nibbles. It frequently sits up on its hindfeet and looks around, to keep watch on its surroundings. If threatened, it rears up on its hindlegs and spits hard at the enemy. During the winter breeding season, females become aggressive towards each other.

- **SIZE** Body length: 22–38 cm (8–15 in). Tail: 7.5–15 cm (3–6 in).
- **OCCURRENCE** W. Chile. In the Andes mountains.

MINIATURE PET
Two pet Chinchillas take a dust-bath (above), a ritual also conducted in the wild. The appealing looks and friendly nature of this small rodent make it a popular pet.

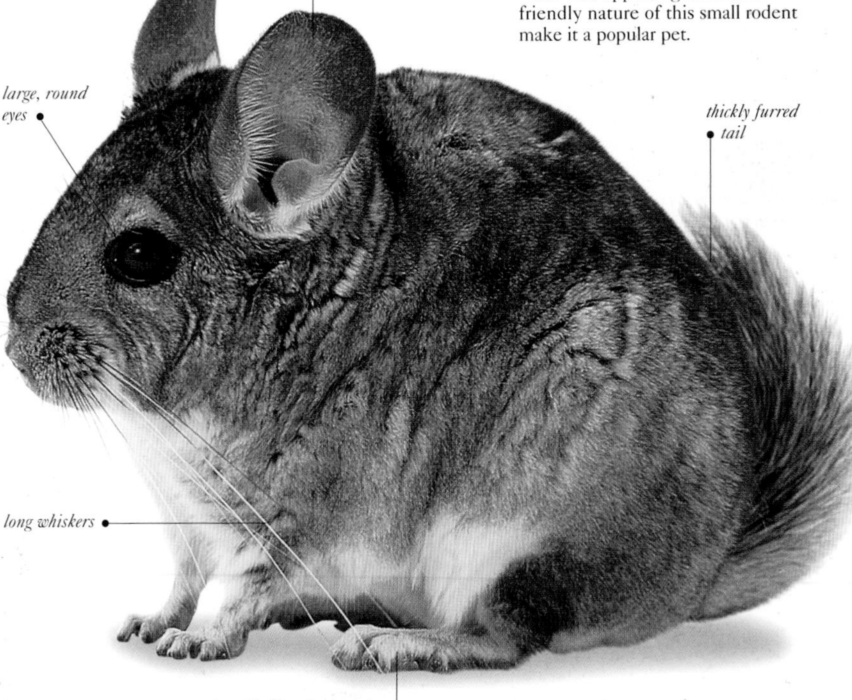

Social unit Social	Gestation 111 days	Young 2–4	Diet

Family MYOCASTORIDAE	Species *Myocastor coypus*	Status Common

COYPU

Resembling a beaver with a rat's tail, this nocturnal rodent has high-set eyes, nostrils, and ears, adapted to its aquatic habits. Its lips close behind the incisors, allowing it to gnaw even under water, and it has a long, rounded tail and webbed hindfeet. This fast swimmer eats mainly water plants, as well as alfalfa, rice, rye grass, and the seedlings of bald cypress, holding and manipulating its food with its forepaws. Highly gregarious, it lives in family groups in bankside tunnels and marks its home range with secretions from its oral and anal glands. It builds platform nests in riverbanks, and the males defend the nest area after the litter is born.

- **SIZE** Body length: 47–58 cm (18½–23 in). Tail: 34–41 cm (13½–16 in).
- **OCCURRENCE** Chile, Argentina, Uruguay, S. Brazil, Paraguay, and Bolivia. Near lakes and rivers.
- **REMARK** Originally from South America, the Coypu has been introduced in North America and Europe, where it is farmed for its dense brown fur. There are now wild populations in several of these countries due to escaped farm animals having set up colonies.

S. AMERICA

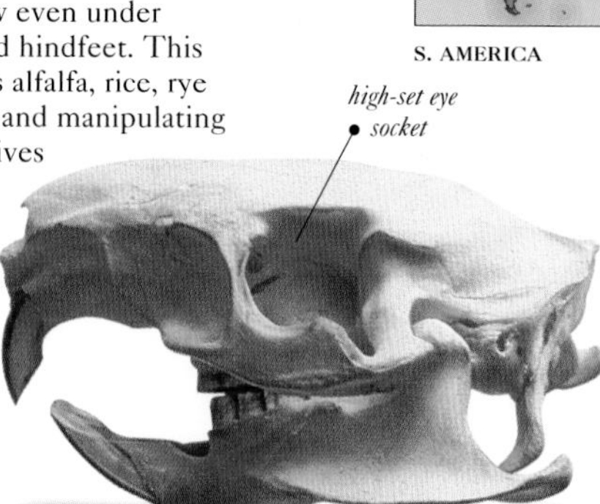

BEAVER-LIKE PROFILE
The Coypu's skull resembles that of the beaver in many ways. The head is flat and heavily set, with a prominent upper jaw and sharp incisors protruding beyond the lips.

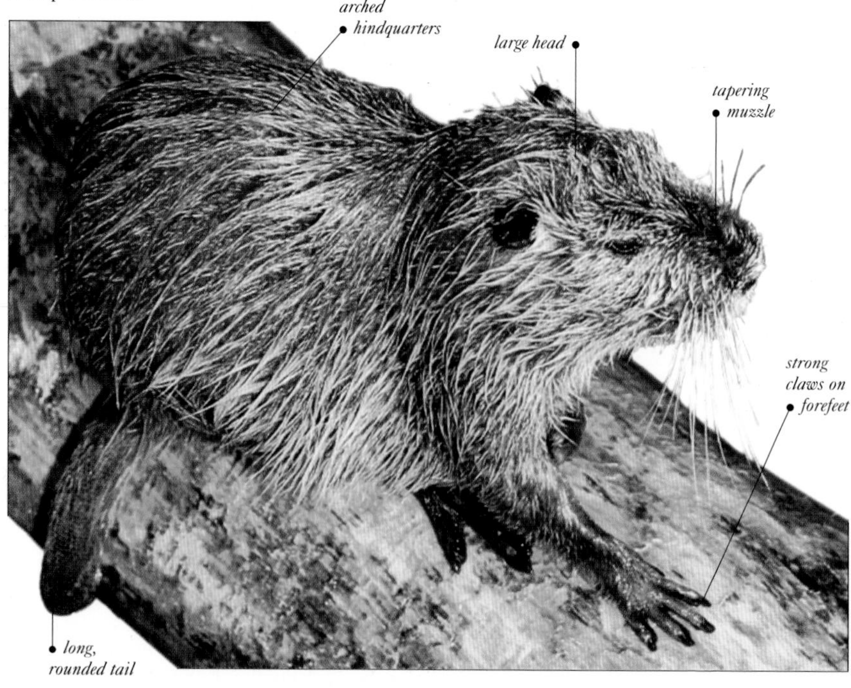

Social unit Social	Gestation 127–139 days	Young 1–12	Diet

Family OCTODONTIDAE	Species *Octodon degus*	Status Locally common

DEGU

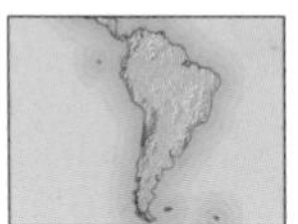

This mountain-dwelling rodent looks like a large, stout mouse and has yellow-brown fur on its back and creamish fur underneath. Usually, it also has a yellow ring around its neck, and yellow furry "lids" above and below the eyes. Found in colonies in elaborate underground burrow systems, it piles up sticks, stones, and animal dung around the entrances. The Degu feeds on a wide variety of vegetation, and is known to eat cattle droppings in the dry season.

- **SIZE** Body length: 25–31 cm (10–12 in). Tail: 7.5–13 cm (3–5 in).
- **OCCURRENCE** W. Chile. In the Andes mountains.
- **REMARK** The tail breaks off easily if grabbed by a predator.

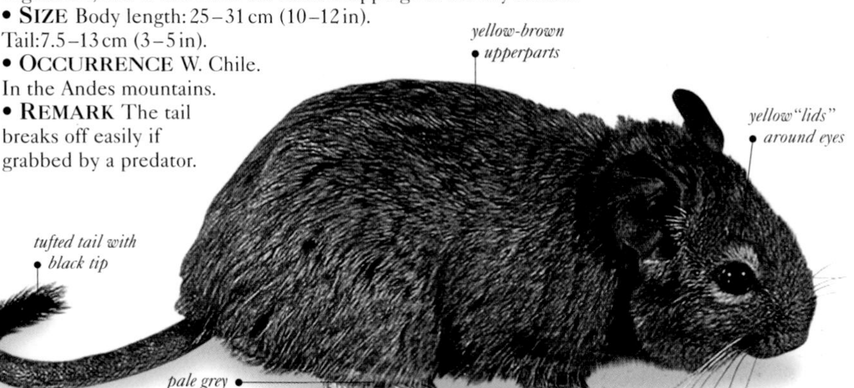

Social unit Variable	Gestation 90 days	Young 4–6	Diet

Family BATHYERGIDAE	Species *Heterocephalus glaber*	Status Locally common

NAKED MOLE RAT

With hairless, loose pinkish grey skin and vestigial ears and minute eyes, this unusual looking, nocturnal rodent has a unique social system. It lives in colonies of 70–80, with a dominant female "queen" that breeds and is tended by several non-workers, while the workers form head-to-tail digging chains in food-gathering galleries that radiate up to 40 m (130 ft) from the central chamber.

- **SIZE** Body length: 8–9 cm (3¼–3½ in). Tail: 3–4.5 cm (1¼–1¾ in).
- **OCCURRENCE** E. Africa. In arid land with a stable temperature.
- **REMARK** This rat lives underground and only surfaces to travel to another colony.

Social unit Social	Gestation 66–74 days	Young 1–12	Diet

CETACEANS

TOOTHED WHALES AND DOLPHINS

TOOTHED WHALES and dolphins make up almost nine-tenths of all cetaceans. They include 71 species of porpoises, river dolphins, dolphins, white whales, and sperm whales.

In addition to the streamlined whale shape, with fins, flippers, and flukes, all members have teeth, in contrast to the baleen whales (see pp.204–205). However, in some species of beaked whales (Ziphiidae), the teeth hardly erupt out of the gums. The nostrils form a single blowhole, usually on the top of the head.

Many toothed whales have elongated jaws forming a "beak" at the front of the mouth, and a bulging forehead known as the "melon". The melon gathers and directs sound waves, which toothed whales use to locate prey and to navigate. The several species of social whales also use these sound waves to communicate.

Family PHOCOENIDAE	Species *Phocoena phocoena*	Status Vulnerable

HARBOUR PORPOISE

A small and stocky animal, this porpoise has a rounded head with no beak, and small, spade-shaped teeth. The flippers are black and a dark stripe extends from them to the lower jaw. The front edge of its dorsal fin sometimes has a row of small, rounded lumps called tubercles (see right). The Harbour Porpoise usually forages alone for food on the seabed, seeking out areas of strong tidal currents to make prey capture easier. At times, however, it is found in hundreds where there is a concentration of prey. Shy of humans, it very rarely approaches boats. The female porpoise is larger than its male counterpart. Although it is found in large groups throughout its wide range, the status of the Harbour Porpoise is considered to be vulnerable. The threat is not so much from its natural predators, such as the Killer Whale and Great White Shark, but more so from human activity. Most fatalities are caused by underwater fishing nets, in which the porpoises become trapped.

PACIFIC, ATLANTIC

round bumps (tubercles) along edge of dorsal fin

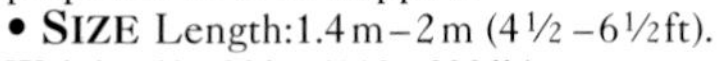

- **SIZE** Length: 1.4 m – 2 m (4½ – 6½ ft). Weight: 50 – 90 kg (110 – 200 lb).
- **OCCURRENCE** N. Pacific and N. Atlantic. In open oceans and along coastlines.
- **REMARK** Despite its small size, this species sometimes dives to depths of over 200 m (655 ft).

Social unit Variable	Gestation 10–11 months	Young 1	Diet

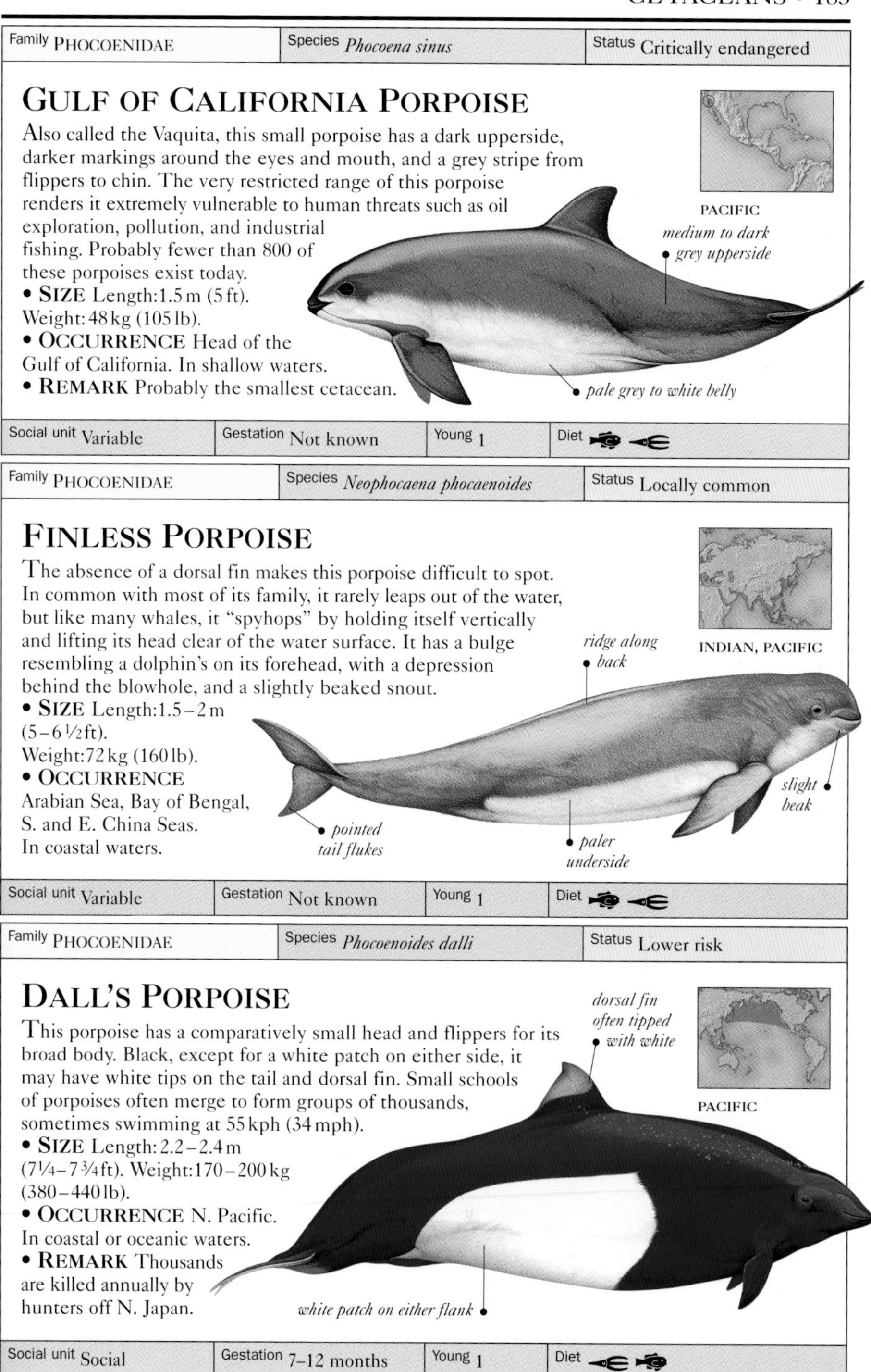

Family PHOCOENIDAE	Species *Phocoena sinus*	Status Critically endangered

GULF OF CALIFORNIA PORPOISE

Also called the Vaquita, this small porpoise has a dark upperside, darker markings around the eyes and mouth, and a grey stripe from flippers to chin. The very restricted range of this porpoise renders it extremely vulnerable to human threats such as oil exploration, pollution, and industrial fishing. Probably fewer than 800 of these porpoises exist today.

- **SIZE** Length: 1.5 m (5 ft). Weight: 48 kg (105 lb).
- **OCCURRENCE** Head of the Gulf of California. In shallow waters.
- **REMARK** Probably the smallest cetacean.

Social unit Variable	Gestation Not known	Young 1	Diet

Family PHOCOENIDAE	Species *Neophocaena phocaenoides*	Status Locally common

FINLESS PORPOISE

The absence of a dorsal fin makes this porpoise difficult to spot. In common with most of its family, it rarely leaps out of the water, but like many whales, it "spyhops" by holding itself vertically and lifting its head clear of the water surface. It has a bulge resembling a dolphin's on its forehead, with a depression behind the blowhole, and a slightly beaked snout.

- **SIZE** Length: 1.5–2 m (5–6½ ft). Weight: 72 kg (160 lb).
- **OCCURRENCE** Arabian Sea, Bay of Bengal, S. and E. China Seas. In coastal waters.

Social unit Variable	Gestation Not known	Young 1	Diet

Family PHOCOENIDAE	Species *Phocoenoides dalli*	Status Lower risk

DALL'S PORPOISE

This porpoise has a comparatively small head and flippers for its broad body. Black, except for a white patch on either side, it may have white tips on the tail and dorsal fin. Small schools of porpoises often merge to form groups of thousands, sometimes swimming at 55 kph (34 mph).

- **SIZE** Length: 2.2–2.4 m (7¼–7¾ ft). Weight: 170–200 kg (380–440 lb).
- **OCCURRENCE** N. Pacific. In coastal or oceanic waters.
- **REMARK** Thousands are killed annually by hunters off N. Japan.

Social unit Social	Gestation 7–12 months	Young 1	Diet

Family INIIDAE	Species *Inia geoffrensis*	Status Vulnerable

AMAZON RIVER DOLPHIN

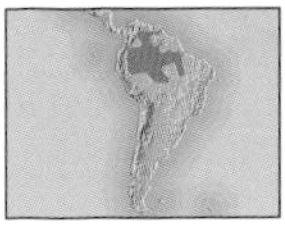

Having extremely small eyes and living in shallow, murky waters, this dolphin–also known as the Boto–pokes in the mud for prey with its long, slim beak, and uses echolocation to find its way and food. It has peg-like front teeth for catching prey and molar-like back teeth for crushing crabs, armoured catfish, and turtles. Plump in shape, it can be vivid pink, bluish grey, or off-white, and has a ridged hump on its back instead of a fin.

- **SIZE** Length: 2–2.6 m (6½–8½ ft). Weight: 100–160 kg (220–350 lb).
- **OCCURRENCE** Amazon and Orinoco basins. In lakes and rivers.

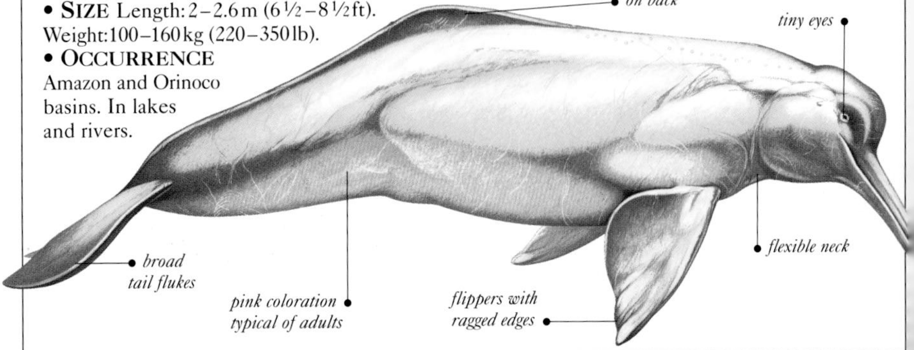

Social unit Variable	Gestation 10–11 months	Young 1	Diet 

Family LIPOTIDAE	Species *Lipotes vexillifer*	Status Critically endangered

CHINESE RIVER DOLPHIN

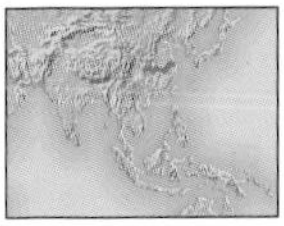

Sometimes called the Whitefin Dolphin, and the "baiji" in the local Chinese dialect, this rare species has a long, thin beak and a flexible neck to dig in the riverbed for prey, mostly at night and early morning. With tiny, underdeveloped eyes, the Chinese River Dolphin navigates by touch and echolocation. Shy and secretive, it lives in groups of 2–6. It faces a number of threats: depletion of prey due to overfishing, obstruction of its migratory routes by dams, chemical pollution, collision with boats, and interference with its echolocation from the noise of boat engines. Despite legal protection since 1949, it is hunted illegally for meat and body parts used in traditional medicines.

- **SIZE** Length: 2.2–2.5 m (7¼–8¼ ft). Weight: 125–160 kg (275–350 lb).
- **OCCURRENCE** China – lower and middle Chang Jiang (Yangtze) River. In rivers and associated lakes.
- **REMARK** Fewer than 150 of these dolphins exist today.

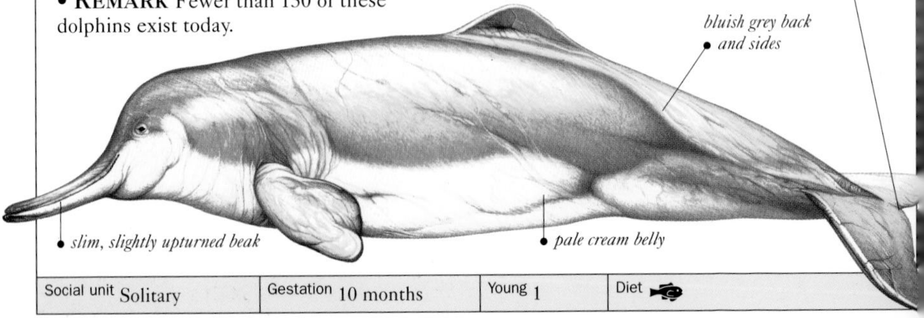

Social unit Solitary	Gestation 10 months	Young 1	Diet

Family PLATANISTIDAE	Species *Platanista gangetica*	Status Endangered

GANGES RIVER DOLPHIN

ASIA

This rare, freshwater dolphin has distinctively broad flippers and a long, thin beak with protruding front teeth that interlock to make a cage for prey. A flexible neck allows it to turn its head at right angles to grub for food and scan the area with echolocating pulses. Although they appear to be identical, the dolphins living in the Indus River and those living in the Ganges–Brahmaputra system are considered to be different subspecies. Both types live in small groups of 4–6, occasionally as many as 30 or more.

- **SIZE** Length: 2.1–2.5m (7–8¼ft). Weight: 85 kg (185 lb).
- **OCCURRENCE** Pakistan and India, throughout the river systems of the Indus, Ganga, and Brahmaputra. In freshwater rivers.
- **REMARK** This is the only cetacean without a crystalline eye lens, making it blind.

large, wide tail flukes
triangular ridge on back
upward curving mouth
paddle-shaped, ragged flippers
pinkish belly

Social unit Variable	Gestation 8–12 months	Young 1	Diet

Family DELPHINIDAE	Species *Lagenorhynchus obscurus*	Status Locally common

DUSKY DOLPHIN

S. AMERICA, AFRICA, NEW ZEALAND

Small and compact in shape, this dolphin has a complex and variable colour pattern. However, it is mainly dark grey to blue-black on its upperside, and pale grey or white below, with a tapering grey stripe separating these two areas. Another forked pale patch embellishes each flank. There are three subspecies: *L. obscurus fitzroyi* found off South America, *L. obscurus obscurus* found off S. Africa, and an unnamed species found off New Zealand. The Dusky Dolphin, active day and night, feeds on small, schooling, open-water fish and squid, taking its prey from the surface as well as the depths.

- **SIZE** Length: 1.7–2.1 m (5½–7 ft). Weight: 70–85 kg (155–185 lb).
- **OCCURRENCE** W. and S. South America, S. Africa, and New Zealand. In coastal and continental shelf waters.

pale grey stripe between black back and white belly
gradually tapering head
tall, curved dorsal fin
forked pale patch on flank
black lips

Social unit Variable	Gestation 13 months	Young 1	Diet

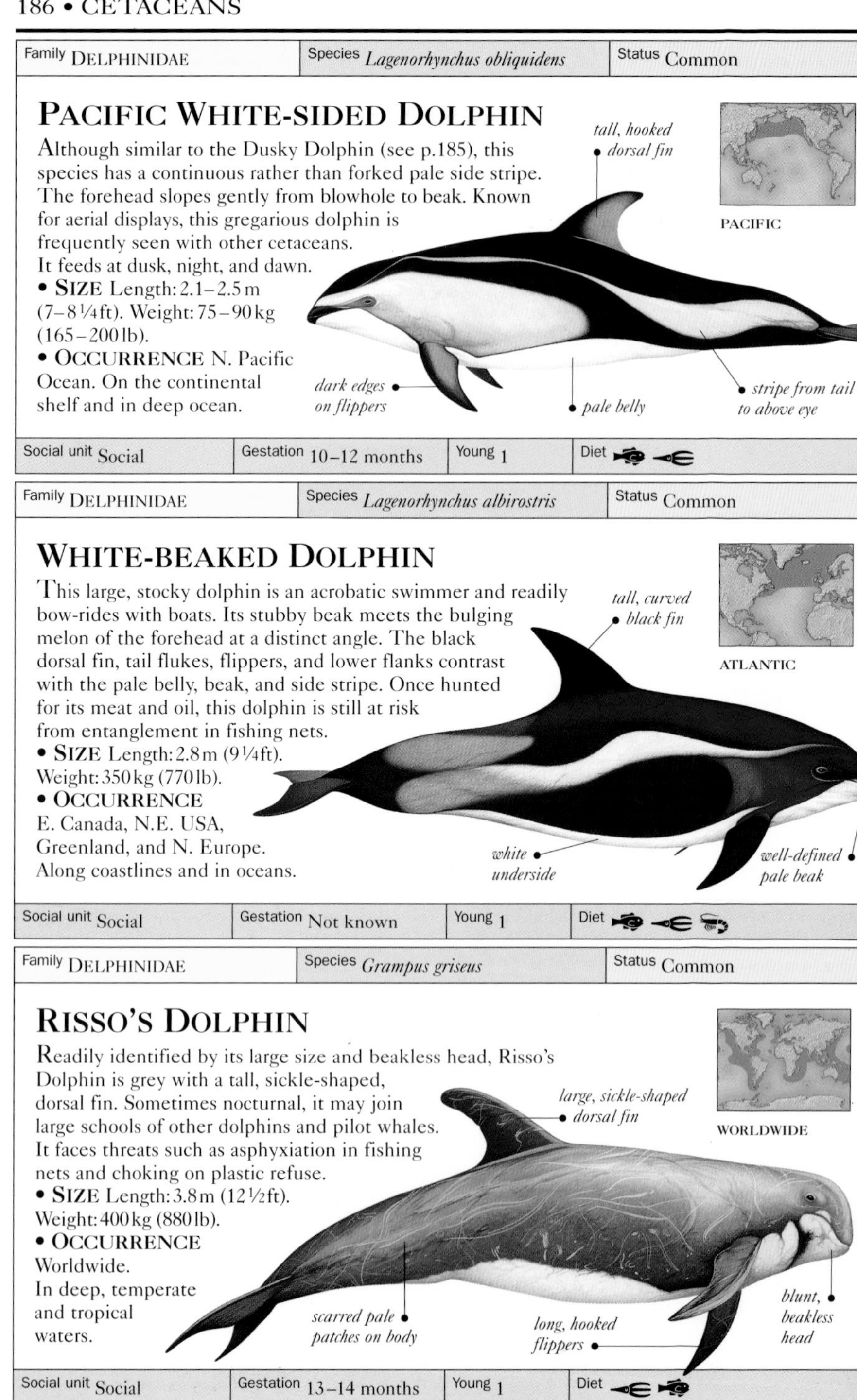

Family DELPHINIDAE	Species *Lagenorhynchus obliquidens*	Status Common

PACIFIC WHITE-SIDED DOLPHIN

Although similar to the Dusky Dolphin (see p.185), this species has a continuous rather than forked pale side stripe. The forehead slopes gently from blowhole to beak. Known for aerial displays, this gregarious dolphin is frequently seen with other cetaceans. It feeds at dusk, night, and dawn.

- **SIZE** Length: 2.1–2.5 m (7–8 1/4 ft). Weight: 75–90 kg (165–200 lb).
- **OCCURRENCE** N. Pacific Ocean. On the continental shelf and in deep ocean.

Social unit Social	Gestation 10–12 months	Young 1	Diet

Family DELPHINIDAE	Species *Lagenorhynchus albirostris*	Status Common

WHITE-BEAKED DOLPHIN

This large, stocky dolphin is an acrobatic swimmer and readily bow-rides with boats. Its stubby beak meets the bulging melon of the forehead at a distinct angle. The black dorsal fin, tail flukes, flippers, and lower flanks contrast with the pale belly, beak, and side stripe. Once hunted for its meat and oil, this dolphin is still at risk from entanglement in fishing nets.

- **SIZE** Length: 2.8 m (9 1/4 ft). Weight: 350 kg (770 lb).
- **OCCURRENCE** E. Canada, N.E. USA, Greenland, and N. Europe. Along coastlines and in oceans.

Social unit Social	Gestation Not known	Young 1	Diet

Family DELPHINIDAE	Species *Grampus griseus*	Status Common

RISSO'S DOLPHIN

Readily identified by its large size and beakless head, Risso's Dolphin is grey with a tall, sickle-shaped, dorsal fin. Sometimes nocturnal, it may join large schools of other dolphins and pilot whales. It faces threats such as asphyxiation in fishing nets and choking on plastic refuse.

- **SIZE** Length: 3.8 m (12 1/2 ft). Weight: 400 kg (880 lb).
- **OCCURRENCE** Worldwide. In deep, temperate and tropical waters.

Social unit Social	Gestation 13–14 months	Young 1	Diet

Family DELPHINIDAE	Species *Stenella longirostris*	Status Lower risk

SPINNER DOLPHIN

WORLDWIDE

Probably no other cetacean displays such wide variations in appearance within the species, as this dolphin. There are three subspecies, as well as a dwarf version found in the Gulf of Thailand. Its body is slender, and its beak long and thin. The male can be identified by its post-anal hump. This highly social animal hunts at night. It swims in schools of tens to thousands, often alongside other dolphins and whales, and shoals of tuna. Its characteristic spinning leaps are probably used, along with various clicks, whistles, and screams, to communicate with others.

- **SIZE** Length: 1.3–2m (4¼–6½ft). Weight: 45–75kg (99–165lb).
- **OCCURRENCE** Worldwide. In tropical oceans.
- **REMARK** Spinner Dolphins are used by fishermen to help locate tuna shoals. Millions have been killed as a result of industrial fishing in the E. Pacific.

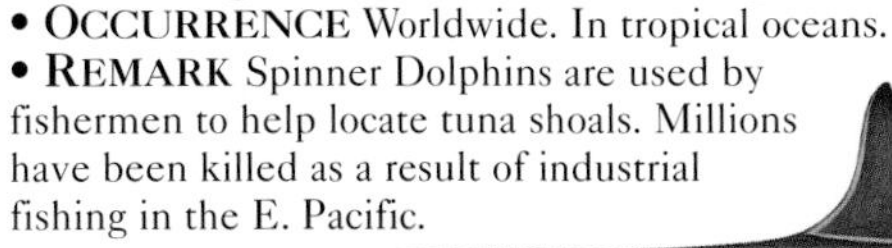
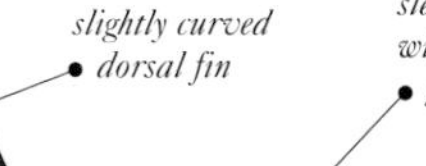
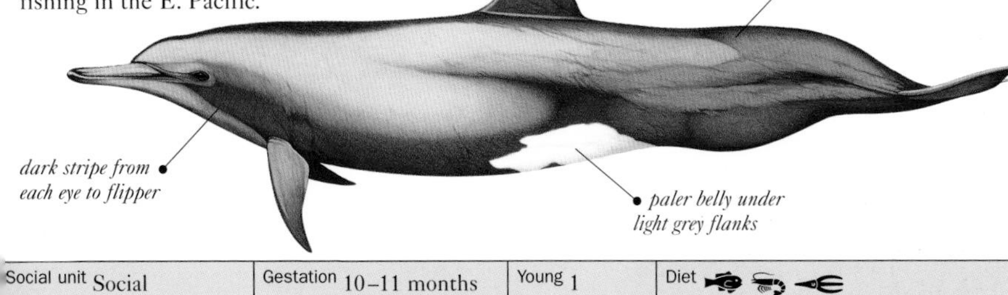

Social unit Social	Gestation 10–11 months	Young 1	Diet

Family DELPHINIDAE	Species *Stenella frontalis*	Status Locally Common

ATLANTIC SPOTTED DOLPHIN

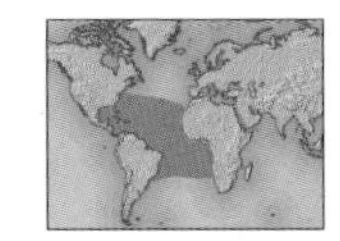
ATLANTIC

This dolphin may be distinguished from the Pantropical Spotted Dolphin (see p.188) by its stouter body and beak, and a light band extending from the shoulder to the dorsal fin. However, like its pantropical relative, it is born without spots: these make their first appearance on the belly and then extend to the back with age. At night, it feeds on fish and squid that come up from deeper waters. During the day the Atlantic Spotted Dolphin feeds on mid-water fish, and may even burrow for prey in the sandy bottom.

- **SIZE** Length: 1.7–2.3m (5½–7½ft). Weight: 140kg (310lb).
- **OCCURRENCE** Atlantic Ocean. In open waters and along coastlines.
- **REMARK** This dolphin is still hunted for meat and bait in the Caribbean Sea.

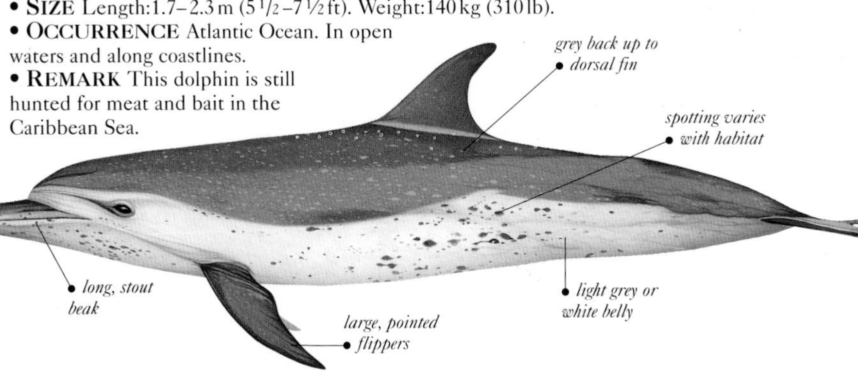

Social unit Variable	Gestation Not known	Young 1	Diet

Family DELPHINIDAE	Species *Stenella coeruleoalba*	Status Lower risk

STRIPED DOLPHIN

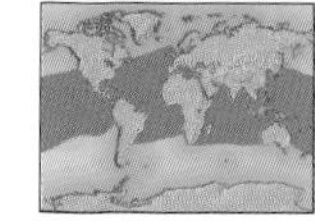
WORLDWIDE

The basic blue-grey colour of this dolphin's body is overlain by a complex pattern of black and white stripes along its back and flanks. A highly social cetacean, usually found in schools of 10–500, the Striped Dolphin is well known for a wide variety of acrobatic leaps and spins while racing across the ocean, and is often spotted riding pressure waves in front of migrating great whales or ships. Sometimes it gathers in thousands, leaping high and whistling to keep in contact.

- **SIZE** Length:1.8–2.5 m (6–8 ¼ft). Weight:110–165 kg (240–360 lb).
- **OCCURRENCE** Worldwide. In tropical and temperate waters.
- **REMARK** Although relatively common, its numbers have declined in recent years. In the early 1990s, the Mediterranean population was devastated by a viral disease.

blue-grey back

black stripe from upper beak to anal region

wide, pale grey stripe

pale cream or pink underside

Social unit Social	Gestation 12–13 months	Young 1	Diet

Family DELPHINIDAE	Species *Stenella attenuata*	Status Lower risk

PANTROPICAL SPOTTED DOLPHIN

WORLDWIDE

One of the commonest cetaceans, this streamlined dolphin is similar to the Atlantic Spotted Dolphin (see p.187), but more slender. It has a dark grey "cape" that runs from its forehead to the dorsal fin, while the belly and lower flanks are pale grey. In adults, these are covered with spots, which vary with habitat and increase with age. Adults may also develop white lips. Large schools of thousands, often segregated into mothers with their young, older juveniles, and other groups, associate with other cetaceans, especially Spinner Dolphins, and with shoals of tuna.

- **SIZE** Length:1.6–2.6 m (5 ¼–8 ½ft). Weight: Up to 120 kg (260 lb).
- **OCCURRENCE** Worldwide. In tropical and temperate waters.
- **REMARK** Like the Spinner Dolphin (see p.187), this species is trapped as a bicatch in industrial tuna fishing operations and many are killed or injured as a result.

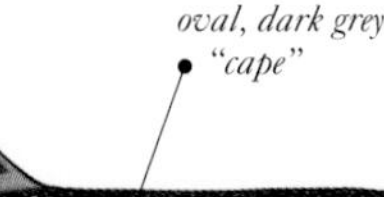

oval, dark grey "cape"

slender, streamlined body

white lips on adults

pale underside

spotting on body increases with age

Social unit Social	Gestation 11–12 months	Young 1	Diet 

Family DELPHINIDAE	Species *Tursiops truncatus*	Status Common

BOTTLENOSE DOLPHIN

WORLDWIDE

A common "performer" at marine life centres, this is the largest of the beaked dolphins. It shows enormous geographic variation in size and colour. In warmer waters it averages 2 m (6 ½ ft) in length, and has relatively large flippers, dorsal fin, and flukes, while in colder open oceans it may be twice this size, but with proportionately smaller extremities. Typically, it has a dark grey or black back and a cream belly. Another possible species of Bottlenose Dolphin, *Tursiops aduncus*, has been identified in the Indian and W. Pacific oceans.

- **SIZE** Length: 1.9–4 m (6 ¼–13 ft). Weight: 500 kg (1,100 lb).
- **OCCURRENCE** Worldwide. In temperate and tropical seas.
- **REMARK** This dolphin can be swum with or stroked by humans.

short and robust beak
large, hooked dorsal fin
large tail flukes
long, pointed flippers

Social unit Variable	Gestation 12 months	Young 1	Diet

Family DELPHINIDAE	Species *Delphinus delphis*	Status Common

COMMON DOLPHIN

WORLDWIDE

A distinctive yellow and grey blaze, shaped like an hourglass, runs along the flank from the face to the tail of this dolphin. Dark stripes run from its chin to its flippers and from the beak to the eyes. A fast swimmer, it makes a variety of sounds – clicks, squeaks, and croaks – that can often be heard from boats as it bow-rides. A second species of Common Dolphin (*Delphinus capensis*) is found in inshore waters. Both species, however, hunt shoaling fish and squid to a depth of 300 m (985 ft).

- **SIZE** Length: 2.3–2.6 m (7 ½–8 ½ ft). Weight: 80 kg (175 lb).
- **OCCURRENCE** Worldwide. In deep tropical and temperate waters.
- **REMARK** The Common Dolphin is hunted in many parts of the world.

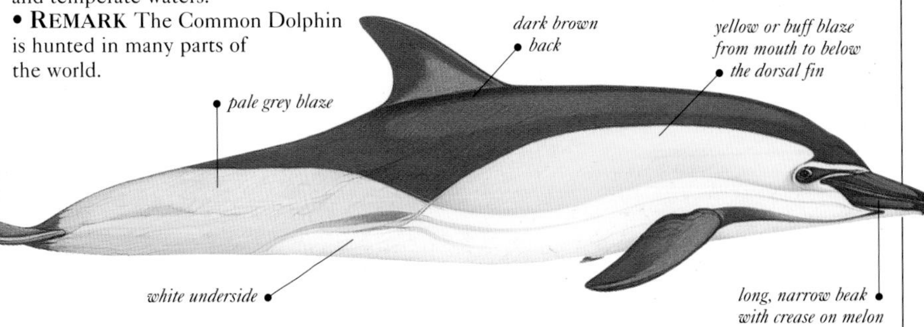

Social unit Social	Gestation 10–11 months	Young 1	Diet

Family DELPHINIDAE	Species *Orcaella brevirostris*	Status Locally common

IRRAWADDY DOLPHIN

ASIA, AUSTRALIA

A rounded, blunt head, ridged lips, bulging forehead, and creased neck characterize this river dolphin, which is capable of a variety of facial expressions due to a complex muscle structure. It is found in small schools of 15 or less, in silt-laden estuaries, but some swim almost 1,500 km (930 miles) upstream, along the waterways of the Irrawaddy and Mekong.

- **SIZE** Length: 2.1–2.8 m (7–9¼ ft). Weight: 90–150 kg (200–330 lb).
- **OCCURRENCE** S.E. Asia and N. Australia. In brackish or fresh water.
- **REMARK** They may sometimes herd fish into fishermen's nets and receive rewards.

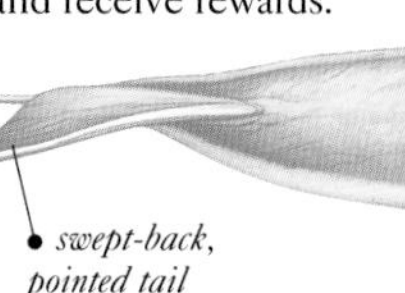

Social unit Variable	Gestation 14 months	Young 1	Diet

Family DELPHINIDAE	Species *Lissodelphis borealis*	Status Common

NORTHERN RIGHT-WHALE DOLPHIN

PACIFIC

Slim and slender, with small flippers and tail flukes, this dolphin's streamlined shape makes it a fast swimmer. It completely lacks a dorsal fin or dorsal ridge, and is named after the Northern Right-whale (see pp. 206–207), which also shares the same feature. This dolphin is almost fully black, except for a white strip below. This is much narrower than that of its cousin, the Southern Right-whale Dolphin (*Lissodelphis peronii*), which is found only in the Southern Hemisphere. Sociable in nature, these cetaceans form schools of 100–200, which merge into gatherings of thousands. Swimming in bouncing movements, they make a graceful sight, often leaping high and riding along with ships. Some dolphins are hunted in Japanese seas, while many are caught in squid driftnets off Japan, Taiwan, and Korea.

- **SIZE** Length: Up to 3 m (9¾ ft). Weight: Up to 115 kg (250 lb).
- **OCCURRENCE** N. Pacific. In deep waters.
- **REMARK** This is the only dolphin without a dorsal fin.

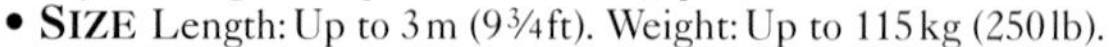
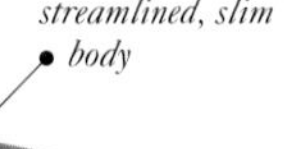
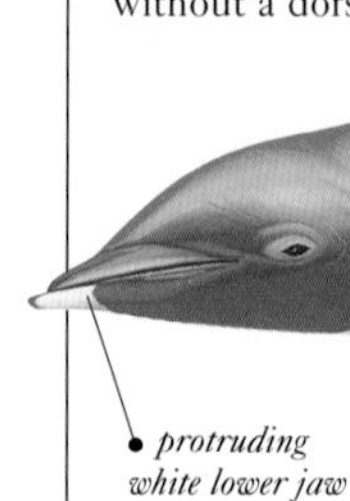

Social unit Social	Gestation Not known	Young 1	Diet

Family DELPHINIDAE	Species *Cephalorhynchus commersonii*	Status Locally common

COMMERSON'S DOLPHIN

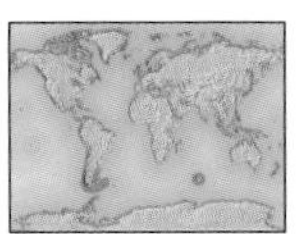

Resembling the Killer Whale (see pp.194–195) in coloration, this dolphin has a smoothly sloping forehead, and rounded fins and tail flukes. It is white, except for black patches on its forehead, flippers, and belly, and between the dorsal fin and tail. The newborn calf is grey and becomes black and white with age. This expert swimmer and leaper forms schools of up to ten, occasionally expanding to 100. Those dolphins found around South America are 25–30 cm (10–12 in) shorter than those inhabiting the Indian Ocean.

- **SIZE** Length: 1.4–1.7 m (4 1/2–5 1/2 ft). Weight: Up to 86 kg (190 lb).
- **OCCURRENCE** S. South America, Falkland Islands, and S. Indian Ocean. In shallow, muddy, coastal waters.

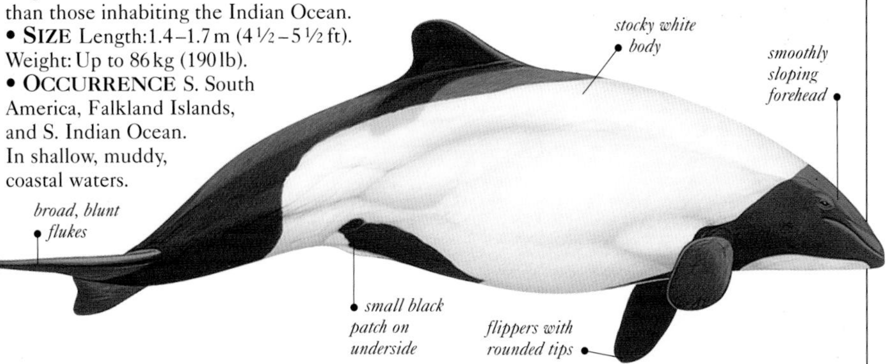

Social unit Social	Gestation 11–12 months	Young 1	Diet

Family DELPHINIDAE	Species *Cephalorhynchus hectori*	Status Endangered

HECTOR'S DOLPHIN

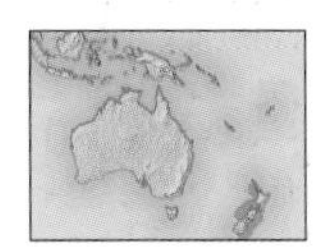

Similar to a porpoise in outline, this small dolphin has a distinctively rounded dorsal fin and complex body pattern of black, white, and grey. Hector's Dolphin has a wide snout, and no beak or melon-like bulge on the forehead. Active and sociable, it swims in small groups of five or less, maintaining high levels of interaction through chasing, flipper-slapping, and touching. As an inshore species, it faces an increased risk from pollution and being trapped and killed in fishing nets.

- **SIZE** Length: 1.2–1.5 m (4–5 ft). Weight: Up to 57 kg (125 lb).
- **OCCURRENCE** New Zealand. In shallow, coastal waters.
- **REMARK** This is one of the rarest marine dolphins.

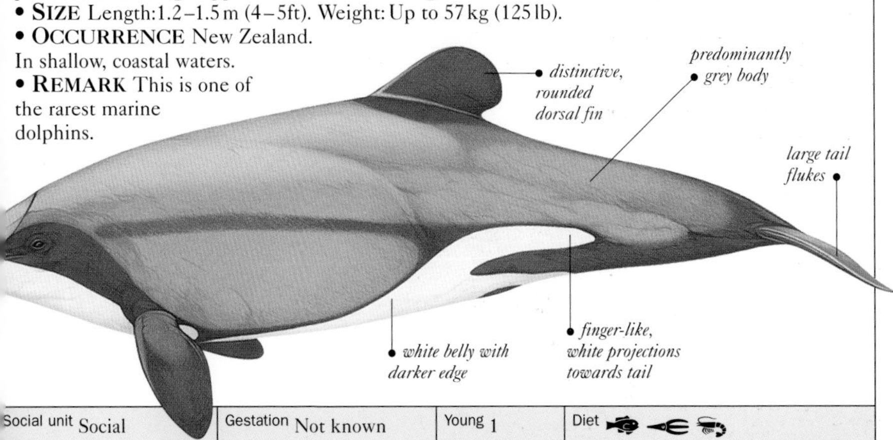

Social unit Social	Gestation Not known	Young 1	Diet

Family DELPHINIDAE	Species *Globicephala macrorhynchus*	Status Lower risk

SHORT-FINNED PILOT WHALE

Distinctive in looks, this toothed whale is uniformly slate-grey or black, with an anchor-shaped pale patch on the throat and chest, and white streaks behind the dorsal fin and eyes. However, at sea it is almost impossible to distinguish it from its close cousin, the Long-finned Pilot Whale (*Globicephala melas*), the only difference being its shorter flippers. The Short-finned Pilot Whale appears to prefer warmer waters, and there is little overlap in the range of the two species. Male whales attain twice the weight of females and live 15 years longer. Scarring on their bodies suggests battles over females. Pilot Whales are highly sociable, forming schools of 10–100. They feed mainly on deep-water squid and octopus at night, often diving below 500 m (1,600 ft), and remaining underwater for more than 15 minutes.

- **SIZE** Length: 5–7m (16–23 ft). Weight: 1–1.8 tonnes (1–1¾ tons).
- **OCCURRENCE** Worldwide. In tropical and warm temperate waters.
- **REMARK** This whale is hunted by being driven into shallow waters.

slender body becomes more robust with age

tail with sharply pointed tips

off-white belly patch

Social unit Social	Gestation 14½–15 months	Young 1	Diet

Family DELPHINIDAE	Species *Pseudorca crassidens*	Status Locally common

FALSE KILLER WHALE

One of the largest dolphins, this uniformly black or slate-grey cetacean has a lighter blaze from its flippers to its belly, a centrally placed, tall, and hooked dorsal fin, and pectoral fins with an elbow-like bend. This slim-bodied whale swims extremely fast for its size and is a formidable hunter. It is equipped with 8–11 pairs of large, conical teeth and catches large, oceanic fish such as Salmon, Tuna, and Barracuda, as well as squid and smaller dolphins. Usually found in schools of 10–20, and rarely even up to 300, it makes various echolocating and communicating sounds, such as clicks and whistles. Known for spectacular leaps, it skilfully surfs breakers and bow waves.

- **SIZE** Length: 5–6 m (16–20 ft). Weight: 1.3–1.4 tonnes (1–1⅜ tons).
- **OCCURRENCE** Worldwide. In deep and offshore, temperate or tropical waters – occasionally along coasts of oceanic islands, especially Japan and Hawaii.
- **REMARK** This whale is extremely susceptible to stranding on the shore, sometimes in large groups of up to 800–1,000.

slender head, tapers to round beak

angled flipper

Social unit Social	Gestation 11–16 months	Young 1	Diet

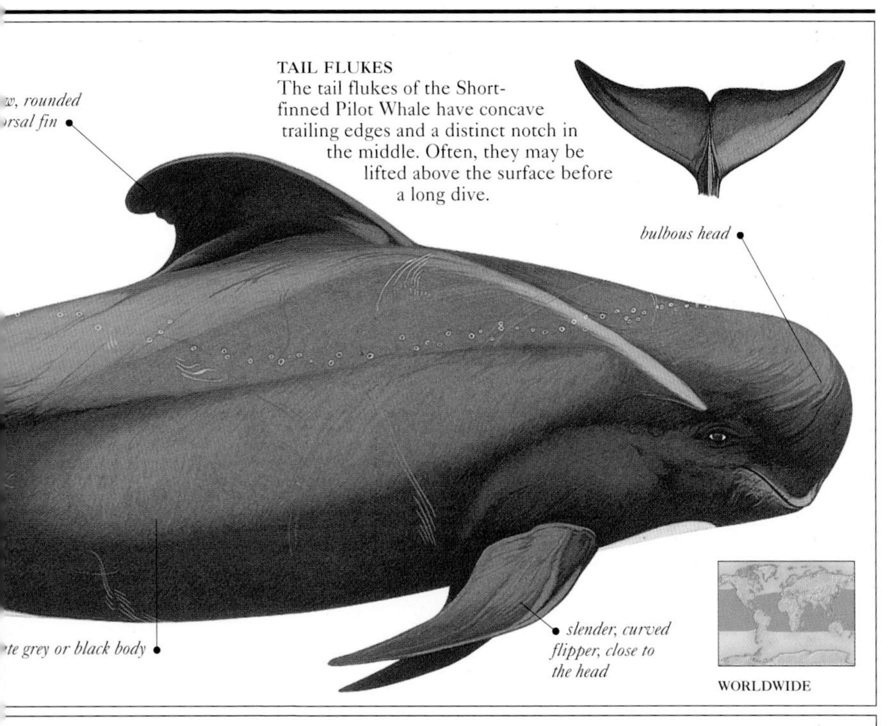
TAIL FLUKES
The tail flukes of the Short-finned Pilot Whale have concave trailing edges and a distinct notch in the middle. Often, they may be lifted above the surface before a long dive.
w, rounded
rsal fin
bulbous head
te grey or black body
slender, curved flipper, close to the head
WORLDWIDE

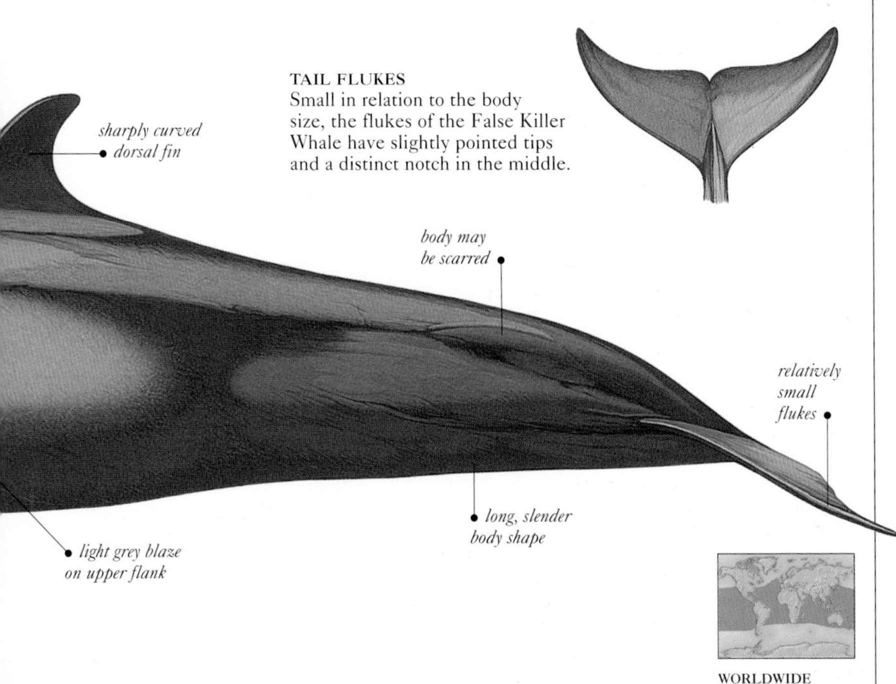
TAIL FLUKES
Small in relation to the body size, the flukes of the False Killer Whale have slightly pointed tips and a distinct notch in the middle.
sharply curved dorsal fin
body may be scarred
relatively small flukes
long, slender body shape
light grey blaze on upper flank
WORLDWIDE

Family DELPHINIDAE	Species *Orcinus orca*	Status Lower risk

KILLER WHALE

The most widely recognized species among toothed whales and dolphins, the Killer Whale is named for its extraordinary and diverse hunting techniques. The food it eats is equally diverse, ranging from herring to Great White Sharks, marine mammals such as small whales and seals, turtles, and seabirds. Its powerful and stocky body is ideally suited to hunting, its broad tail flukes propel it at high speeds, and the tall dorsal fin and paddle-like flippers provide stability. The distinctive black-and-white markings provide effective camouflage underwater, seen from above and below. Highly social in nature, the Killer Whale lives in long-lasting matriarchal pods or family groups. Male and female calves stay with the mother for life. As the young begin to reproduce, their offspring build up a multi-generation family around the matriarch. The average pod size is 30, but pods often merge to form "superpods" of 150. The pods travel in tight formation, with the females and calves in the centre and the males around the fringes, or spread across distances of up to 1 km (¾ mile). Communication between whales is by highly distinctive cries and screams, which also act as social signals to reinforce group identity. The Killer Whale displays a variety of surface movements including spyhopping (rising slowly vertically, until its head is above the water), tail- and flipper-slapping, and breaching.

WORLDWIDE

dorsal fin forms isosceles triangle

conspicuous white eyepatches

white chin

paddle-shaped flippers

flippers may be one-fifth body length

• **SIZE** Length: Up to 9 m (30 ft). Weight: Up to 10 tonnes (10 tons).
• **OCCURRENCE** Worldwide. From estuaries to open oceans and icefields; common along coastlines and areas rich in marine life.
• **REMARK** Despite its name, the Killer Whale is approachable, rarely harming humans.

Social unit Social	Gestation 12–17 months	Young 1	Diet

ACE HUNTER
The Killer Whale is an ingenious hunter, working in coordinated groups to capture a wide variety of prey. It herds schools of fish together before attacking them from different angles. It tips over ice floes, dropping seals and penguins into the water to capture them, and intentionally beaches itself to pounce on unsuspecting seals.

SPYHOPPING
The Killer Whale rises slowly out of the water, raising its head and most of its flippers above the surface, before sinking out of sight.

TAIL FLUKES
The Killer Whale's wide tail flukes work as a propeller when it swims at high speed. They have a distinct notch in the middle and are white below.

POD COORDINATION
Pods may travel closely together or spread apart, surfacing and diving in one coordinated movement.

Family MONODONTIDAE	Species *Delphinapterus leucas*	Status Vulnerable

BELUGA

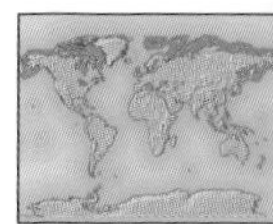
ARCTIC

The only all-white cetacean, the Beluga has a long, muscular body, no beak, an extremely flexible neck, and a fibrous dorsal ridge. The absence of a dorsal fin and its white skin are adaptations to life around and under floating ice. Tinged dark grey at birth, the skin gradually fades to become pure white. In an apparently unique departure from other cetaceans, the Beluga may moult in summer, when the skin may acquire a yellowish tinge. Well-known for emitting a wide variety of sounds, including squeaks, whistles, mews, clicks, and belches, the Beluga was nicknamed "sea canary" by whalers. It keeps to the upper edges of the Arctic icefields and has been radio-tracked diving to over 300 m (990 ft), to find prey.

- **SIZE** Length: 4–5.5 m (13–18 ft). Weight: 1–1.5 tonnes (1–1½ tons).
- **OCCURRENCE** Circumpolar in the Northern Hemisphere. In temperate to Arctic waters.
- **REMARK** Severely depleted by whaling operations, the Beluga population is now threatened by chemical pollution.

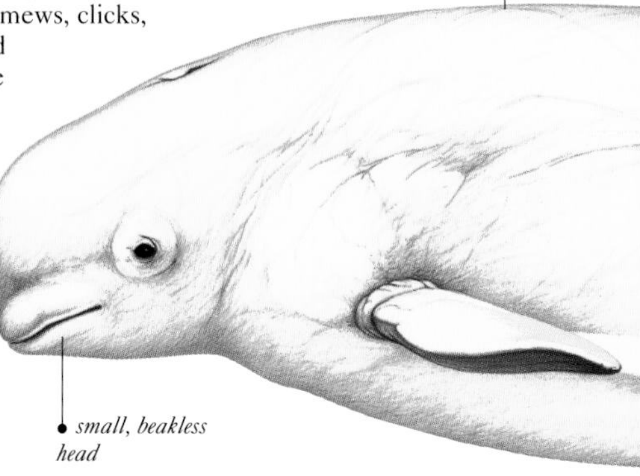

Social unit Social	Gestation 14 months	Young 1	Diet

Family MONODONTIDAE	Species *Monodon monoceros*	Status Locally common

NARWHAL

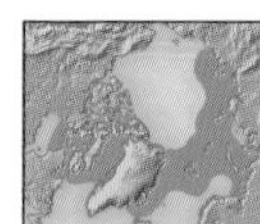
ARCTIC

With the most northerly range of any mammal, the Narwhal rarely strays south of 60° North. The most famous feature of the male is its tusk, which is actually a modified upper left tooth that grows in a clockwise spiral through its upper lip. The tusk is used as a weapon, but also to break open breathing holes in ice floes or for grubbing up the seabed. The mouth itself is toothless and the bulbous melon has no beak. The Narwhal's stocky body has no dorsal fin, and its small pectoral fins are upturned at the tip in adults. The C-shaped tail flukes of the adult are shaped like a fan, composed of two, joined semi-circles (see right). Calves are born uniformly grey, but the adults have pale grey skin speckled black.

Like the Belugas, with which they are often found, Narwhals form big schools segregated by age and sex, and use a wide variety of sounds, either for communicating with each other and warning against predators, or for locating their prey .

- **SIZE** Length: 4–4.5 m (13–15 ft). Weight: 0.8–1.6 tonnes (¾–1⅝ tons).
- **OCCURRENCE** Circumpolar. In oceans and along coastlines.
- **REMARK** The Narwhal is hunted for meat, oil, skin, and tusks.

Social unit Social	Gestation 14–15 months	Young 1	Diet 

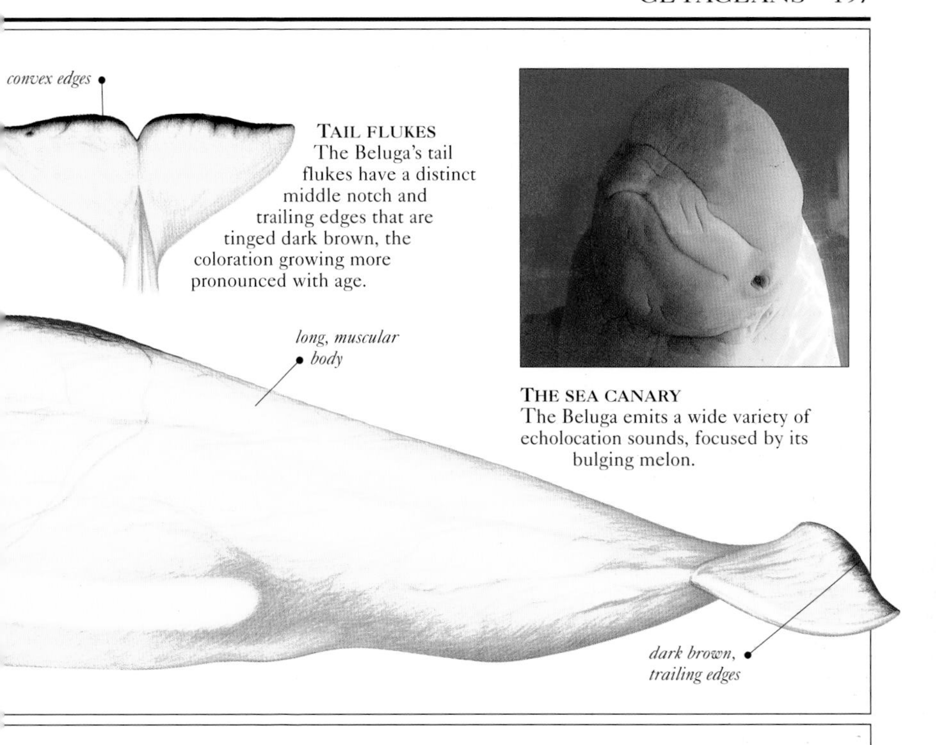

TAIL FLUKES
The Beluga's tail flukes have a distinct middle notch and trailing edges that are tinged dark brown, the coloration growing more pronounced with age.

THE SEA CANARY
The Beluga emits a wide variety of echolocation sounds, focused by its bulging melon.

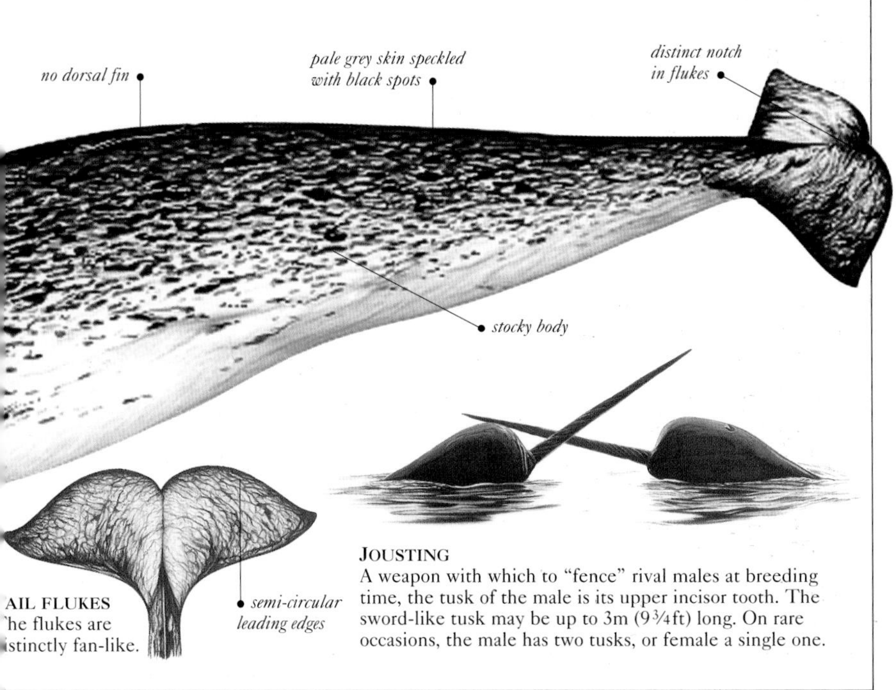

AIL FLUKES
'he flukes are istinctly fan-like.

JOUSTING
A weapon with which to "fence" rival males at breeding time, the tusk of the male is its upper incisor tooth. The sword-like tusk may be up to 3m (9¾ft) long. On rare occasions, the male has two tusks, or female a single one.

Family ZIPHIIDAE	Species *Hyperoodon ampullatus*	Status Lower risk

NORTHERN BOTTLENOSE WHALE

One of about 19 species of beaked whales, this cetacean has a steep, almost box-like forehead with a distinct beak protruding from below. Two tusks erupt from the tip of its lower jaw, which also has a pair of throat grooves. An erect dorsal fin is found about two-thirds down its streamlined body, whose colour varies from brown or orange to grey on the back. Despite their large size, these whales are capable of spectacular leaps, but they are mostly seen breathing at the surface between long dives to the seabed, where they suck up their prey.

• **SIZE** Length: 6–10 m (20–33 ft). Weight: Not recorded.

• **OCCURRENCE** N. Atlantic down to the Spanish coast. In open waters.

• **REMARK** This whale is now protected, although many are known to choke to death on plastic refuse ingested while sucking sediment from the seabed.

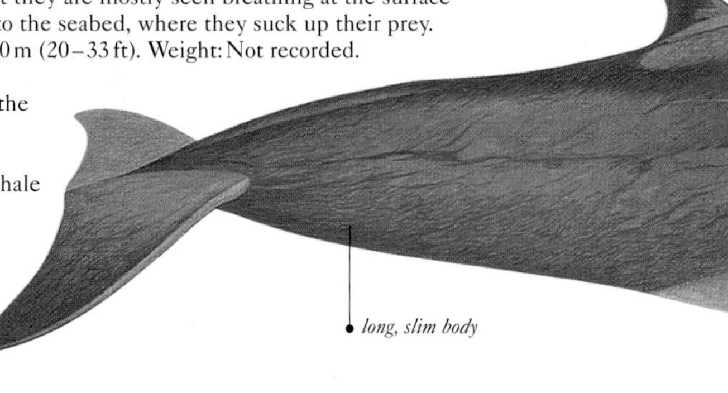

Social unit Social	Gestation 12 months	Young 1	Diet

Family ZIPHIIDAE	Species *Ziphius cavirostris*	Status Locally common

CUVIER'S BEAKED WHALE

The jawline of this whale curves up towards the tip of its snout and then down. Combined with the gradual slope of its relatively smooth forehead, this has given rise to an alternative name, Goosebeak. The grey-blue to tan body is covered with pale scars from shark wounds, and with teeth marks from other Cuvier's Whales. Its small flippers fit into indentations in the body, so that this whale is streamlined for fast swimming and for diving to great depths, using only its tail flukes. There are two throat grooves under its lower jaw and two conical teeth erupt from the lower jaw of the male, showing even when the mouth is closed. The female whales have no teeth. Older Cuvier's males tend to live alone or in small schools.

tan body coloration

broad tail flukes

• **SIZE** Length: 7–7.5 m (23–25 ft). Weight: 3–4 tonnes (3–3⅞ tons).

• **OCCURRENCE** Worldwide. In deep temperate and tropical oceans.

Social unit Variable	Gestation Not known	Young 1	Diet

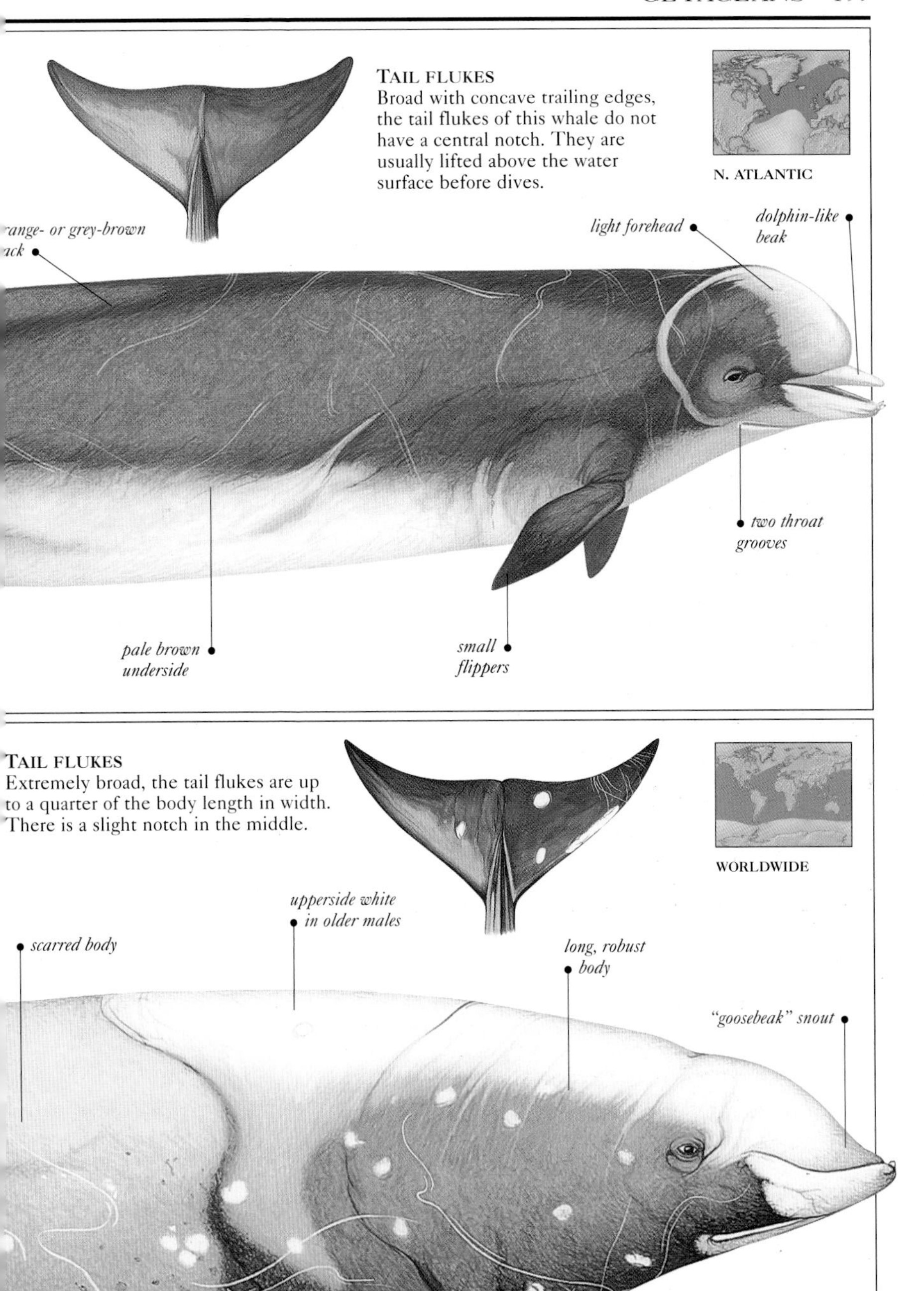

TAIL FLUKES
Broad with concave trailing edges, the tail flukes of this whale do not have a central notch. They are usually lifted above the water surface before dives.

TAIL FLUKES
Extremely broad, the tail flukes are up to a quarter of the body length in width. There is a slight notch in the middle.

Family ZIPHIIDAE	Species *Mesoplodon layardii*	Status Unconfirmed

STRAP-TOOTHED WHALE

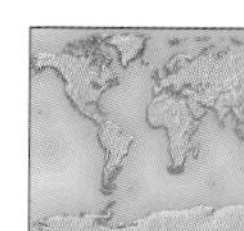
SOUTHERN OCEAN

One of the largest of the beaked whales, the male Strap-toothed Whale can be easily identified by a pair of extraordinary teeth that grow from its lower jaw, curling upwards and backwards over the top of its upper jaw – in adult whales they may grow to a length of 30 cm (12 in). These teeth do not erupt in female whales, while in juvenile males they are smaller and more triangular in shape. The Strap-toothed Whale has distinct black-and-white markings, with a black "face mask". It has a long, slender beak and slightly bulging melon, and a relatively small dorsal fin and flippers for its size. Rarely seen in the wild, and difficult to approach, especially in large vessels, this whale may sometimes be seen basking at the surface on calm, sunny days.

• **SIZE** Length: 5–6.2 m (16½–20¼ ft). Weight: 1–3 tonnes (1–3 tons).

• **OCCURRENCE** Chile, Argentina, Uruguay, Falkland Islands, Namibia, South Africa, Australia, Tasmania, New Zealand. In cold, temperate, offshore waters.

• **REMARK** The Strap-toothed Whale is the most commonly reported species of *Mesoplodon* in the Southern Hemisphere.

scarring on body

small, narrow flippers

Social unit Variable	Gestation Not known	Young 1	Diet

Family ZIPHIIDAE	Species *Mesoplodon bidens*	Status Unconfirmed

SOWERBY'S BEAKED WHALE

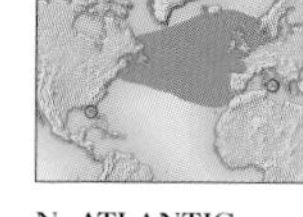
N. ATLANTIC

With one of the most northerly distributions of all the beaked whales, Sowerby's Beaked Whale is seldom sighted, in spite of being a commonly stranded species. Males are characterized by a pair of teeth mid-way on the beak, visible when the mouth is shut. The upper body of this whale is slate-grey or bluish grey, while the underside is lighter. The beak is long and slender, and the blowhole is situated behind the forehead bulge.

• **SIZE** Length: 4–5 m (13½–16½ ft). Weight: 1–1.3 tonnes (1–1.3 tons).

• **OCCURRENCE** E. and W. North Atlantic. In temperate and sub-Arctic waters.

• **REMARK** The first beaked whale to be discovered (off the Scottish coast in 1800), it was painted by the English artist, James Sowerby – hence its common name.

concave trailing edges

TAIL FLUKES
There is no notch between the tail flukes, which are dark on both sides.

teeth visible when mouth is shut

small, curved dorsal fin

relatively long flippers

belly may have pale spots

Social unit Solitary	Gestation Not known	Young 1	Diet

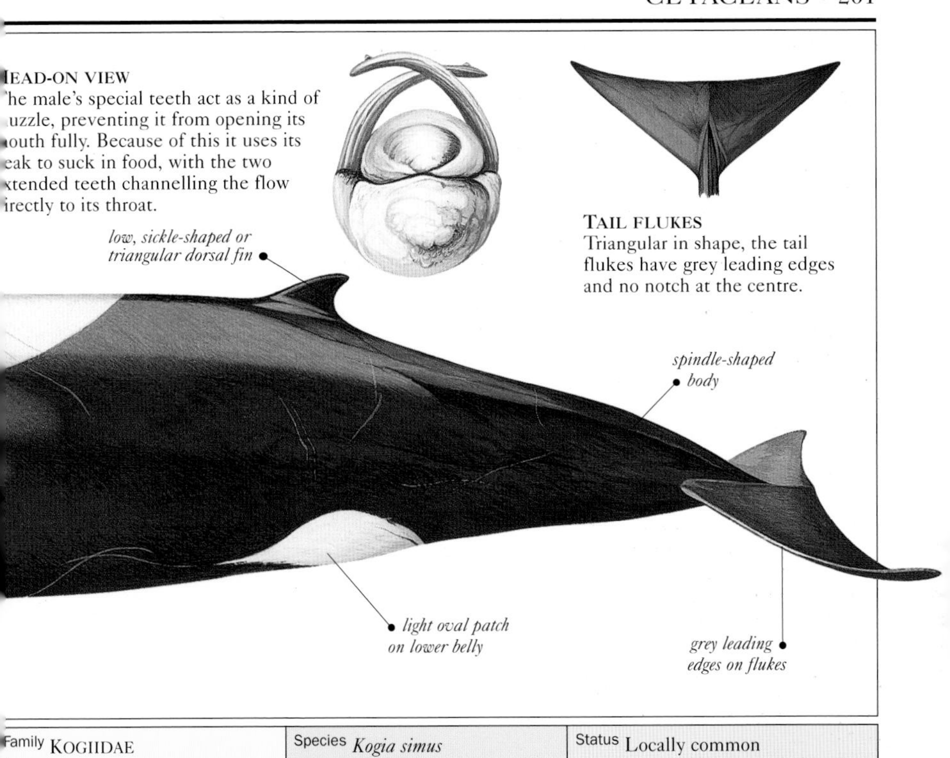

HEAD-ON VIEW
The male's special teeth act as a kind of muzzle, preventing it from opening its mouth fully. Because of this it uses its beak to suck in food, with the two extended teeth channelling the flow directly to its throat.

TAIL FLUKES
Triangular in shape, the tail flukes have grey leading edges and no notch at the centre.

Family KOGIIDAE	Species *Kogia simus*	Status Locally common

DWARF SPERM WHALE

The smallest of the three sperm whales, the Dwarf Sperm Whale has a blue-grey back, fin, flippers, and flukes, while its belly is creamish white. A distinctive pale crescent behind its eyes and mouth creates the false impression of a gill slit. Its lower jaw, with 7–13 pairs of sharp teeth, is slung in shark-like fashion under its large, bulbous head, while the upper jaw has only three pairs of teeth. This whale dives to 300 m (1,000 ft) for its prey. A shy animal, it lives alone or in small schools of fewer than ten, and releases a cloud of excreta to repel predators.

WORLDWIDE

- **SIZE** Length: Up to 2.7 m (8¾ ft). Weight: 135–270 kg (300–600 lb).
- **OCCURRENCE** Worldwide. Around the tropical and temperate continental shelf and adjoining coasts.
- **REMARK** This whale is prone to group strandings.

TAIL FLUKES
The broad flukes have concave edges and pointed tips.

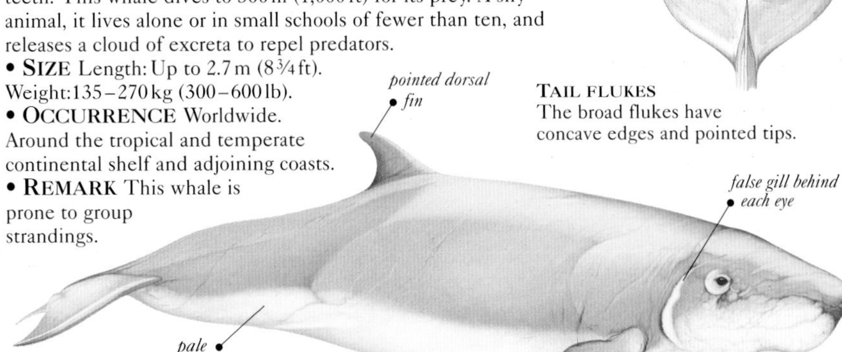

Social unit Variable	Gestation 9 months	Young 1	Diet

Family PHYSETERIDAE	Species *Physeter macrocephalus*	Status Vulnerable

SPERM WHALE

This whale is known for its huge, box-like head, a quarter to a third of its body size, in which is found the unique spermaceti organ that helps it make impressively deep dives. Between dives, it lies log-like on the surface of the water, exhaling a misty plume of air at 45° from its single blowhole. Dark grey or brown in colour, it has a paler underside and a cream patch on its narrow lower jaw. This contains 50 pairs of conical teeth; there are no visible upper teeth. Male whales are twice the weight of the females and migrate further towards the poles for summer feeding. They form loose bachelor pods (groups) when young, and become solitary with age. Females stay nearer to the tropics in mixed groups containing young and juveniles. The single calf is born in summer or autumn, with the next birth following in 3–15 years. Sperm Whales swim close, touching and caressing each other. They produce loud clicks and banging sounds, which may aid individual recognition.

- **SIZE** Length: 11–20 m (36–65 ft).
Weight: 20–57 tonnes (19½–56 tons).
- **OCCURRENCE** Worldwide. In deep waters, at and beyond the continental shelf.
- **REMARK** The Sperm Whale is the largest carnivore in the world.

HEAD-ON VIEW The Sperm Whale's head contains the spermaceti organ, a mass of waxy oil that works as a buoyancy aid during deep dives, and changes density according to the water pressure and temperature.

body colour varies from dark grey to light brown

head larger in male than female

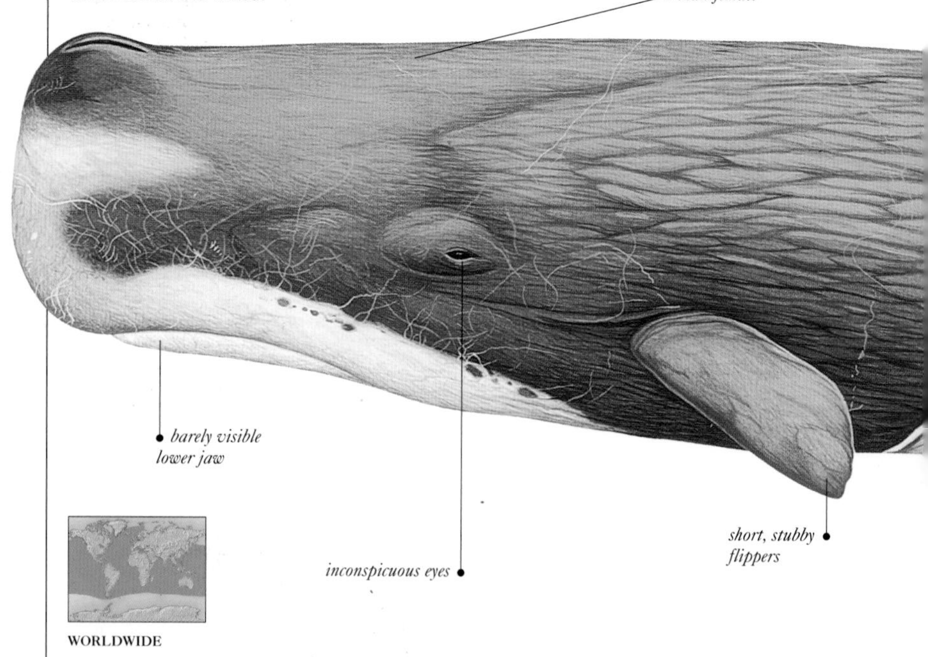

WORLDWIDE

Social unit Variable	Gestation 14–15½ months	Young 1	Diet

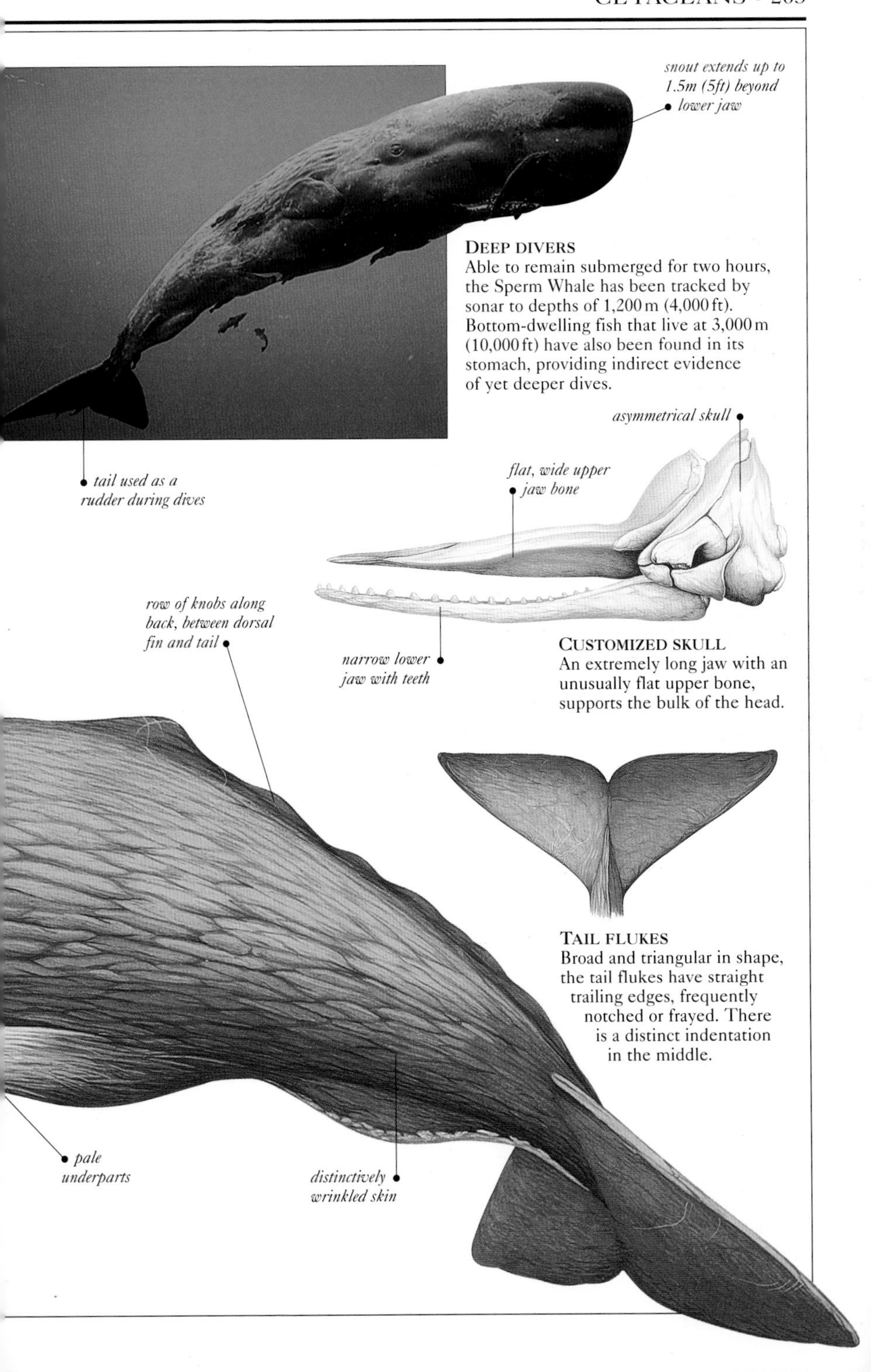

DEEP DIVERS
Able to remain submerged for two hours, the Sperm Whale has been tracked by sonar to depths of 1,200 m (4,000 ft). Bottom-dwelling fish that live at 3,000 m (10,000 ft) have also been found in its stomach, providing indirect evidence of yet deeper dives.

CUSTOMIZED SKULL
An extremely long jaw with an unusually flat upper bone, supports the bulk of the head.

TAIL FLUKES
Broad and triangular in shape, the tail flukes have straight trailing edges, frequently notched or frayed. There is a distinct indentation in the middle.

CETACEANS

BALEEN WHALES

THE ANATOMY OF the great whales differs from that of a typical mammal perhaps more than any species. They are gigantic, almost hairless, possess front limbs modified as flippers, and have lost their rear limbs. They also lack teeth. Instead, hanging from either side of the upper jaw is a row of plates made of baleen, or whalebone, which is not true bone, but a springy material similar to cartilage. These "curtains" are used to sieve small food items from sea water.

The 12 species of baleen whales include the Gray Whale, the Right and Bowhead Whales, and the Rorquals. Members of this last family vary from the Minke, about 10 m (33 ft) in length, to the Blue Whale, the biggest living animal, at up to 30 m (98 ft) long.

Family BALAENA	Species *Balaena mysticetus*	Status Endangered

BOWHEAD WHALE

The massive head of the Bowhead Whale accounts for about one-third of its body weight and makes up about 40 per cent of its length. The skin, unusually clear of barnacles and whale lice, is black in adults, with white around the chin and lower jaw, and at the base of the tail. The brownish or bluish black baleen plates are the longest of any whale and can be up to 4.6 m (15 ft) long. Each side of its strongly curved or "bowed" upper jaw has between 240 and 340 baleen plates. The Bowhead Whale has two blowholes and can blow up to 6 m (20 ft). It is known to produce intense low-frequency underwater sounds, which range from simple tonal moans to elephant trumpetings and screeching, and may last up to 7 seconds or more.

- **SIZE** Length: 14–18 m (45–59 ft). Weight: 50–60 tonnes (49½–59 tons).
- **OCCURRENCE** Northern Hemisphere, especially N. Canada, Alaska, and N. Russia. In Arctic and subarctic waters.
- **REMARK** Tens of thousands of Bowhead Whales were killed by early whalers, so that populations in the N. Atlantic are now perilously small. Current threats include oil exploration.

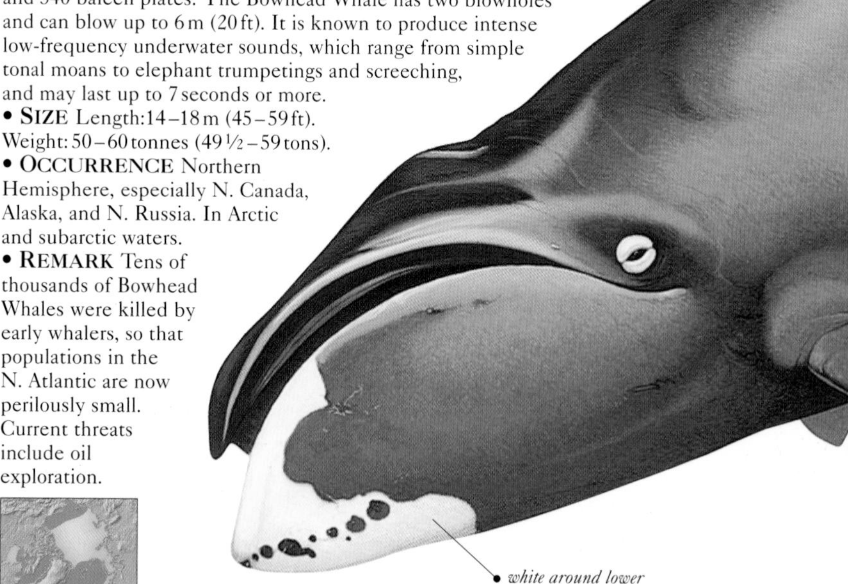

ARCTIC

Social unit Social	Gestation 12–14 months	Young 1	Diet

UNDERWATER NAVIGATORS
Since it inhabits polar waters, the Bowhead Whale spends part of the year in continuous night. Capable of navigating under ice in pitch dark, it can break through ice sheets more than 20 cm (8 in) thick. It uses echolocation to investigate the ice conditions ahead.

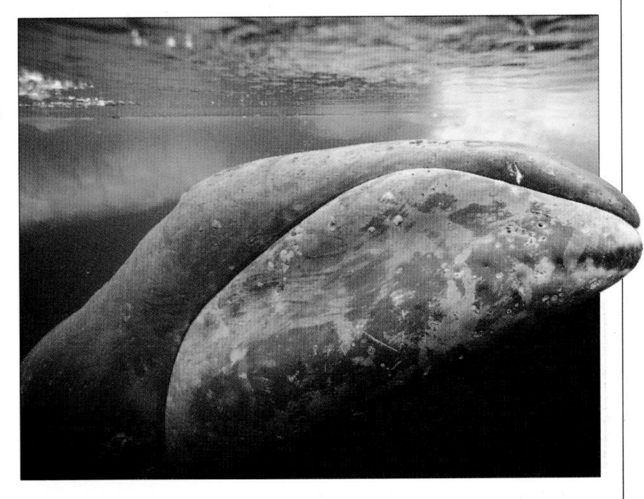

large, stocky body

pale patch on tail stock

very wide tail flukes

paddle-shaped flippers

twin blowholes

BIRD'S-EYE VIEW OF HEAD
Seen directly from above, the Bowhead Whale has a narrow rostrum (snout), with a white chin. The curvaceous mouthline cuts a distinctive arch through the head, and the two separated blowholes are at the centre.

slightly concave edges

TAIL FLUKES
The width of the flukes may be almost half the total body length. They have pointed tips and may be white on the upper trailing edges.

Family BALAENIDAE	Species *Eubalaena glacialis*	Status Endangered

NORTHERN RIGHT WHALE

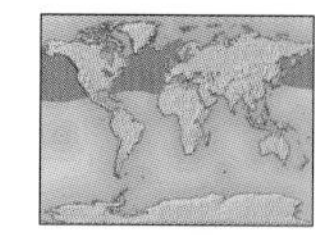

WORLDWIDE

Perhaps the most endangered of the large whales, the Northern Right Whale was so named by whalers since it has all the "right" features to make it the ideal catch – it lives and feeds close to the shore, is easy to approach, and is a valuable source of meat, blubber, and bone. Hunted indiscriminately in the nineteenth century to near extinction, the whale has been protected since 1937, but its numbers are recovering extremely slowly. Being a slow surface swimmer that dives only for a few minutes, it is susceptible to fatalities from ship collisions or being trapped in fishing equipment. This whale's body is black with scattered white patches. Similar in shape to the Bowhead Whale (see pp. 204–205), the Northern Right Whale has a head that is up to a quarter of its body length, with a downward-curving jawline and narrow baleen plates up to 3 m (9¾ ft) long. The head is encrusted with fibrous growths, or callosities, which are covered with whale lice that often add a pink, yellow, or orange tinge, and its twin blowholes create a bushy, V-shaped blow. Found singly or in small groups, it migrates to the far north or south in summer, returning to warmer mid-latitude waters in winter to breed. Despite being a sluggish swimmer, it is surprisingly acrobatic, and may be seen flipper-slapping, tail-slapping, breaching, or even doing a "headstand" by raising its tail flukes high above the surface of the water, almost at a right angle.

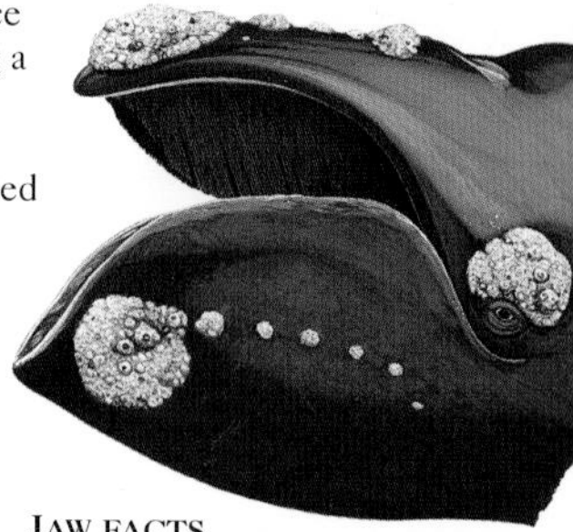

JAW FACTS
The baleen plates of this whale are exceptionally long and narrow. Densely fringed with fine bristles, they number 200–270 on each side of the upper jaw. The deep, bowed lower jaw is used for closing the mouth.

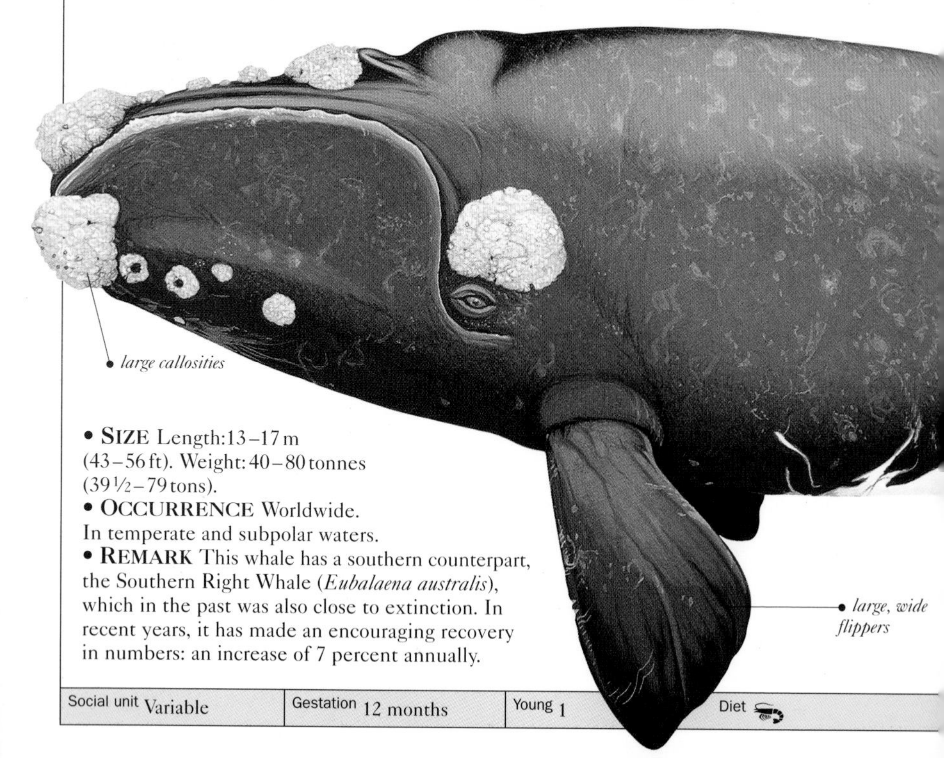

• **SIZE** Length: 13–17 m (43–56 ft). Weight: 40–80 tonnes (39½–79 tons).
• **OCCURRENCE** Worldwide. In temperate and subpolar waters.
• **REMARK** This whale has a southern counterpart, the Southern Right Whale (*Eubalaena australis*), which in the past was also close to extinction. In recent years, it has made an encouraging recovery in numbers: an increase of 7 percent annually.

Social unit Variable	Gestation 12 months	Young 1	Diet

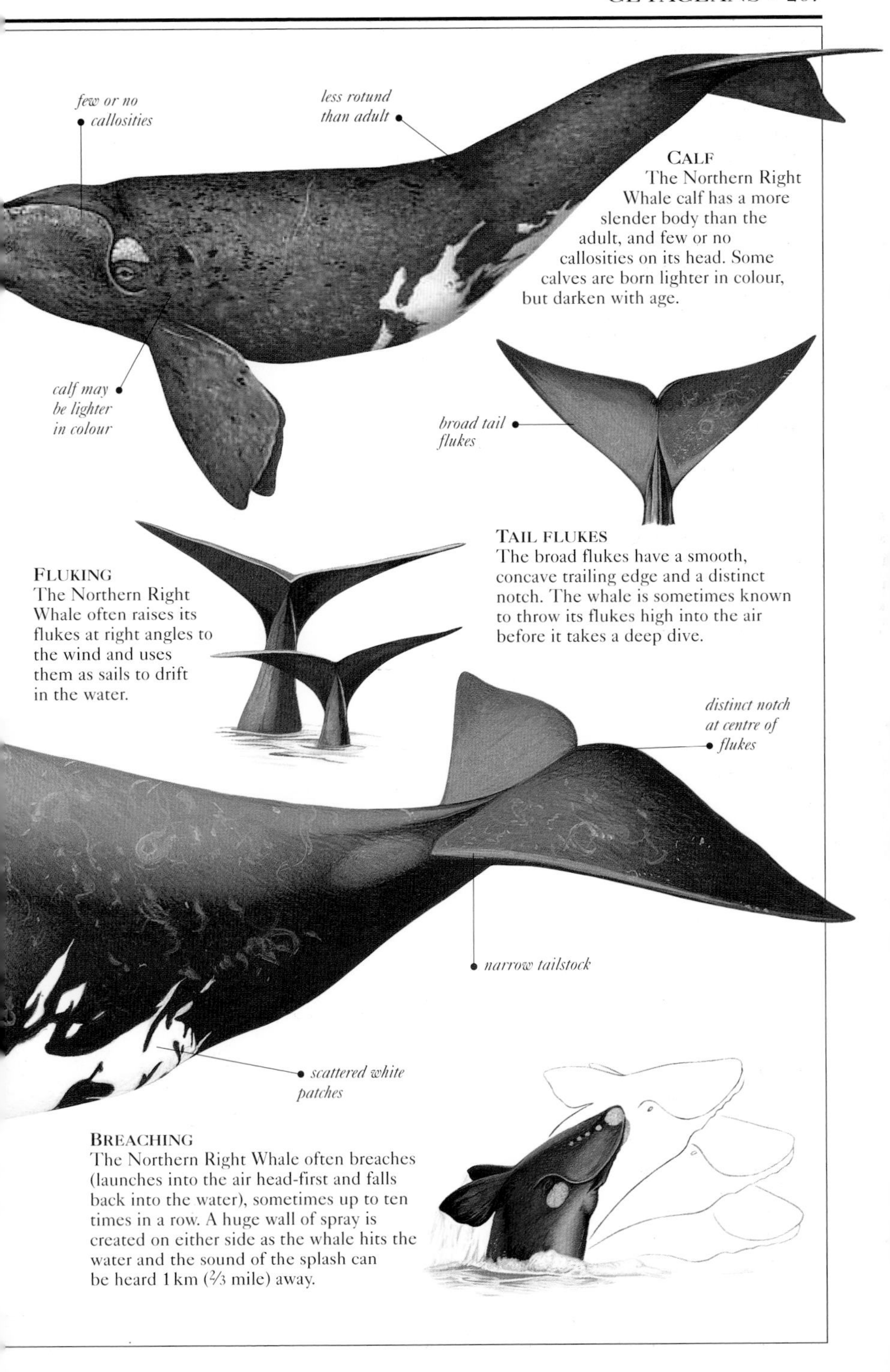

CALF
The Northern Right Whale calf has a more slender body than the adult, and few or no callosities on its head. Some calves are born lighter in colour, but darken with age.

TAIL FLUKES
The broad flukes have a smooth, concave trailing edge and a distinct notch. The whale is sometimes known to throw its flukes high into the air before it takes a deep dive.

FLUKING
The Northern Right Whale often raises its flukes at right angles to the wind and uses them as sails to drift in the water.

BREACHING
The Northern Right Whale often breaches (launches into the air head-first and falls back into the water), sometimes up to ten times in a row. A huge wall of spray is created on either side as the whale hits the water and the sound of the splash can be heard 1 km (2/3 mile) away.

Family ESCHRICHTIIDAE	Species *Eschrichtius robustus*	Status Endangered

GRAY WHALE

N. PACIFIC

Among the most active of all large whales, the Gray Whale makes one of the longest migrations of any mammal. Groups of up to ten travel to the Arctic for summer feeding, returning south to warm-water lagoons to rest and produce calves in winter. When migrating, this whale typically blows 3–6 times before diving for 3–5 minutes. It enjoys surf-riding, and is frequently found in surf (especially in Baja California) in very shallow water, sometimes lying on its side and waving a flipper in the air. Apart from filter-feeding, it also dives to the shallow seabed, scoops up huge mouthfuls of mud, and filters it for worms, starfish, shrimps, and other small creatures, making it a unique bottom-feeder among whales. Feeding near the shore, it is accessible to watchers, especially in the east Pacific. The Gray Whale swims in a coordinated fashion with other whales, staying in line or arching out of the water together. Like many other baleen whales, it spyhops, or raises its head vertically out of the water. Its mottled grey skin is encrusted with barnacles and sea lice. This whale has no dorsal fin, but a series of eight or nine bumps runs along the last third of its dorsal ridge.

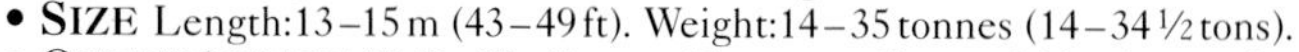

• **SIZE** Length: 13–15 m (43–49 ft). Weight: 14–35 tonnes (14–34½ tons).
• **OCCURRENCE** N. Pacific Ocean. Along coastlines, within the continental shelf, up to the surf zone; breeding in sheltered lagoons in the tropics.
• **REMARK** Atlantic stocks were wiped out by excessive whaling before this whale was declared a protected species in 1946. The Californian population has since made a remarkable recovery. However, Siberian and Alaskan Inuit people are still allowed to kill a certain number every year.

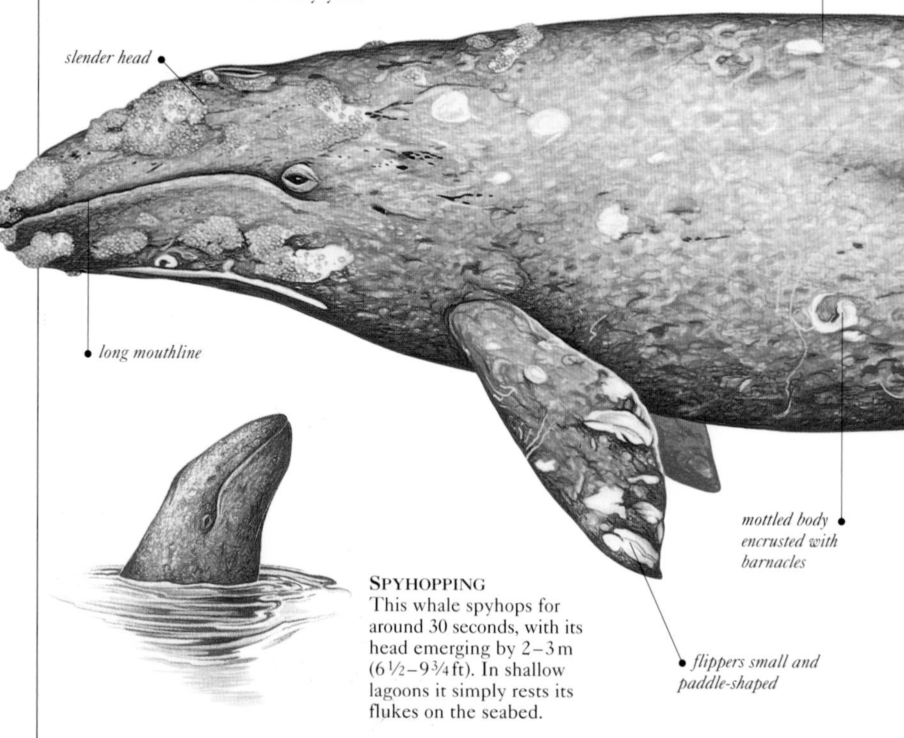

SPYHOPPING
This whale spyhops for around 30 seconds, with its head emerging by 2–3 m (6½–9¾ ft). In shallow lagoons it simply rests its flukes on the seabed.

Social unit Social	Gestation 13½ months	Young 1	Diet

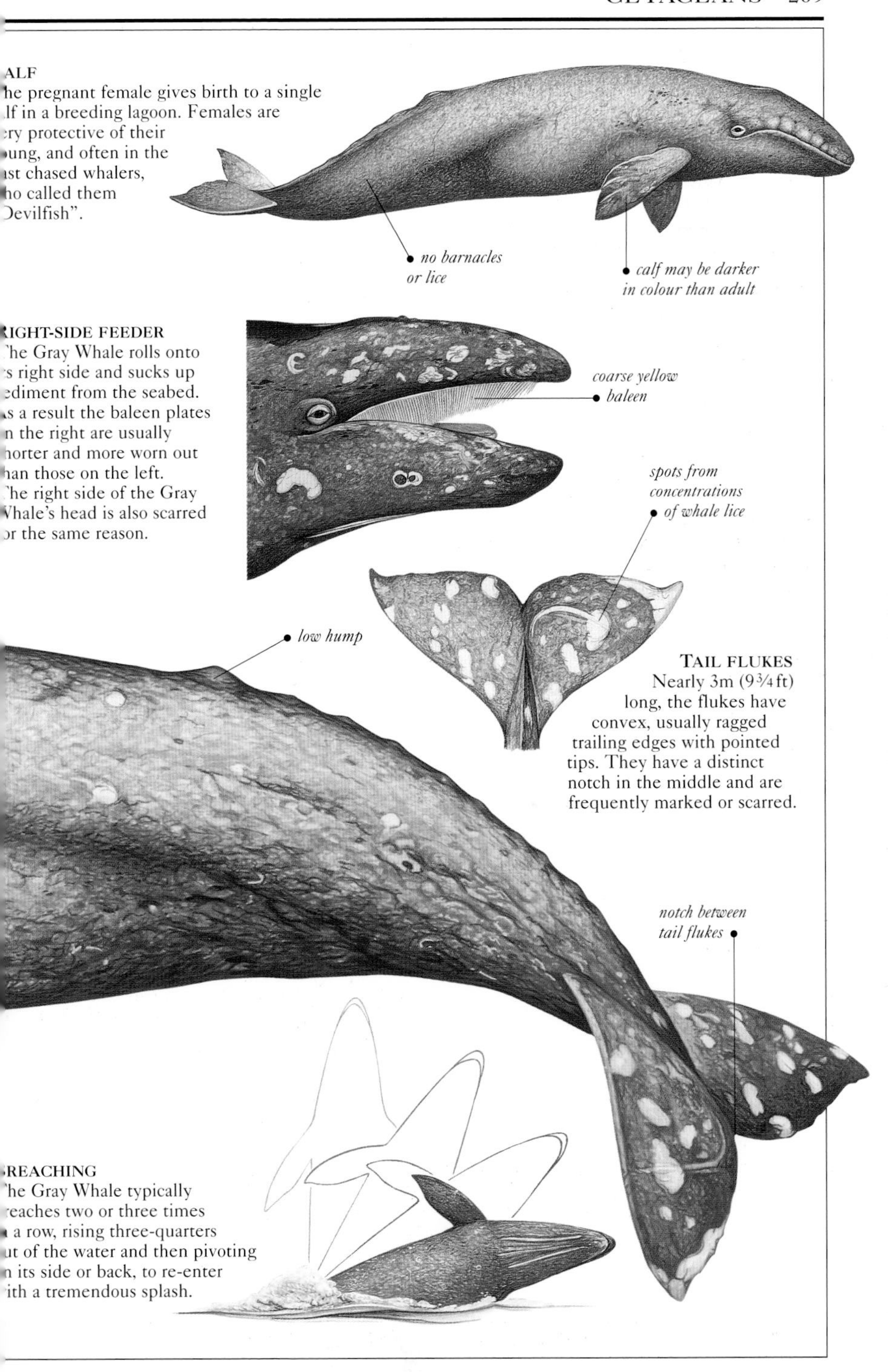

ALF
he pregnant female gives birth to a single
lf in a breeding lagoon. Females are
ry protective of their
ung, and often in the
st chased whalers,
ho called them
)evilfish".

IGHT-SIDE FEEDER
he Gray Whale rolls onto
s right side and sucks up
diment from the seabed.
s a result the baleen plates
n the right are usually
horter and more worn out
han those on the left.
he right side of the Gray
Vhale's head is also scarred
r the same reason.

TAIL FLUKES
Nearly 3m (9¾ft) long, the flukes have convex, usually ragged trailing edges with pointed tips. They have a distinct notch in the middle and are frequently marked or scarred.

REACHING
he Gray Whale typically
reaches two or three times
a row, rising three-quarters
ut of the water and then pivoting
n its side or back, to re-enter
ith a tremendous splash.

Family BALAENOPTERIDAE	Species *Balaenoptera physalus*	Status Endangered

FIN WHALE

The second largest whale and one of the fastest, this species can sometimes reach a speed of 30 kph (19 mph). The Fin Whale has a grey back, flippers, and tail flukes, and its dorsal fin, set about two-thirds of the way along the spine, has a concave rear edge. A distinct ridge runs from its dorsal fin to the flukes, which accounts for an alternative name: Razorback. This whale is always found swimming on its right side, the left side of its mouth is black, and the right side is white. Such asymmetrical coloration is highly unusual in mammals. Like other great whales, it undertakes long migrations from high latitudes in summer towards the tropical regions for winter, when it breeds. In addition to hums and squeals, it produces an immensely deep, loud moan that can be heard hundreds of kilometres away.

• **SIZE** Length: 19–22 m (62–72 ft). Weight: 45–75 tonnes (44½–74 tons).

• **OCCURRENCE** Worldwide, except in the Mediterranean, Baltic, and Red Seas, and Arabian Gulf. In open oceans, particularly in areas of seasonal plankton.

• **REMARK** Heavily exploited by industrial whaling, particularly in the Southern Hemisphere where around three-quarters of its population has been wiped out, the Fin Whale is currently protected. However, there is a growing concern that low-frequency sound produced by military equipment and shipping activities may disrupt the normal communication patterns of this whale reducing, its ability to navigate or find mates.

BREACHING
When breaching, the Fin Whale leaves the water at an angle, usually with its tail-end submerge As its body begins to fall, it may twist in mid-air, re-entering the water with a loud splash, typically landing on its belly, and less commonly on its back.

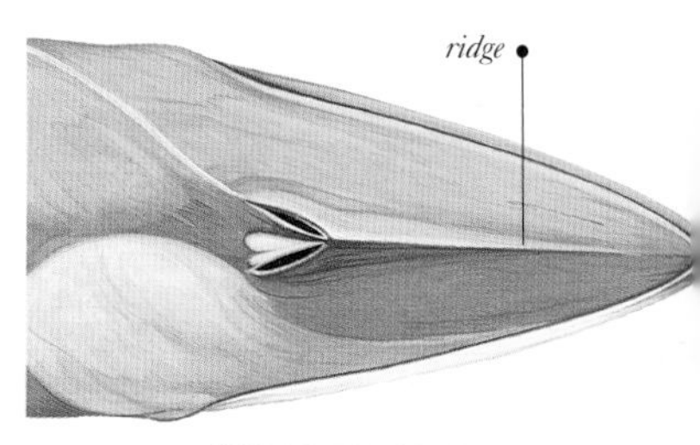

RIDGE ON HEAD
The head of this whale is betwee one-fifth and one-quarter of its to body length. It usually has a singl longitudinal ridge along its rostrur a few whales having two extra rid

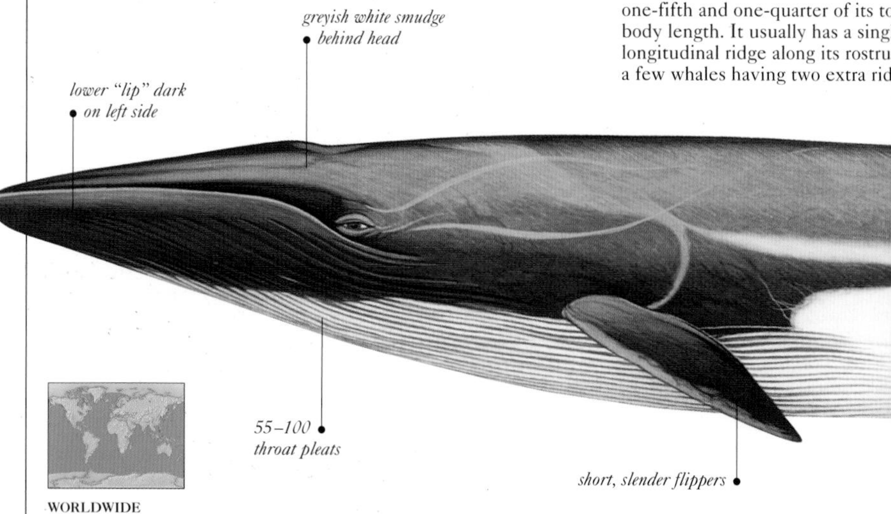

WORLDWIDE

Social unit Variable	Gestation 11 months	Young 1	Diet

:IGHT-SIDED FEEDING
,wimming only on its right de, the Fin Whale lunges : high-speed at krill and fish ıch as capelin or herring. ynchronizing its attacks ith those of other Fin /hales, it takes in huge olumes of water, then closes s mouth and forces out the ater to trap fish in its aleen plates. The baleen lates on the right side are hite in up to a quarter or ıird of the mouth (the left de is dark grey), and when wimming below the water's ırface, the white right "lip" often clearly visible.

broad, flat rostrum, flatter than in Blue Whale

TAIL FLUKES
The Fin Whale rarely shows its broad, slightly triangular flukes. These have a distinct notch in the middle, and the slightly concave and trailing edges may be notched or frayed. The undersides of the flukes are white.

ASYMMETRICAL HEAD
The uneven coloration of the 'in Whale's head can be extensive, vith the white patch on the right xtending anywhere from the ıpper "lip" to the neck.

dorsal fin slopes backwards

distinct ridge from dorsal fin to flukes

thick tailstock

white underside

Family BALAENOPTERIDAE	Species *Balaenoptera musculus*	Status Endangered

BLUE WHALE

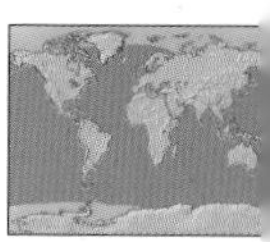

The largest animal in the world, the Blue Whale is surprisingly streamlined, with its pointed head, long body, slender flippers, and large, narrow tail flukes. It is mainly grey-blue in colour, but the skin on the belly may be tinted yellow-brown by algae. Its lower jaw has 55–68 grooves running up to the navel, making the skin extremely distensible. The Blue Whale engulfs shoals of krill, its exclusive diet, with the throat swelling to four times its normal size. In a day it can consume around 6 tonnes (5 7/8 tons) of food. Feeding occurs mainly in summer around the krill-rich polar waters. The Blue Whale is known to migrate south to warmer latitudes in winter, when the female gives birth. Usually found alone, or in mother–calf pairs, this whale may gather in loose groups to feed. Its grunts, hums, and moans, sometimes at volumes over 180 decibels, are the loudest of any animal sounds, and can be heard by other whales 1,000 km (625 miles) away. It blows every 10–20 seconds for a total of 2–6 minutes and then dives for 5–20 minutes (although it is able to stay underwater longer). Adult whales rarely breach clear of the water, but younger animals have been observed breaching at an angle of 45°, landing on their stomach or side.

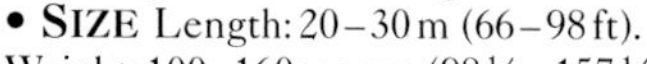

- **SIZE** Length: 20–30 m (66–98 ft). Weight: 100–160 tonnes (98 1/2–157 1/2 tons).
- **OCCURRENCE** Worldwide, except the Mediterranean, Baltic, and Red Seas, and Arabian Gulf. In deep, open, mainly cold waters.
- **REMARK** The Blue Whale was hunted close to extinction by whalers, and some populations may never recover.

pale grey or white mottling, especially behind head

broad, flat head

long, slender flippers

pale patch on underside

DISTINCTIVE HEAD

The head of the Blue Whale is about a quarter of its body length and broad and flat compared to that of other rorquals. Basically U-shaped, it is often likened to a Gothic arch and has a single vertical ridge along the top of the rostrum, and a fleshy splashguard.

black baleen plates

throat grooves

two distinct blowholes

splashguard

vertical ridge along centre

Social unit Solitary	Gestation 11 months	Young 1	Diet

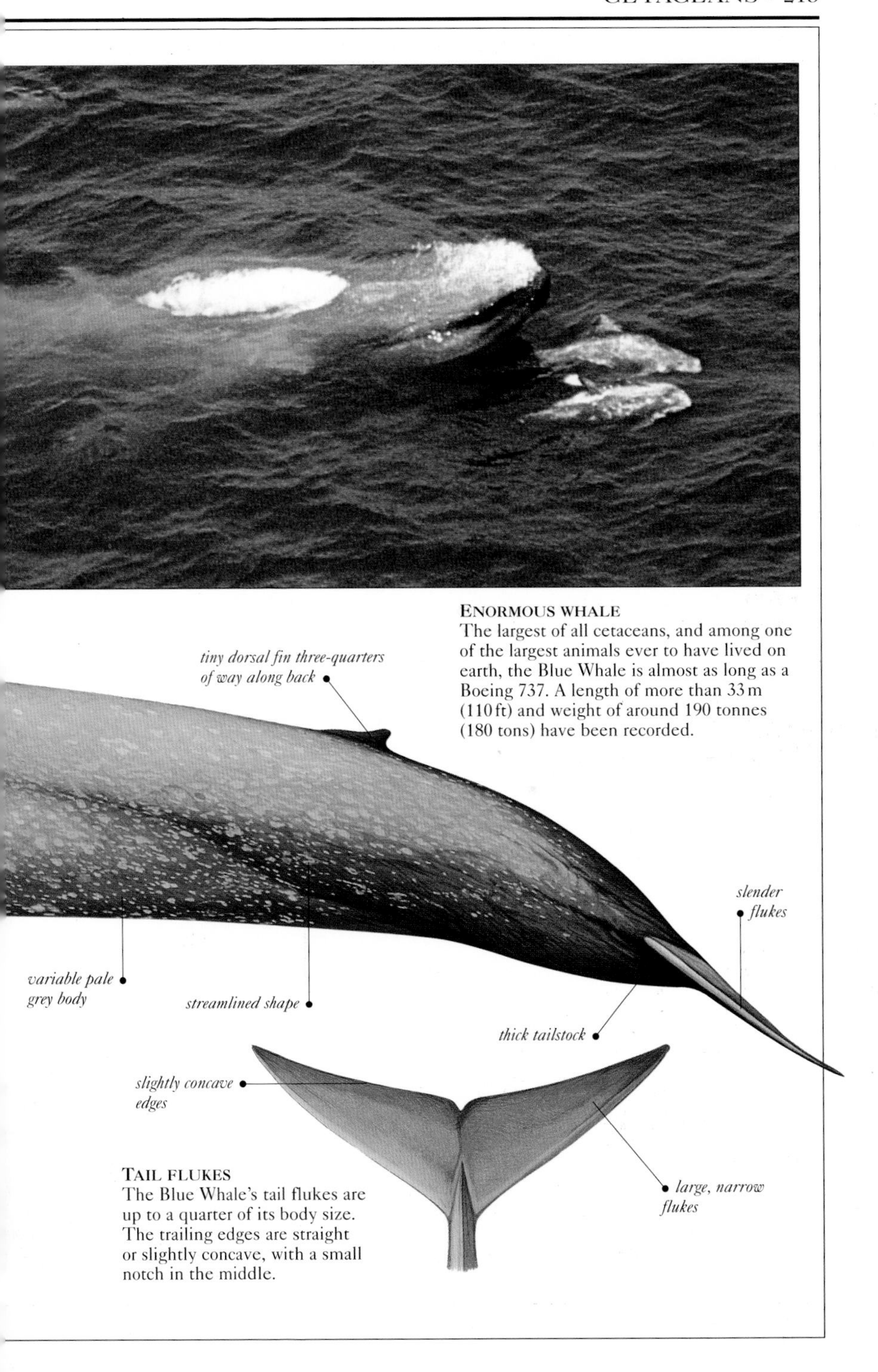

ENORMOUS WHALE
The largest of all cetaceans, and among one of the largest animals ever to have lived on earth, the Blue Whale is almost as long as a Boeing 737. A length of more than 33 m (110 ft) and weight of around 190 tonnes (180 tons) have been recorded.

TAIL FLUKES
The Blue Whale's tail flukes are up to a quarter of its body size. The trailing edges are straight or slightly concave, with a small notch in the middle.

Family BALAENOPTERIDAE	Species *Balaenoptera acutorostrata*	Status Lower risk

MINKE WHALE

Resembling a dolphin in shape, this species is the smallest of the baleen whales. Its back is black, and its belly, lower jaw, and throat are white, with smoky grey patterns called chevrons blurring the border between them. Its sickle-shaped dorsal fin is large compared to that of other baleen whales. The baleen plates are up to 30 cm (12 in) long, creamish in front and grey at the rear. A fast and agile swimmer, it sometimes makes spectacular leaps from the water. It is generally solitary, but is sometimes found feeding with other whales. It also hunts where other predators such as birds and fish are active at a particularly dense supply of food. This whale is not shy and often approaches stationary boats. It communicates with grunts, clicks, ratchets, and a variety of other sounds.

- **SIZE** Length: 8–10 m (26–33 ft). Weight: 8–13 tonnes (7 ½–13 tons).
- **OCCURRENCE** Worldwide (except E. Mediterranean). In oceans and along coastlines.
- **REMARK** A smaller form of the Minke Whale has recently been recognized as a separate species – the Antarctic Minke Whale (*Balaenoptera bonaerensis*).

relatively lar dorsal fin

white belly

Social unit Solitary	Gestation 10 months	Young 1	Diet

Family BALAENOPTERIDAE	Species *Megaptera novaeangliae*	Status Vulnerable

HUMPBACK WHALE

This highly vocal whale is recognized by its rather dumpy shape and huge pectoral fins. These are normally used to channel prey towards its mouth, but aggressive whales can beat adversaries to death with them. The upper and lower jaws are lined with tubercles that are often encrusted with barnacles, and there is a slight ridge from the blowholes to the tip of the snout. The back is black or blue-black, with paler patches below. These colour patterns vary from whale to whale and can be used to identify individuals. The dorsal fin, located on a pad of fat, may be triangular to sharply hooked. Humpback Whales migrate from warm breeding grounds in winter to fertile feeding grounds, at higher latitudes, in summer. They specialize in rounding up prey by blowing underwater "bubble curtains", with sometimes a dozen or more whales hunting together.

triangular dorsal fin

ridge from dorsal fin to tail

WORLDWIDE

- **SIZE** Length: 13–14 m (43–46 ft). Weight: 25–30 tonnes (24 ⅝–29 ½ tons).
- **OCCURRENCE** Worldwide. Along coasts or in deep seas.
- **REMARK** This species is known to sing for over 22 hours. Each male whale has its signature tune, which it changes over the years. It may be used to attract a mate or ward off rivals.

Social unit Solitary	Gestation 11 months	Young 1	Diet

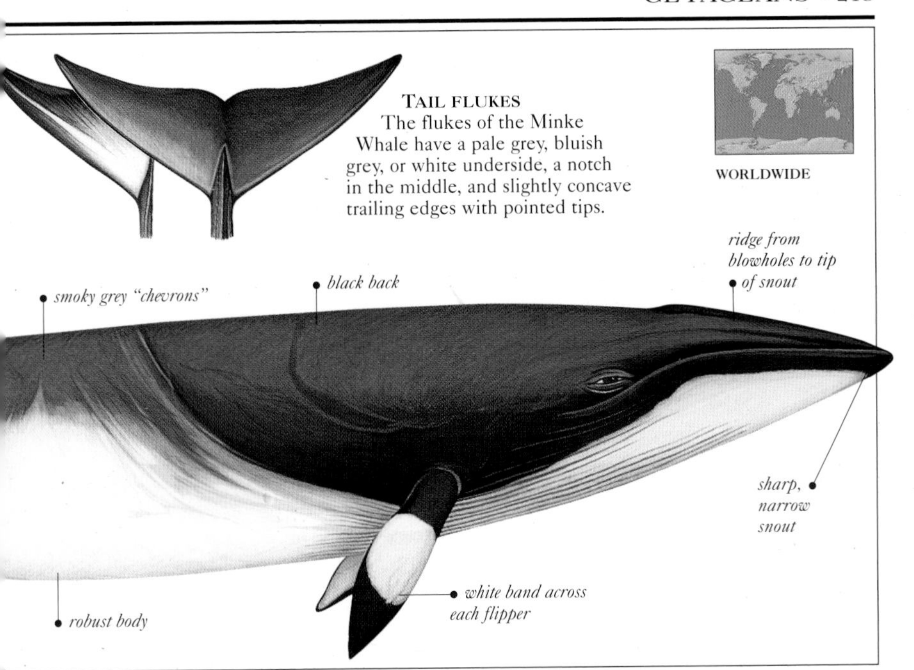

TAIL FLUKES
The flukes of the Minke Whale have a pale grey, bluish grey, or white underside, a notch in the middle, and slightly concave trailing edges with pointed tips.

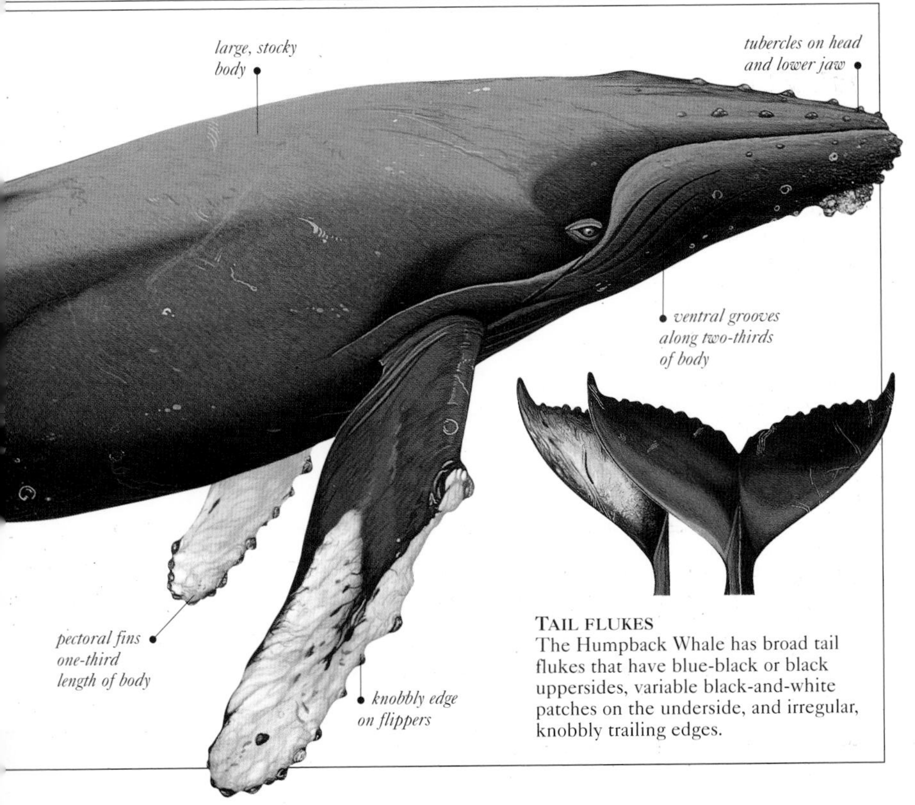

TAIL FLUKES
The Humpback Whale has broad tail flukes that have blue-black or black uppersides, variable black-and-white patches on the underside, and irregular, knobbly trailing edges.

CARNIVORES

DOGS AND FOXES

TWO OF THE most familiar carnivores, the Grey Wolf (the ancestor of the domestic dog) and the Red Fox, belong to the Canid or dog family. So do another 34 species among them dingoes, coyotes, and jackals.

Canids are typified by a muscular but slender body, powerful legs, a long and bushy tail, a long muzzle and sensitive nose, large ears for keen hearing, and strong, large-toothed jaws.

Found worldwide, generally in ope habitats, they have an opportunisti approach to diet. Smaller specie especially foxes, feed on insects an rodents, and live alone or in pair Larger species like wolves and Africa wild dogs live and hunt in packs.

Several canids have faced persecutio over the ages, and a few, such as th Ethiopian and red wolves, are now o the critical list.

Family CANIDAE	Species *Vulpes zerda*	Status Unconfirmed

FENNEC FOX

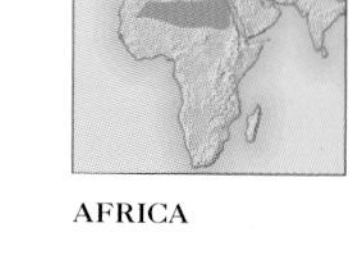

The smallest of all foxes, the Fennec is readily identified by its huge ears. It has a cream to yellowish coat with white underparts and a black-tipped tail. The soles of its feet are furry for protection when walking on hot, soft sand. Usually nocturnal, this omnivorous fox has a wide-ranging diet which includes fruits, seeds, small rodents, birds, eggs, reptiles, and insects. It is unusual in associating in groups of up to ten, but the relationships between the members are not clear. Each member digs a den several metres deep, the male marking his territory with urine and becoming aggressive during the breeding season. Mating takes place in mid- and late winter, this fox often mating again if the litter is lost. The female alone defends the nest site, and the cubs remain in the den for two months, protected by the mother. During this time the male does not enter the den.

- **SIZE** Body length: 24–41 cm (9½–16 in). Tail: 18–31 cm (7–12 in).
- **OCCURRENCE** Throughout the Sahara. In sandy desert.
- **REMARK** The Fennec Fox is trapped and sold as a pet, as well as extensively hunted for its pelt.

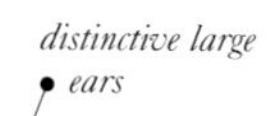

Social unit Social	Gestation 50–52 days	Young 2–5	Diet

Family CANIDAE	Species *Vulpes cana*	Status Locally common

BLANFORD'S FOX

Cat-like in its appearance and movements, this small fox has a mottled coat that is black, grey, and white, dark hindlegs, and almost white underparts. This species is also characterized by its large ears and a bushy tail which often has a dark tip. Wholly nocturnal, it eats more fruits than other foxes and is often found near orchards.

- **SIZE** Body length: 42 cm (16½ in). Tail: 30 cm (12 in).
- **OCCURRENCE** W. and S. Asia. In grassland and mountains.
- **REMARK** This fox is hunted heavily for its skin.

Social unit Solitary	Gestation 50–60 days	Young 1–3	Diet

Family CANIDAE	Species *Vulpes velox*	Status Unconfirmed

SWIFT FOX

This fox is closely related to the Kit Fox (*Vulpes macrotis*) from which it can be distinguished by its widely spaced ears, rounded, dog-like head, and more graceful form. Its coat is greyish on the head, back, and flanks, and its long, bushy tail has a black tip and is tinged redder in summer. Usually nocturnal, it may lie in the sun by its den during the day.

- **SIZE** Body length: 38–53 cm (12½–21 in). Tail: 18–26 cm (7–10 in).
- **OCCURRENCE** C. North America. In prairie and grassland.

widely spaced ears

greyish red coat

black-tipped tail

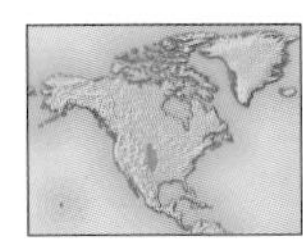

Social unit Pair	Gestation 50–60 days	Young 3–6	Diet 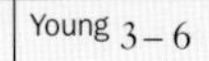 

Family CANIDAE	Species *Vulpes vulpes*	Status Common

RED FOX

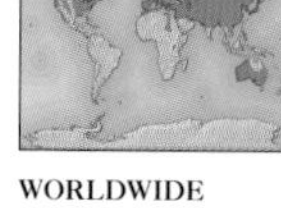

This fox has a coat that varies from grey and rust to flame-red, and a large, bushy, white-tipped tail. The backs of its ears are usually black and, often, so are the lower limbs and feet. However, the Red Fox has been bred extensively for its pelt and several colour morphs, including completely black or white, have been produced. This stealthy hunter is active by day and night, and feeds chiefly on rabbits and young hares in grassy or farmed areas. It stalks its prey and then makes a dash for the victim before it can reach its burrow or speed away. The fox then carries the prey by the neck to a secluded spot where it eats at leisure. The Red Fox also feeds on a variety of other food, including carrion and refuse. It hoards food by burying it, using its keen memory to locate it afterwards. It shelters in an earthen den in rabbit burrows, crevices, or in outbuildings. The primary social unit consists of a vixen (female) and a dog (male), who sometimes share their territory with non-breeding kin. Mating takes place in late winter or early spring, when the female makes eerie shrieks (known as the "vixen's scream").

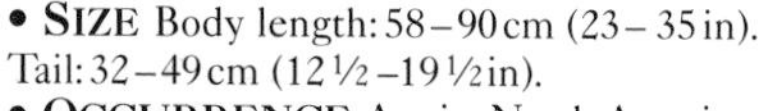

- **SIZE** Body length: 58–90 cm (23–35 in). Tail: 32–49 cm (12 ½–19 ½ in).
- **OCCURRENCE** Arctic, North America, Europe, W. Asia, N. Africa, and Australia. In desert, forest, mountains, tundra, and urban areas.
- **REMARK** Trapped extensively for its fur, this fox has also been killed in enormous numbers during rabies control exercises in North America and Europe.

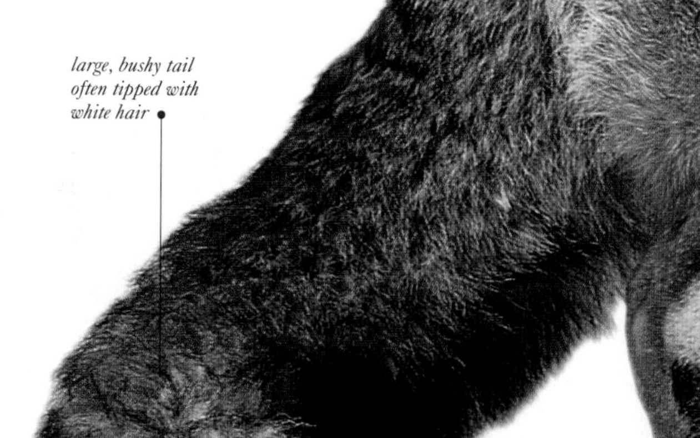

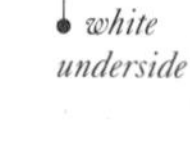

Social unit Pair	Gestation 49–55 days	Young 3–12	Diet

PINNING DOWN PREY

The Red Fox often hunts in dense vegetation for rodents and earthworms. Using its sharp hearing it first pinpoints the location of the prey. Then making a high vertical leap, the fox lands on its frontpaws, and catches its victim by pinning it down.

pointed ears black on upperside

pointed muzzle

EARLY DAYS

Red Fox cubs are usually cared for by both the parents and "helper" or non-breeding females. They remain in or near the den for the first three months, when they are vulnerable to predators.

Family CANIDAE	Species *Vulpes rueppelli*	Status Unconfirmed

RUEPPELL'S FOX

AFRICA, ASIA

Sometimes called the Sand Fox, this species is slighter in build than the Red Fox (see pp. 218–219) and has a soft, dense, sandy or silver-grey coat to blend in with its arid habitat. It has black patches on its face, wide ears, short legs, and a conspicuously white tail-tip. In some regions (such as Oman), it appears to live in territorial pairs, but usually this gregarious fox is found in what are probably extended family groups of up to 15 individuals. During the day, Rueppell's Fox shelters in crevices and burrows, changing its den every few days.

- **SIZE** Body length: 40–52 cm (16–20 ½ in). Tail: 25–39 cm (10–15 ½ in).
- **OCCURRENCE** N. Africa and W. Asia. In stony or sandy desert.
- **REMARK** Human encroachment into its habitat is a threat to this fox, which is also killed by Bedouins for food.

Social unit Social/Pair	Gestation Not known	Young 2–3	Diet

Family CANIDAE	Species *Alopex lagopus*	Status Common

ARCTIC FOX

This fox is unusual in having two colour "phases" or types. Those of the "white" phase become almost completely white in winter for camouflage in snow, and grey or brown in summer to blend into plains and grassy hillocks. Foxes of the "blue" phase, found more often in coastal or shrubby habitats, are grey-brown with a blue tinge in winter, and dark brown in summer. Adapted to the freezing cold, this fox has small ears, a blunt muzzle, and short legs and tail, since these areas lose heat fastest.

- **SIZE** Body length: 53–55 cm (21–22 in). Tail: 30 cm (12 in).
- **OCCURRENCE** Alaska, N. Canada, Greenland, N. Europe, and N. Asia. In Arctic tundra and coastal habitats.
- **REMARK** Human persecution, hunting for fur, and land development leading to habitat destruction are serious threats.

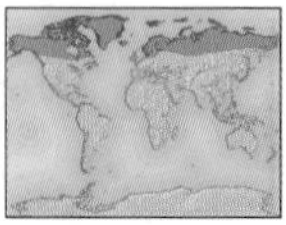

N. AMERICA, GREENLAND & EURASIA

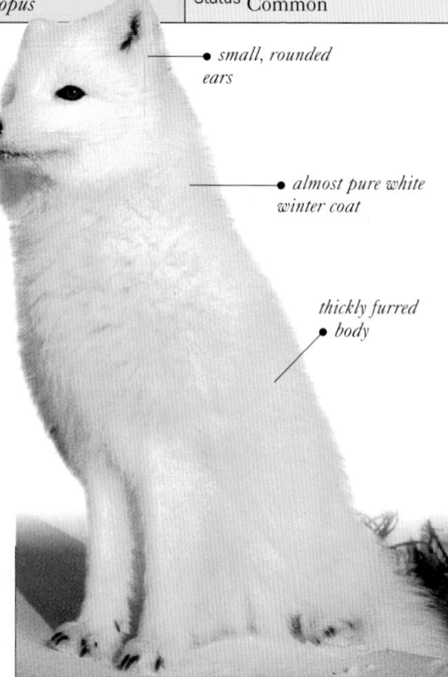

Social unit Social	Gestation 51–54 days	Young 6–16	Diet

Family CANIDAE	Species *Urocyon cinereoargenteus*	Status Common

Grey Fox

Also known as the Tree Fox, this species is often found in woodland, climbing branches and leaping up tree trunks like a cat. This nocturnal animal eats a variety of insects and small mammals, but may seasonally feed more on fruits and seeds. It rarely digs its own den but usually makes it in a tree-hole, crevices in rocks and logs, or on a building ledge or roof of a house. Most Grey Foxes form breeding pairs. The newborn cubs, usually four in a litter, are helpless and blind. In about 9–12 days, their eyes open, and within four weeks the cubs venture out of the den and begin to climb, accompanied by a parent. The young disperse in the first year and have been known to wander up to 85 km (50 miles).

• **Size** Body length: 53–81 cm (21–32 in). Tail: 27–44 cm (10½–17½ in).

• **Occurrence** S. Canada to N. South America. In temperate woodland and deciduous forest, abandoned oil fields, and urban areas.

Grizzled grey
The Grey Fox's coat owes its mottled appearance to individual hairs that are banded white, grey, and black. It has a small, dark grey mane, and parts of its neck, flanks, and legs are reddish, whereas its chin and belly are white or buff.

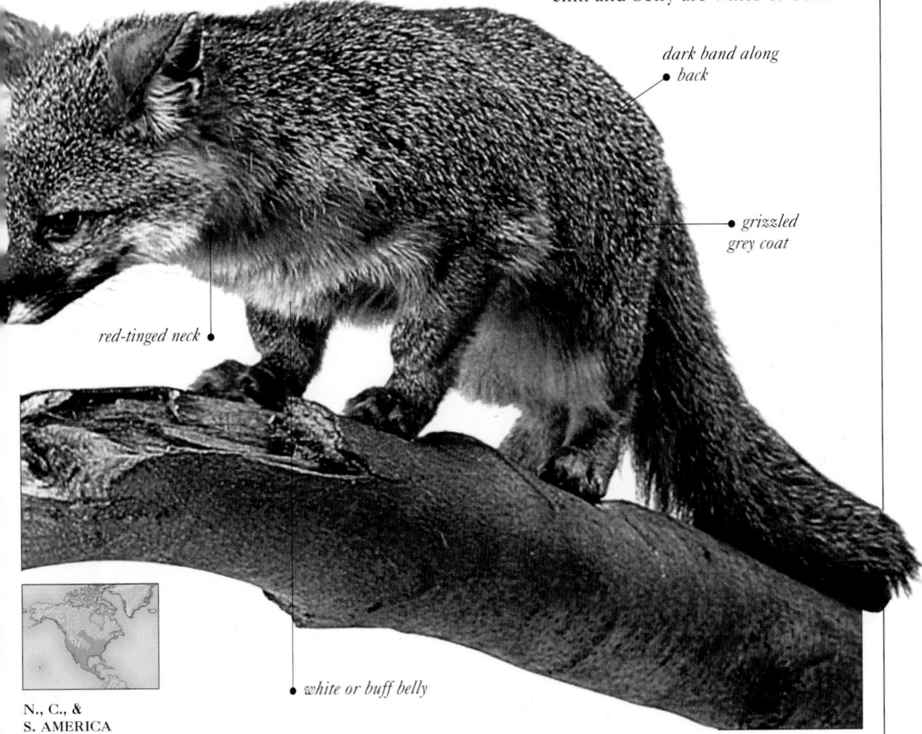

N., C., & S. AMERICA

Social unit Pair	Gestation 51–63 days	Young 1–10	Diet

Family CANIDAE	Species *Chrysocyon brachyurus*	Status Lower risk

MANED WOLF

S. AMERICA

Sometimes described as a Red Fox on stilts because of its characteristic long, dark limbs, the Maned Wolf has black feet which look like they are clad in stockings. With a reddish yellow coat and a darker coloured stripe running from the nape of the neck to its back, it has an erect mane and a dark muzzle. The long, bushy tail is usually dark, but may be lighter, sometimes even white. An inhabitant of open grassland or scrub, this wolf may be found peering over vegetation for prey or potential danger. It is active at twilight and night and has a varied diet which includes rabbits, birds, mice, grubs, and ants, as well as plant matter such as fruits and berries. The Maned Wolf is known to kill small livestock, especially poultry, and is hunted in some areas as a pest, although it is also kept as a pet. Female and male wolves form monogamous pairs, sharing a territory and mating every year, usually in May or June. However, they are hardly found together at other times of the year.

reddish yellow coat

- **SIZE** Body length: 1.2–1.3 m (4–4 1/4 ft). Tail: 28–45 cm (11–18 in).
- **OCCURRENCE** C. and E. South America. In a variety of open habitats, including grassland, scrub, and agricultural areas.
- **REMARK** The Maned Wolf probably benefits from the initial stages of forest clearance, but intensive use of land for cultivation leads to the loss of its habitat, adversely affecting this species. Disease is another major threat to its survival.

MOVING THROUGH GRASS
This wolf is not a very fast runner. Its long legs are more an adaptation to its habitat, which includes areas of tall grassland.

VOCAL WOLVES
Maned wolves make several sounds that are usually heard at night. Disputes over territory evoke typical dog-like growls. Guttural barks also warn away intruders and probably help the territorial pair to keep in audible contact.

black "stockings" on lower limbs

Social unit Solitary	Gestation 62–66 days	Young 1–5	Diet

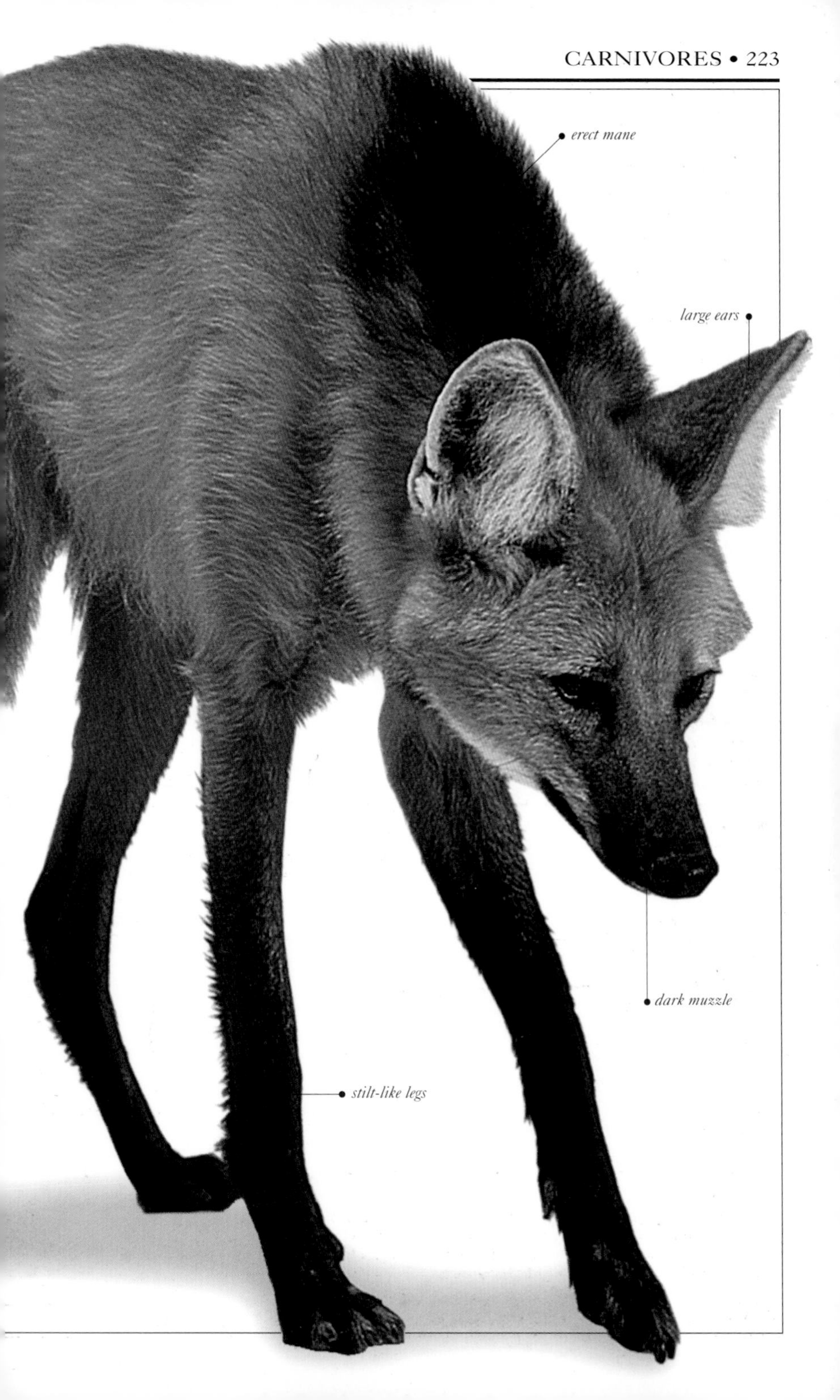
erect mane
large ears
dark muzzle
stilt-like legs

Family CANIDAE	Species *Pseudalopex culpaeus*	Status Locally common

CULPEO FOX

A large, powerful fox with a grizzled grey coat and reddish head, neck, and ears, this species is extensively trapped and hunted for its pelt, and to prevent it preying on livestock, such as lambs and poultry. The Culpeo Fox also feeds on rodents, rabbits, birds and their eggs, as well as seasonal fruits and berries, and stores food when it is easily available, burying the excess or wedging it under logs and rocks. This fox is thought to have suffered from large-scale forest clearance, with some other species, such as the Grey Fox, benefiting at its cost.

- **SIZE** Body length: 60–120 cm (23 ½–47 in). Tail: 30–45 cm (12–18 in).
- **OCCURRENCE** Andean and Patagonian regions of W. South America. In mountains and pampas.

S. AMERICA

Social unit Pair	Gestation 55–60 days	Young 3–8	Diet 

Family CANIDAE	Species *Cerdocyon thous*	Status Common

CRAB-EATING FOX

Widespread in a variety of habitats, this species shows much variation across its range, although it is usually greyish brown above and white below. With a reddish face, ears, and legs, it is black on the tail-tip, ear-tips, and backs of the legs. Usually found in monogamous pairs, this nocturnal predator feeds on crabs–both coastal and freshwater–as well as a variety of other food. It is occasionally shot by ranchers and farmers, and is also hunted for its pelt.

- **SIZE** Body length: 64 cm (25 in). Tail: 29 cm (11 ½ in).
- **OCCURRENCE** N. and E. South America. In grassland, temperate forest, and lowland tropical forest.

S. AMERICA

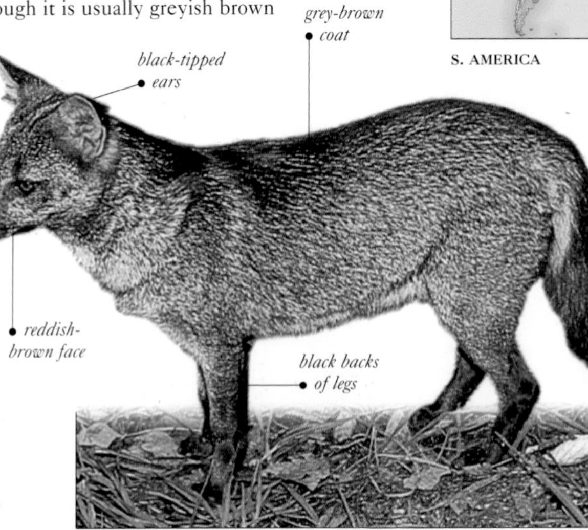

Social unit Social	Gestation 52–59 days	Young 3–6	Diet

Family CANIDAE	Species *Otocyon megalotis*	Status Common

BAT-EARED FOX

This fox derives its name from its large ears, which may be up to 12 cm (4¾in) long. It has a distinctive small, black-masked face with a pointed muzzle and black ear-tips. Its teeth, much smaller than those of most canids, have eight extra molars, and may number 48: more than any other non-marsupial mammal. However, it is a typical fox in its breeding and social habits.

• **SIZE** Body length: 46–66 cm (18–26 in). Tail: 23–34 cm (9–13½ in).

• **OCCURRENCE** E. and S. Africa. In open grassland, semi-desert areas, and forest margins.

• **REMARK** Probably the only canid to have largely abandoned mammal prey, it feeds chiefly on termites and dung beetles.

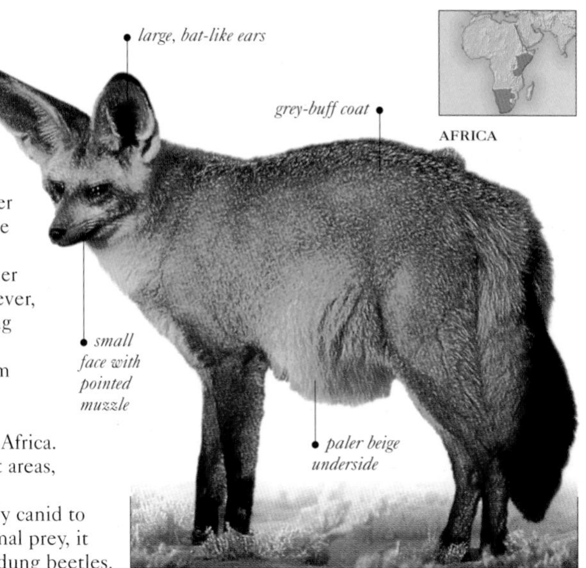

Social unit Variable	Gestation 60–75 days	Young 1–6	Diet

Family CANIDAE	Species *Atelocynus microtis*	Status Locally common

SMALL-EARED DOG

Rather like the Raccoon Dog (see p. 226), with its short, rounded ears, this species has much shorter and more velvety fur, which is dark grey to black on top and varying shades of tawny grey on the underside. Also called the Small-eared Zorro, this chiefly nocturnal canid is reportedly solitary, moving stealthily in a cat-like manner across the forest floor. It probably eats mainly rodents and some plant matter.

• **SIZE** Body length: 72–100 cm (28–39 in). Tail: 25–35 cm (10–14 in).

• **OCCURRENCE** Amazon Basin. In tropical forest, up to an altitude of 1,000 m (3,300 ft).

Social unit Solitary	Gestation Not known	Young Not known	Diet

Family CANIDAE	Species *Nyctereutes procyonoides*	Status Locally common

RACCOON DOG

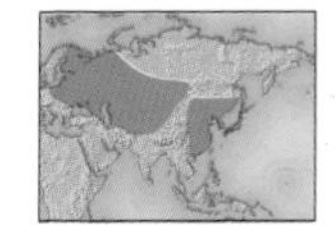

Resembling both a raccoon and a dog, this unusual member of the canid family has a black "robber's mask" on its face, a white muzzle, sleek black legs, and a bushy tail. It has a variable yellow-tinged, brown-black coat, and the upperside of its tail is usually black. The nocturnal Raccoon Dog forages along riverbanks, lakesides, and the seashore, on a huge range of food, from birds and small mammals to fruits. Found in pairs or temporary family groups, it is abundant in Japan, but extinct in areas of China; it has spread rapidly in parts of Europe, where it has been introduced. Unlike most members of the dog family, it can climb well. It is also unique among canids in that it hibernates in winter, feasting in autumn to increase its body weight by up to 50 per cent.

- **SIZE** Body length: 50–60 cm (20–23½ in). Tail: 18 cm (7 in).
- **OCCURRENCE** Europe, C., N., and E. Asia. In temperate woodland and forested river valleys.

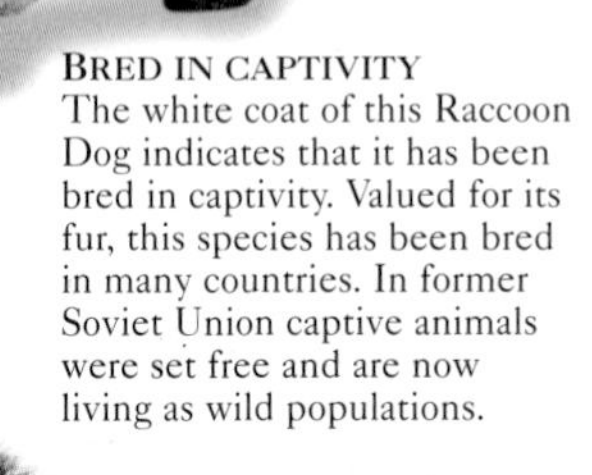

BRED IN CAPTIVITY
The white coat of this Raccoon Dog indicates that it has been bred in captivity. Valued for its fur, this species has been bred in many countries. In former Soviet Union captive animals were set free and are now living as wild populations.

brown-black fur on body

black face mask

white muzzle with black nose

black, short-furred legs

Social unit Variable	Gestation 60–65 days	Young 4–12	Diet

Family CANIDAE	Species *Canis dingo*	Status Locally common

DINGO

Variously regarded as a subspecies of the domestic dog or of its ancestor the Grey Wolf, or as a species in its own right, the Dingo probably descended from the domestic dog in the last 10,000 years, and is now able to survive in the wild. Its coat varies from sandy brown to deep reddish ginger and it has irregular white patches on its chest, feet, and tail-tip. Breeding adults often form settled packs, older members teaching the young their place in the hierarchy by nips and other rebuffs. The dominant female may kill the young of subordinates.

- **SIZE** Body length: 72–110 cm (28–43 in). Tail: 21–36 cm (8 1/2–14 in).
- **OCCURRENCE** Australia. In desert, grassland, tropical and temperate forest, and forest edges.
- **REMARK** The Dingo breeds readily with domestic dogs, and in parts of Australia one-third of individuals are hybrids. It is regarded as a pest as it kills livestock and spreads rabies.

Social unit Social	Gestation 63 days	Young 1–10	Diet

Family CANIDAE	Species *Canis latrans*	Status Common

COYOTE

black shoulder saddle

long muzzle

grey or white underparts

black tail

The grizzled buff coat of the Coyote is yellowish on the outer ears, legs, and feet, but grey or white on the underparts. The shoulders, back, and tail may be tinged black. Highy adaptable in habitat and diet, this opportunistic stalk-and-pounce predator feeds on Pronghorns and deer, as well as Mountain Sheep, livestock, carrion, and refuse. It often hunts jackrabbits by rapidly sprinting after them. The Coyote was once believed to be solitary, but may form breeding pairs or, when larger prey is at hand, gather as a small hunting pack. The distinctive night-time howl of the Coyote usually announces the whereabouts of an individual or its territory to its neighbours.

- **SIZE** Body length: 70–97 cm (28–38 in). Tail: 30–38 cm (12–15 in).
- **OCCURRENCE** North America to N. Central America. In grassland, temperate forest, tundra, mountains, and urban areas.

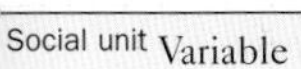

Social unit Variable	Gestation 63 days	Young 6–18	Diet

Family CANIDAE	Species *Canis simensis*	Status Critically endangered

ETHIOPIAN WOLF

The distinctive reddish yellow fur of this wolf, once known as the Simian Jackal, is lighter in females and juveniles, and has white patches on the throat, neck, and chest. Foraging alone by day, it gathers in noisy groups of 2–12 in the morning, noontime, and evening, to greet other wolves.

• **SIZE** Body length: 1 m (3 ¼ ft). Tail: 33 cm (13 in).

• **OCCURRENCE** Ethiopian Highlands. In open moorland and grassland.

• **REMARK** A key species in the food chain of the Ethiopian Highlands, the three remaining populations of this species are vulnerable to disease and competition from domestic dogs and livestock. Reduction of its rodent prey, mainly due to overgrazing, is another factor leading to its rapid decline.

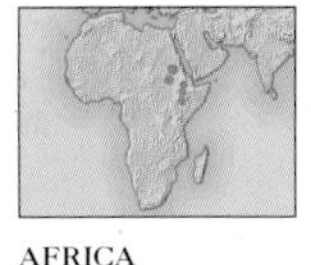

AFRICA

Social unit Social	Gestation 60 days	Young 2–7	Diet

Family CANIDAE	Species *Canis aureus*	Status Common

GOLDEN JACKAL

The coat of this jackal varies with season and region, but is usually a pale golden brown or brown-tipped yellow, with rufous sides, and a ginger or nearly white underside. These omnivorous foragers are usually found in territorial breeding pairs, cooperation between the two leading to highly successful hunting. Strictly nocturnal near human habitations, packs of up to 20 are found in areas where food is plentiful. The pups are cared for in a secure den, by parents, older siblings, and other young adults.

• **SIZE** Body length: 60–110 cm (23 ½–43 in). Tail: 20–30 cm (8–12 in).

• **OCCURRENCE** S.E. Europe, N. and E. Africa, W. to S.E. Asia. In grassland and desert.

EURASIA, AFRICA

Social unit Pair	Gestation 63 days	Young 1–9	Diet

Family CANIDAE	Species *Canis mesomelas*	Status Common

BLACK-BACKED JACKAL

Sometimes called the Silver-backed Jackal, this ginger-coloured canid has a distinctive black saddle over its shoulders and back, and a bushy black tail. Occurring in a wide variety of habitats, from the suburbs of large cities to the Namibian Desert, male–female pairs feed during the day and at twilight, or at night near human settlements, on a variety of food. This jackal is known to attack sheep or young cattle and is viewed as a pest in parts of Africa. Cubs are reared by parents and siblings in dens, which are sometimes built in old termite mounds or abandoned Aardvark holes.

• **SIZE** Body length: 45–90 cm (18–35 in). Tail: 26–40 cm (10–16 in).

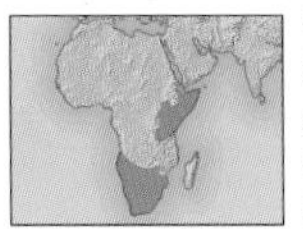
AFRICA

• **OCCURRENCE** E. and S. Africa. In grassland, desert, and urban areas.

• **REMARK** Female and male jackals mate for life and hunt together as a team.

Social unit Pair	Gestation 60 days	Young 1–8	Diet

Family CANIDAE	Species *Canis adustus*	Status Unconfirmed

SIDE-STRIPED JACKAL

Greyish yellow with paler undersides and a white tail-tip, the Side-striped Jackal often has indistinct white and black stripes down its sides. More omnivorous than other jackals, it feeds on rodents, birds, eggs, lizards, insects, refuse, carrion, and plant matter. Like the Red Fox (see pp. 218–219), it is not averse to foraging near urban areas and has been sighted near forest edges and in mixed farmland.

• **SIZE** Body length: 65–81 cm (26–32 in). Tail: 30–41 cm (12–16 in).

• **OCCURRENCE** C., E., and S. Africa. In moist savanna, tropical forest, and farmland.

• **REMARK** This species died in large numbers in the early twentieth century because of distemper.

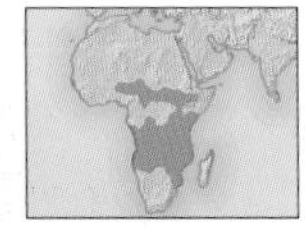
AFRICA

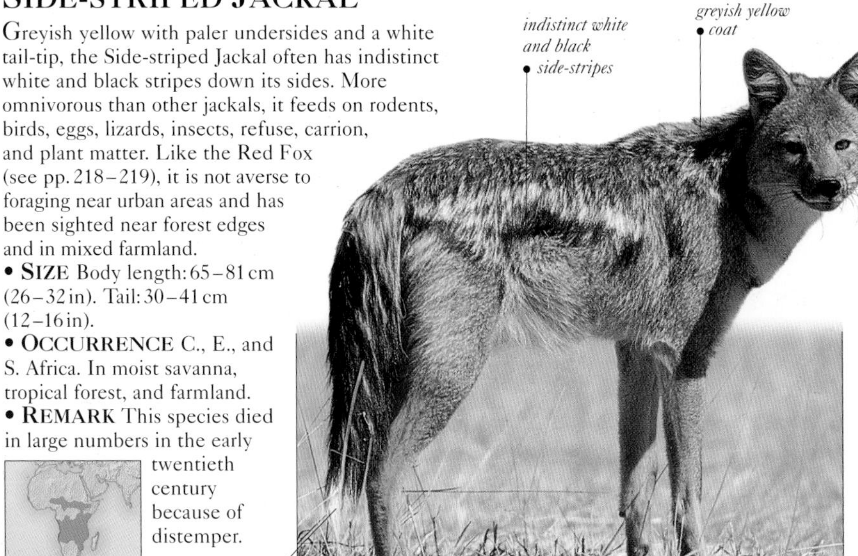

Social unit Pair	Gestation 57–70 days	Young 3–6	Diet

Family CANIDAE	Species *Canis lupus*	Status Vulnerable

GREY WOLF

N. AMERICA, GREENLAND, EURASIA

The ancestor of the domestic dog, the Grey Wolf is the largest member of the canid family. Its great success as a hunter is due to the complex social organization into packs or family groups of 8–12 animals. The clearly defined hierarchy within a group centres around a dominant breeding pair that usually mates for life. Packs patrol large territories, marking them with scent. They announce their presence by howling, which is heard at distances of up to 10 km (6 miles), and warns rival packs to stay well separated and so avoid confrontation. A strong, stocky build, and acute senses of hearing and smell also help in making this wolf a good hunter. Its thick fur is usually grey, but can vary from nearly pure white to red, brown, and black. Although a major part of the Grey Wolf's diet comprises large ungulates, such as deer, Moose, Caribou, and Elk, it will also eat livestock, carrion, and garbage. Pups are weaned in one month and then fed on scraps of food regurgitated by adults. A well-fed pup is strong enough to travel along with the pack by 3–5 months, and a juvenile may choose to leave the pack within a year.

long, sharp teeth

• **SIZE** Body length: 1–1.5 m (3 1/4–5 ft). Tail: 30–51 cm (1–1 3/4 ft).

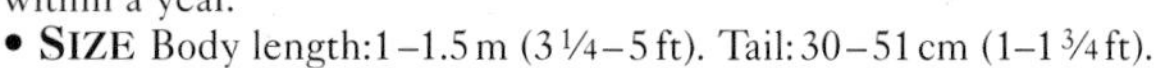

• **OCCURRENCE** N. America, Greenland, Europe, and Asia: above 15°N latitude. In wilderness or remote areas, but may live close to human habitations if food is abundant.

• **REMARK** Originally one of the world's most widely distributed wild mammals, Grey Wolves have been exterminated by humans across much of their range, due to an exaggerated view of their potential ferocity.

STRENGTH IN NUMBERS
Communal hunting allows the Grey Wolf to attack a wide range of prey, which may be ten times its own weight. After the kill, other pack members wait behind the dominant pair for access to the carcass.

pack member awaiting its turn to feed

Social unit Social	Gestation 61–63 days	Young 1–11	Diet

:ANIS FAMILIARIS

'he modern domestic dog (*Canis familiaris*), and the Grey Volf share a similar set of genes. The domestic dog differs rom the wolf in only two significant ways – its teeth are maller and more crowded, and its brain is about a iird smaller than that of the wolf. The learning entres in the brain of the wolf used for mapping erritories lost their relevance in animals that ettled down in human habitations. On the other and, the brain centres responsible for learning to dapt to other species, namely humans, icreased in size and efficiency.

Family CANIDAE	Species *Speothos venaticus*	Status Vulnerable

BUSH DOG

Rather like a weasel in appearance, the Bush Dog is found in family groups of up to ten members. This diurnal predator feeds individually on ground-dwelling birds and rodents such as the Agouti, while the pack may tackle prey as large as the Capybara, hunting them even in water. At night it shelters in burrows and the male brings food to the suckling female in its den.

- **SIZE** Body length: 57–75 cm (22½–30 in). Tail: 12.5–15 cm (5–6 in).
- **OCCURRENCE** Central America to N. and C. South America. In tropical forest and wet savanna.
- **REMARK** The Bush Dog is the most social of small canids.

weasel-like face
long, slender body
small, thick tail
short, sturdy legs

Social unit Social	Gestation 67 days	Young 1–6	Diet

Family CANIDAE	Species *Cuon alpinus*	Status Vulnerable

DHOLE

Sometimes called the Asian Wild Dog, this large carnivore is tawny or dark red with a darker, bushy tail. This territorial dog lives in extended family units of up to 25 individuals, sheltering in dens with their pups, which are fed on regurgitated food even after they leave the den.

- **SIZE** Body length: 90 cm (35 in). Tail: 40–45 cm (16–18 in).
- **OCCURRENCE** S., E., and S.E. Asia. In dense tropical, temperate, and montane forest.
- **REMARK** A widespread species, it is facing a shrinking range and declining numbers.

Social unit Social	Gestation 60–62 days	Young 3–9	Diet

Family CANIDAE	Species *Lycaon pictus*	Status Endangered

AFRICAN WILD DOG

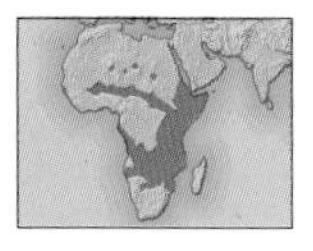

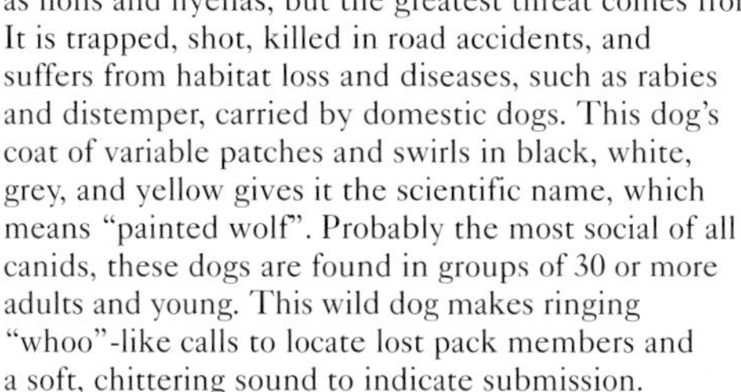

Once found throughout Africa in a variety of habitats, the African Wild or Hunting Dog has now been reduced to a few fragmented populations. It is preyed upon by larger animals such as lions and hyenas, but the greatest threat comes from humans. It is trapped, shot, killed in road accidents, and suffers from habitat loss and diseases, such as rabies and distemper, carried by domestic dogs. This dog's coat of variable patches and swirls in black, white, grey, and yellow gives it the scientific name, which means "painted wolf". Probably the most social of all canids, these dogs are found in groups of 30 or more adults and young. This wild dog makes ringing "whoo"-like calls to locate lost pack members and a soft, chittering sound to indicate submission.

- **SIZE** Body length: 76–110 cm (30–43 in). Tail: 30–41 cm (12–16 in).
- **OCCURRENCE** Sub-Saharan Africa, especially in Tanzania and South Africa. In bush habitats and mountains, and along coastlines.
- **REMARK** The survival of this species depends heavily on active conservation.

PROTECTING THE WILD DOG Scientists have gathered information on the African Wild Dog, such as size, weight, and blood samples for genetic analysis, in an effort to save it from extinction. This is done by anaesthetizing the dog and putting on a radio collar to track its movements.

Social unit Social	Gestation 69–73 days	Young 10–12	Diet

CARNIVORES

BEARS

THE EIGHT BEAR species, known as Ursids, are similar in form and body proportions. They are strong and heavily built, with stout, powerful, large-clawed limbs, a bulky head, small ears, small eyes with poor sight, and a long muzzle with a keen sense of smell.

Although classified as carnivores, most bears are omnivorous. There are some exceptions, however: Polar Bears are almost exclusively flesh-eaters; the Spectacled Bear and Asiatic Black Bea are mainly herbivorous, and the Gian Panda is almost totally so.

Most species are solitary and, i northern regions, hibernate throug winter. Their size, power, occasiona attacks on livestock, and tendency t aggression, especially when raising cubs, have long made bears targets o persecution. All but two species ar listed as in some way threatened.

Family URSIDAE	Species *Ursus americanus*	Status Lower risk*

AMERICAN BLACK BEAR

This bear actually varies in colour from black to cinnamon, brown, or blonde, and even a unique grey-blue occurring on the Pacific coast. It adapts to various habitats quite easily but is usually found in heavily forested areas. An exceptional climber, the American Black Bear plucks fruits, buds, berries, and nuts with its prehensile lips. It opens up old logs and flips over stones with its powerful front limbs and curved claws to feed on insects, and may even break into outbuildings or vehicles to obtain food left by humans. This species uses a complex set of body postures to communicate. Yawning, averted eyes, or a lowered head may be used to determine social hierarchy and to minimize physical contact.

• **SIZE** Length: 1.3–1.9 m (4 1/4–6 1/4 ft). Weight: 55–300 kg (120–660 lb).

• **OCCURRENCE** Alaska, Canada, USA, and Mexico. In temperate and coniferous forest, and mountains.

• **REMARK** The American Black Bear has adapted to human presence successfully and is rarely known to attack humans as it prefers to avoid any confrontation.

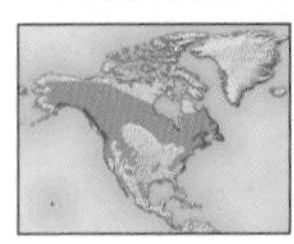

N. AMERICA

Social unit Solitary	Gestation 6 weeks	Young 1–5	Diet

Family URSIDAE	Species *Ursus maritimus*	Status Lower risk

POLAR BEAR

One of the largest carnivores, the Polar Bear is an excellent swimmer and diver which has adapted to its aquatic life in various ways– hollow, air-filled guard hairs aid buoyancy, while its nostrils close underwater, allowing it to hold its breath before pouncing on unwary prey from beneath. This bear's diet consists chiefly of Ringed, Bearded, and Harp Seals, which it may hunt by stalking or by waiting motionlessly in front of seal breathing holes. Capable of fasting for extremely long periods when food is scarce, especially when forced ashore by melting sea ice, it can survive on morsels such as seaweeds, moss, and berries for as long as five months.

WHALE HUNTING
The Polar Bear chiefly feeds on seals, but can also effectively hunt larger aquatic mammals such as Walruses and even Narwhals. Here, it is shown attempting to catch a Beluga Whale.

- **SIZE** Length: 2.1–3.4 m (7–11 ft). Weight: 400–680 kg (880–1,500 lb).
- **OCCURRENCE** Circumpolar across the Arctic in North America, Greenland, Norway, and Russia. In polar ice, near oceans.
- **REMARK** The Polar Bear probably has the most advanced sense of smell among all bears. It can locate seal breathing holes, covered by 90 cm (35 in) of ice, from a kilometre (3/4 mile) away.

Social unit Solitary	Gestation 9 weeks	Young 1–4	Diet

Family URSIDAE	Species *Ursus arctos*	Status Lower risk*

BROWN BEAR

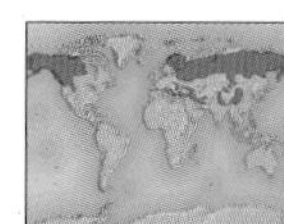

N. AMERICA, EUROPE, ASIA

Large and powerfully built, the Brown Bear is the most widely distributed of all the bear species and varies greatly in size depending on its food and habitat. Several races have been recognized, among them the Grizzly Bear, Kodiak Bear, Alaskan Bear, Eurasian Brown Bear, Syrian Bear, Siberian Bear, Manchurian Bear, and Hokkaido Bear. All share the distinctive shoulder "hump" of muscle, long foreclaws that help to dig for roots and bulbs, concave facial profile, and small ears. Most commonly dark brown in colour, it can vary from blonde to black. Its long guard hairs are tipped with white, giving it a "grizzled" appearance. Although there is not much difference in body length between the sexes, the male Brown Bear may be twice as heavy as the female. This bear is chiefly herbivorous, with 95 per cent of its diet consisting of grasses, roots, bulbs, tubers, berries, and nuts. However, it also savours fish and insects. It feeds intensively during the warmer months, putting on weight in preparation for its long winter sleep.

- **SIZE** Length: 2–3 m (6½–9¾ ft). Weight: 100–1,000 kg (220–2,200 lb).
- **OCCURRENCE** North America, N. Europe, and Asia. In grassland, dry desert, dense temperate and coniferous forest, and mountains.
- **REMARK** Large tracts of wilderness are necessary for the survival of this bear and habitat destruction has led to a drastic decline in its numbers. It is also poached for its body parts, especially the gall bladder, used in making traditional medicines.

dense, dark brown coat may vary from blonde to black

SURVEYING THE SCENE
The Brown Bear sometimes stands upright on its rear limbs to identify a possible source of food or danger. Contrary to popular belief this is not an aggressive posture. Bears only rarely attack humans, as the result of a sudden encounter.

A FISHY DIET
Although the Brown Bear is mainly herbivorous, some coastal populations feed on large quantities of salmon, killing them with a blow or sharp bite, as the fish swim upstream to spawn.

short, powerful limbs

Social unit Solitary	Gestation 9 weeks	Young 1–4	Diet

small
ears
large, broad
head
non-retractable
front claws

Family URSIDAE	Species *Ursus thibetanus*	Status Vulnerable

ASIATIC BLACK BEAR

Resembling the American Black Bear (see p. 234) in appearance and habits, this species is often found in trees, foraging for nuts and fruits. Black, with a yellow-white chest patch, it has tufted ears and a short tail. It raids crops and sometimes kills humans as much of its habitat has been replaced by farmland.

- **SIZE** Length: 1.3–1.9 m (4¼–6¼ ft). Weight: 100–200 kg (220–440 lb).
- **OCCURRENCE** S. and S.E. Asia, N.E. China, far eastern Russia, and Japan. In forested hills.
- **REMARK** It is hunted for its body parts, used in local medicines.

muzzle lighter in colour

whitish yellow chest patch

strong hindlegs allow it to walk upright

ASIA

Social unit Solitary	Gestation 8 months	Young 1–3	Diet

Family URSIDAE	Species *Helarctos malayanus*	Status Endangered

SUN BEAR

Elusive and nocturnal, this stocky and dog-like species is also known as the Dog or Honey Bear. Largely arboreal, it hauls itself up trees by hugging the trunks with its arms and gripping them with its teeth and long claws. The Sun Bear makes a rough nest out of bent branches in which to sleep. Digging into old logs and bees' nests with its claws for grubs and honey, it also pushes its 25 cm (10 in) long tongue into crevices to lick up similar food. Sometimes, it thrusts its paws alternately into termite nests, taking them out to lick off the insects. Its sleek, short fur varies from black to grey or rusty, with a U-shaped or circular creamish patch on the chest and a short, pale muzzle. If grabbed from behind by an attacker, its loose neck skin allows it to turn around and fight.

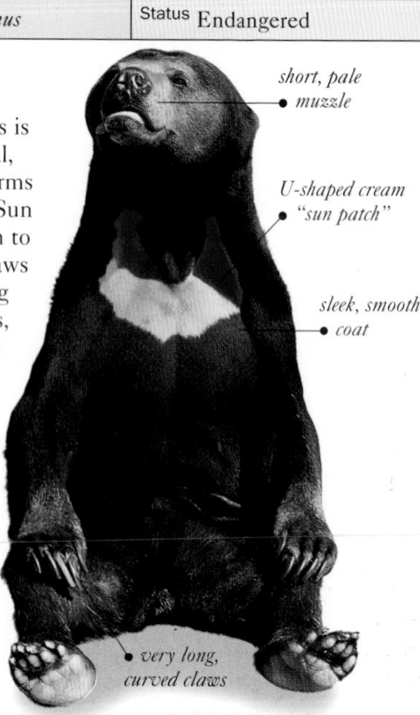

- **SIZE** Length: 1.1–1.4 m (3½–4½ ft). Weight: 50–65 kg (110–145 lb).
- **OCCURRENCE** S.E. Asia. In lowland hardwood rainforest.
- **REMARK** The Sun Bear is the smallest of the eight bear species and best-adapted to the tropics.

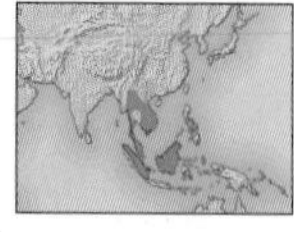

ASIA

Social unit Solitary	Gestation 96 days	Young Not known	Diet

Family URSIDAE	Species *Melursus ursinus*	Status Endangered

SLOTH BEAR

One of the most distinctive looking bears, with its untidy, shaggy coat and long hair around the ears, neck, and shoulders, the Sloth Bear frequents habitats that provide its essential diet of ants, termites, and fruits. It tears open insect nests with its foreclaws, which are 8 cm (3 in) long; then, shutting its nostrils and pursing its lips, it slurps up its prey through a gap formed by missing incisors. Like other bears, it stands on its hindfeet to assess its surroundings.

• **SIZE** Length: 1.4–1.8 m (4½–6 ft). Weight: 55–190 kg (120–420 lb).

• **OCCURRENCE** Indian subcontinent. In grassland, thorny scrub, and tropical forest.

ASIA

• **REMARK** Threats to this species include habitat loss, hunting for body parts, and cub capture for performance.

Social unit Variable	Gestation 6–7 months	Young 2–3	Diet

Family URSIDAE	Species *Tremarctos ornatus*	Status Vulnerable

SPECTACLED BEAR

South America's only bear, the Spectacled or Andean Bear plays a prominent role in local folklore and mythology. Once found in a wide variety of habitats, this excellent climber is now increasingly restricted to cloud forest, where food is most abundant. The massive jaws of this mainly herbivorous bear enable it to chew the toughest plants. It is named after the pale, circular patches around its eyes.

S. AMERICA

• **SIZE** Length: 1.5–2 m (5–6½ ft). Weight: 140–175 kg (310–390 lb).

• **OCCURRENCE** Andes. In coastal areas and steppe, as well as forest.

• **REMARK** The largest land mammal in South America after the tapirs, the Spectacled Bear is a valuable disperser of seeds.

Social unit Solitary	Gestation 7–8 months	Young 1–4	Diet

Family URSIDAE	Species *Ailuropoda melanoleuca*	Status Critically endangered

GIANT PANDA

Immediately recognizable by its striking two-toned colouring, this stocky, barrel-shaped bear is white, except for its black ears, oval eye-patches, nose, shoulder saddle, and limbs. It has a large head, erect ears, and small, beady eyes. The Giant Panda, called "Daxiongmao" (giant bear-cat) or "Baixiong" (white bear) in Chinese, has one of the most restricted diets of all mammals, feeding almost entirely on bamboo. It eats different parts of more than 30 species of bamboo – new shoots in spring, leaves in summer, and stems in winter. The panda has the single stomach and short intestines of a carnivorous mammal and is known sometimes to eat carrion. Normally solitary, it feeds at dusk and dawn, and sleeps in the bamboo thicket that constitutes its feeding area. Not strictly territorial, the Giant Panda prefers to avoid conflict, and marks its home range with scent, urine, and claw scratches. The female moans, bleats, and barks (about 11 panda calls have been identified) during the breeding season; and males chase and fight each other for a receptive female. Sometimes, the male kills the newborn cubs.

- **SIZE** Length:1.6–1.9 m (5 1/4–6 3/4 ft). Weight:70–125 kg (155–280 lb).
- **OCCURRENCE** W. China (occurs in 25 separate populations in three provinces). In temperate, tropical, and mountainous bamboo forest.
- **REMARK** Unlike most bears, pandas do not hibernate.

"HAND" WITH SIX FINGERS
The pad-like false thumb flexes and opposes the true thumb (first digit), helping the Panda to grip bamboo stems and leaves.

coarse, oily guard hair, up to 10 cm (4 in) long

PANDAS IN DANGER
More than 100 pandas are kep in captivity, but only 30 perce of the young born to them survive. Most wild population are small, scattered, and genetically unfit. They are poached despite stringent law (death sentence). Creating huge reserves may be the only answer to the Panda's survival

Social unit Solitary	Gestation 97–181 days	Young 1–2	Diet

CHINA
black ears
white face
muscular front limbs
hindquarters less powerful than forelimbs

CARNIVORES

RACCOONS

WITH A "BANDIT" face mask, ringed tail, bold and busy lifestyle, opportunistic habits, and immensely adaptable diet, the Common Raccoon typifies the family of Procyonids. The 20 species also include coatis, ringtails, kinkajous, and olingos, all found in the forests of the Americas. An exception is the Lesser Panda of Asia, sometimes included with the Giant Panda in the two-member family, Ailuridae.

Members of the Procyonidae fami are generally nocturnal, and rema solitary outside the breeding seaso They have a long body and tail, pointed muzzle, broad face, an smallish, rounded or pointed ears. The legs are relatively short, and their stanc is plantigrade: the soles of their feet re on the ground (as in bears and humans However, all procyonids have sha claws and can climb skilfully.

Family PROCYONIDAE	Species *Ailurus fulgens*	Status Endangered

LESSER PANDA

Also called the Red Panda, this small carnivore is reddish brown to deep chestnut with whitish cheeks, muzzle, and spots above the eyes. It has a round, heavy head and large, pointed ears with white edges. The Lesser Panda lives mainly on the ground, but uses its partly retractable claws to climb up branches, sunning itself high up in trees in winter. It is solitary by nature, but the cub may stay with the mother for up to a year. Mainly nocturnal, it marks its territory with droppings, urine, and musk-like secretions from its anal glands.

- **SIZE** Body length: 50–64 cm (20–25 in). Tail: 28–50 cm (11–20 in).
- **OCCURRENCE** S. and S.E. Asia. In remote high-altitude bamboo forest.
- **REMARK** Habitat destruction and hunting are serious threats to this species.

ASIA

Social unit Solitary	Gestation 114–145 days	Young 1–5	Diet

Family PROCYONIDAE	Species *Bassariscus astutus*	Status Unconfirmed

RINGTAIL

This slim-bodied, agile carnivore, sometimes called the Ringtailed Cat or Cacomistle, is greyish brown or buff, with a striking black-and-white ringed tail. It has black eye-rings and a white muzzle and eyebrows. An effective nocturnal hunter, the Ringtail is known to use barks and long, high-pitched screams to communicate.

• **SIZE** Body length: 30–42 cm (12–16½ in). Tail: 31–44 cm (12–17½ in).

• **OCCURRENCE** C. and W. USA to S. Mexico. In desert, forest, and mountains.

• **REMARK** Extremely dextrous, this animal has hindfeet that rotate through at least 180 degrees, allowing it great flexibility.

N. AMERICA

greyish brown or buff upper coat

slim body

bold, black-and-white rings on tail

Social unit Solitary	Gestation 51–60 days	Young 1–4	Diet

Family PROCYONIDAE	Species *Procyon lotor*	Status Common

COMMON RACCOON

Amazingly adaptable, the Common Raccoon or Mapache is a familiar sight in different kinds of habitat, ranging from desert to woodland and urban areas. The black "robber's mask" seems to reflect the opportunistic habits of this noisy animal. It can climb and dig, and even open doors with its forepaws to enter livestock enclosures.

• **SIZE** Body length: 40–65 cm (16–26 in). Tail: 25–35 cm (10–14 in).

• **OCCURRENCE** S. Canada to Central America. In desert, tropical, temperate, and coniferous forest, near lakes and rivers, and in urban areas.

• **REMARK** Its hands are adapted to manipulate food.

N. & C. AMERICA

grey to blackish long fur

rounded, short ears

black eye-patches

Social unit Solitary	Gestation 60–73 days	Young 1–7	Diet

Family PROCYONIDAE	Species *Procyon cancrivorus*	Status Common

CRAB-EATING RACCOON

About the size of a large cat, this raccoon has short, coarse fur, and forward-pointing hair on its neck. This nocturnal omnivore forages near streams, marshes, lakes, and seashores, feeling for prey with its fingers. It is well-adapted to searching for prey along water edges, going in and out of water with ease. Although terrestrial, it dens in tree-holes.

- **SIZE** Body length: 45–90 cm (18–35 in). Tail: 20–56 cm (8–22 in).
- **OCCURRENCE** E. Costa Rica and adjacent Panama to N. Argentina and Uruguay. In areas near water.

Social unit Solitary	Gestation 60–73 days	Young 2–4	Diet

Family PROCYONIDAE	Species *Nasua nasua*	Status Locally common

SOUTHERN RING-TAILED COATI

Also called the Coatimundi, Tejón, or Pizote, the Southern Ring-tailed Coati may range from reddish brown through yellowish to greyish brown. By day, it forages in noisy groups of 10–20, using its long, white-tipped snout to search for anything edible, while "sentries" around the pack's fringes watch out for predators. Highly vocal, this procyonid uses soft calls, barks, whistles, and squeaks, as well as tail movements, to communicate.

short, rounded ears

long, tapering ringed tail

elongated snout

N., C., & S. AMERICA

- **SIZE** Body length: 40–70 cm (16–28 in). Tail: 32–70 cm (12½–28 in).
- **OCCURRENCE** S.W. USA, Mexico, Central and South America. In desert, forest, near water, and in mangrove swamps.

Social unit Variable	Gestation 10–11 weeks	Young 2–7	Diet

Family PROCYONIDAE	Species *Potos flavus*	Status Endangered*

KINKAJOU

Buff golden brown or tawny to dark greyish brown in colour, this highly agile climber uses its strongly prehensile tail and clawed feet, with naked, padded soles, to clamber up trees. A nocturnal carnivore, it moves about the forest canopy carefully searching for food, and nests in tree hollows or thickets. The Kinkajou has a wide repertoire of calls, used variously to proclaim territory, attract a mate, or warn off predators.

- **SIZE** Body length: 39–76 cm (15 ½–30 in). Tail: 39–57 cm (15 ½–22 ½ in).
- **OCCURRENCE** S. Mexico to South America. In tropical forest, near wetlands and mangrove swamps.

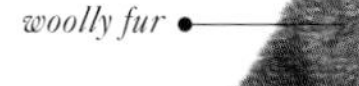

broad, rounded head

buffish gold to grey coat

clawed feet with padded soles

prehensile tail for gripping

N., C., & S. AMERICA

Social unit Solitary	Gestation 112–120 days	Young 1	Diet

Family PROCYONIDAE	Species *Bassaricyon gabbii*	Status Lower risk*

BUSHY-TAILED OLINGO

This slim, cat-like, greyish brown to pale brown procyonid inhabits a variety of forest areas, especially cloud forest. Moving with great agility through the trees, the Bushy-tailed Olingo uses its naked-soled, clawed paws to grasp branches, and its faintly banded, non-prehensile tail for balance. Chiefly active at night, it rarely comes to the ground, foraging for food in the tree canopy, and curling up on a branch or in a tree-hole to rest. It is solitary, except in the breeding season, when males and females call out loudly and can be seen in pairs.

- **SIZE** Body length: 36–42 cm (14–16 ½ in). Tail: 37–49 cm (14 ½–19 ½ in).
- **OCCURRENCE** Central to N. South America. In tropical rainforest and mountains.

C. & S. AMERICA

- **REMARK** Particularly sensitive to deforestation and forest disturbance, the Bushy-tailed Olingo has not adapted well to cleared spaces or secondary forest.

greyish brown to pale brown coat

large brown eyes

long, fluffy tail

Social unit Solitary	Gestation 73–74 days	Young 1	Diet

CARNIVORES

MUSTELIDS

A LONG, SLIM, and flexible or sinuous body, shortish legs, small ears, beady eyes, and very sharp teeth and senses typify the mustelid family.

The 67 species range in size from the Least Weasel, the smallest of all carnivores, which could curl up in a person's cupped hands, through stoats and skunks, to the Giant Otter and Wolverine, which are the size of a very large dog. The different species vary in their lifestyles, from mainly terrestria polecats, to arboreal martens, burrowin badgers, semi-aquatic minks, an almost fully aquatic otters.

Unusually for carnivores, mustelid have five clawed toes on all the fee Glands near the tail produce mus odours, chiefly to mark territories Mustelids are found worldwide apar from Australia, Antarctica, and some o Southeast Asia.

Family MUSTELIDAE	Species *Mustela putorious*	Status Locally common

EUROPEAN POLECAT

Considered to be the ancestor of the domestic ferret, the European Polecat was once valued for its pelt. It has long, buff to black hair, with creamish yellow underfur visible beneath, and a black mask across its eyes. Its head is small and flat, its ears rounded, and snout blunt. Long and sinuous in shape, it glides through small openings with ease, and runs, climbs, and swims agilely after a variety of prey. Since its vision is poor, this night-predator hunts by smell and hearing. Rabbits are its principal prey, which it often takes from their nests. Other prey include small rodents, birds, and amphibians. In turn, it is preyed upon by foxes and raptors, and trapped by gamekeepers as it is considered a threat to game and poultry. Male polecats are considerably larger than females and they defend exclusive territories, although these may overlap with the territories of females. If threatened, this polecat may release an unpleasant odour from its anal glands.

• **SIZE** Body length: 35 – 51 cm (14 – 20 in). Tail: 12 – 19 cm (4¾ – 7½ in).

• **OCCURRENCE** Europe. In woodland, forest, plantations, farmland, marsh, and riverbanks.

EUROPE

FRIENDLY FERRET
The domestic ferret (shown in its light winter coat), is kept as a pet, and has descended from the European Polecat.

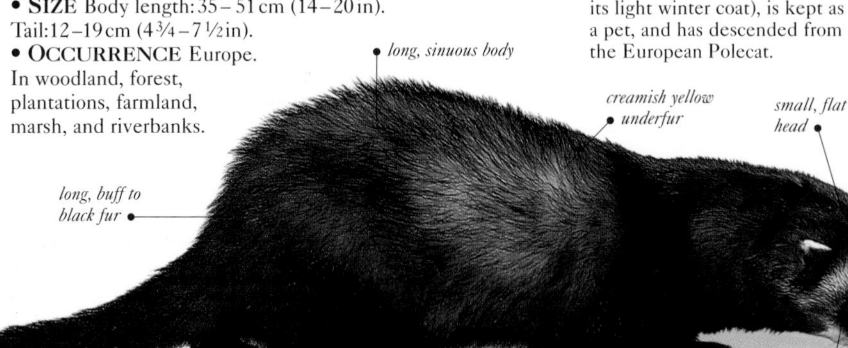

Social unit Solitary	Gestation 40 – 43 days	Young 5 – 8	Diet

Family MUSTELIDAE	Species *Mustela nigripes*	Status Critically endangered

BLACK-FOOTED FERRET

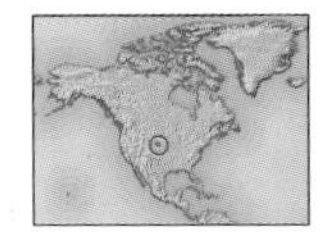

One of the world's rarest mammals, the Blackfooted Ferret has a yellow-buff coat with a distinctive black mask, and black feet and tail-tip. It nearly became extinct in the wild in the 1990s, due to the extermination of its main prey, the prairie dog. As well as feeding on prairie dogs, this nocturnal ferret also lives in its burrows.

- **SIZE** Body length: 38–41 cm (15–16 in). Tail: 11–13 cm (4½–5 in).
- **OCCURRENCE** W. North America. In open grassland, steppe, and scrub, within the prairie dog's range.
- **REMARK** A small population of about 190 ferrets has been bred in captivity and released in Wyoming, USA. Some are breeding in the wild again.

Social unit Solitary/Pair	Gestation 42–45 days	Young 3–6	Diet

Family MUSTELIDAE	Species *Mustela erminea*	Status Locally common

STOAT

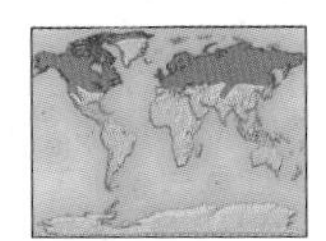

This extremely widespread, typical mustelid has a slim body, pointed muzzle, small eyes and ears, and short legs. Active day and night, it needs only cover and prey to thrive. In warmer climates, the Stoat's upper coat is russet to ginger-brown with a sharply demarcated cream belly. However, those found in the north turn almost completely white in winter for camouflage in the snow.

- **SIZE** Body length: 17–24 cm (6½–9½ in). Tail: 9–12 cm (3½–4¾ in).
- **OCCURRENCE** North America and Eurasia. In tundra, temperate and coniferous forest, and mountain areas.
- **REMARK** The Stoat is called the Ermine in its white winter coat.

Social unit Solitary	Gestation 28 days	Young 4–9	Diet

Family MUSTELIDAE	Species *Mustela nivalis*	Status Locally common

WEASEL

One of the smallest and most widespread mustelids, the adaptable Weasel is found over an enormous range of habitats. Its body is long and slender, and its coat is chocolate-brown or russet on the back, limbs, and tail, and white on the underparts. Its head is small, its rounded ears are relatively large, and it has a long neck. The overall size of the Weasel, however, varies greatly across its range. In northern areas it is frequently larger, and turns white in the winter months. Active by day and night, in bouts of 10–45 minutes, alternating with rest periods, the Weasel must consume one-third of its body weight in food a day in order to survive. A specialist predator of mice and voles, and occasionally birds, it hunts chiefly by sight and smell, delivering an accurate and fatal bite on the neck of its victim. The Weasel hisses or screams when it is threatened, and emits a powerful, unpleasant scent from its anal glands, which is also believed to be used in communication with others of its kind.

- **SIZE** Body length:16.5–24 cm (6½–9½ in). Tail: 3–9 cm (1¼–3½ in).
- **OCCURRENCE** North America, Europe to N., C., and E. Asia. In a variety of habitats: tundra, steppe, semi-desert, open forest, farmland, and meadow.
- **REMARK** The Weasel survives the harsh northern winters by often hunting for prey under a thick blanket of snow. This serves as insulation in extremely low temperatures.

Social unit Solitary	Gestation 34–37 days	Young 1–7	Diet

SOLITARY DWELLER
The Weasel lives alone, making several nests, lined with the fur or feathers of its prey, in a crevice, tree root, or burrow abandoned by some other animal.

FLATTENED SKULL
The Weasel has a long, flattened head. Its front incisors are small and its sharp-ridged molars slice the gristle and sinew of its prey.

BABY WEASEL
Weasel young are looked after by the mother alone for 9–12 weeks following their birth. The female gives birth to an average litter of 1–7 in a breeding nest lined with soft material. Male Weasels grow to be around a quarter longer than females, and twice as heavy.

Family MUSTELIDAE	Species *Mustela vison*	Status Locally common

AMERICAN MINK

This mink is dark brown to almost black in colour, but about one in ten is silvery grey, and a number of colour variations have also been bred on fur farms. An opportunistic predator, it hunts at night, dusk, or dawn for a variety of small creatures such as rats, rabbits, frogs, fish, crayfish, and shore crabs. Although found close to water, the American Mink is not a powerful swimmer and since its eyesight is not well-adapted to underwater vision, it locates its prey on the surface before pursuing it. This mustelid has a repertoire of sounds, which includes screams and hisses, and chuckling calls between males and females during the breeding season. Mating is often preceded by very aggressive encounters between the two sexes. Males first seek out those females whose territories overlap their own, before searching for others. Typical territories are 1–3 km (0.6–1.8 miles) across for the females, and 2–5 km (1.2–3.1 miles) for the males. The female alone takes care of the young, suckling them for 5–6 weeks in her nesting den, built among roots or rocks.

• **SIZE** Body length: 30–54 cm (12–21 ½ in). Tail: 14–21 cm (5 ½–8 ½ in).

• **OCCURRENCE** North America, S. South America, Europe, and Asia. In lakes and rivers, and along coastlines.

• **REMARK** Trapped for its luxuriant fur over hundreds of years, the American Mink was taken to Europe in the early part of the 20th century for farming. Escaped animals have established wild populations and are regarded as pests and a threat to native species, especially the European Water Vole (see p.164).

N. & S. AMERICA, EURASIA

small ears

white patch on chin

dark brown coat

thick underfur

darker, coarse guard hairs

NEAR WATER
Semi-aquatic in lifestyle, the American Mink has a coat of dark guard hairs that waterproof its fur while swimming. Its partially webbed feet enable it to hunt on land as well as in water.

Social unit Solitary	Gestation 40–75 days	Young 3–6	Diet

Family MUSTELIDAE	Species *Mustela lutreola*	Status Endangered

EUROPEAN MINK

Although it is not closely related to the American Mink (see opposite), the slightly smaller European Mink resembles it in habits and appearance. Its fur is usually dark brown to almost black, and the upper and lower lips are ringed with a thin band of white fur. The males are known to be up to 85 per cent larger than the females. This predominantly nocturnal predator hunts under water and on land for water voles, fish, frogs, crayfish, and water birds. The European Mink has very few natural predators, although it is under threat from introduced species such as the American Mink. It releases a noxious secretion from its anal glands when threatened.

- **SIZE** Body length: 30–40 cm (12–16 in). Tail: 12–19 cm (4¾–7½ in).
- **OCCURRENCE** N. Spain, W. France, Belarus, and Russia. In lakes and slow-moving rivers, especially in or near woods.

EUROPE

- **REMARK** This is Europe's most endangered mammal, and is now being bred in captivity.

Social unit Solitary	Gestation 40 – 43 days	Young 4–5	Diet

Family MUSTELIDAE	Species *Poecilogale albinucha*	Status Lower risk*

AFRICAN STRIPED WEASEL

AFRICA

The long, sinuous body of this animal is black apart from a white patch running from the forehead to the neck, where it splits into two white stripes, each stripe dividing into two again along the back and sides of its body. All four stripes come together at the tail, which is bushy and white. The front claws, longer than those on the hindfeet, are used for burrowing. The African Striped Weasel feeds at night, chiefly on small rodents, killing those up to its own size by means of a well-directed neck bite. When faced with predators such as foxes, cats, or owls, this weasel arches its back and squirts noxious fluid from its anal glands.

- **SIZE** Body length: 25–35 cm (10–14 in). Tail: 15–23 cm (6–9 in).
- **OCCURRENCE** C. to S. Africa. In grassland areas with rainfall.

Social unit Solitary	Gestation 32 days	Young 1–3	Diet

Family MUSTELIDAE	Species *Vormela peregusna*	Status Vulnerable

MARBLED POLECAT

As its name suggests, the Marbled Polecat has a black coat, mottled with white or yellow spots and stripes. A distinctive black mask over its eyes characterizes its face, while its underparts are black. It has a long, sinuous body with short legs, a small, flat head with a blunt snout, and rounded ears. When threatened, it curls its tail to display its warning coloration and emits an unpleasant odour. The Marbled Polecat enlarges rodent burrows to make its den, and hunts at night, dawn, or dusk.

- **SIZE** Body length: 33–35 cm (13–14 in). Tail: 12–22 cm (4¾–9 in).
- **OCCURRENCE** S.E. Europe to W. China. In dry, open steppe and semi-arid regions.
- **REMARK** This species is threatened by habitat loss and depletion of steppe rodents, one of its main foods.

Social unit Solitary	Gestation 56–63 days	Young 4–8	Diet

Family MUSTELIDAE	Species *Martes foina*	Status Common

BEECH MARTEN

Adapted to human presence, this marten scavenges on refuse and hunts around farmhouses, and also eats rodents, birds, and fruits. Short in stature but long-legged, it has a dense brown coat with a "bow tie" patch of white on its throat. Also known as the Stone Marten, this typically solitary animal makes its den in a rocky crevice, tree-hole, old rodent burrow, or outhouse near human settlements. The Beech Marten is strictly nocturnal by nature.

- **SIZE** Body length: 42–48 cm (16¼–19 in). Tail: 26 cm (10 in).
- **OCCURRENCE** S. and C. Europe, and C. Asia. In deciduous woodland, rocky hillsides, and near human habitations.

Social unit Solitary	Gestation 30 days	Young 12	Diet

Family MUSTELIDAE	Species *Martes flavigula*	Status Endangered

YELLOW-THROATED MARTEN

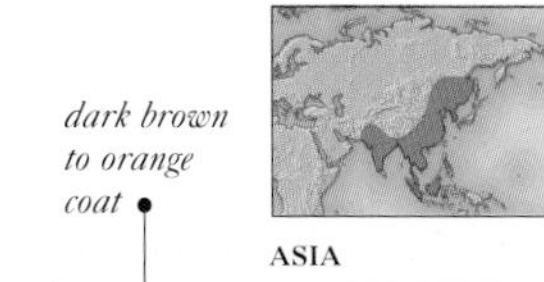

An agile climber that uses its claws to grip branches and its tail for maintaining balance, the Yellow-throated Marten also bounds along the ground in long leaps. This medium-sized carnivore has orange-yellow to dark brown fur, with a creamy yellow throat "bib" and a bushy black tail. It has a wedge-shaped head and large, round ears. Hunting both on trees and on the ground at night, it also climbs trees to escape from danger, and releases a noxious smell when alarmed. It is preyed upon by eagles and other carnivores, and is occasionally hunted by humans for its pelt.

- **SIZE** Body length: 48–70 cm (19–28 in). Tail: 35–45 cm (14–18 in).
- **OCCURRENCE** India to S.E. Asia. In temperate and coniferous forest.

Social unit Solitary	Gestation 5–6 months	Young 2–5	Diet

Family MUSTELIDAE	Species *Martes zibellina*	Status Endangered*

SABLE

Extensively hunted over centuries for its pelt, the Sable is now a protected species in many countries. It has dense brown-black fur, with an indistinct, paler brown patch on the throat, long legs with sharp, partly retractable claws, and a bushy tail. Like others of its family, it is fast and agile on the ground. However, it rarely climbs, except to escape from danger, in spite of being well-adapted for this activity. Occupying old burrows for its main nest, the territory of this species is considerably larger in larch forest than in pine forest. The Sable is active both during the day and at night.

- **SIZE** Body length: 32–46 cm (12½–18 in). Tail: 14–18 cm (5½–7 in).
- **OCCURRENCE** N.E. Asia and islands of N. Japan. In temperate and coniferous forest (especially larch and pine).
- **REMARK** The Sable is almost extinct in Europe, but was reintroduced in European Russia for fur-farming and hunting.

Social unit Solitary	Gestation 25–45 days	Young 3–4	Diet

Family MUSTELIDAE	Species *Martes pennanti*	Status Lower risk*

FISHER

Belying its name, the Fisher hunts terrestrial prey such as rodents, porcupines, squirrels, and hares, both during the day and night. This medium-sized carnivore has a wedge-shaped nose and round ears; its dark brown fur has a sheen on the head and shoulders, and its bushy tail and legs are black. An agile creature, the Fisher is well-adapted to climbing and prefers to raise its young on high tree branches, although it also makes its den among rocks, roots, stumps, and bushes.

• **SIZE** Body length: 47–75 cm (18 ½ –30 in). Tail: 30–42 cm (12–16 ½ in).

• **OCCURRENCE** Canada and N. USA. In coniferous and hardwood forest.

• **REMARK** Some 50,000–130,000 animals are trapped each year for their pelts. Habitat loss due to logging also remains a major threat.

Social unit Solitary	Gestation 11–12 months	Young 1–5	Diet

Family MUSTELIDAE	Species *Galictis vittata*	Status Lower risk*

GREATER GRISON

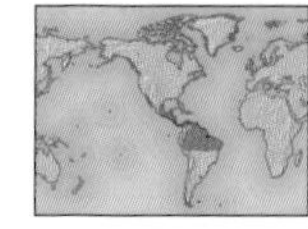

Long and sinuous in shape, this mustelid has a slim, pointed head and a flexible neck, and relatively short legs and tail. The fur on its pale grey upperparts is grizzled, but its muzzle, throat, chest, and forefeet are black. The grey crown has a broad white band on the forehead above the eyes, across the ears, and down the side of its neck. Usually found living alone or in pairs, this agile runner, climber, and swimmer produces a noxious smell when threatened.

• **SIZE** Body length: 47–55 cm (18 ½ –22 in). Tail: 14–20 cm (5 ½ –8 in).

• **OCCURRENCE** S. Mexico, and Central and South America. In tropical forest and grassland.

• **REMARK** A related species, the Lesser Grison (*Galictis cuja*), occurs in southern temperate latitudes and at higher elevations. It is sometimes tamed to control rodents.

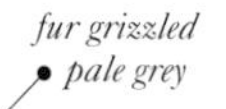

Social unit Solitary/Pair	Gestation Not known	Young 2–4	Diet

Family MUSTELIDAE	Species *Ictonyx striatus*	Status Locally common

ZORILLA

Jet black, with four white stripes fanning out across its back, the Zorilla (also known as the African or Striped Polecat) resembles a small skunk (see pp. 256–257). Its behaviour is similar too: when threatened by predators, it sprays noxious fumes from its anal glands, lifting up its tail or standing on its hindlegs, hissing and screaming. It uses its rounded muzzle to poke into leaf litter, and then uses its long front claws to dig up burrowing insect prey. A nocturnal animal, it usually shelters in tree crevices or disused animal holes, or digs its own burrow in soft earth. It is mainly terrestrial, but may occasionally climb trees.

- **SIZE** Body length: 28–38 cm (11–15 in). Tail: 20–30 cm (8–12 in).
- **OCCURRENCE** Africa, south of the Sahara Desert. In grassland, desert, and tropical forest.

AFRICA

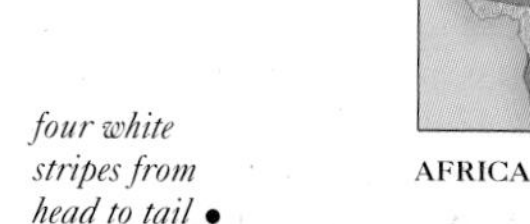

Social unit Solitary	Gestation 36 days	Young 2–3	Diet

Family MUSTELIDAE	Species *Gulo gulo*	Status Vulnerable

WOLVERINE

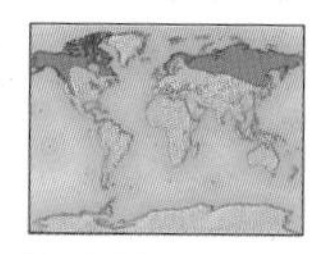

N. AMERICA, EURASIA

Stocky and bear-like, this large mustelid lopes across the snow, hunting prey ranging from deer and Elk to hares, birds, and mice. Also known as the Glutton, it scavenges on carcasses of reindeer, crunching the frozen meat and bones with it powerful jaws. It lives alone in a den made among roots and rocks, and is active during both day and night.

- **SIZE** Body length: 65–105 cm (26–41 in). Tail: 17–26 cm (6½–10 in).
- **OCCURRENCE** Canada, N.W. USA, N. Europe, and N.E. Asia. In tundra and coniferous forest.
- **REMARK** The Wolverine is the largest mustelid after the Giant Otter (see p. 265).

Social unit Solitary	Gestation 30–50 days	Young 1–6	Diet 

Family MUSTELIDAE	Species *Mephitis mephitis*	Status Common

STRIPED SKUNK

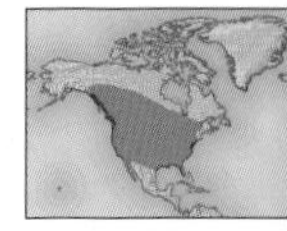
N. AMERICA

With a small, pointed snout, short legs, and a fluffy tail, the Striped Skunk is a typical example of its genus. Its black coat has a thin white stripe on the muzzle and two wider stripes on the back, stretching from head to tail. Like all skunks, it is sluggish in its movements and relies largely on its warning coloration and spray (see right) to protect itself from predators, mostly horned owls, hawks, coyotes, bobcats, foxes, and dogs. Highly adaptable by nature, the Striped Skunk prospers in places that have been cleared for farming, where larger predators have been driven out by humans. The diet of this nocturnal animal often depends on habitat and availability of food. Generally solitary, families and individuals often gather together in communal winter dens located in abandoned burrows, buildings, or rock piles. A wide repertoire of sounds such as hisses, growls, squeals, and soft cooing notes are used to communicate. Males use their scent spray to communicate with females during the mating season. Females build nests of dried grass and weeds; the young stay with the mother for over a year.

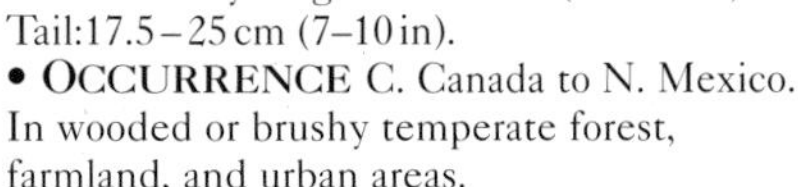

- **SIZE** Body length: 55–75 cm (22–30 in). Tail: 17.5–25 cm (7–10 in).
- **OCCURRENCE** C. Canada to N. Mexico. In wooded or brushy temperate forest, farmland, and urban areas.
- **REMARK** This skunk is a major carrier of rabies in the USA, and its proximity to humans and domestic animals is a cause for concern.

Social unit Solitary	Gestation 60–77 days	Young 5–6	Diet

ON THE PROWL
The omnivorous Striped Skunk may be seen ambling along in search of food such as insects, small mammals, fish, crustaceans, fruits, grain, carrion, and refuse. It will also dig out bee and wasp nests, and may even rummage through rubbish bins.

UP IN THE AIR
The scientific name of the Striped Skunk is derived from a Latin word that means "poisonous vapour". When threatened by predators, it stands on its forefeet and ejects a noxious fluid over its head, for up to 3 m (9 3/4 ft), towards the enemy.

Family MUSTELIDAE	Species *Conepatus humboldti*	Status Vulnerable*

PATAGONIAN HOG-NOSED SKUNK

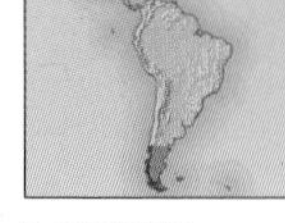

A white stripe runs along each side of this skunk up to the tail. Its black or reddish brown body is stocky, with a small head and bushy tail. A broad nose pad helps it root out food at night, primarily insects, but virtually anything that is edible. Like other skunks, it occupies a secure shelter under rocks, in a burrow, or among bushes, and ejects a noxious fluid from its anal scent glands when threatened. It uses a variety of sounds, from soft chittering to squeals and growls, to communicate.

- **SIZE** Body length: 25–37 cm (10–14 ½ in). Tail: 30–57 cm (12–22 ½ in).
- **OCCURRENCE** S. Chile and Argentina. In wooded, brushy areas and farmland.
- **REMARK** Like others in its family, this skunk does not spray its own kind even in the fiercest fight.

Social unit Solitary	Gestation 42 days	Young 2–4	Diet

Family MUSTELIDAE	Species *Spilogale putorius*	Status Lower risk*

EASTERN SPOTTED SKUNK

The striking black and white coloration of this skunk warns its predators – owls, coyotes, and foxes – of the foul-smelling fluid that it discharges, sometimes when doing a handstand. The white markings on the body differ from one individual to another, but all usually have a white forehead patch and white tail-tip. This slow-moving omnivore feeds on a variety of small animals, insects, and plant matter, even rummaging in rubbish heaps for food. The Eastern Spotted Skunk is more active, alert, and nocturnal by nature than the Striped Skunk (see pp. 256–257). Although usually solitary, up to eight skunks may share an underground den in winter. They are also known to climb and take shelter in trees.

- **SIZE** Body length: 30–34 cm (12–13 ½ in). Tail: 17–21 cm (6 ½–8 ½ in).
- **OCCURRENCE** E. to C. USA, and N.E. Mexico. In grassland and temperate forest.
- **REMARK** When threatened, it sometimes contorts its body in a horseshoe shape towards the aggressor in order to squirt its noxious fluid.

Social unit Solitary	Gestation 42 days	Young 3–6	Diet

Family MUSTELIDAE	Species *Taxidea taxus*	Status Common

AMERICAN BADGER

Similar in appearance to the Eurasian Badger (see p. 261) but smaller, the American Badger has long, grizzled grey fur and yellowish underparts, and a prominent white crescent on either side of its blackish face. A solitary, nocturnal hunter and an expert digger, it uses its stout claws to unearth burrowing rodents such as Prairie Dogs and other ground squirrels. Although it does not hibernate, it may spend several days underground during inclement weather. The American Badger has few natural predators and displays very ferocious behaviour when attacked.

• **SIZE** Body length: 42–72 cm (16½–28 in). Tail: 10–16 cm (4–6½ in).

• **OCCURRENCE** S.W. Canada and C. USA to N. Mexico. In areas with loose soil: grassland, shrub, and temperate forest.

• **REMARK** It has a transparent third inner eyelid to protect its eyes while digging.

N. AMERICA

Social unit Solitary	Gestation 42 days	Young 2–3	Diet

Family MUSTELIDAE	Species *Melogale personata*	Status Lower risk

BURMESE FERRET BADGER

With a long, flexible body rather like a ferret's, this bushy-tailed badger is much smaller than others, and is dark brown or grey in colour. Its face has white or yellow cheek patches and a pale band between the eyes and from the top of the head to the shoulders. Unlike other badgers, the Burmese Ferret Badger is sometimes found in trees, the ridges on the pads of its feet providing a good grip. It feeds at night, dawn, and dusk, on insects, snails, birds, small mammals, and vegetable matter, using its massive teeth to crush mollusc shells and insects. When threatened, the Burmese Ferret Badger digs into the ground with its long claws and bites its aggressor or sprays it with a foul-smelling fluid.

• **SIZE** Body length: 33–43 cm (13–17 in). Tail: 15–23 cm (6–9 in).

• **OCCURRENCE** N.E. India, Nepal, Burma, Thailand, and S.E. Asia. In wooded hillsides, as well as open grassland.

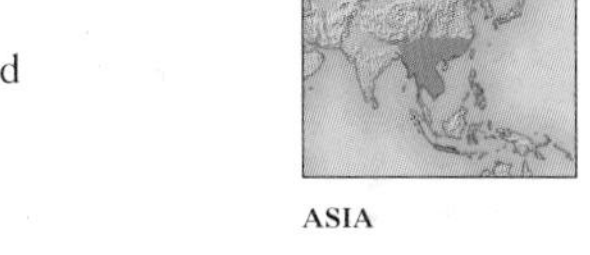

Social unit Solitary	Gestation 57–80 days	Young 1–5	Diet

Family MUSTELIDAE	Species *Mellivora capensis*	Status Lower risk*

HONEY BADGER

AFRICA, ASIA

Also called the Ratel, this heavily built badger is characterized by silvery grey upperparts, but is black or dark brown elsewhere. The Honey Badger feeds on honey and bee larvae, as well as a wide range of prey from worms, termites, and scorpions, to hares and porcupines. It excavates large burrows, but may also live in rock crevices and holes in tree roots. It fearlessly fights off predators, sometimes producing an offensive smell in defence.

- **SIZE** Body length: 60–77 cm (23 ½–30 in). Tail: 20–30 cm (8–12 in).
- **OCCURRENCE** Sub-Saharan and W. Africa, Middle East, and India. In grassland, desert, forest, and mountains.
- **REMARK** The Honey Badger has a unique symbiotic relationship with the Honeyguide Bird (*Indicator indicator*), which leads it to beehives and then waits for the badger to break them open, so that both can feed on grubs and honey.

Social unit Variable	Gestation 5–6 months	Young 1–4	Diet

Family MUSTELIDAE	Species *Arctonyx collaris*	Status Lower risk*

HOG-BADGER

ASIA

This badger takes its name from its pig-like snout, with protruding incisors and canines on the lower jaw, with which it tends to root in the ground for grubs and plant matter. Grey to yellowish in colour, the Hog-badger has a white face and ears, and a distinctive black stripe extending from the nose through the eye to the ear, on each side. It digs elaborate burrow systems, sometimes using crevices under rocks and boulders as shelters. Feeding according to seasonal availability, it uses its strong sense of smell to locate fruits, tubers, and small animals. This badger is known to fight savagely when cornered, and it sometimes escapes underground. Hunted by tigers and leopards, its black-and-white face stripe probably acts as a warning to predators.

- **SIZE** Body length: 55–70 cm (22–28 in). Tail: 12–17 cm (4¾–6 ½ in).
- **OCCURRENCE** N.E. India to China, and S.E. Asia. In lowland jungle and low, wooded hills.

Social unit Social	Gestation 6 weeks	Young 2–4	Diet

Family MUSTELIDAE	Species *Meles meles*	Status Locally common

EURASIAN BADGER

Unusually among mustelids, the Eurasian Badger lives in clans – probably the most efficient way of hunting for irregularly distributed food. This stocky, short-limbed badger has a small, pointed head, short neck, strong limbs, and a small tail. A nocturnal omnivore, its diet varies with season and availability, but it feeds mostly on earthworms, sucking them out when they emerge from their holes on wet nights. It also eats insects, lizards, frogs, small mammals, birds and their eggs, fruits, and carrion, and sometimes digs out wasps from their nests and rabbits from their burrows with its powerful front claws. Since its vision is poor, it locates its food by means of its well-developed senses of smell and hearing. This badger vigorously defends its territory, which may range over 50–150 hectares (125–370 acres), against other clans. Badger cubs are hunted by eagles, owls, wolves, and wolverines, while adult animals also face persecution by humans.

• **SIZE** Body length: 56–90 cm (22–35 in). Tail: 12–20 cm (4¾–8 in).

• **OCCURRENCE** Europe to E. Asia. In woodland and steppe.

• **REMARK** The striped face of the Eurasian Badger varies slightly between individuals, perhaps allowing clan members to recognize each other, or acting as camouflage.

BADGER COLONY
A badger clan has about six individuals, with a dominant boar (male), one or more sows (females), and cubs. The extensive system of underground chambers and pathways (the "sett") is kept scrupulously clean, and enlarged over generations.

Social unit Social	Gestation 7 weeks	Young 2–6	Diet

Family MUSTELIDAE	Species *Lutra lutra*	Status Vulnerable

EUROPEAN OTTER

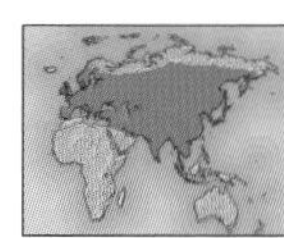

Uniformly brown with a paler throat, this mustelid, sometimes called the Eurasian River Otter, has a flat head with a broad muzzle, and small eyes and ears. Its elongated body, waterproof coat, webbed feet, and stout, flattened tail (used to aid propulsion and as a rudder) are adaptations to its aquatic lifestyle. Although usually active at night, dawn, and dusk, the otters found along coastlines are more active during the day. The holt (burrow), well hidden in bank vegetation or under overhanging tree roots, is located within a bankside territory, 4–20 km (2½–12 miles) long, that is marked by scent and droppings. Mostly solitary, this species is known to form temporary pairs for two or three months during the breeding season. The cubs are suckled for three months, but may remain with the mother for more than a year. This otter communicates by a variety of sounds, and through scent emitted from its glands, signifying identity and status.

- **SIZE** Body length: 57–70 cm (22½–28 in). Tail: 35–40 cm (10–16 in).
- **OCCURRENCE** Eurasia, south of the tundra. Near riverbanks, lakes, and coastlines.
- **REMARK** The European Otter is a protected species. It has been hunted for fur, fishery protection, and sport. It faces additional threats from water pollution and from activities such as clearance of banks, irrigation, and water sports.

Social unit Solitary	Gestation 60–70 days	Young 2–3	Diet

QUATIC DIET
he European Otter is truly an mphibious predator. It feeds chiefly n fish, which it locates underwater ith the help of its stiff, sensitive hiskers, adapted to feeling currents r prey movement. Otters also eat els, other aquatic animals and birds.

ADEPT DIVER
An excellent swimmer and diver, this otter makes short dives lasting about 5–30 seconds. Sometimes killed in fishing nets, the European Otter is particularly vulnerable to the loss of its riverine habitat

Family MUSTELIDAE	Species *Lontra canadensis*	Status Lower risk*

NORTH AMERICAN RIVER OTTER

This truly amphibious otter is well-adapted for swimming and diving, with eyes adapted for underwater vision, an elongated and sinuous body, webbed feet, and a flattened tail, acting as rudder. It has grey-brown or red to black, velvety fur above, whereas its underparts are silvery or greyish brown, and the throat and cheeks are paler.

- **SIZE** Body length: 66–110 cm (26–43 in). Tail: 32–46 cm (12–18 in).
- **OCCURRENCE** Canada and USA. In rivers, streams, lakes, and coastal marshes.

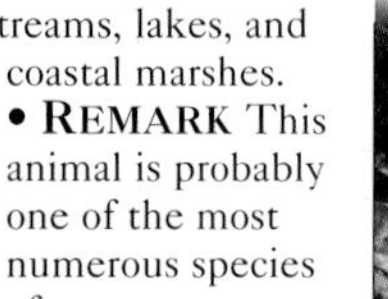

N. AMERICA

- **REMARK** This animal is probably one of the most numerous species of otters.

Social unit Solitary	Gestation 60–70 days	Young Not known	Diet

Family MUSTELIDAE	Species *Aonyx capensis*	Status Lower risk*

AFRICAN CLAWLESS OTTER

An expert swimmer, this otter has light to dark brown fur on its long, lithe body, with off-white patches on its chest. Its rear feet are webbed and have small claws on the third and fourth toes, while its forefeet have no claws and the digits are used to handle prey.

- **SIZE** Body length: 73–95 cm (20–37 in). Tail: 41–67 cm (16–26 in).
- **OCCURRENCE** Sub-Saharan Africa, excluding desert. In fresh- or seawater habitats.
- **REMARK** It is the largest of the two otter species in Africa.

AFRICA

flattened tail
dark brown fur
clawless digits resembling fingers

Social unit Social	Gestation Not known	Young 2	Diet 

Family MUSTELIDAE	Species *Aonyx cinerea*	Status Lower risk*

ORIENTAL SHORT-CLAWED OTTER

Unusually for an otter, this species has short claws that do not extend beyond the fleshy pads of its webbed feet. Equally unusually, fish are not an important part of its diet and its broad teeth are adapted to crushing molluscs. These playful and social otters form groups of about 12, keeping in touch through noises and scents.

ASIA

- **SIZE** Body length: 45–61 cm (18–24 in). Tail: 25–35 cm (10–14 in).
- **OCCURRENCE** From India to Malaysia and S.E China. In rivers, creeks, estuaries, and along coastlines.
- **REMARK** This is the smallest otter.

short limbs
lithe body
flattened, webbed feet

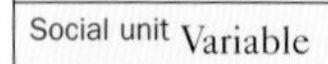

Social unit Variable	Gestation 60–64 days	Young 1–6	Diet 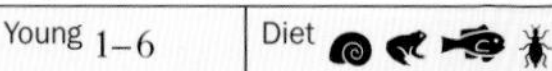

Family MUSTELIDAE	Species *Pteroneura brasiliensis*	Status Endangered

GIANT OTTER

Once widespread throughout the Amazon Basin, this otter is now very reduced in numbers. It has short legs, well-webbed toes, and a flattened, wide-based tail, which make it adept at swimming and diving. Its stout whiskers and sensitive eyes also help it to spot prey while underwater. Like other otters, this species has a sleek coat with short, dense fur. Its upperparts are rich brown, with cream spots on the chin, throat, and chest that may merge into a "bib". These noisy otters form groups of 5–9 individuals that hunt together to provide food for their young. Extremely vocal when alarmed, Giant Otters also attack in a group to defend their young against predators.

• **SIZE** Body length: 1–1.4 m (3½–4½ ft). Tail: 45–65 cm (18–26 in).
• **OCCURRENCE** N. and C. South America. In lowland tropical forest, rivers, and lakes.
• **REMARK** The largest mustelid, it faces the major threats of habitat destruction and pollution.

S. AMERICA

Social unit Social	Gestation 65–72 days	Young 2	Diet

Family MUSTELIDAE	Species *Enhydra lutris*	Status Endangered

SEA OTTER

The smallest marine mammal, this otter both lives and feeds in the ocean and has a specialized diet of abalone, sea urchins, and clams. It has excellent underwater eyesight, a rudder-like tail, and luxuriant fur (the densest of all animals) to insulate it effectively. Its lungs, twice the size of similar-sized land animals, enable it to dive up to 30 m (98 ft).

• **SIZE** Body length: 55–130 cm (22–51 in). Tail: 13–33 cm (5–13 in).
• **OCCURRENCE** N. Pacific (Kamchatka Peninsula to C. California). Along coastlines.
• **REMARK** The Sea Otter was once hunted for its pelt. It is now a legally protected species.

N. PACIFIC

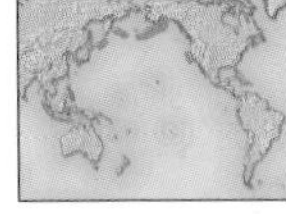

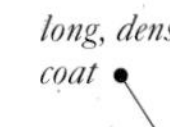

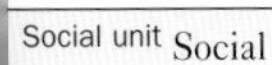

Social unit Social	Gestation 4 months	Young 1	Diet

CARNIVORES

VIVERRIDS

THE 76 SPECIES of civets, linsangs, genets, mongooses, meerkats, and other species in the Viverridae family resemble a combination of felid (cat) and mustelid (stoat). They are, however, more primitive, and have a longer snout and extra teeth.

The typical viverrid is long and slim, from the tapering snout to the lengthy tail. It has sharp senses, moves with agile stealth, and eats a mixed diet. However, some species tend towards meat-eating, and they stalk their victims in cat-like fashion. Viverrids are found from southern Europe, across Africa to southern Asia, in forest, desert, and savanna. They are generally nocturnal, and mainly terrestrial, but are good climbers. However, the Binturong is almost wholly arboreal, while the otter genets are semi-aquatic.

Civets and genets tend to have spots in longitudinal rows along their bodies for camouflage, while the mongoose species are either plain or striped. Another feature that distinguishes this family is scent glands in the anal region, used to mark territories. In civets, the gland secretions are collected by people to use as a base for perfumes.

Family VIVERRIDAE	Species *Genetta genetta*	Status Common

SMALL SPOTTED GENET

Also called the Common Genet, this slender, cat-like animal has a strikingly marked, pointed face, spotted body, and a banded tail. An excellent climber, it has semi-retractable claws and a crest along its back that can be erected. This swift, omnivorous predator hunts at night, dawn, and dusk, and is considered a pest for its raids on poultry farms. The male has a larger home range than the female, and marks its territory using scent, urine, and faeces. It builds its den in a hole or tangle of roots in thick bushes.

EUROPE, AFRICA

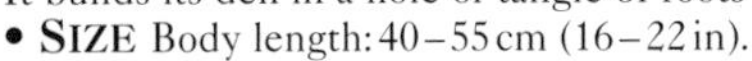

- **SIZE** Body length: 40–55 cm (16–22 in). Tail: 40–51 cm (16–20 in)
- **OCCURRENCE** W. Europe and W., E., and S. Africa. In woodland, savanna, and grassland.

Social unit Solitary	Gestation 70 days	Young 2–3	Diet

Family VIVERRIDAE	Species *Prionodon pardicolor*	Status Rare

ORIENTAL LINSANG

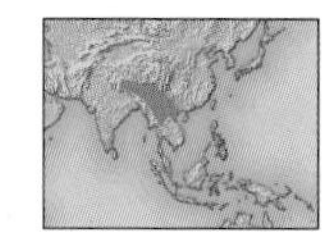

This slender carnivore "flows" like quicksilver through the branches of trees with amazing grace and agility, using its retractable claws to grip, and its tail to balance and brake while climbing. Solitary and nocturnal, the large eyes of this viverrid are adapted for night vision, and it has dark spots on its brownish orange coat and a banded tail. It is found equally on the ground and on trees, and winds its tail around its body while sleeping. The Oriental Linsang stalks its prey and kills it with a bite on the neck.

• **SIZE** Body length: 37–43 cm (14½–17 in). Tail: 30–36 cm (12–14 in).
• **OCCURRENCE** S., E., and S.E. Asia. In hill and mountain forest, also shrub and forest at lower elevations.
• **REMARK** Male linsangs are twice the size of the females.

Social unit Solitary	Gestation Not known	Young 2–3	Diet

Family VIVERRIDAE	Species *Viverra tangalunga*	Status Common

MALAYAN CIVET

Like other civets, this species has many dark spots that form lines along its coat, but is distinct in having a black-and-white neck collar, white underside, black legs and feet, and about 15 bands on its tail. It has a blackish crest along its spine, which it sometimes erects, and semi-retractable claws that enable it to climb trees. Although it occasionally takes to the trees, this nocturnal animal forages on the forest floor searching for creatures including millipedes, giant centipedes, scorpions, and small mammals such as mice.

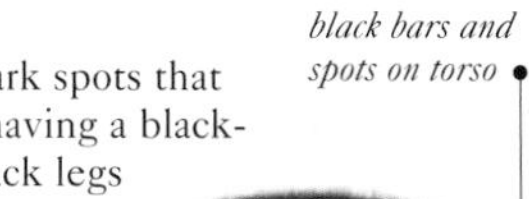

• **SIZE** Body length: 62–66 cm (24–26 in). Tail: 28–35 cm (11–14 in).
• **OCCURRENCE** Indonesia, Philippines, Malayasia, and Borneo. In lowland tropical forest and neighbouring cultivated land.

Social unit Solitary	Gestation Not known	Young Not known	Diet

Family VIVERRIDAE	Species *Paradoxurus hermaphroditus*	Status Common

PALM CIVET

This civet has a greyish brown coat with darker spots and black stripes on its back. It has a large, bushy tail, and a face mask of dark and light patches, like that of the European Polecat (see p. 246). Nocturnal by nature, this adaptable mustelid is an excellent climber, staying mainly in trees, but sometimes resting on rooftops. Fond of fruits, especially figs, it also eats buds, grasses, insects, small animals, and even poultry.

face mask of dark and light patches

black stripes on back

large, bushy tail

• **SIZE** Body length: 43–71 cm (17–28 in). Tail: 40–66 cm (16–26 in).

• **OCCURRENCE** S. and S.E. Asia, S. China. In forested areas, as well as around human settlements.

• **REMARK** This animal is known to be particularly fond of fermented palm tree juice or toddy, hence its other name, the Toddy Cat.

ASIA

Social unit Solitary	Gestation Not known	Young 2–4	Diet

Family VIVERRIDAE	Species *Arctictis binturong*	Status Common

BINTURONG

Moving cautiously among branches in search of fruits, shoots, small animals, birds, and insects, the Binturong is the only carnivore with a prehensile tail, apart from the Kinkajou (see p. 245). It has distinctive long-tufted ears and a shaggy black coat with lighter buffish tips. This animal has semi-retractable, short, and slightly curved claws, and moves through the forest canopy on the soles of its feet, which are naked up to the heels. Often seen curled up on a secluded branch, the Binturong may even continue to feed in this position. Chiefly nocturnal, it uses scent to mark its territory.

ASIA

shaggy black coat

long tufts on ear-tips

small, pointed muzzle

• **SIZE** Body length: 61–96 cm (24–38 in). Tail: 56–89 cm (22–35 in).

• **OCCURRENCE** N.E. India, Bhutan, Nepal, and S.E. Asia. In dense tropical, semi-evergreen, or deciduous forest.

Social unit Solitary	Gestation 92 days	Young 1–3	Diet

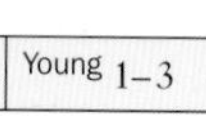

Family HERPESTIDAE	Species *Cynictis penicillata*	Status Common

YELLOW MONGOOSE

The colour of this medium-sized mongoose is variable: yellowish in the south of its range to greyish in the north. It has a grizzled appearance, and a tail that is tipped white. Equipped with long claws, it digs its own burrow or takes over the burrow system of meerkats and ground squirrels, even coexisting with these species. The Yellow Mongoose usually lives in small family groups of a breeding pair, their offspring, and young non-breeding adults. It feeds mainly on insects such as termites, ants, and beetles, but its diet also includes birds, eggs, frogs, as well as rodents.

- **SIZE** Body length: 23–33 cm (9–13 in). Tail: 18–25 cm (7–10 in).
- **OCCURRENCE** S.W. and S. Africa. In open grassland and semi-desert scrub.

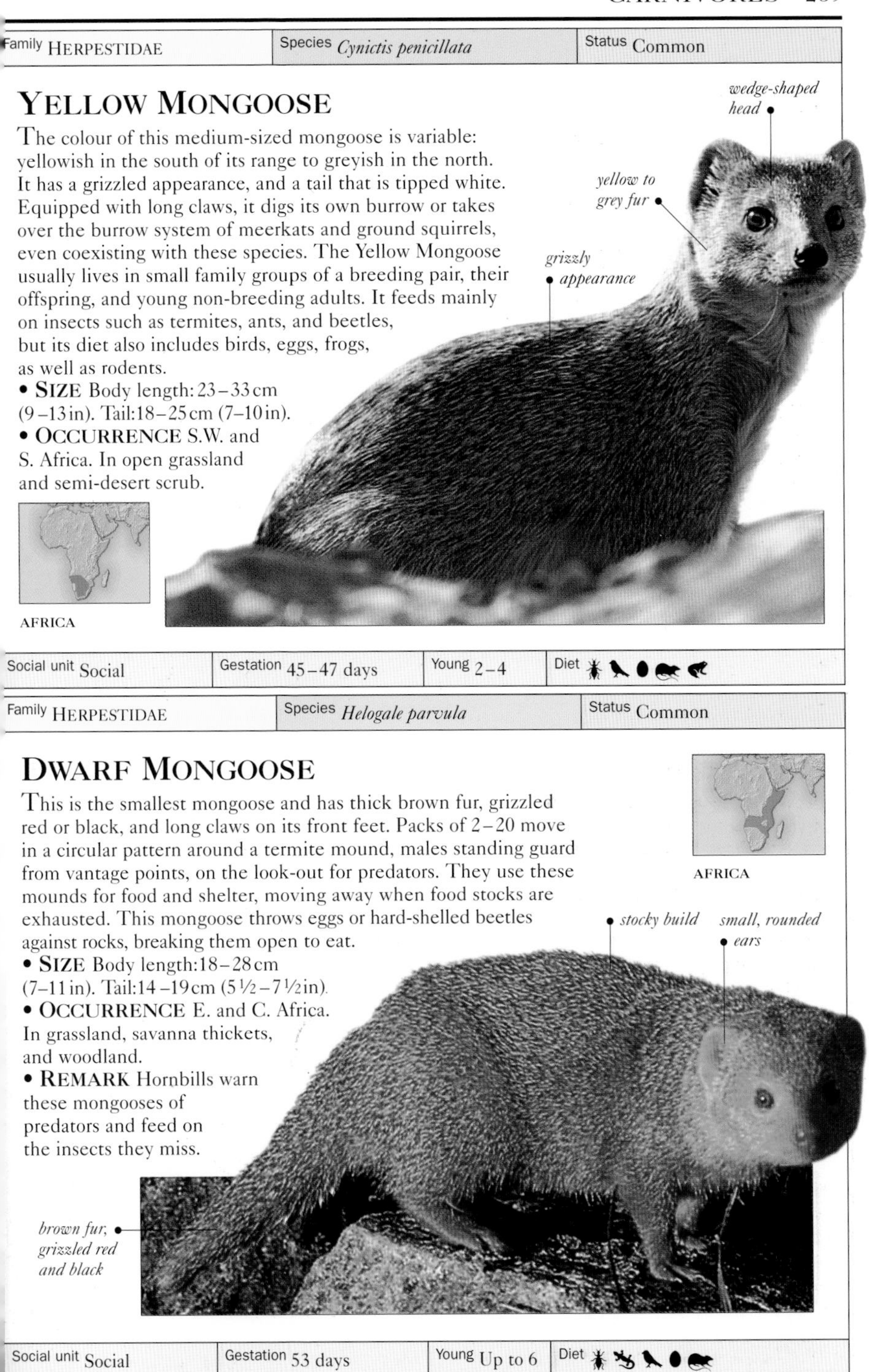

Social unit Social	Gestation 45–47 days	Young 2–4	Diet

Family HERPESTIDAE	Species *Helogale parvula*	Status Common

DWARF MONGOOSE

This is the smallest mongoose and has thick brown fur, grizzled red or black, and long claws on its front feet. Packs of 2–20 move in a circular pattern around a termite mound, males standing guard from vantage points, on the look-out for predators. They use these mounds for food and shelter, moving away when food stocks are exhausted. This mongoose throws eggs or hard-shelled beetles against rocks, breaking them open to eat.

- **SIZE** Body length: 18–28 cm (7–11 in). Tail: 14–19 cm (5½–7½ in).
- **OCCURRENCE** E. and C. Africa. In grassland, savanna thickets, and woodland.
- **REMARK** Hornbills warn these mongooses of predators and feed on the insects they miss.

Social unit Social	Gestation 53 days	Young Up to 6	Diet

Family VIVERRIDAE	Species *Mungos mungo*	Status Common

BANDED MONGOOSE

This lively, opportunistic mongoose has a stocky, brownish grey body with a striking pattern of dark bands across its back. The populations from moist habitats are darker in colour than those from drier areas. Found in packs of 15–40, they often live in enlarged cavities of termite mounds. Individuals make twittering, chirping sounds to communicate with others of the pack as they forage for food, primarily insects, during the day. This mongoose also eats eggs, breaking them open by throwing them against rocks.

• **SIZE** Body length: 30–45 cm (12–18 in). Tail: 15–30 cm (6–12 in).

• **OCCURRENCE** E. and C. Africa, south of Sahara. In woodland and savanna.

• **REMARK** It is sometimes kept as a pet.

Social unit Social	Gestation 2 months	Young 1–4	Diet

Family VIVERRIDAE	Species *Cryptoprocta ferox*	Status Endangered

FOSSA

The largest carnivore found in Madagascar, the reddish to dark brown Fossa has a cat-like head with prominent, forward-pointing eyes and rounded ears. Its short, sharp claws can be retracted and its anal glands are known to produce a strong, disagreeable odour. The Fossa has the predator's typical strong jaws and teeth. Active during the day and night, this agile leaper and climber once specialized in hunting all species of lemur, but now also stalks and kills pigs, poultry, and other domesticated animals. A solitary carnivore, it occupies a large territory of more than 4 square km (1 ½ square miles) in area, and so population densities are low. The Fossa has been successfully bred in zoos.

• **SIZE** Body length: 60–76 cm (23 ½–30 in). Tail: 55–70 cm (22–28 in).

• **OCCURRENCE** Madagascar. In undisturbed areas of tropical forest .

• **REMARK** Confined to the tiny island of Madagascar, the Fossa is threatened by the rapid depletion of its forest habitat. It is also persecuted and hunted by humans because of its frequent attacks on livestock.

Social unit Solitary	Gestation 3 months	Young 2–3	Diet

Family VIVERRIDAE	Species *Suricata suricatta*	Status Common

MEERKAT

A silvery brown, grizzled coat with eight distinctive dark bands on the lower half of the body distinguishes the Meerkat, or Suricate as it is sometimes called. This carnivore has small, pointed ears, dark circles around its eyes, and a pointed nose. Its slender tail ends in a dark tip. Adapted to burrowing, the Meerkat can close its ears underground and has long foreclaws that it uses for digging up soil and rooting for food items. Active by day, this sociable animal is found in colonies of up to 30, which live in enlarged burrow systems of ground squirrels. The male marks the home range of the group, chasing off all rival pack members. At dawn, the Meerkat emerges from its burrow and warms itself by standing up on its haunches and facing the sun. While most pack members forage, mainly for insects, some act as sentries, watching out for hawks and other predatory birds. Sentries stand at vantage points, on mounds or in bushes, and cheep or cluck out warnings. Sharper barks or growls indicate more urgent threats, upon which the whole pack dives for cover. Mating is preceded by mock fights, and the young are born during the rains, when food is available in plenty.

- **SIZE** Body length: 25–35 cm (10–14 in). Tail: 17–25 cm (6½–10 in).
- **OCCURRENCE** S.W. Africa. In semi-desert scrub and woodland, especially stony, open country.

Social unit Social	Gestation 11 weeks	Young 2–5	Diet

CARNIVORES

HYENAS AND AARDWOLF

THE THREE SPECIES of hyenas and the single species of Aardwolf superficially resemble dogs. But they are in fact more closely related to other members of the carnivore family such as civets, genets, and cats.

All four species of Hyaenidae have a distinctive body profile with a back that slopes down, due to long front legs and shorter rear ones. Other common features include a large head, big ears, muscular body and limbs, and a man[e] that extends along the back, except i[n] the Spotted Hyena.

Hyenas have immensely strong jaw[s] and teeth, usually live in families, an[d] are hunter-scavengers. The smalle[r] aardwolf is solitary and licks up ants an[d] termites. All species are nocturnal an[d] found in savanna and semi-arid habitat[s] in Africa, with the Striped Hyena'[s] range extending to southern Asia.

Family HYAENIDAE	Species *Hyaena hyaena*	Status Lower risk

STRIPED HYENA

This medium-sized, dog-like carnivore is grey or pale brown with a dark brown or black patch on its throat and five or six vertical stripes down its flanks. The mane on its neck merges with its bushy, black-and-white tail and is erected when this hyena is threatened by predators. It has well-developed forequarters and its body slopes down towards its back. The Striped Hyena essentially scavenges on kills of other predators. It crushes bones by using its massive molars and powerful jaws. It also hunts dogs, sheep, goats, and poultry, and eats invertebrates, vegetables, and fruits.

• **SIZE** Body length: 1.1 m (3 ¼ ft). Tail: 20 cm (8 in).

• **OCCURRENCE** W., N., and E. Africa and W. to S. Asia. In open habitat or sparse woodland savanna: avoids extreme desert, high altitudes, and forest.

AFRICA, ASIA

erectile neck mane

dark brown or black throat patch

horizontal black stripes on forelegs

well-developed, powerful hindlegs

Social unit Solitary	Gestation 84 days	Young 1–5	Diet

Family HYAENIDAE	Species *Parahyaena brunnea*	Status Lower risk*

BROWN HYENA

Ranging further into desert than other hyenas, this shaggy, dark brown to black species can smell carrion from a distance of 14 km (8 ½ miles). A typical hyena with its powerful jaws and shearing teeth, it feeds on almost any carcass, from those of seal pups along the Namib Desert coast to those of Springhares in the Kalahari Desert. It forms loose territorial colonies, and the size of territories varies according to the availabilty of food.

• **SIZE** Body length: 1.3 m (4 ¼ ft). Tail: 21 cm (8½ in).

• **OCCURRENCE** S. Africa – chiefly below the Kunene–Zambezi river system. In remote habitats: arid grassland, desert, and mountains.

AFRICA

dark brown to black coat

pale, tawny neck mantle

barred legs

Social unit Variable	Gestation 97 days	Young 1–5	Diet

Family HYAENIDAE	Species *Proteles cristatus*	Status Lower risk*

AARDWOLF

Smaller than hyenas, the Aardwolf is unique within its family in having a diet consisting almost wholly of termites, particularly the surface-foraging Nasute (snouted) Harvester Termites. It has a pale buff to yellowish white, down-sloping body with a shoulder mane, which it erects under stress in order to appear larger. Three stripes run down each side, while diagonal stripes mark the fore- and hindquarters. The front teeth resemble those of a hyena, but the molars are small pegs, the food being ground into smaller particles in the stomach.

• **SIZE** Body length: 67 cm (26 in). Tail: 24 cm (9 ½ in).

AFRICA

• **OCCURRENCE** E. and S. Africa. In woodland savanna and desert.

• **REMARK** Aardwolf meat is regarded as a delicacy by humans in some areas.

erect ears

three vertical black stripes on body

buff to yellowish white body

Social unit Solitary	Gestation 90 days	Young 2–4	Diet 

Family HYAENIDAE	Species *Crocuta crocuta*	Status Lower risk

SPOTTED HYENA

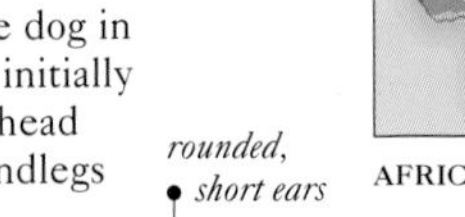

The largest species of hyena, the Spotted Hyena is like a large dog in appearance, with a sandy to greyish brown, spotted coat. The initially black spots turn brown, and finally fade out with age. A large head with powerful jaws, a downward-sloping back, with shorter hindlegs than forelegs, and a short tail that ends in a bushy, black tip further characterize this carnivore. Its fur is coarse and bristl. The hair in its mane slopes forwards and is raised when the hyena is excited. The Spotted Hyena lives in female-dominated clans which vary from up to five animals in desert to 50 or more in the fertile savanna. The female hyena is about 10 percent larger than the male, and its large external genitals look deceptively like male organs. The clan occupies a communal den and uses communal latrines, jointly defending a territory of 40–1,000 square km (15–390 square miles) by calls, and scent-markings, as well as border patrols. The Spotted Hyena is a powerful hunter and may form a pack to kill large prey such as ungulates. When alone, it preys on hares, ground birds, and fish from swamps. Gorging its food, it can consume up to one-third of its total body weight in a single meal.

- **SIZE** Body length:1.3 m (4½ft). Tail: 25 cm (10 in).
- **OCCURRENCE** W. to E., and S. Africa. In semi-desert, savanna, and woodland.
- **REMARK** The notorious hyena "laugh" signifies submission to a senior member of the clan.

rounded, short ears

powerful jaws

dog-like black muzzle

sandy to greyish brown, spotted coat

long front legs

blunt claws that are non-retractable

MOTHER AND CUB
The female Spotted Hyena is solely responsible for rearing her young. The cubs are born black and change colour after a few months.The dominant cub controls access to the mother's milk. After two or three months the cubs are transferred to a communal den and suckled by any lactating female.

Social unit Social	Gestation 110 days	Young 1–3	Diet 

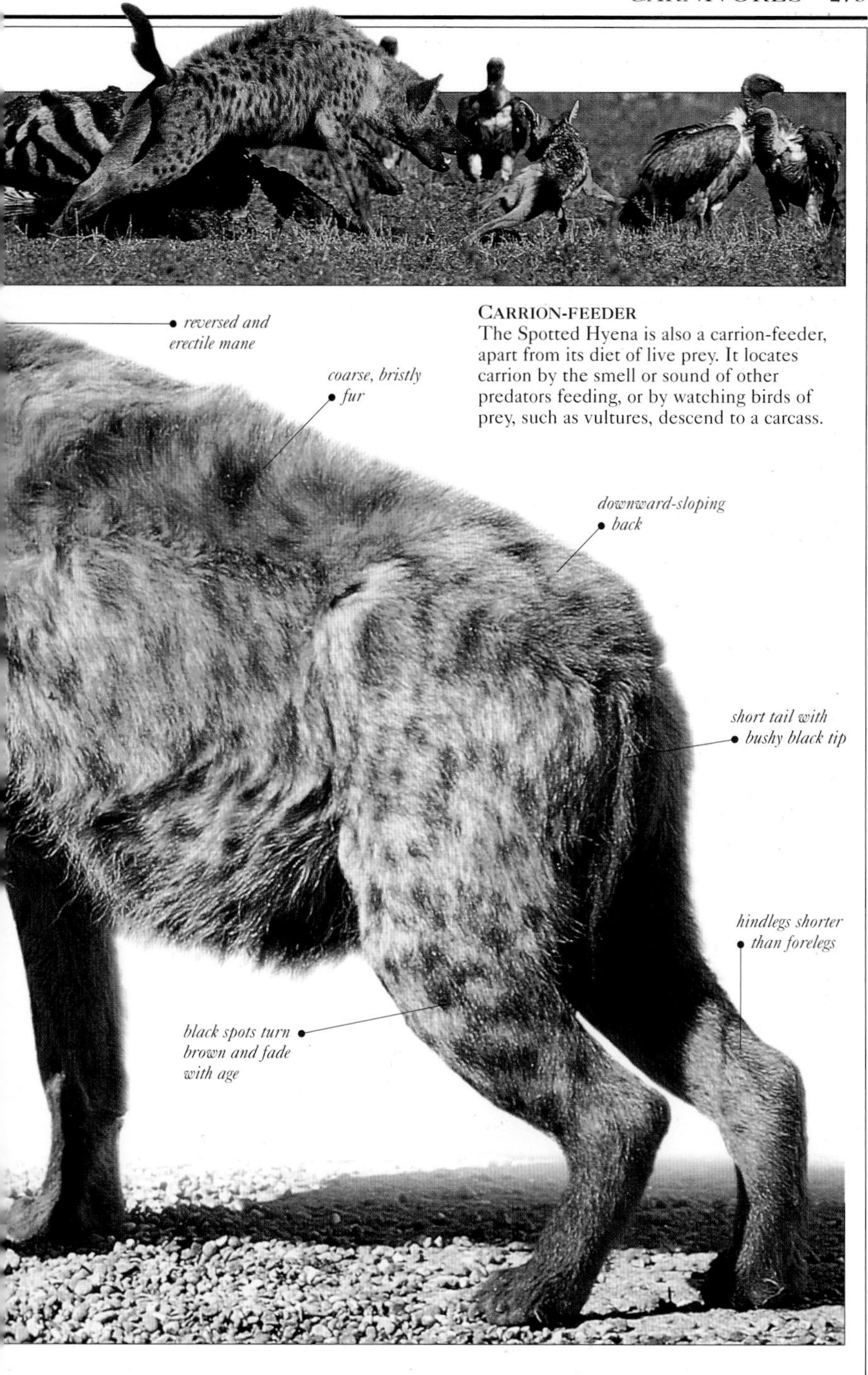

CARRION-FEEDER
The Spotted Hyena is also a carrion-feeder, apart from its diet of live prey. It locates carrion by the smell or sound of other predators feeding, or by watching birds of prey, such as vultures, descend to a carcass.

CARNIVORES
CATS

FEW MAMMAL FAMILIES display as much similarity among their members as is shown by the 38 species of cats (Felidae). All are generally solitary, stealthy, and nocturnal, and are model hunters with lightning reflexes.

A typical cat has a round face, short muzzle, wide gape, large eyes, pointed ears, sharp teeth, and a lithe body, with powerful limbs and acute senses.

Cats are found all over Eurasia, Africa and the Americas, from mountains to deserts, and swamps. The seven "big" cats are the Tiger, Lion (a rare social species), Cheetah, Jaguar, Leopard, Snow Leopard, and Clouded Leopard.

Traditionally all "small" cats have been grouped in one genus, *Felis*, although some authorities divide them into several different genera.

Family FELIDAE	Species *Felis silvestris*	Status Lower risk*

WILD CAT

Resembling a slightly larger, longer-furred (especially in winter) version of the domestic tabby cat, the Wild Cat varies from sandy to grey-brown. Active at night, dawn, and dusk, it has several dens in tree hollows, thickets, or rock crevices, where it rests by day.

longitudinal stripes on forehead

long, dense fur

EUROPE, ASIA, AFRICA

- **SIZE** Length: 50–75 cm (20–30 in). Weight: 3–8 kg (6½–18 lb).
- **OCCURRENCE** Europe, W. and C. Asia, and Africa. In mixed forest.
- **REMARK** The African subspecies, *Felis silvestris libyca*, is presumed to be the ancestor of the domestic cat.

Social unit Solitary	Gestation 63–68 days	Young 1–8	Diet

Family FELIDAE	Species *Felis chaus*	Status Lower risk*

JUNGLE CAT

This species is more aptly called the Swamp or Reed Cat, since it hunts around marshes, riverbanks, and ponds – often near human habitations. Slender and long-legged, it has an unpatterned coat varying from yellow- or brownish grey to tawny red, and its tail has black rings and a black tip. This cat rests in thick vegetation or abandoned burrows, and is active by day and night.

- **SIZE** Length: 50–94 cm (20–37 in). Weight: 4–16 kg (8¾–35 lb).
- **OCCURRENCE** N.E. Africa, W. to S.E. Asia. In dense woodland, near water.

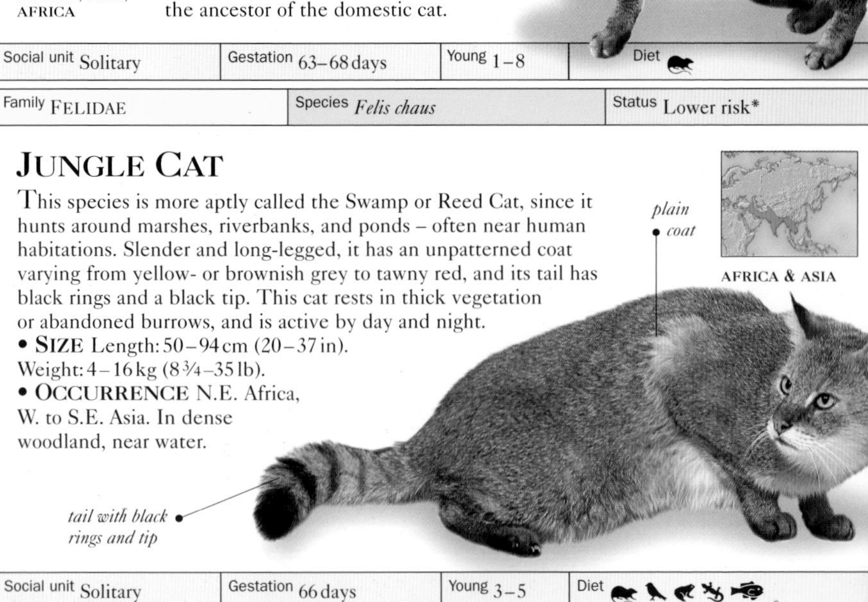

AFRICA & ASIA

Social unit Solitary	Gestation 66 days	Young 3–5	Diet

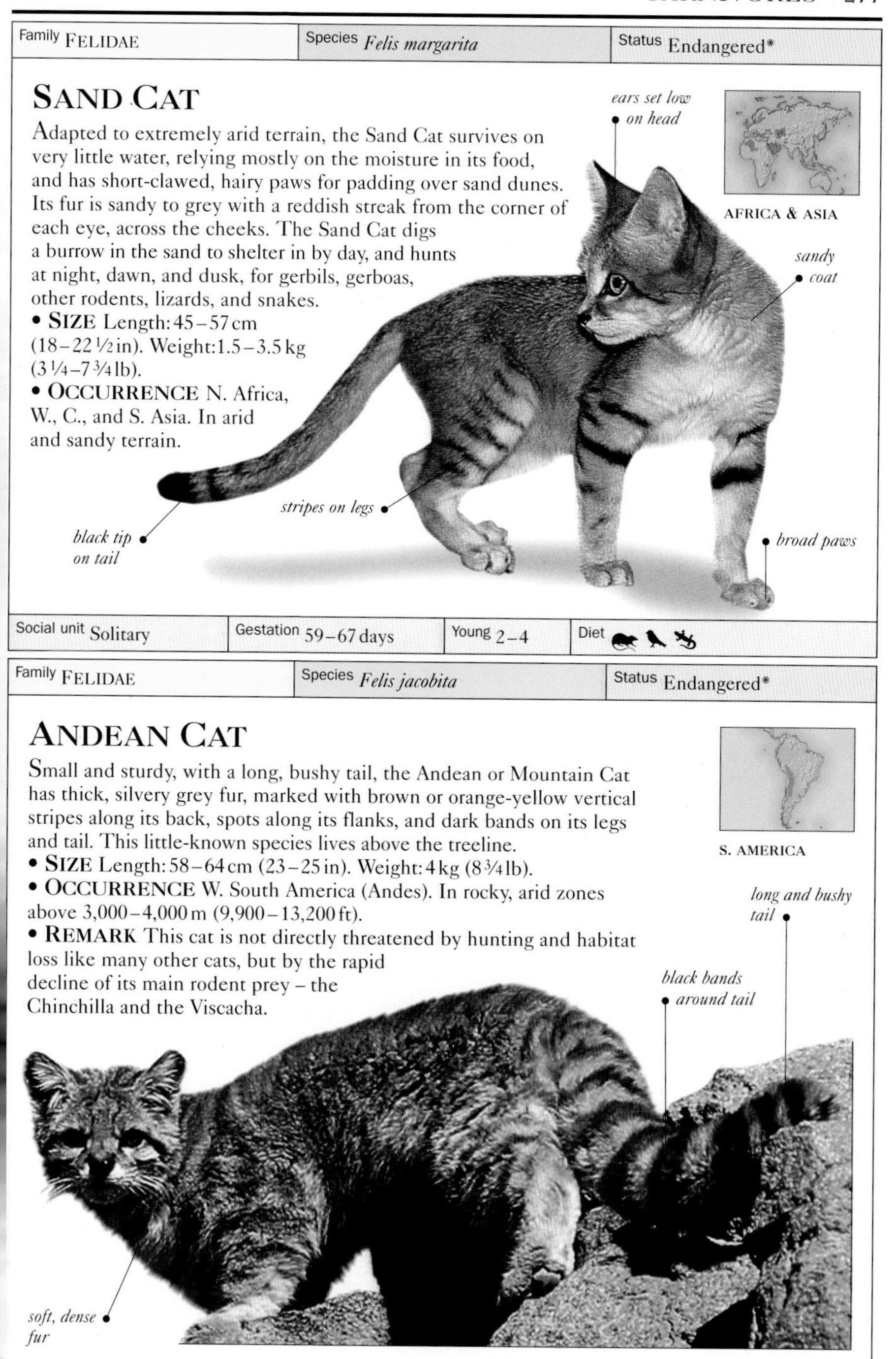

Family FELIDAE	Species *Felis margarita*	Status Endangered*

SAND CAT

Adapted to extremely arid terrain, the Sand Cat survives on very little water, relying mostly on the moisture in its food, and has short-clawed, hairy paws for padding over sand dunes. Its fur is sandy to grey with a reddish streak from the corner of each eye, across the cheeks. The Sand Cat digs a burrow in the sand to shelter in by day, and hunts at night, dawn, and dusk, for gerbils, gerboas, other rodents, lizards, and snakes.

- **SIZE** Length: 45–57 cm (18–22½ in). Weight: 1.5–3.5 kg (3¼–7¾ lb).
- **OCCURRENCE** N. Africa, W., C., and S. Asia. In arid and sandy terrain.

Social unit Solitary	Gestation 59–67 days	Young 2–4	Diet

Family FELIDAE	Species *Felis jacobita*	Status Endangered*

ANDEAN CAT

Small and sturdy, with a long, bushy tail, the Andean or Mountain Cat has thick, silvery grey fur, marked with brown or orange-yellow vertical stripes along its back, spots along its flanks, and dark bands on its legs and tail. This little-known species lives above the treeline.

- **SIZE** Length: 58–64 cm (23–25 in). Weight: 4 kg (8¾ lb).
- **OCCURRENCE** W. South America (Andes). In rocky, arid zones above 3,000–4,000 m (9,900–13,200 ft).
- **REMARK** This cat is not directly threatened by hunting and habitat loss like many other cats, but by the rapid decline of its main rodent prey – the Chinchilla and the Viscacha.

Social unit Solitary	Gestation Not known	Young Not known	Diet

Family FELIDAE	Species *Felis serval*	Status Lower risk*

SERVAL

Resembling a small Cheetah, with its slim body and long limbs, the Serval has a yellow coat with dark spots that tend to merge into longitudinal stripes on the back and head. Its large ears are rounded, and its tail has several dark rings and a black tip. The male cats are generally larger than females. Preferring to live among reeds fringing wetlands, this cat hunts rats, birds, fish, frogs, and large insects such as locusts. It usually hunts at dusk, its long neck and legs raising its head above the tall grass, allowing it to spot its prey. Once the prey is located, the Serval leaps up to 4 m (13 ft) horizontally, and 1 m (3 1/4 ft) vertically, to strike its victim with its forepaws. It uses a shrill cry to communicate, and also growls and purrs.

- **SIZE** Length: 60–100 cm (23 1/2–39 in). Weight: 9–18 kg (20–40 lb).
- **OCCURRENCE** W., C., and E. Africa. Along vegetated riverbanks and streams.
- **REMARK** Servals help farmers, as they prey on locusts and rats, and rarely hunt for livestock. Due to their concentration along riverbanks, they are exposed to the dangers of being hunted, and habitat loss.

HUNTED
Mercilessly hunted in farming areas of South Africa, the Serval is today considered to be rare in that country. It no longer occurs in populated areas.

Social unit Solitary	Gestation 73 days	Young 2	Diet

Family FELIDAE	Species *Felis viverrinus*	Status Lower risk

FISHING CAT

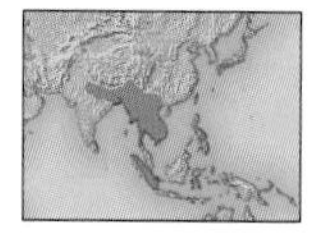

Like a civet in proportions, with a long, stocky body and relatively short legs, the Fishing Cat is a semi-aquatic hunter of fish, frogs, snakes, water insects, crabs, crayfish, and shellfish. However, its adaptations to water are largely behavioural, as its toes are only slightly webbed, and its teeth are not suitable for grasping slippery prey. It scoops prey from the water with its paws, often diving, and sometimes it suddenly surfaces under a waterbird. The Fishing Cat's coat is grizzled olive-grey with dark brown spots running along its body.

- **SIZE** Length: 75–86 cm (30–34 in). Weight: 8–14 kg (18–31 lb).
- **OCCURRENCE** India to S.E. Asia. Along lakes, riverbanks, and creeks; in mangrove swamps and marshes.
- **REMARK** Drainage of wetlands for agriculture is adversely affecting this species, which is dependent on water-edge habitats.

dark stripes from forehead to neck

short legs

dark brown spots on fur

Social unit Solitary	Gestation 63 days	Young 1–4	Diet

Family FELIDAE	Species *Felis planiceps*	Status Vulnerable

FLAT-HEADED CAT

Slightly smaller than a domestic cat, the Flat-headed Cat lives around rivers, lakes, swamps, and canals. A semi-aquatic predator of fish, it also takes shrimps, frogs, rodents, and small birds. Its toes are partially webbed and the claws cannot be retracted fully, and its upper premolars are large and sharp, helping it to grip slippery prey. The most outstanding feature of this cat, however, is its long and narrow skull with a flattened forehead and small, low-set ears. Its coat is dark brown, tinged silver.

- **SIZE** Length: 41–50 cm (16–20 in). Weight: 1½–2 kg (3¼–4½ lb).
- **OCCURRENCE** S.E. Asia. In riverine forest, swamps, and around irrigation canals.

eyes set close together

small ears

brown fur tinged silver

Social unit Solitary	Gestation 56 days	Young Not known	Diet

Family FELIDAE	Species *Felis marmorata*	Status Vulnerable*

MARBLED CAT

This cat resembles a small Clouded Leopard (see p. 288) in appearance. Varying from brownish grey to bright yellow or reddish brown, it has large dark blotches, outlined in black, along its sides. Its limbs and underparts have solid black dots, and its long, bushy tail is spotted and tipped with black. Little is known about the lifestyle of this cat, but it is thought to be nocturnal and partly arboreal. It preys mainly on birds, squirrels, and rats, and probably lizards and frogs.

- **SIZE** Length: 45 – 53 cm (18 – 21 in). Weight: 2 – 5 kg (4 ½ – 11 lb).
- **OCCURRENCE** N.E. India to S.E. Asia. In tropical forest.
- **REMARK** This cat is affected by human disturbance and habitat loss.

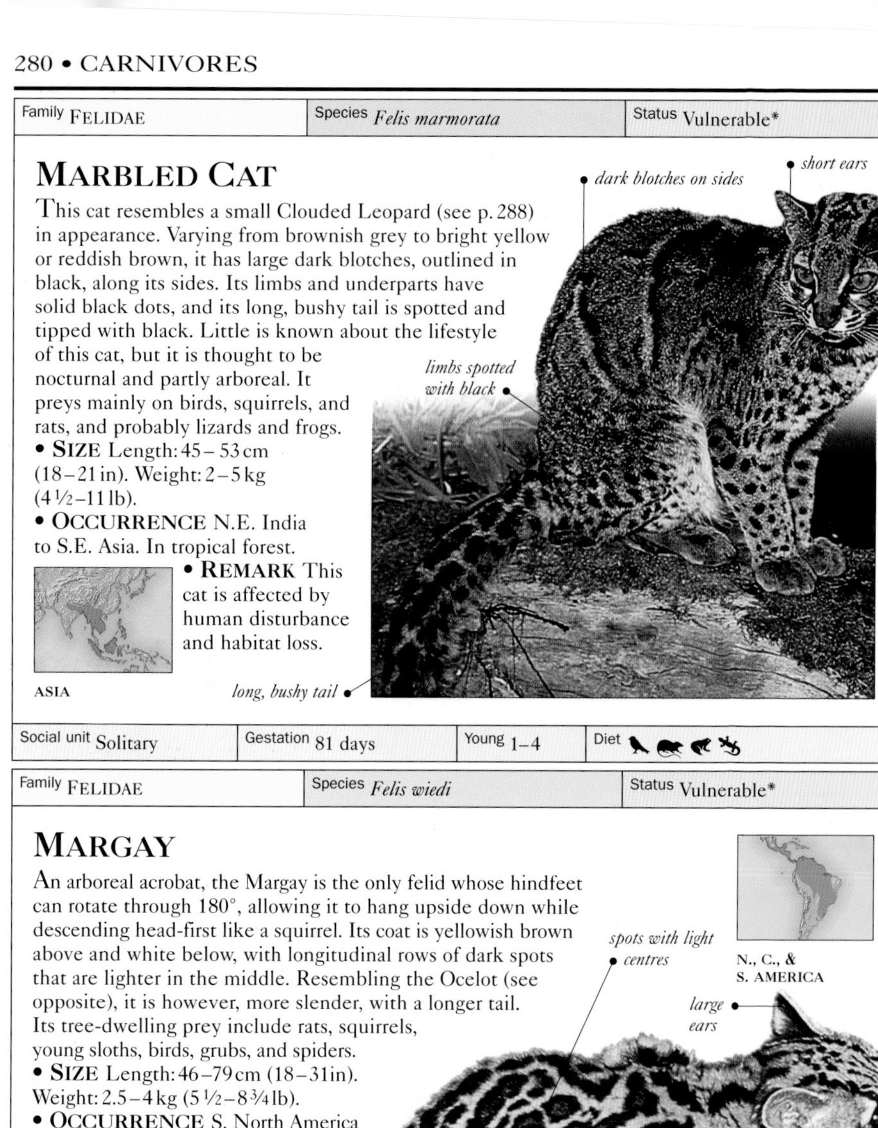

Social unit Solitary	Gestation 81 days	Young 1–4	Diet

Family FELIDAE	Species *Felis wiedi*	Status Vulnerable*

MARGAY

An arboreal acrobat, the Margay is the only felid whose hindfeet can rotate through 180°, allowing it to hang upside down while descending head-first like a squirrel. Its coat is yellowish brown above and white below, with longitudinal rows of dark spots that are lighter in the middle. Resembling the Ocelot (see opposite), it is however, more slender, with a longer tail. Its tree-dwelling prey include rats, squirrels, young sloths, birds, grubs, and spiders.

- **SIZE** Length: 46–79 cm (18–31 in). Weight: 2.5–4 kg (5 ½–8 ¾ lb).
- **OCCURRENCE** S. North America to Central and South America. In tropical rainforest.
- **REMARK** With the declining availability of Ocelot pelts, the Margay has become a highly sought-after small cat in the fur trade.

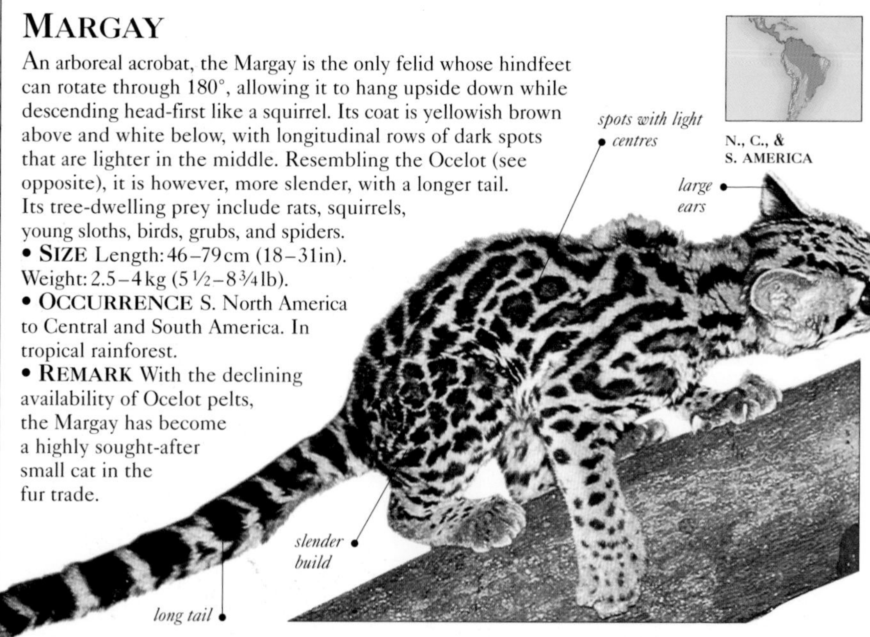

Social unit Solitary	Gestation 76–85 days	Young 1	Diet

Family FELIDAE	Species *Felis pardalis*	Status Lower risk*

OCELOT

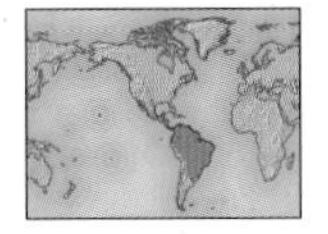

Typically cat-like in its lifestyle, the Ocelot is extremely adaptable by nature. It inhabits a wide range of habitats from humid forests to dry scrubland – as long as there is dense cover provided by vegetation. Mainly terrestrial, it also climbs, jumps, and swims well, and rests by day in a tree hollow or on a branch. A nocturnal hunter, the Ocelot takes a variety of prey such as rodents, birds, lizards, fish, bats, young deer, monkeys, armadillos, anteaters, and turtles.

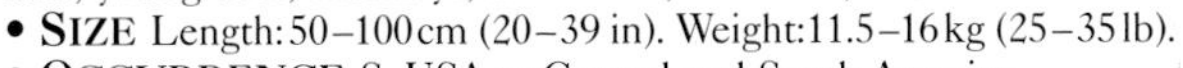

• **SIZE** Length: 50–100 cm (20–39 in). Weight: 11.5–16 kg (25–35 lb).
• **OCCURRENCE** S. USA to Central and South America. In tropical rainforest, grassland, and swamp.
• **REMARK** In the 1960s and 1970s, around 200,000 Ocelots were lost to the fur trade each year. Protected in much of its range, the Ocelot's numbers have risen – although it is now threatened by deforestation.

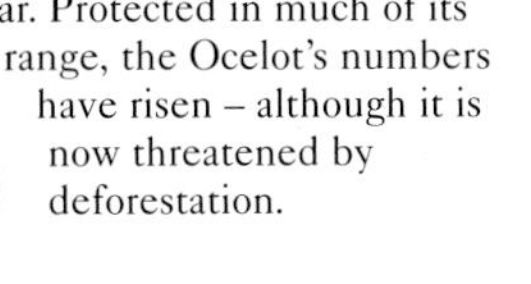

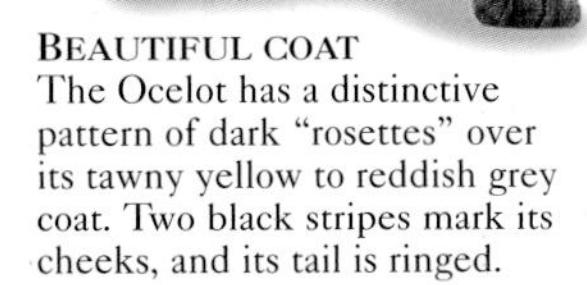

BEAUTIFUL COAT
The Ocelot has a distinctive pattern of dark "rosettes" over its tawny yellow to reddish grey coat. Two black stripes mark its cheeks, and its tail is ringed.

chain-like pattern of "rosettes"

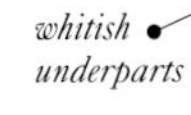

whitish underparts

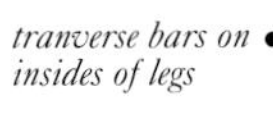

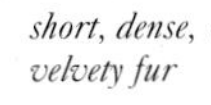

Social unit Solitary	Gestation 79–85 days	Young 1–3	Diet

Family FELIDAE	Species *Felis rufus*	Status Lower risk*

BOBCAT

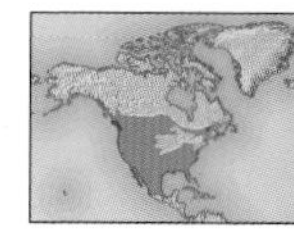

N. AMERICA

A short, "bobbed" tail gives this cat its name. Mainly varying between shades of buff and brown, its coat is spotted and lined with dark brown and black. The Bobcat resembles the Eurasian Lynx (see opposite), but is smaller, has more slender legs, and its ears (distinctly marked with black on the back) are less conspicuously tufted, or not at all. As with the Lynx, it has a ruff of fur extending from its ears to the jowls. The Bobcat is seen in a greater variety of habitats than the Lynx. It is mainly nocturnal and terrestrial, although it can climb with ease, and shelters by day in a thicket, hollow tree, or crevice. It usually stalks its prey with great stealth and seizes it after a swift leap. Its diet consists mainly of rabbits and birds, although it may hunt larger mammals such as deer in winter.

• **SIZE** Length: 65–110 cm (26–43 in).
Weight: 4–15.5 kg (8¾–34 lb).

• **OCCURRENCE** S. Canada, USA, and Mexico. In desert, brush, mixed woodland, coniferous forest.

• **REMARK** Occasionally trapped and hunted by humans, it has been exterminated over much of the Ohio Valley, upper Mississippi Valley, and the southern Great Lakes region.

MIXED SPOTS
The spots on the coat vary in density: they are either prominent all over, or only on the underside.

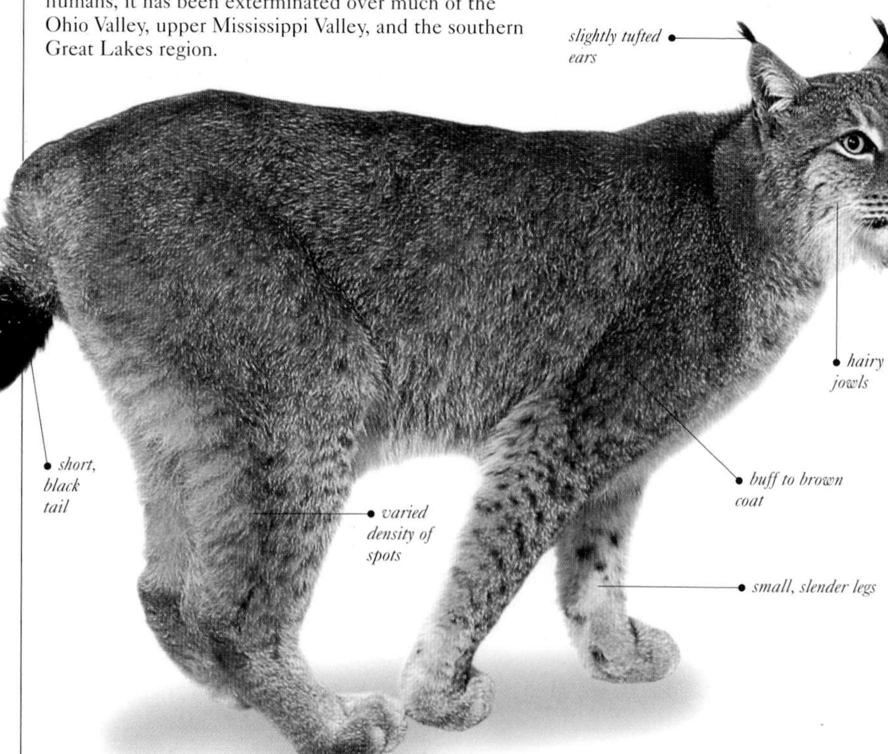

Social unit Solitary	Gestation 60–70 days	Young 2–3	Diet

Family FELIDAE	Species *Felis lynx*	Status Lower risk*

EURASIAN LYNX

Primarily an inhabitant of mixed forest, the Eurasian Lynx has now been driven to open woods and rocky mountains, due to human presence and persecution. Its exceedingly dense coat is more variable than that of any other cat. There are three predominant patterns: mainly striped, mostly spotted (as shown below), and plain. The background colour ranges from reddish brown or yellowish grey, to almost white. Its underparts are usually white, its ears have prominent black tufts, and its tail is tipped black. Large feet like "snowshoes", which are thickly haired in winter, enable it to walk on snow. In summer, the Eurasian Lynx's coat is short with prominent markings, while in winter it grows longer. The Eurasian Lynx is able to kill prey three or four times its own size and hunts mainly ungulates such as deer, goats, and sheep. However, if these are scarce, it also preys on pikas, hares, rodents, and birds.

• **SIZE** Length: 0.8–1.3 m (2 1/2–4 1/4 ft). Weight: 8–38 kg (18–84 lb).

• **OCCURRENCE** N. Europe to E. Asia. In mixed forest and steppe.

• **REMARK** The Eurasian Lynx remains rare in W. Europe despite conservation efforts, as it is hunted by farmers and frequently killed in road accidents. It has also been observed that male cubs have a low survival rate, probably for genetic reasons.

EURASIA

IBERIAN LYNX
Around half the size of the Eurasian Lynx, and an endangered species, the Iberian Lynx *(Felis pardina)* is found in S.W. Europe.

Social unit Solitary	Gestation 67–74 days	Young 1–4	Diet

Family FELIDAE	Species *Felis caracal*	Status Lower risk*

CARACAL

AFRICA & ASIA

Sometimes called the Desert Lynx because of the arid, scrubby areas that it inhabits, the Caracal is usually reddish brown, with a white chin, throat, and belly, and a narrow black band from the eye to the nose. Its distinctive pointed ears are black on the outside, and have long black tufts. The Caracal is well known for its ability to spring vertically and "bat" flying birds with its paws, often leaping as high as 3 m (10 ft). Largely nocturnal, this cat climbs and jumps well and is also believed to be the fastest feline of its size. It stalks and kills its prey after a quick dash or leap. The Caracal is territorial and marks its home range with urine. Usually solitary, it sometimes forms small groups comprising adults and young. It makes its dens in porcupine burrows, rocky crevices, or dense vegetation.

- **SIZE** Length: 60–91 cm (23½–36 in). Weight: 6–19 kg (13–42 lb).
- **OCCURRENCE** Africa, W., C., and S. Asia. In woodland, savanna, and scrubland.

HUNTER'S FRIEND
Despite its ferocious appearance, the Caracal is easily tamed and is sometimes used to assist hunters in India and Iran.

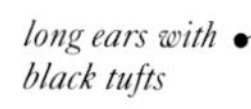

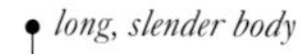

white chin

Social unit Solitary	Gestation 69–81 days	Young 1–6	Diet 

Family FELIDAE	Species *Felis aurata*	Status Lower risk*

AFRICAN GOLDEN CAT

This little-known, medium-sized inhabitant of tropical rainforest varies from grey to reddish brown, and may be faintly spotted or uniformly coloured. Its cheeks, chin, and underparts are white, and it has long legs, a small head, and large paws. The male is usually larger than the female. Active during the day and night, it is mainly terrestrial but prefers areas near water. It feeds on prey such as rodents, hyraxes, small antelopes, and birds, which it usually stalks and kills with a sudden rush.

- **SIZE** Length: 61–100 cm (24–39 in). Weight: 5.5–16 kg (12–35 lb).
- **OCCURRENCE** W. and C. Africa. In forest as well as mountains.

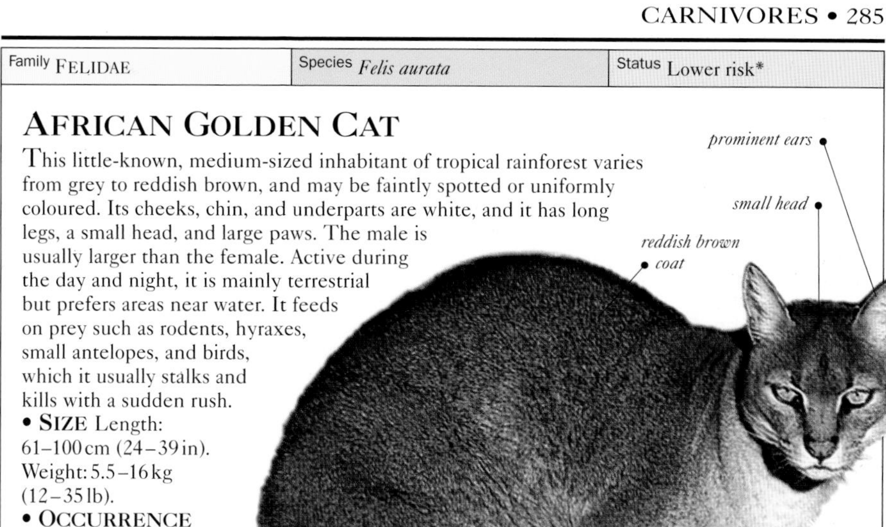

AFRICA

Social unit Solitary	Gestation Not known	Young Not known	Diet

Family FELIDAE	Species *Felis yagouaroundi*	Status Lower risk

JAGUARUNDI

More mustelid than felid in appearance, with its pointed snout, long body, and short legs, the Jaguarundi varies from black in forested areas to pale grey-brown or red in dry shrubland. Much less nocturnal than most cats, it hunts in the morning and evening and forages mainly on the ground.

- **SIZE** Length: 55–77 cm (22–30 in). Weight: 4.5–9 kg (10–20 lb).
- **OCCURRENCE** S. USA to South America. In lowland forest and thickets.
- **REMARK** It is hunted and subject to habitat destruction.

N., C., & S. AMERICA

Social unit Solitary	Gestation 70–75 days	Young 1–4	Diet

Family FELIDAE	Species *Puma concolor*	Status Lower risk

PUMA

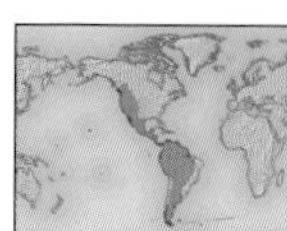

Also known as the Panther, Cougar, or Mountain Lion, this species larger than some "big" cats (it is about the same size as the Leopard), but is thought to be closer to "small" cats in classification. The Puma has an elongated body, small head, and short face, the Puma has uniformly buff-coloured fur. It has powerful and muscular limbs, and its hindlegs are longer than its forelegs. Throughout its range, the Puma's most important food is deer – especially Mule Deer and Elk. It stealthily stalks its prey, leaping upon the victim's back or seizing it after a rapid dash. It drags the carcass to a sheltered spot and eats its fill, then covers the remains with leaves and debris, for later consumption. This agile cat can leap to a height of 5.5 m (18 ft) from the ground and is also a good swimmer, although it prefers not to enter water. Usually solitary, individuals deliberately avoid each other except for a brief period of courtship. There is no fixed den, except when females are rearing their young. However, temporary shelter is taken in dense vegetation, rocky crevices, and caves. The Puma cannot roar like other big cats. Instead, it communicates by growls, hisses, and bird-like calls, and emits an eerily human-like scream during courtship.

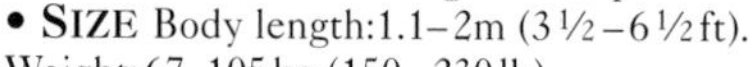

• **SIZE** Body length: 1.1–2m (3½–6½ ft).
Weight: 67–105 kg (150–230 lb).

• **OCCURRENCE** W. and S. North America, Central America, and South America. In montane coniferous forest, lowland tropical forest, swamps, grassland, dry brush country, or any other area with adequate cover and prey.

• **REMARK** Amazingly adaptable, the Puma has the greatest natural distribution of any indigenous mammal in the Western Hemisphere, with the exception of humans.

coat uniformly buff-coloured

SPOTTED KITTENS
Puma kittens have spots on their body until they are about six months old. They begin to feed on meat at six weeks and, if born in spring, are able to go hunting with the mother by autumn, making their own kills by winter. However, they remain with their mother for several more months.

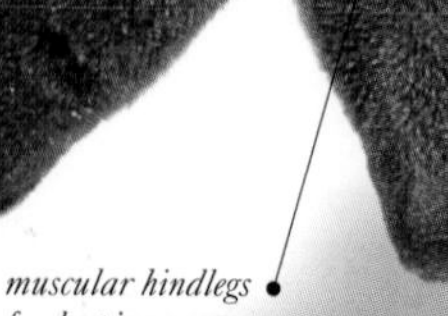

muscular hindlegs for leaping

Social unit Solitary	Gestation 90–96 days	Young 1–6	Diet

THREATENED
Extensively hunted since the arrival of the European colonists, the Puma has also declined in numbers in places where agricultural interests have led to the destruction of forests.

Family FELIDAE	Species *Neofelis nebulosa*	Status Vulnerable

CLOUDED LEOPARD

The smallest of the big cats, this species derives its name from the "cloudy", black-bordered, dark patches, on its tawny, grey, or silver coat. Its forehead and legs are spotted, and the tail banded. Little is known about this elusive cat, but it is believed to be mainly arboreal, avoiding danger, resting, and stalking prey, from overhead branches. Its climbing skill rivals that of many small cats: it runs down trees head first or moves about among horizontal branches with its back to the ground.

- **SIZE** Length: 60–110 cm (23 ½–43 in). Weight: 16–23 kg (35–51 lb).
- **OCCURRENCE** S., S.E., and E. Asia. In tropical and temperate forest, mountains, and grassland.
- **REMARK** It is threatened by habitat loss and hunted excessively.

ASIA

dark "clouds"

banded tail with spotted base

Social unit Solitary	Gestation 93 days	Young 1–5	Diet

Family FELIDAE	Species *Uncia uncia*	Status Endangered

SNOW LEOPARD

Large dark rosettes on a ground of pale grey or creamy smoke-grey characterize this woolly cat. Like the Leopard (see pp. 290–291), it feeds on a wide range of prey. Often active by day, especially in the early morning and late afternoon, the Snow Leopard dens in a cavern or crevice.

- **SIZE** Length: 1–1.3 m (3 ¼–4 ¼ ft). Weight: 25–75 kg (55–165 lb).
- **OCCURRENCE** C., S., and E. Asia. In mountains and alpine meadows.
- **REMARK** Hunted for its fur, it is also threatened by the scarcity of its natural prey, and the encroachment of livestock on pastures.

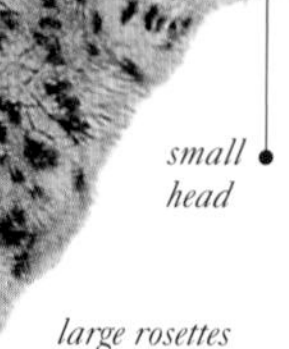

solid spots on neck

small head

large rosettes on body

rings on tail

whitish below

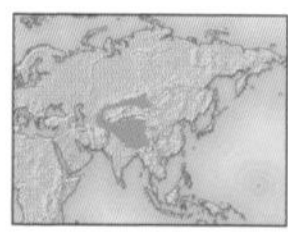

ASIA

Social unit Solitary	Gestation 90–103 days	Young 1–5	Diet 

Family FELIDAE	Species *Panthera onca*	Status Lower risk

JAGUAR

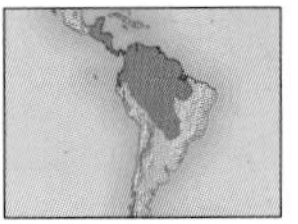

C. & S. AMERICA

The only "big" cat found in the Americas, the Jaguar looks like the Leopard (see pp. 290–291), but its coat is patterned with dark-centred rosettes and it is more squat and powerfully built, with a large, broad head, and heavily muscled quarters. Along the midline of its back is a row of elongated spots that may merge to form a solid line. Melanistic individuals are common; the spots of such animals may still be seen in bright light. Living in watery habitats – permanent swamps and seasonally flooded forest – the Jaguar is an excellent swimmer, even feeding on aquatic prey such as crocodilians. However, it hunts mostly on the ground, stalking or ambushing its prey and dragging it away to a sheltered spot to eat. Solitary and territorial, the Jaguar marks its range with urine and tree scrapes, communicating with others of its kind through a variety of sounds, including roars, grunts, and mews.

- **SIZE** Length: 1.1–1.9 m (3½–6¼ ft). Weight: 36–160 kg (79–350 lb).
- **OCCURRENCE** Central to N. South America. In tropical forest, savanna, scrub, and wetland.
- **REMARK** Despite legal protection and reduced hunting for fur, the Jaguar is increasingly at risk from habitat loss and persecution as a predator, especially on cattle ranches.

elongated spots along midline of body

ROSETTES AND SPOTS
Varying from pale yellow through reddish yellow to reddish brown, the Jaguar has black rings or rosettes on its shoulders, back, and flanks. The head, neck, limbs, and underparts have black spots.

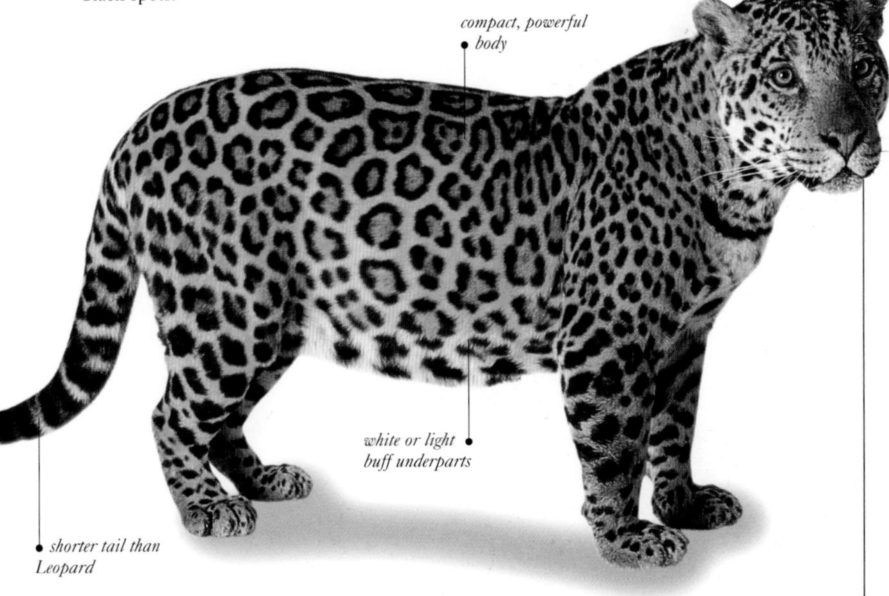

Social unit Solitary	Gestation 93–105 days	Young 1–4	Diet

Family FELIDAE	Species *Panthera pardus*	Status Lower risk

LEOPARD

AFRICA, ASIA

Widely distributed over different habitats, the Leopard is extremely varied both in appearance and in prey preference. Varying in colour with habitat, these large cats may be pale yellow in deserts and deeper yellow in grassland. The Leopard's diet ranges from tiny creatures such as dung beetles to animals much larger than itself, such as antelopes. A large victim may provide enough food for two weeks, although such kills are usually made every three days, and twice as often by a female with cubs. The leopard's large head has powerful jaws that enable it to kill and dismember prey. An adept climber with immensely strong shoulders and forelimbs, the Leopard often drags its prey up into trees, for immediate consumption as well as for caching (hiding for future use). Up in the branches, it can eat undisturbed, the meat remaining safe from scavengers such as hyenas and jackals. Individual Leopards usually keep to a specific area, defending it against others, although the range of a male may include that of one or more females. The Leopard is a solitary animal, but there are reports of males remaining with females after mating and helping to rear the cubs. More adaptable than the Tiger to the presence of humans, the Leopard often hunts for prey within a few kilometres of large towns. It has survived well, despite numerous threats to its existence.

- **SIZE** Length: 0.9–1.9 m (3–6 ¼ ft). Weight: 37–90 kg (82–200 lb).
- **OCCURRENCE** W., C., S., E., and S.E. Asia, and Africa. In lowland forest, mountains, grassland, brush country, and semi-arid desert.
- **REMARK** Persecuted as a predator, and for its beautiful spotted fur, the Leopard is also threatened by the loss of habitat and prey.

straw or greyish buff, to ochre and chestnut coat

dark spots arranged as rosettes on flanks

ringed tail

Social unit Solitary	Gestation 90–105 days	Young 1–6	Diet

BLACK PANTHER
The Leopard sometimes exhibits melanism, or an excessive amount of dark pigment in its skin and fur, but its spots are usually faintly visible.

LIFE IN TREES
Usually nocturnal, the Leopard rests by day on the branches of a tree or remains hidden in dense vegetation. Large trees are important for a female with cubs, both to escape from danger and as a place to hide food for her young.

Family FELIDAE	Species *Panthera tigris*	Status Endangered

TIGER

The largest member of the cat family, the Tiger is instantly recognizable by its orange coat, vividly patterned with black stripes and white markings. Although eight subspecies of tiger have been recognized, three have become extinct since the 1950s. The remaining five subspecies are all endangered, some critically so. The Tiger's size, coat, colour, and markings vary according to the subspecies. Once found as far west as eastern Turkey, the Tiger now exists in small, scattered populations from India to Vietnam, and in Siberia, China, and Sumatra. It lives in a variety of habitats, ranging from tropical forest to the freezing steppe, but its basic requirements are the same: dense cover, access to water, and sufficient prey. Tigers may require as much as 40 kg (88 lb) of meat at a time and return to a large kill for 3–6 days. Animals taken are mainly deer and pigs, cattle in certain regions, as well as monkeys, birds, reptiles, and fish. Although solitary, tigers also travel in groups, with a male occasionally resting and feeding with a female and her cubs. The Bengal Tiger (*Panthera tigris tigris* – shown here) is the most common subspecies and exhibits the classic tiger coat: deep orange with white undersides, cheeks, and eye areas, and distinctive black markings that help camouflage it in the tall jungle grass. The other existing subspecies are the Indo-Chinese Tiger *(P. t. corbetti)*, the Chinese Tiger *(P. t. amoyensis)*, the Sumatran Tiger *(P. t. sumatrae)*, and the Siberian Tiger *(P. t. altaica)*.

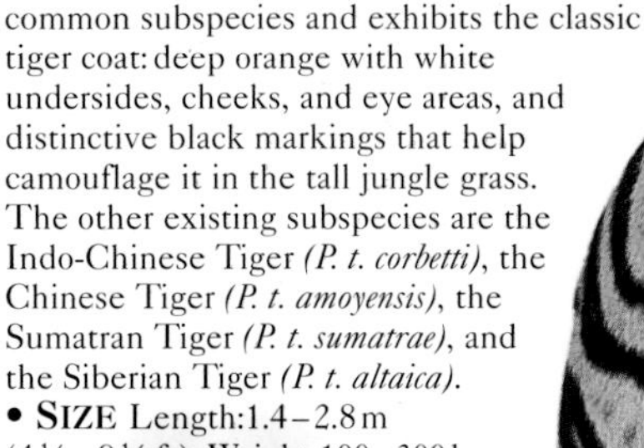

- **SIZE** Length: 1.4–2.8 m (4½–9¼ ft). Weight: 100–300 kg (220–660 lb).
- **OCCURRENCE** S. and E. Asia. In tropical and evergreen forest, mangrove swamps, grassland, savanna, and rocky country.
- **REMARK** Protected in most areas, the Tiger continues to be hunted illegally for its skin and its body parts. Several conservation projects now exist to monitor and safeguard Tiger populations.

Social unit Solitary	Gestation 93–111 days	Young 1–6	Diet

SIBERIAN TIGER
Found in Siberia and Manchuria, the Siberian Tiger (*Panthera tigris altaica*) is the biggest living cat. Its coat is the lightest among all tigers, and its fur is long and dense to protect it against the cold. Critically endangered, its numbers may be as low as 150–200.

SUMATRAN TIGER
The Sumatran Tiger *(Panthera tigris sumatrae)* is classified as critically endangered. Its numbers have decreased from 1,000 in the 1970s to about 400 in the wild, and around 194 in captivity.

Family FELIDAE	Species *Panthera leo*	Status Vulnerable

LION

This "big" cat is unique among the usually solitary felids in forming long-term social bonds with related females and their young, forming a group called a pride, and remaining together over several generations. Around two or three unrelated males or four or five related males form a loose "coalition" and defend a large area against other coalitions of males. They mate with the pride occupying this territory for a period of two or three years, until they are driven out by a contending group of males. The Lion usually hunts by a low stalk, alternately creeping and freezing, making the best use of cover. Pride members may hunt cooperatively, fanning out to close in on large quarry such as Zebra, Wildebeest, Impala, or Buffalo. Males living with a pride let the females hunt, but have first access to the kill. Varying widely in colour from light buff and silvery grey, to yellowish red and dark ochre-brown, the Lion is instantly recognizable by the male's magnificent mane.

- **SIZE** Length: 1.7–2.5 m (5 ½ – 8 ¼ ft). Weight: 150 – 250 kg (330 – 550 lb).
- **OCCURRENCE** Africa, S. Asia. In grassy plains, savanna, open woodland, and scrub.
- **REMARK** Seriously threatened by the expansion of human activities and persecution, the Lion has become extinct in Asia except in the Gir Forest in N.W. India.

AFRICA & ASIA

relatively large head

LION CUBS
Sometimes born with their eyes open, Lion cubs follow their mother after three months, beginning to participate in kills at about 11 months. Females of the pride tend to give birth around the same time, and suckle each other's young.

Social unit Social	Gestation 110–119 days	Young 1–6	Diet

Family FELIDAE	Species *Acinonyx jubatus*	Status Vulnerable

CHEETAH

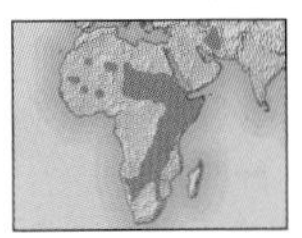

AFRICA & ASIA

The fastest land animal in the world, the Cheetah can reach a speed of over 100 kph (60 mph), for a burst of 10–12 seconds, before it begins to overheat. If its prey, consisting chiefly of small and medium-sized ungulates, can stay ahead for longer than this, it invariably escapes. Once the Cheetah overtakes, it knocks the animal down by the sheer speed of its charge, seizes it by the throat, and strangles it. A female with cubs may make a kill every day, while lone adults hunt every 2–5 days. More sociable than any of the other "big" cats, except Lions, siblings leave their mother at 13–20 months, but may stay together for longer; brothers sometimes live with each other for years. The Cheetah has a slim body, long legs, and a rounded head with distinctive black stripes running down its face. Its coat is yellowish, with small black spots – the cats found in the desert being paler with smaller spots. The King Cheetah from southeast Africa has the largest spots.

- **SIZE** Length: 1.1–1.5 m (3 ½–5 ft). Weight: 21–72 kg (46–160 lb).
- **OCCURRENCE** Africa and W. Asia. In grassland and arid bush.
- **REMARK** The Cheetah seems to be more vulnerable than other big cats to environmental changes brought about by human activity. Originally found from the Arabian Peninsula to C. India, and throughout most of Africa, it has now virtually disappeared from Asia, except the western parts.

HUNTING
The Cheetah is unusual among cats in stalking rather than ambushing its prey, and then charging from about 70–100 m (230–330 ft) away.

very distinctive black face stripes

small black spots on coat

tawny to pale buff or greyish white coat

tail with black rings

very long legs

pale buff to white underside

Social unit Solitary/Pair	Gestation 90–95 days	Young 1–8	Diet

SEALS AND SEA LIONS

THE ORDER PINNIPEDIA (meaning "flipper-feet"), which is sometimes incorporated as a suborder of Carnivora, contains 34 species in three families: seals (Phocidae), sea lions and fur seals (Otariidae), and the single species of walrus (Odobenidae).

Nearly all have a tapering muzzle, large eyes, a short neck that merges into a smooth, torpedo-shaped body, and four flipper-like limbs. Seals have no external ear flaps, and their rear flippers point backwards; they wriggle, or "hump", on land. The sea lion family has visible ear flaps, and rear flippers that can rotate, for waddling on land. The Walrus has almost no hair and possesses long tusks.

Pinnipeds are swift, agile underwater predators, but return to land or floating ice during the annual breeding season. There are species of seals and sea lions in all the world's oceans, while the Walrus is found in the Arctic region.

Family OTARIIDAE	Species *Zalophus californianus*	Status Vulnerable*

CALIFORNIAN SEA LION

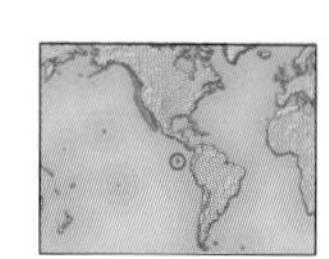

N. AMERICA, GALAPAGOS ISLANDS

Rarely straying away from the coast, this sea lion often enters harbours and estuaries for food and shelter. The male has a peaked head and is dark brown; females and juveniles are a uniform tan. Its main prey is shoaling fish, caught in two-minute dives of about 75 m (245 ft) deep. During the breeding season (May to July), males fight for small territories on the beach and around rock pools. However, after a fortnight, they swim away to feed, and on returning, must battle again for a new territory.

- **SIZE** Length: Up to 2.4 m (7¾ ft). Weight: 275–390 kg (610–860 lb).
- **OCCURRENCE** California and Galapagos Islands. Along coastlines.
- **REMARK** The common performing seal at marine parks, it can apparently be taught to understand artificial language as with dolphins and chimpanzees.

Social unit Variable	Gestation 11 months	Young 1	Diet

Family OTARIIDAE	Species *Arctocephalus pusillus*	Status Locally common

CAPE FUR SEAL

There are two subspecies of the Cape or Australian Fur Seal: the one found off South Africa is darker grey-brown and dives twice as deep as its counterpart found off the Australian coast. Males fight for territory before the females come ashore, and the cows too establish their own territory for pupping. The pups play together in nursery pools, while the mothers go to sea to feed for a few days at a time.

• **SIZE** Length: 1.8–2.3 m (6–7 ½ ft). Weight: 200–360 kg (440–790 lb).

• **OCCURRENCE** S. Africa, S.E. Australia, Tasmania, and islands off the Bass Straits. In open oceans and along coastlines.

• **REMARK** Namibia allows limited commercial hunting of this seal each year, while Australia has banned its hunting since 1975.

Social unit Variable	Gestation 11 ¾ months	Young 1	Diet

Family OTARIIDAE	Species *Phocarctos hookeri*	Status Vulnerable

NEW ZEALAND SEA LION

Also known as Hooker's Sea Lion, this species is restricted to a few islands south of New Zealand. The male is dark brown with silvery grey hindquarters and a shoulder mane, while the females and juveniles are silvery or brownish grey above and yellowish tan beneath. Foraging up to 150 km (95 miles) out at sea, it travels 1 km (⅔ mile) inland to rest among cliffs or trees. An opportunistic feeder and expert diver, its diet includes fish, squid, crustaceans, penguins, and even seal pups. Although it forms dense breeding colonies, it is solitary at sea.

• **SIZE** Length: 2–3.3 m (6 ½–11 ft). Weight: 300–450 kg (660–990 lb).

• **OCCURRENCE** Islands south of New Zealand. In open ocean and along coastlines.

• **REMARK** Now protected all over its range, this seal has been hunted since prehistoric times for its meat, skin, and oil.

Social unit Variable	Gestation 11 ¾ months	Young 1	Diet

Family OTARIIDAE	Species *Otaria byronia*	Status Vulnerable*

SOUTH AMERICAN SEA LION

A huge, heavy head, and a thick shoulder and chest mane in males, characterize this sea lion. It has brown fur that is paler or yellow on the underside, a thickened neck and chest, and a broad, upturned muzzle. Since it does not migrate, it uses its breeding areas (rookeries) around the year for resting. The mother coaxes her pups into the water at the early age of one or two months.

• **SIZE** Length: 2.3–2.8 m (7½–9¼ ft). Weight: 300–350 kg (660–770 lb).

• **OCCURRENCE** W., S., and E. South America and Falkland Islands. Along coastlines and over the continental shelf, as well as in deeper waters, rivers, and around glaciers.

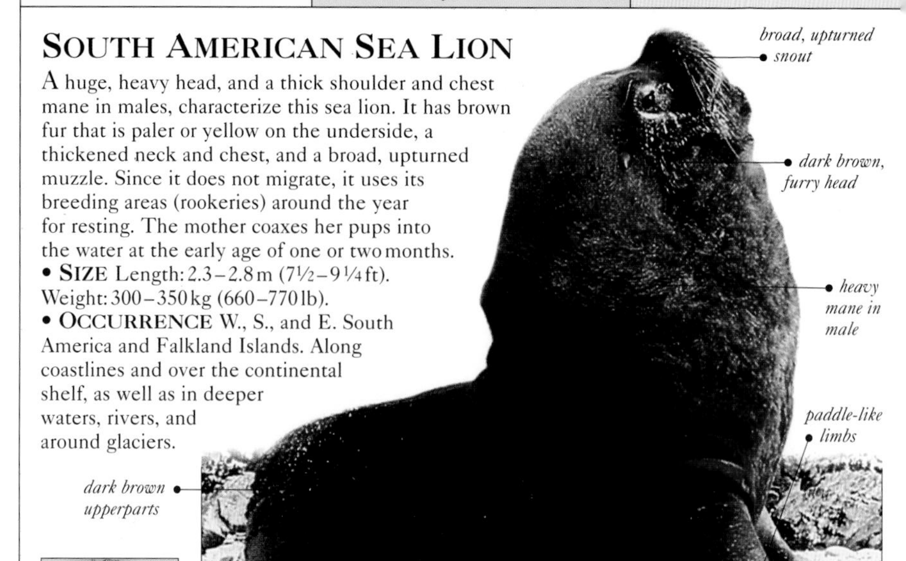

Social unit Variable	Gestation 11¾ months	Young 1	Diet

Family OTARIIDAE	Species *Callorhinus ursinus*	Status Vulnerable

NORTHERN FUR SEAL

The male of this species is brown-grey, while females and juveniles are silver-grey above, red-brown below, with a grey-white chest patch. The front flippers are long and the fur on them appears to be cut off at the wrist, and the back flippers can rotate, allowing this seal a walking motion. Most populations are migratory (except the seals on San Miguel Island), with the adult males heading south in August and staying at sea for nine or ten months of the year. Females and juveniles follow in November.

• **SIZE** Length: Up to 2.1 m (7ft). Weight: 180–270 kg (400–600 lb).

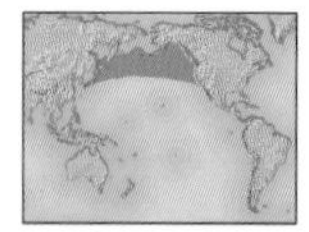

• **OCCURRENCE** N. Pacific (Bering Sea to California). In sub-Arctic oceans and coastal waters.

Social unit Variable	Gestation 11¾ months	Young 1	Diet

Family ODOBENIDAE	Species *Odobenus rosmarus*	Status Vulnerable*

WALRUS

Large and bulky, the Walrus has a blunt, thickly whiskered muzzle that widens rapidly to a broad body, and then tapers again at the tail, which is embedded in a web of skin. Its front flippers resemble those of sea lions, whereas the back flippers resemble those of seals; its rough, creased skin is grey to cinnamon-brown. The male Walrus is twice the size of the female, and its upper canines grow longer than the female's, to almost 1 m (3¼ ft) in length. The Walrus dives more than 100 m (330 ft) deep, for over 25 minutes, and locates its prey with sensory nerves in its whiskers and snout. It sucks up food from the seabed, first rooting for it with its nose, aided by jets of water squirted from its mouth. Walruses remain huddled on land or ice floes in groups of 100, breaking up into smaller groups of 10 when at sea. During breeding, males establish small territories in water next to females perched on ice floes, using their tusks to defend their territories. Walruses migrate to warmer waters in April and May.

- **SIZE** Length: 3–3.6 m (9¾–12 ft). Weight: 1.2–2 tonnes (1⅛–2 tons).
- **OCCURRENCE** Arctic Ocean (circumpolar). In shallow, continental shelf water and along coastlines.

COMMUNAL SUNBATHING
The skin colour of Walruses basking in the sun shows through the animals' short fur and flushes rose-red, as the blood vessels dilate to absorb the maximum heat.

Social unit Social	Gestation 15 months	Young 1	Diet

Family PHOCIDAE	Species *Monachus monachus*	Status Critically endangered

MEDITERRANEAN MONK SEAL

MEDITERRANEAN, BLACK SEA, ATLANTIC

This extremely rare species has a dark brown coat, which is said to resemble a friar's robe – hence the name Monk Seal. It has a small head and robust body. Since its back flippers do not rotate, it cannot walk on land. One of the least social seals out of water, it is found in small, widely separated groups or mother–pup pairs. Diet comprises fish such as sardines, tuna, eels, and mullet, as well as lobster and octopus.

- **SIZE** Length: 2.4–2.8 m (7¾–9¼ ft). Weight: 250–400 kg (550–880 lb).
- **OCCURRENCE** Mediterranean, Black Sea, and Atlantic (N.W. Africa). Along coastlines.
- **REMARK** Sensitive to disturbance caused by tourist activity, this seal shelters in sea caves. However, cave collapses, overfishing, pollution, and exposure to viral infection pose serious threats to its existence.

robust body

compressed muzzle

rear flippers have several branches

Social unit Variable	Gestation 11 months	Young 1	Diet

Family PHOCIDAE	Species *Lobodon carcinophagus*	Status Common

CRABEATER SEAL

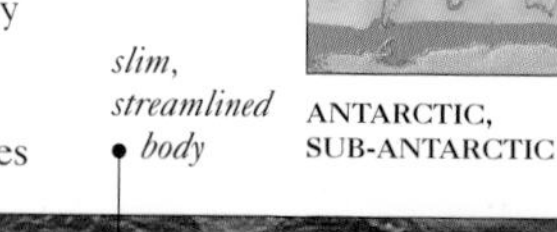

ANTARCTIC, SUB-ANTARCTIC

Long and lithe, with a silver-grey to yellow-brown coat dotted with irregular darker spots and rings, the Crabeater Seal is oddly named, since it eats mainly krill. Its small head tapers to a thin muzzle that has short, inconspicuous whiskers. A typical seal, it has no external ears, and cannot walk on land. To feed, it dives down to 40 m (130 ft) for around five minutes.

- **SIZE** Length: 2.2–2.6 m (7¼–8½ ft). Weight: 220 kg (490 lb).
- **OCCURRENCE** Antarctica. In open ocean and along coastlines.
- **REMARK** This seal is one of the most abundant and fastest-swimming seal species.

Social unit Variable	Gestation 11 months	Young 1	Diet

Family PHOCIDAE	Species *Hydrurga leptonyx*	Status Locally common

LEOPARD SEAL

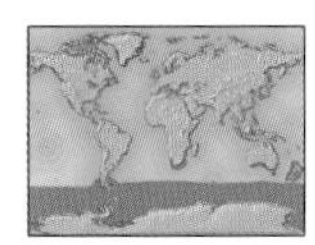

This seal is silver to dark grey with variable spots. It has a sinuous body and massive, reptile-like head that lacks a forehead, and a wide, deep lower jaw. It is widest around the shoulders and its entire body, including the flippers, is furred. Unusually for a seal, the Leopard Seal swims with its front flippers which are tipped with claws. Its large canine teeth enable it to feed on smaller seals, penguins, and other birds.

• **SIZE** Length: 2.5–3.2 m (8¼–10 ft). Weight: 200–455 kg (440–1,000 lb).

• **OCCURRENCE** Antarctica, especially high southern latitudes. In polar and subpolar waters, and on pack ice and islands.

• **REMARK** The Leopard Seal is the largest of the four Antarctic seal species.

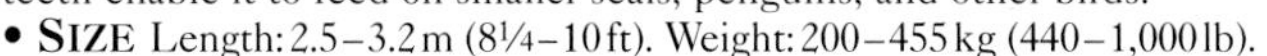

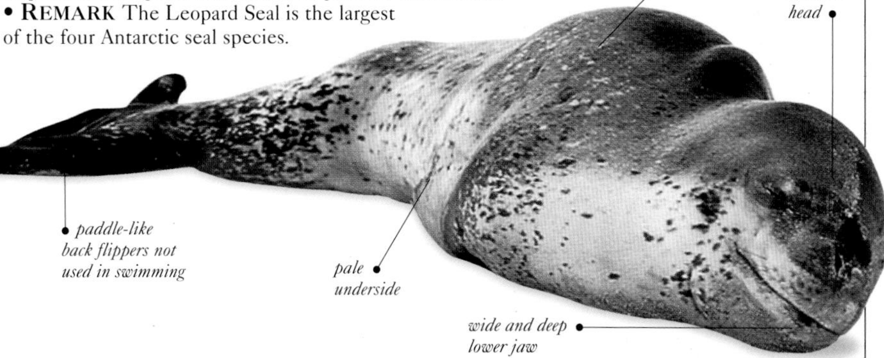

Social unit Solitary	Gestation 11 months	Young 1	Diet

Family PHOCIDAE	Species *Leptonychotes weddelli*	Status Locally common

WEDDELL SEAL

Bulky, with a small head and flippers, the Weddell Seal has a blunt muzzle with a few short whiskers, close-set eyes, and an upturned mouthline. Its coat is dark silvery grey above and off-white below, with variable dark and light patches. An expert diver, it reaches depths of 500 m (1,600 ft), staying underwater for over an hour. Its modified upper incisors are used to cut breathing holes in ice. This seal does not migrate, but moves with ice floes.

• **SIZE** Length: 2.5–2.9 m (8¼–9½ ft). Weight: 400–600 kg (880–1,320 lb).

• **OCCURRENCE** Antarctica (circumpolar). In open ocean and along coastlines.

• **REMARK** Named after the sealer, Captain James Weddell, who wrote about an encounter with this seal in 1820, today the Weddell Seal is protected from commercial hunting.

ANTARCTIC

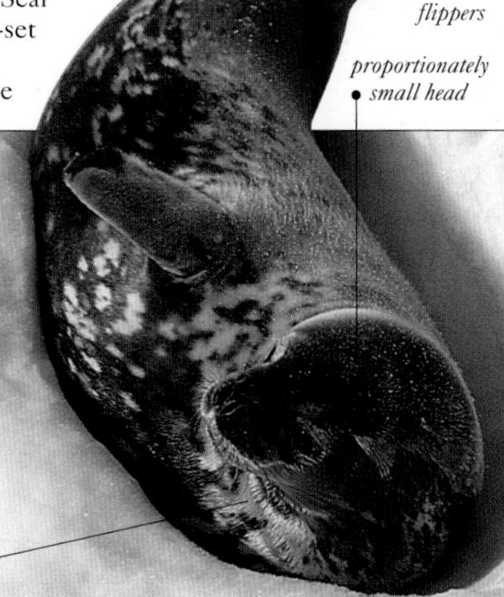

Social unit Variable	Gestation 10¼ months	Young 1	Diet

Family PHOCIDAE	Species *Mirounga leonina*	Status Locally common

SOUTHERN ELEPHANT-SEAL

The male Elephant-seal has an inflatable trunk-like nose from which it earns its name. Both sexes have a uniform light to dark silver-grey coat, with a broad head, muzzle, jaw, and neck. Males are four or five times heavier than the females. They fight rivals during the two-month mating season by rearing up on their hindquarters and roaring loudly, inflating their noses in display. This is followed by a butting and slapping fight that may last from a few seconds to half an hour, leaving most animals battle-scarred. The victorious male or "beachmaster" has a harem of 20–40 cows, although even larger ones of 100 are known to occur. A creature of the open seas, this seal spends ten months (apart from breeding and moulting time) foraging over wide areas. Diving continuously day and night, each dive lasting an average of 20–22 minutes, it spends 90 per cent of its time underwater.

- **SIZE** Length: 4.2–6 m (14–20 ft). Weight: 2.2–5 tonnes (2 ⅛–4 ⅞ tons).
- **OCCURRENCE** Antarctica. North of seasonally shifting ice and around sub-Antarctic islands.
- **REMARK** The male Elephant Seal is the largest member of its order (Pinnipedia).

FEMALE FEATURES
The female seal has a fleshy, blunt nose. She nurses her pup for 19–23 days, during which time she does not leave the beach and loses 35 per cent of her total weight.

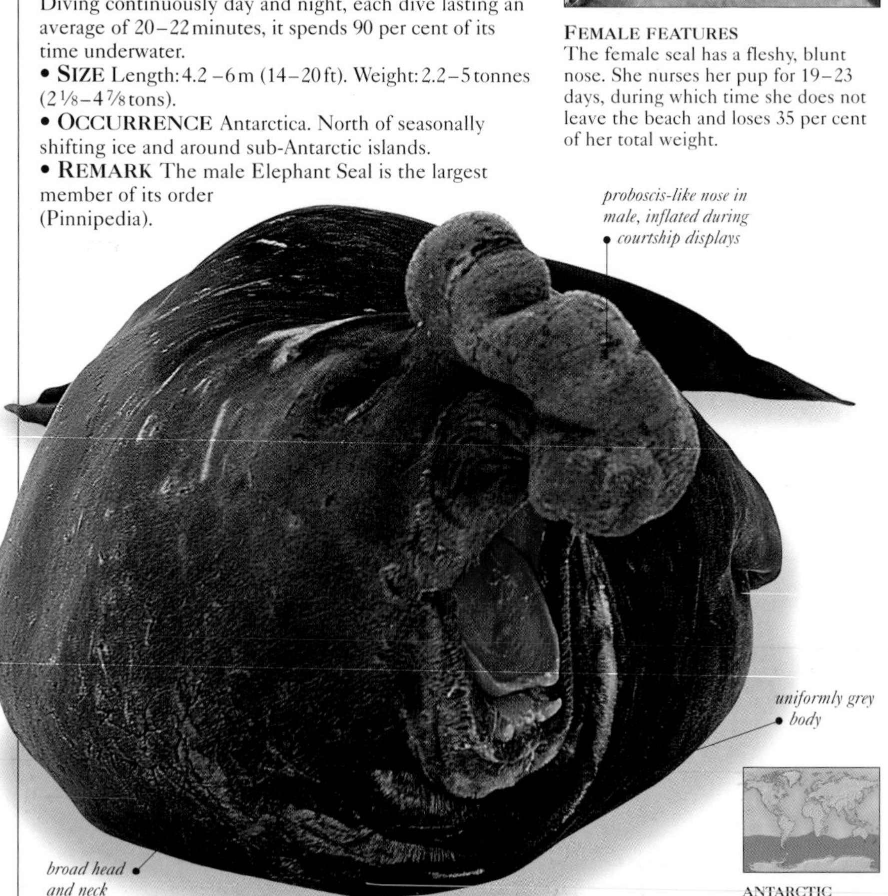

Social unit Variable	Gestation 11 ¼ months	Young 1	Diet

Family PHOCIDAE	Species *Ommatophoca rossii*	Status Vulnerable

ROSS SEAL

With the shortest fur of any seal, the Ross Seal is dark grey to chestnut-brown, with broad, dark bands along its body. It has a distinctively blunt muzzle, wide head, and long rear flippers. Less social than other seals, on ice it is found alone or in mother–pup pairs. The smallest of Antarctic seals, it is also one of the rarest, and little is known about its lifestyle.

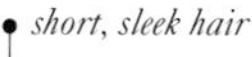
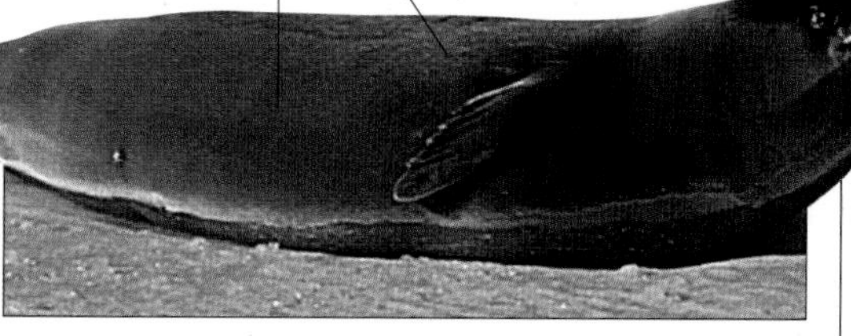

- **SIZE** Length: 1.7–3 m (5½–9¾ ft). Weight: 130–215 kg (290–470 lb).
- **OCCURRENCE** Antarctica, especially Ross Sea. In open waters and along coastlines.

Social unit Solitary	Gestation 11 months	Young 1	Diet

Family PHOCIDAE	Species *Cystophora cristata*	Status Locally common

HOODED SEAL

This seal has a fleshy muzzle that droops over its mouth. The male intimidates rivals by inflating its nasal cavity to double the size of its head and by extruding a membrane from the left nostril that also blows up like a red balloon.

- **SIZE** Length: 2.5– 2.7 m (8¼–8¾ ft). Weight: 300–410 kg (660–900 lb).
- **OCCURRENCE** N. Atlantic to the Arctic Ocean. In open waters and on pack ice.
- **REMARK** The pup is weaned earlier than any other mammal (four or five days).

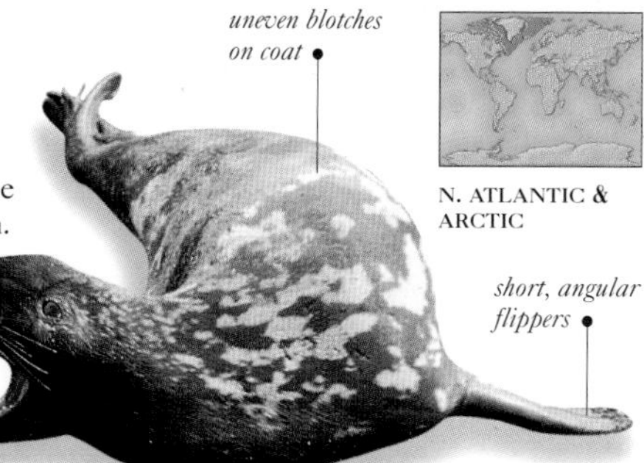

Social unit Variable	Gestation 11½ months	Young 1	Diet

Family PHOCIDAE	Species *Halichoerus grypus*	Status Common

GREY SEAL

There are three populations of Grey Seals: those found in the N.W. Atlantic are heavier and breed from December to February; those in the N.E. Atlantic breed from July to December; whereas those from the Baltic Sea breed from February to April. The male is grey-brown, and the female pale grey.

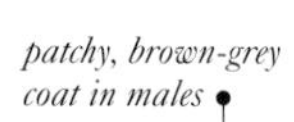

- **SIZE** Length: 2–2.5 m (6½ –8¼ ft). Weight: 170–310 kg (370–680 lb).
- **OCCURRENCE** N. Atlantic and Baltic Sea. In open waters and along coasts.

Social unit Variable	Gestation 11¼ months	Young 1	Diet

Family PHOCIDAE	Species *Pagophilus groenlandicus*	Status Common

HARP SEAL

This species has a wide face with close-set eyes, strong black claws, and silver-white fur with curved dark bands on the back that form the pattern of a harp. Feeding on cod, capelin, herring, and other fish, it migrates along edges of pack ice in noisy groups. The male courts the female by calling out, blowing bubbles underwater, and chasing her across the ice. Breeding takes place on rough, hummocky ice, which provides adequate shelter for the pups.

- **SIZE** Length: 1.7 m (5½ ft). Weight: 130 kg (290 lb).
- **OCCURRENCE** N. Atlantic to the Arctic Ocean. In polar regions – open ocean and along coastlines.
- **REMARK** Long hunted for fur and oil, Harp Seals have been protected from commercial hunters in Canada since 1987.

N. ATLANTIC, ARCTIC

black markings on body

long, flattened head

small back flippers

silvery white fur

furred flippers

Social unit Variable	Gestation 11½ months	Young 1	Diet

Family PHOCIDAE	Species *Phoca sibirica*	Status Lower risk

BAIKAL SEAL

Smaller than many other seals and the only wholly freshwater pinniped species, the Baikal Seal nevertheless resembles its marine relatives in most ways. With a dark, brownish grey coat that fades to yellowish grey on the sides, this seal has large, powerful foreflippers and claws. It is chiefly solitary and the female mates with the same male over several years. Mating takes place in water, and the single pup moults its woolly white coat to the silvery grey adult coat after 6–8 weeks. The Baikal Seal's lifespan of 50–55 years is longer than that of many seals. Evenly distributed across Lake Baikal during summer, these seals migrate to the northern part of the lake in winter, remaining there until the beginning of spring.

- **SIZE** Length: 1.2–1.4 m (4–4½ ft). Weight: 80–90 kg (175–200 lb).
- **OCCURRENCE** Russia (Lake Baikal and its feeder streams). In fresh water.
- **REMARK** Hunted since prehistoric times, it continues to be commercially exploited for its meat and skin.

ASIA

brownish grey body

dark patches near eyes

strong front flippers and claws

Social unit Solitary	Gestation 9½ months	Young 1	Diet

Family PHOCIDAE	Species *Phoca vitulina*	Status Common

COMMON SEAL

The most widespread pinniped, the Common or Harbour Seal has a plump body and a small, cat-like head with a slight forehead and nostrils that form a V-shaped pattern. Its coat ranges from light to dark grey or brown, and has spots, rings, and blotches all over. The breeding season varies regionally, and the pups are able to crawl and swim almost within an hour of birth. There are at least five subspecies of Common Seal. One of them, the Ungava Seal, lives in fresh water in N. Quebec, Canada.

• **SIZE** Length:1.4–1.9 m (4½–6¼ ft). Weight: 55–170 kg (120–370 lb).

• **OCCURRENCE** N. Pacific and N. Atlantic, from the polar to temperate region. In lakes and rivers, and along coastlines.

• **REMARK** The Common Seal feeds on many fish that are exploited commercially, leading to seal deaths in fishing nets. In several countries it is legal to shoot these seals to protect fish farms.

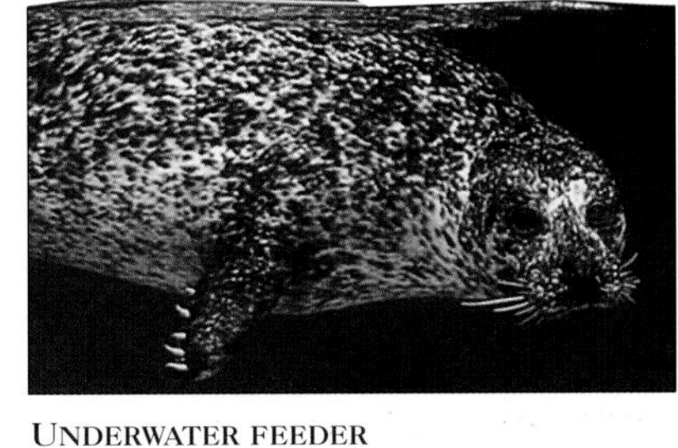

UNDERWATER FEEDER
The Common Seal feeds near the shore, diving to depths of less than 100 m (330 ft) for about 3–5 minutes. It is an opportunistic forager, feeding on herring, sandeels, sandgobies, hake, whiting, and crustaceans.

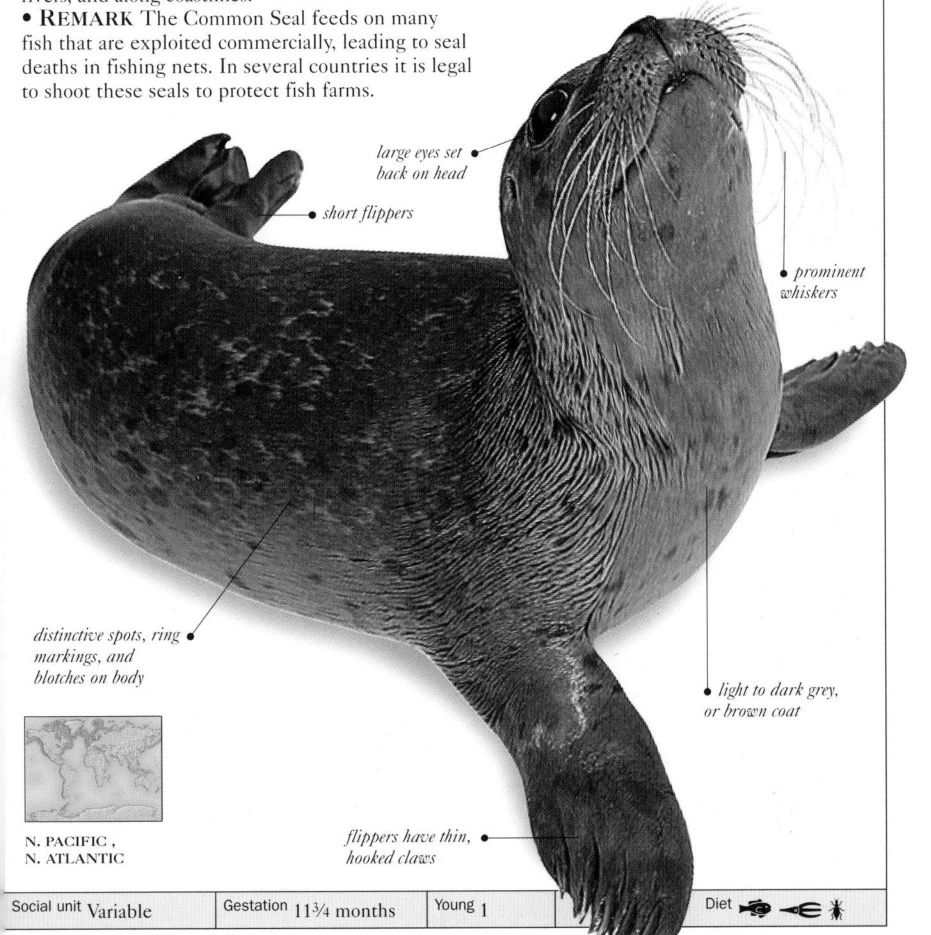

N. PACIFIC,
N. ATLANTIC

Social unit Variable	Gestation 11¾ months	Young 1		Diet

ELEPHANTS

FOR MANY YEARS the order Proboscidea was considered to contain one living family, Elephantidae, which had two herbivorous species, the African and Asian (Indian) Elephants.

However, the African Forest Elephant, found deep in thickly wooded areas of west-central Africa, has now been differentiated as a separate species from the African Elephant, which frequents bush, savanna, and scattered forest from the east to the south.

All elephants are instantly recognize by their great size, thick and almos hairless skin, huge head and ear flaps a very long, mobile trunk (the elongate nose and upper lip), bulky body, an four pillar-like legs.

The upper incisors gradually grow int tusks in all species, but in adult Asia females, rarely protrude from th mouth. Elephants live in family-base groups dominated by a senior female the matriarch.

Family ELEPHANTIDAE	Species *Elephas maximus*	Status Endangered

ASIAN ELEPHANT

Unlike its African counterpart (see pp. 308–309), the Asian Elephant has small ears and a single finger-like, gripping extremity at the tip of its trunk. Its tusks are relatively small and in females are absent or do not protrude beyond the lips. Some males can also be tuskless (called "makhnas" locally). Elephants locate, select, and pluck food with the trunk-tip, in conjunction with the tusks and forefoot, before placing it in the mouth. Young elephants learn the various techniques involved in food gathering by imitation and practice. Highly intelligent animals, elephants are found in complex, matriarchal societies, made up of mothers with their offspring and juvenile females, all sharing close bonds. Young males roam in bachelor herds, while a tusker (dominant male) mates with the adult females of the herd.

- **SIZE** Length: Up to 3.5 m (11 ft). Weight: 2–5 tonnes (2–4 7/8 tons).
- **OCCURRENCE** S. and S.E. Asia. In forest, close to water.
- **REMARK** The elephant population is adversely affected by habitat loss and the disruption of traditional herd routes by human activity.

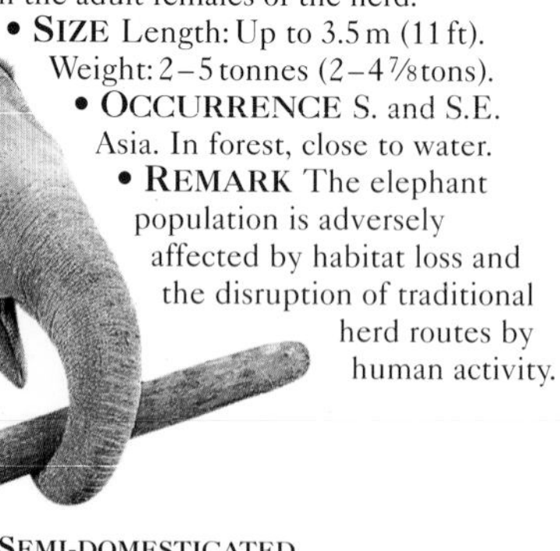

convex back

tail with dark hair at tip

SEMI-DOMESTICATED
Although the Asian Elephant has lived quite closely with humans for centuries, it has never been wholly domesticated. Until very recently, each new generation of working elephants has been captured from the wild.

Social unit Social	Gestation Up to 22 months	Young 1	Diet

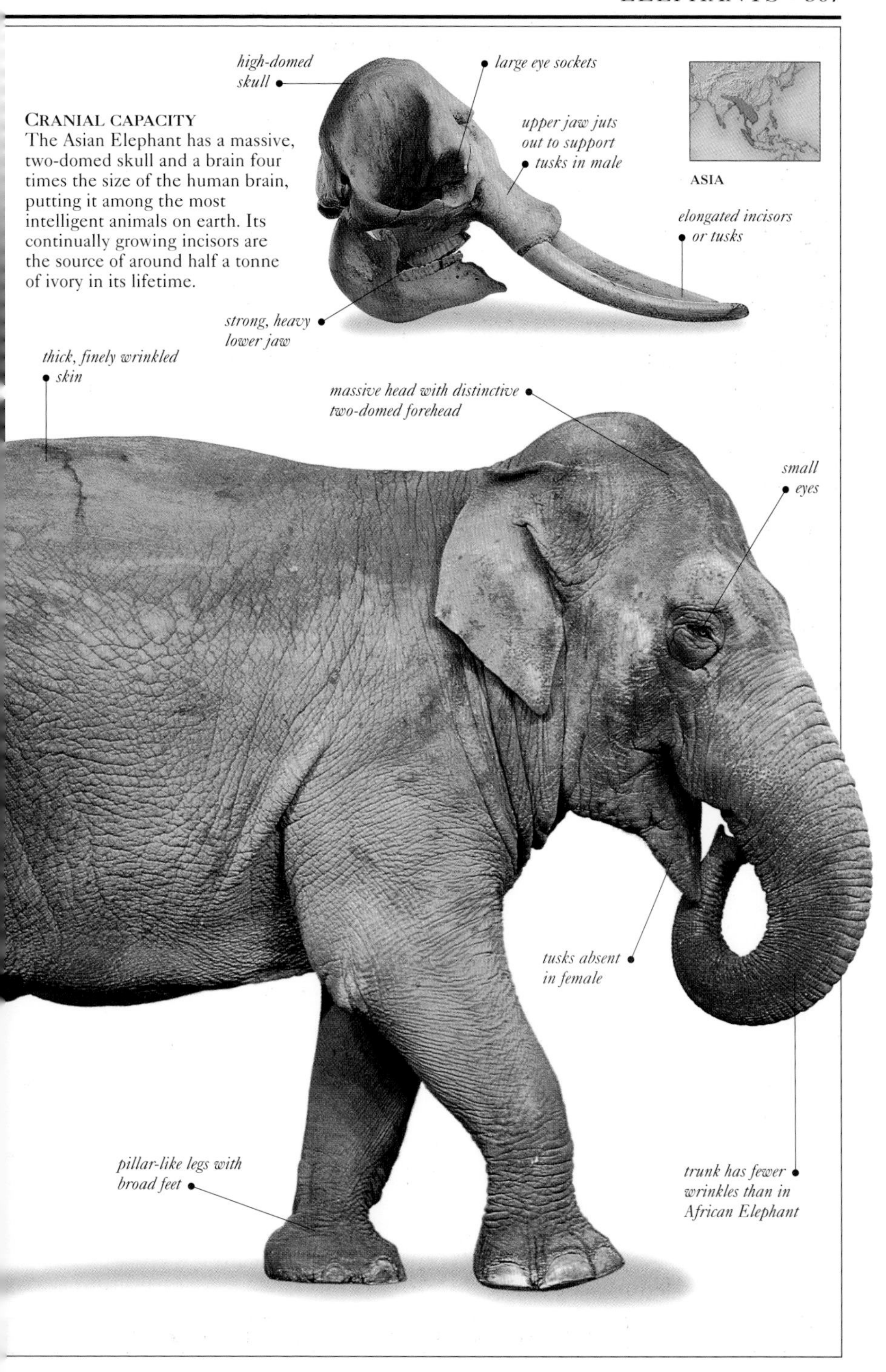

CRANIAL CAPACITY
The Asian Elephant has a massive, two-domed skull and a brain four times the size of the human brain, putting it among the most intelligent animals on earth. Its continually growing incisors are the source of around half a tonne of ivory in its lifetime.

Family ELEPHANTIDAE	Species *Loxodonta africana*	Status Endangered

AFRICAN ELEPHANT

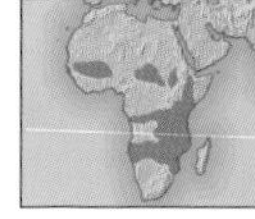
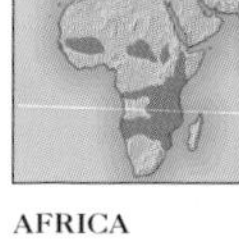
AFRICA

The largest living land animal, the African, Bush, or Savanna Elephant, as it is variously called, lives in a range of habitats, from desert to high rainforest. Its huge, prominent ears are much larger than those of its Asian cousin, and both sexes have forward-pointing tusks, which are sometimes used to loosen the mineral-rich soil that it eats as a dietary supplement. The male may be nearly twice the weight of the female, and its tusks are thicker at the base. Feeding for 20 hours each day, this elephant consumes vegetation up to about 5 per cent of its body weight. It also visits a water source each day to drink, bathe, and wallow. Foraging over large areas, a herd of elephants may cause dramatic changes to the ecosystem, especially during drought. Like the Asian Elephant (see pp. 306–307), the African Elephant lives in matriarchal societies.

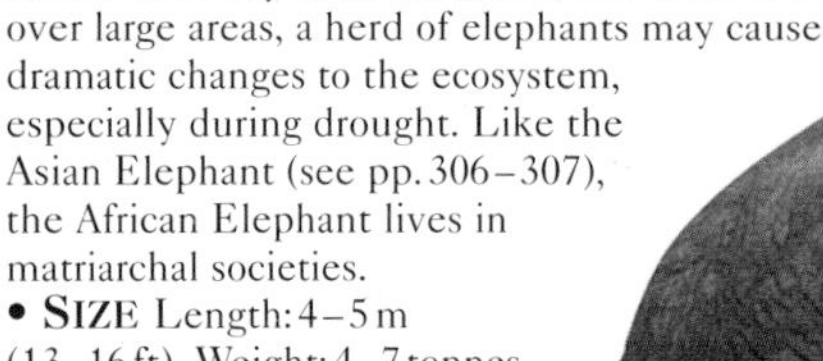

- **SIZE** Length: 4–5 m (13–16 ft). Weight: 4–7 tonnes ($3\frac{7}{8}$–$6\frac{7}{8}$ tons).
- **OCCURRENCE** Sub-Saharan Africa. In grassland, desert, tropical forest, and wetlands; near lakes and rivers.
- **REMARK** The African Elephant (especially the male tusker) is widely hunted for its tusks, which are a major source of ivory.

huge ears, the biggest of any living being

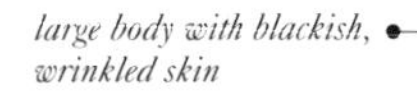

large body with blackish, wrinkled skin

Social unit Social	Gestation Up to 22 months	Young 1	Diet

PROTECTING THE YOUNG
Vulnerable to large predators such as lions, the elephant calf may remain with its mother for the first three or four years of its life. It is also protected from danger by other females (an older sister or female cousin) in the herd.

AFRICAN FOREST ELEPHANT
Formerly regarded as a subspecies of the African Elephant, the smaller and lighter African Forest Elephant (*Loxodonta cyclotis*), found deep within dense rainforest vegetation, was recently accorded the status of a separate species.

HYRAXES

IN A CURIOUS twist of evolution, hyraxes resemble small-eared rabbits in size and shape, but their detailed anatomy and genetics seems to suggests that they are more closely related to primitive hoofed mammals. The eight species in the order Hyracoidea, some of which are also known as dassies or conies, are found in Africa and the Middle East.

Certain species are arboreal and tend to live singly or in small groups. Others prefer rocky outcrops and are more gregarious. All hyraxes are opportunistic herbivores, switching diet with the season and plant availability. Their climbing abilities are aided by secretions from glands under their feet, which improve grip on smooth surfaces. Another hyrax feature is their poor control of body temperature, compared to most mammals. They are often seen warming themselves in the sun or cooling down in the shade, like reptiles.

Family PROCAVIIDAE	Species *Procavia capensis*	Status Locally common

ROCK HYRAX

Also called the Rock Dassie, the Rock Hyrax has a stout body with short, dense, grey or grey-brown fur above, and a paler underside. It lives in colonies of 4–40, which consist of one dominant male, other males, females, and young. Found in a variety of habitats, it favours rocky outcrops and crags, where it makes a grass-lined nest. It is hunted for its meat and pelt by local people.

small, rounded ears
grey or greyish brown coat
paler underside

AFRICA & ASIA

- **SIZE** Body length: 30–58 cm (12–23 in). Tail: 20–31 cm (8–12 in).
- **OCCURRENCE** S. and E. Africa and W. Asia. In grassland, desert, forest, and hills.

Social unit Social	Gestation 7–8 months	Young 1–6	Diet

Family PROCAVIIDAE	Species *Dendrohyrax arboreus*	Status Vulnerable

TREE HYRAX

Grey-brown with buff underparts, this hyrax has a yellowish patch on its rump. Its head, legs, and tail are relatively small for its stout body. Also known as the Tree Dassie, this species lives among trees, shrubs, and creepers, nesting in tree-holes and rarely feeding on the ground.

AFRICA

- **SIZE** Body length: 40–70 cm (16–28 in). Tail: 1–3 cm (3/8–1 1/4 in).
- **OCCURRENCE** E. and S. Africa. In tropical forest and mountains.
- **REMARK** It is hunted for its meat and pelt.

Social unit Variable	Gestation 7–8 months	Young 1–3	Diet

AARDVARK

THE ONLY SPECIES in its order, the African Aardvark has a specialized diet of ants and termites. Traditional anatomical studies have long suggested that, like the hyrax, its closest relatives may be hoofed mammals such as the elephant. However, newer genetic studies are challenging this viewpoint.

The Aardvark's sight is poor and it hunts mainly by smell. It has powerful limbs with long, straight, shovel-like claws for rapid digging. There are four toes on each front foot and five on each rear foot.

The Aardvark's teeth are unique. Only 20 premolars and molars are present, and these are seldom used, since most of the prey are swallowed whole. The teeth lack the usual enamel, being covered by a bone-like substance called cementum. Inside the tooth is dentine with numerous cylindrical cavities, giving the name for the order, Tubulidentata, or "tubule-tooth".

Family ORYCTEROPODIDAE	Species *Orycteropus afer*	Status Unconfirmed

AARDVARK

One of the most powerful diggers among mammals, this nocturnal species, also known as the Ant-bear or Ant-pig, digs burrows up to 10 m (33 ft) long around its home range of 2–5 square km (3/4–2 square miles). It prefers ants as food, especially in summer, when they are available in plenty. However, it also eats termites when ants are not available. The Aardvark is unusual in chewing one species of ant with its molar teeth, but swallowing other species of ants and termites whole, and grinding them in its muscular stomach. It has a distinctive curved back, and long tapering snout, ears, and tail. A dense mat of hair around its nostrils helps to filter out the dust while it is digging for food.

• **SIZE** Body length: 1.6 m (5 1/4 ft). Tail: 55 cm (22 in).

• **OCCURRENCE** Africa, south of the Sahara. In open woods and grassland.

AFRICA

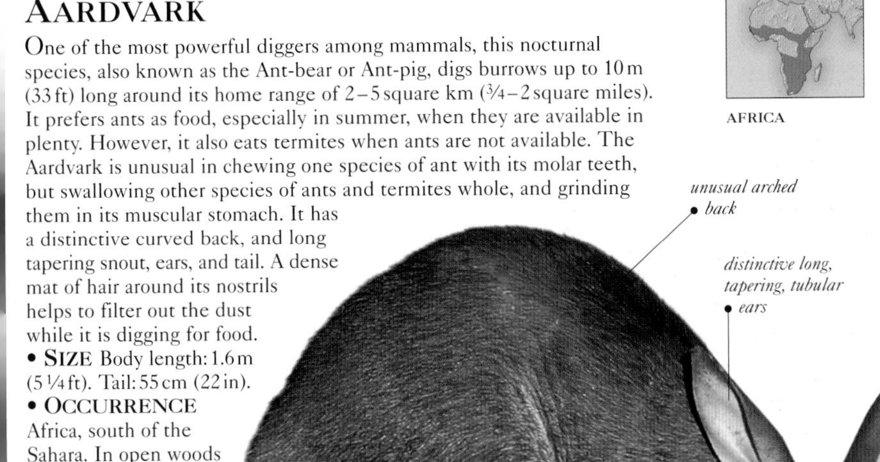

Social unit Solitary	Gestation 243 days	Young 1	Diet

SIRENIANS

THE THREE SPECIES of manatees and the Dugong make up the order Sirenia – also called "sea cows" because of their bulky bodies, slow movements, and habit of grazing on sea grasses and other aquatic plants.

Despite their outward resemblance to seals, sirenians are the only purely herbivorous marine mammals (they also live in rivers and lagoons). They have very thick skin, paddle-like front limbs, a flattened tail, small eyes, and a fleshy, rounded muzzle with a mobile upper lip with which they pluck food.

The three manatee species occur in the West Indies, the Amazon region, and West Africa, whereas the Dugong lives principally in the Indian Ocean. The slow, unaggressive lifestyle of sirenians, foraging in coastal shallows, has made them vulnerable to humans. With a total population of perhaps 130,000, this is possibly the least numerous of all orders of mammals.

Family DUGONGIDAE	Species *Dugong dugon*	Status Vulnerable

DUGONG

With a torpedo-shaped, grey to brown-grey body, and a crescent-shaped tail, the Dugong is adapted to aquatic life in various ways: the "feet" have evolved into flippers and the tail is modified for propulsion. The thick, heavy bones provide diving weight for easier locomotion underwater. Usually diurnal, it moves daily between on- and offshore islands depending on the tide and supply of food. In certain areas, it makes longer seasonal migrations, of sometimes hundreds of kilometres, to follow sea grass and algae growth, and to avoid cold-water currents. Hunted by Killer Whales, sharks, and saltwater crocodiles, members of a group may congregate to intimidate and butt large predators.

- **SIZE** Length: 2.5–4 m (8¼–13 ft). Weight: 250–900 kg (550–1,985 lb).
- **OCCURRENCE** E. Africa, W., S., and S.E. Asia, Australia, and Pacific islands. Along shallow tropical coasts.
- **REMARK** It is widely hunted for meat, oil, leather, teeth, and bones.

AFRICA, ASIA, AUSTRALIA

crescent-shaped tail

tail used as paddle

grey to brown-grey, generally hairless skin

short, paddle-shaped flippers

Social unit Social	Gestation 13–14 months	Young 1	Diet

Family TRICHECHIDAE	Species *Trichechus manatus*	Status Vulnerable

WEST INDIAN MANATEE

N., C., & S. AMERICA

Probably the best known of the manatees (the other two are the Amazonian and the West African), the West Indian Manatee has grey-brown skin, paler below, which may harbour growths of algae. Like other sirenians, it has tiny eyes and no external ear flaps. Groups of 2–20 manatees may swell to 100 in warm waters during winter, or in areas where food is abundant. These groups are fluid and individuals come and go, ranging widely along shallow coastlines and through fresh and salt water. Most communication is tactile and includes touching, nuzzling, and rubbing. Manatees make high-pitched squeals and whistles underwater, usually during mother–calf bonding, or to warn others of danger. Auditory and tactile signals are also used during courtship when several males may compete for a single female. Only the mother cares for the calf – probably with the help of older offspring and female relatives.

- **SIZE** Length: 2.5–4.5 m (8 ¼–15 ft). Weight: 200–600 kg (440–1,320 lb).
- **OCCURRENCE** S.E. USA to N.E. South America. In shallow tropical coasts, coastal rivers and estuaries, and freshwater springs.
- **REMARK** Although protected by law, the West Indian Manatee continues to be threatened by hunting, habitat degradation, and pollution.

FEEDING
This manatee feeds from the surface down to 4 m (13 ft). It holds food with its flippers, directing it into its mouth, using flexible lips. Daily food intake is about one-quarter of its body weight.

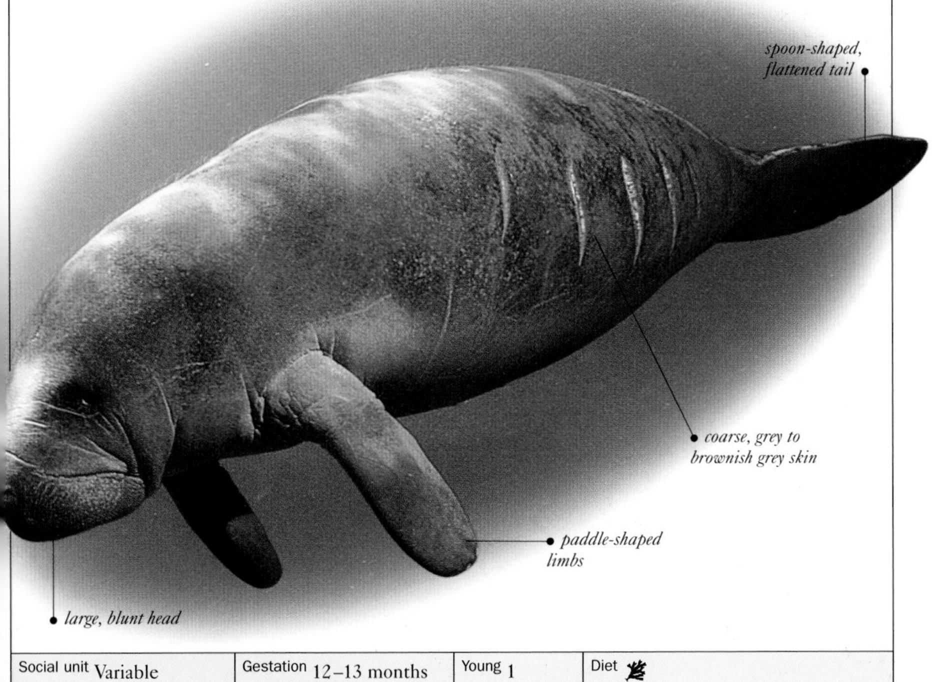

Social unit Variable	Gestation 12–13 months	Young 1	Diet

ODD-TOED UNGULATES

HORSES

IN ADDITION TO domesticated and wild horses, this group includes zebras and asses, such as the Onager and Kiang. There are altogether ten species of the family Equidae, which in turn forms part of the order Perissodactyla, the odd-toed ungulates.

Equids are the minimalists of the order, with just one toe on each foot, capped by a hard, horny hoof. The head is large with long jaws and batteries of cheek teeth for chewing grass and other plants. The neck and body are long and powerful, with relatively long, slim limbs for great stamina and speed, to outrun predators.

Most equids live in family groups or larger herds, in open habitats such as grassland and dry scrub, across Africa and Asia. Various species have been bred and interbred in captivity, and widely introduced around the world.

Family EQUIDAE	Species *Equus ferus przewalskii*	Status Extinct in wild

PRZEWALSKI'S WILD HORSE

Also known as the Mongolian Wild Horse, this member of the horse family is heavily built, with a thick neck, large head, and short legs. Its dun-coloured coat, short in summer, becomes longer and paler in winter. Like other feral horses, it lives in herds that wander across large distances. A typical group is led by a senior mare, with 2–4 other mares, their young, and a single stallion that remains on the fringes.

- **SIZE** Length: 2.2–2.6 m (7 ¼–8 ½ ft). Weight: 200–300 kg (440–660 lb).
- **OCCURRENCE** Unknown in the wild since 1968. Was native to C. Asia and Mongolia. In steppe grassland.
- **REMARK** This horse exists in zoos, parks, and field stations. Several attempts have been made to reintroduce it in Mongolia.

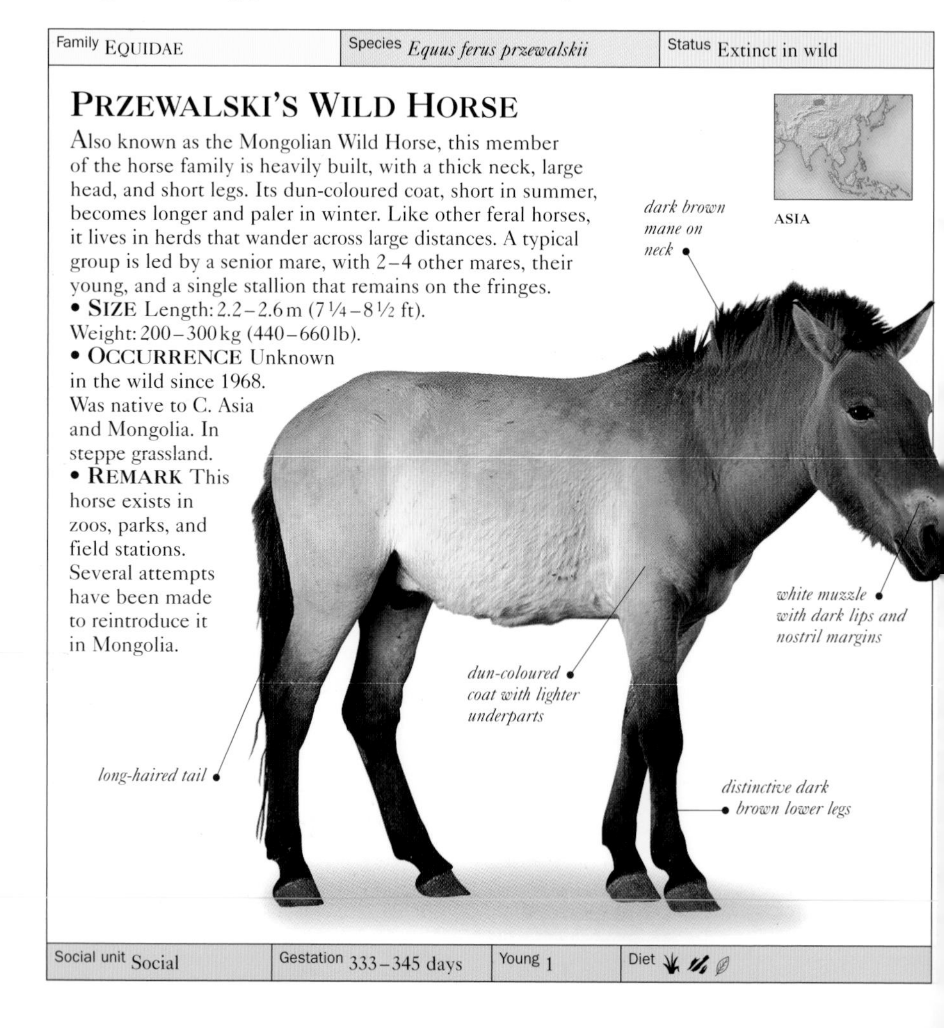

Social unit Social	Gestation 333–345 days	Young 1	Diet

Family EQUIDAE	Species *Equus africanus*	Status Critically endangered

AFRICAN WILD ASS

Grey with variable stripes across its legs, this wild ass has a coat that changes from buff-grey in summer to iron-grey in winter. It survives in rocky deserts where the ground temperature exceeds 50°C (122°F), and is adapted to its habitat in various ways. Its narrow hooves give it surefootedness rather than speed and help in climbing over rocks. It feeds on almost any plant food, from grasses to thorny acacia, and can survive without water for several days.

- **SIZE** Length: 2–2.3 m (6 ½–7 ½ ft). Weight: 200–230 kg (440–510 lb).
- **OCCURRENCE** E. Africa. In desert.

AFRICA

short, sparse mane permanently erect

white underparts

variable transverse leg stripes

Social unit Social	Gestation 360–370 days	Young 1	Diet

Family EQUIDAE	Species *Equus hemionus*	Status Vulnerable

ONAGER

The Onager or Asian Wild Ass, has a tawny, yellowish, or grey coat that has a dark stripe with a white border down its back. The elongated lower portion of its limbs enables it to run swiftly across long distances. Females and young form loose nomadic herds, whereas immature males gather in bachelor groups. Mature males may kick and bite rivals to occupy breeding territory.

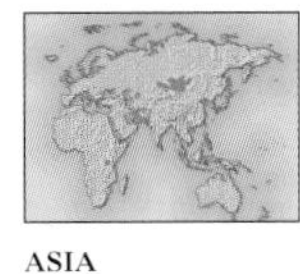

ASIA

- **SIZE** Length: 2–2.5 m (6 ½–8 ¼ ft). Weight: 200–260 kg (440–570 lb).
- **OCCURRENCE** W., C., and S. Asia. In stony desert.

tawny to grey coat

slender head with dark mane

long legs

Social unit Social	Gestation 11–12 months	Young 1	Diet 

Family EQUIDAE	Species *Equus burchelli*	Status Lower risk*

BURCHELL'S ZEBRA

Also called the Common Zebra, this equid is characterized by a striped pattern which may have faint "shadow" bands between its larger flank stripes. It forms long-term social units consisting of a stallion, his harem, and several offspring, and is found grazing with other species such as Grevy's Zebra, Wildebeest, Roan Antelope, and Hartebeest.

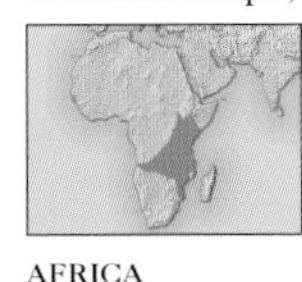

AFRICA

- **SIZE** Length: 2.2–2.5 m (7 ¼–8 ¼ ft). Weight: 175–385 kg (390–850 lb).
- **OCCURRENCE** E. and S. Africa. In savanna, light woodland, and grassland.

broad black stripes along back and belly

Social unit Social	Gestation 360–396 days	Young 1	Diet 

Family EQUIDAE	Species *Equus grevyi*	Status Endangered

GREVY'S ZEBRA

AFRICA

The largest species of zebra, Grevy's Zebra is also the biggest wild equid. It has a black-and-white erect mane and a characteristic pattern of very narrow, densely distributed black-and-white stripes, which remain distinct all the way down to the hooves; the belly and base of the tail are white. Unlike most members of the equid family (horses and asses), which are uniformly coloured, the zebra's strikingly patterned coat is believed to serve some special function: it may allow recognition of herd members, or regulate body temperature, or create a "dazzle" effect to confuse predators. Grevy's Zebra also has characteristic large, furry ears, and a tuft of long hair at the tip of its tail. The male is slightly larger than the female and has larger canine teeth. The female alone cares for the young, often roaming freely, accompanied by a foal and older offspring. Sometimes, she may have to leave the foal alone while she searches for water, making it particularly vulnerable to predators. Grevy's Zebra forms small, loose groups to graze, often along with Burchell's Zebra. It communicates by a range of sounds, including a series of deep grunts, punctuated by a whistle-like squeal, and through various body postures. Hunted by large carnivores such as lions and hyenas, Grevy's Zebra flees swiftly from predators, kicking and biting if cornered.

• **SIZE** Body length: 2.5–3 m (8 1/4–9 3/4 ft). Tail: 38–60 cm (15–23 1/2 in).

• **OCCURRENCE** S. Ethiopia, Somalia, S. Sudan, and N. Kenya. In dry desert and open grassland.

blackish nose

distinct pattern of black-and-white stripes on body

white belly, unlike Burchell's Zebra

AUSTERE DIET
With a diet consisting of tough, fibrous grasses and other plants, Grevy's Zebra uses its upper and lower canines to crop grass. Poor quality food and this animal's small stomach means that it spends a lot of time eating. Grevy's Zebra may make seasonal migrations from arid localities where conditions are harsh, to areas where food and water are more easily available.

Social unit Social	Gestation 390 days	Young 1	Diet

SOCIAL GROUPS
Grevy's Zebras live in fluid social groups without any fixed hierarchy. Breeding stallions defend territories, which may be as large as 15 square km (6 square miles), one of the largest among herbivores. They mate with mares that range through their territory.

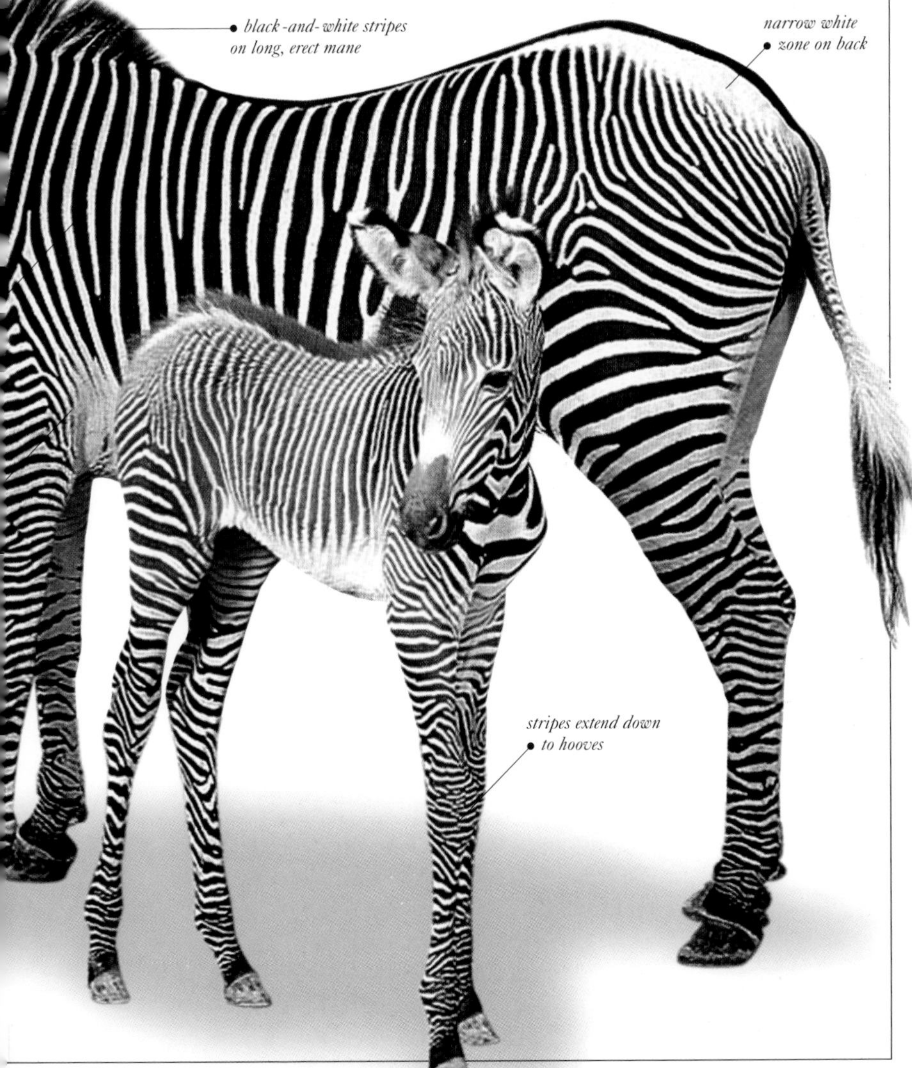

ODD-TOED UNGULATES

RHINOCEROSES

ALL FIVE RHINOCEROS species are endangered: three of them critically so. With little variation across the family, all rhinoceroses are massive-bodied, with mostly bare, extremely thick skin with folds. They have short limbs, tubular ears, tiny eyes (sight is poor), and nasal horns.

The horns are not made of true horn or bone, but of matted keratin (which also forms hair and hooves). The African rhinoceros species have two horns; of the three Asian species, only the Sumatran grows two horns.

The Indian and White Rhinoceroses are grazers of swampy grassland and savanna, while the other three species live in forest and browse. Rhinoceroses tend to be solitary and mainly nocturnal, and most males are territorial.

Family RHINOCEROTIDAE	Species *Ceratotherium simum*	Status Lower risk

WHITE RHINOCEROS

The most numerous rhinoceros, this once widespread inhabitant of the African savanna saw a rapid decline in numbers over the last century, but has made a spectacular recovery since, due to conservation efforts. It is also the largest rhinoceros; its thick grey hide has few wrinkles, apart from at the foreleg joint with the body and on the flank. Almost exclusively a grazer, it has a wide, straight, and hard upper lip pad, allowing it to graze very close to the ground, and giving it the other name of Square-lipped Rhinoceros. Males weigh up to 500 kg (1,100 lb) more than females, have a more pronounced nuchal crest or hump at the shoulders, and a larger front horn that may reach 1.3 m (4 ¼ ft) in length. Placid and sociable by nature, these rhinoceroses are found in small herds of mother-calf pairs and up to seven juveniles. Mature males tend to be solitary and defend their territory of around 1 sq km (½ sq mile) by ritualized displays, fighting with the horn if necessary. Only the dominant male mates within his territory, and is successful only after several attempts.

larger front horn

hard, square lip pad

elongated head

- **SIZE** Length: 3.7–4 m (12–13 ft). Weight: Up to 2.3 tonnes (2 ¼ tons).
- **OCCURRENCE** N.E. and S. Africa. In wooded savanna in N.E. Africa and in dry savanna in S. Africa.
- **REMARK** Although dependent on conservation, there are several secure populations of the Southern White Rhinoceros *(Ceratherium simum simum)*, totalling over 8,500, and the most numerous of all rhinos. The Northern White Rhinoceros (*C. simum cottoni)*, on the other hand, probably numbers fewer than 30 and is on the critical list.

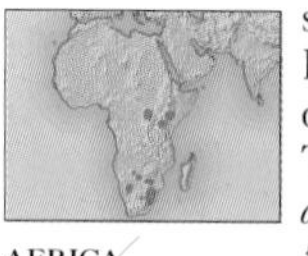

AFRICA

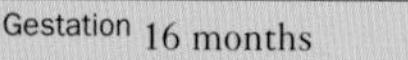

Social unit Social	Gestation 16 months	Young 1	Diet 

MARKING BY URINE
The male rhinoceros sprays urine between his hindlegs to mark a territory that may be about 1 sq km (½ sq mile) in area.

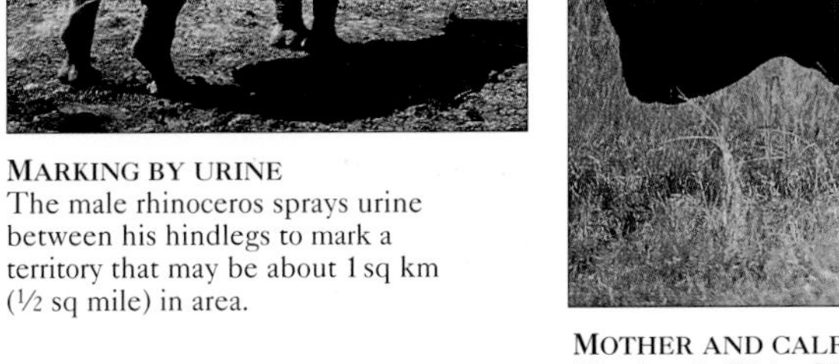

MOTHER AND CALF
The female White Rhinoceros gives birth to her single young, which remains with the mother for 3 years until the next birth. The mother and young use squeaks to communicate with each other.

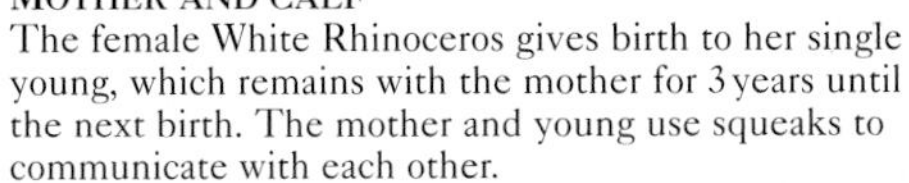

Family RHINOCEROTIDAE	Species *Diceros bicornis*	Status Critically endangered

BLACK RHINOCEROS

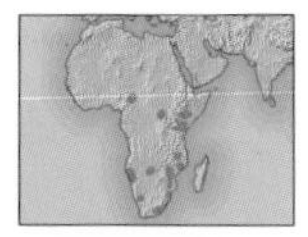

AFRICA

This rhinoceros has two horns, the front one being larger than the back. Unlike the White Rhinoceros (see pp.318 – 319), it does not have a nuchal crest (shoulder hump). Its relatively smooth, grey skin is largely hairless except for the eyelashes, ear-tips, and tail-tip. In common with all rhinoceroses, it has a sharp sense of smell and good hearing, but poor eyesight. Feeding on a wide variety of bushes and low trees, especially at night and in the early morning, this grazer spends the day dozing in shade or wallowing in mud. Usually solitary, the Black Rhinoceros marks its home range plentifully with urine and piles of dung. It sometimes tolerates intruders – of its own species or humans – but is known suddenly to charge or jab with its horns when provoked. Mating couples remain together only for a while, but the calf stays with the mother until she gives birth to another young.

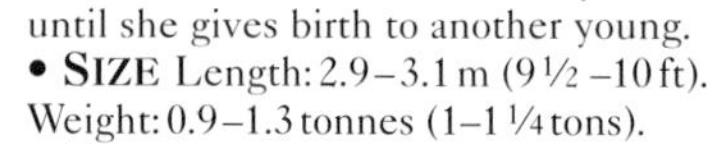

• **SIZE** Length: 2.9 – 3.1 m (9½ – 10 ft). Weight: 0.9 – 1.3 tonnes (1 – 1¼ tons).

• **OCCURRENCE** Sub-Saharan Africa, except the Congo Basin. In a wide range of habitats, from desert to mountain, but primarily in woodland savanna.

• **REMARK** Trade in rhinoceros horns, used in local medicines and for dagger handles, has caused a massive decline in Black Rhinoceros numbers, which fell from 65,000 in 1970 to just 2,500 in the 1990s.

CURLED UPPER LIP

Sometimes called the Hook-lipped Rhinoceros, this species uses its pointed, prehensile upper lip to draw twigs and shoots into its mouth, then bites them off with its strong molar teeth.

Social unit Solitary	Gestation 15 months	Young 1	Diet

Family RHINOCEROTIDAE	Species *Rhinoceros unicornis*	Status Endangered

INDIAN RHINOCEROS

The hairless skin of this one-horned rhinoceros has tubercles on the sides and rear that resemble rivets. The pink skin within the folds attracts parasites, which are often removed by egrets and tick birds. It feeds at night, dawn, and dusk.

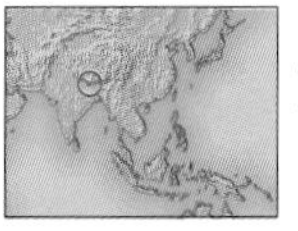

ASIA

• **SIZE** Length: Up to 3.8 m (12 ft). Weight: Up to 2.2 tonnes (2 1/8 tons).

• **OCCURRENCE** India (Brahmaputra Valley). In grassland.

Social unit Solitary	Gestation 16 months	Young 1	Diet

Family RHINOCEROTIDAE	Species *Rhinoceros sondaicus*	Status Critically endangered

JAVAN RHINOCEROS

Largely hairless except for its ears and tail-tip, the Javan Rhinoceros is probably the rarest large mammal in the world. It feeds on leaves and bamboo, uprooting saplings or using its weight to push over vegetation.

• **SIZE** Length: 3–3.5 m (9 3/4–11 ft). Weight: Up to 1.4 tonnes (1 1/2 tons).

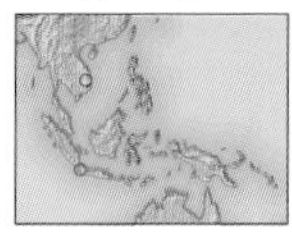

S.E. ASIA

• **OCCURRENCE** S.E. Asia. In tropical forest, mangrove swamps, and bamboo groves.

• **REMARK** Habitat loss and poaching have led to a massive decline in numbers.

Social unit Solitary	Gestation 16 months	Young 1	Diet

Family RHINOCEROTIDAE	Species *Dicerorhinus sumatrensis*	Status Critically endangered

SUMATRAN RHINOCEROS

Also known as the Asiatic Two-horned Rhinoceros, this species is the smallest and hairiest of all rhinoceros species. It browses actively at night and during the day it keeps cool by wallowing in mud, which also coats its skin, protecting it from flies and other insects.

• **SIZE** Length: 2.5–3.2 m (8 1/4–10 ft). Weight: Up to 800 kg (1,760 lb).

ASIA

• **OCCURRENCE** S.E. Asia. In montane rainforest, especially primary forest on slopes.

• **REMARK** The high-altitude habitat of this rhinoceros is now subject to logging; horn poachers are also a major threat.

larger front horn

prominent shoulder fold

skin has few wrinkles

long toes

Social unit Solitary	Gestation 7–8 months	Young 1	Diet

ODD-TOED UNGULATES

TAPIRS

THE FOUR SPECIES of tapirs (family Tapiridae), from Central and South America and Southeast Asia, can be viewed as "living fossils", since their basic anatomy has hardly changed for close to 30 million years.

The pig-like body is supported by relatively slender legs, the ears are large and erect, and the eyes small and deep-set. The flexible, trunk-like snout is used for rooting up various plant foods.

All species are found in forest, close to water, and the tapir's streamlined, compact shape allows it to push through the undergrowth. It can remain almost submerged in water for hours, to stay cool or to avoid predators, using its "trunk" as a snorkel for breathing.

Although tapirs are perissodactyls (odd-toed ungulates), with three toes on each rear foot, the front foot has an extra toe.

Family TAPIRIDAE	Species *Tapirus pinchaque*	Status Endangered

MOUNTAIN TAPIR

The smallest of all tapirs, the Mountain Tapir is also the furriest. Covered with thick, woolly, dark brown to coal-black fur, it has furry white fringes around the lips and on the tips of its rounded ears. Sheltering in forest or thickets by day, it feeds at night, dawn, and dusk, on a variety of dwarf trees and shrubs. The Mountain Tapir is an important disperser of seeds for cloud forest trees and plants, which it ejects whole in its droppings.

- **SIZE** Length:1.8 m (6 ft). Weight:150 kg (330 lb).
- **OCCURRENCE** N.W. South America. In mountains and grassland.
- **REMARK** This tapir is hunted for meat and its body parts are used in traditional medicine.

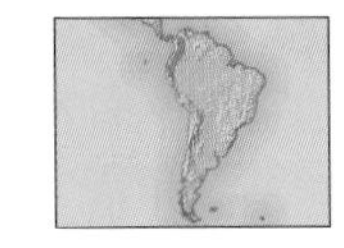

S. AMERICA

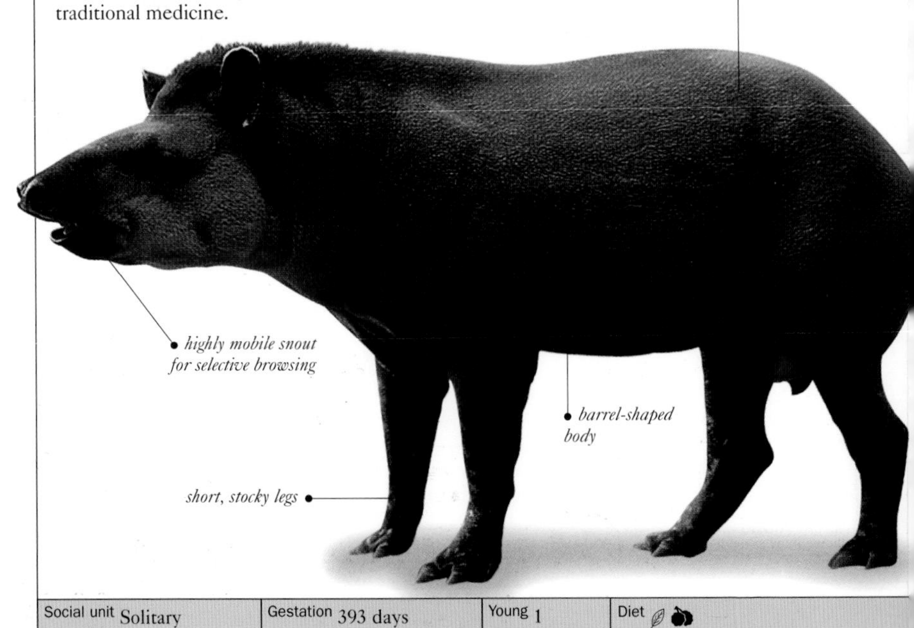

Social unit Solitary	Gestation 393 days	Young 1	Diet

Family TAPIRIDAE	Species *Tapirus bairdii*	Status Vulnerable

BAIRD'S TAPIR

The largest American tapir, this dark brown mammal has pale grey-yellow cheeks and throat, and white-fringed ears. It whistles to communicate with its young or warn other adults away from its territory.

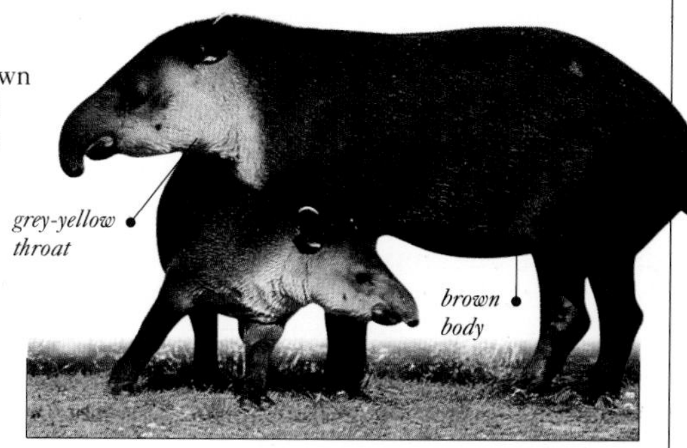

• **SIZE** Length: 2 m (6 ½ ft). Weight: 240–400 kg (530–880 lb).

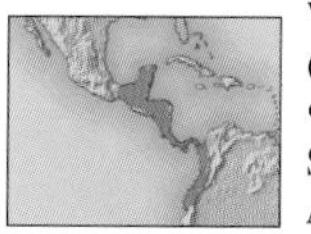
N., C., & S. AMERICA

• **OCCURRENCE** S. Mexico to N. South America. In forest, marshes, and swamps.

Social unit Solitary	Gestation 390–400 days	Young 1	Diet

Family TAPIRIDAE	Species *Tapirus terrestris*	Status Lower risk

SOUTH AMERICAN TAPIR

This tapir has a bristly grey coat and a short, stiff, narrow mane. Its throat and chest may be paler. A good swimmer, it often dives into water to escape predators such as Pumas or Jaguars. The young, as in all tapirs, are born with spots and stripes for camouflage in vegetation.

• **SIZE** Length: 1.7–2 m (5 ½–6 ½ ft). Weight: 225–250 kg (500–550 lb).

S. AMERICA

• **OCCURRENCE** N. and C. South America. In tropical rainforest, gallery forest, and occasionally, open grassy habitat.

Social unit Solitary	Gestation 13 months	Young 1	Diet

Family TAPIRIDAE	Species *Tapirus indicus*	Status Vulnerable

MALAYAN TAPIR

The largest and only Old World tapir, this species has a striking black-and-white coat, which helps to break up its body outline in shady forest for effective camouflage. The Malayan Tapir feeds on soft vegetation and fallen fruits.

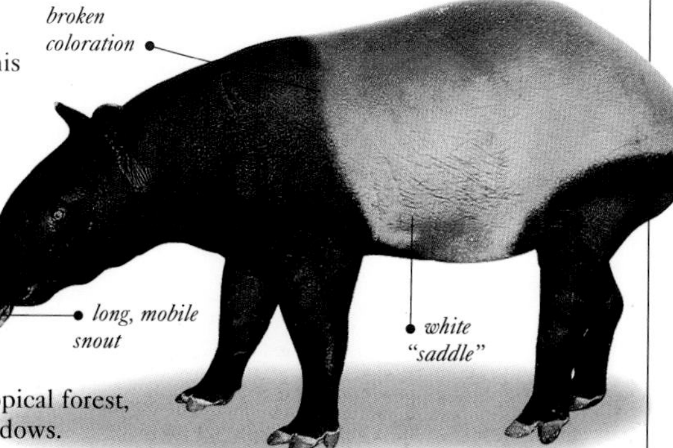

• **SIZE** Length: 1.8–2.5 m (6–8 ¼ ft). Weight: 250–540 kg (550–1,190 lb).

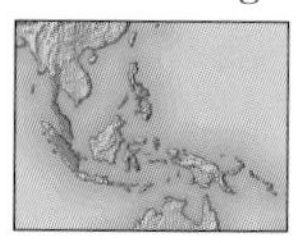
ASIA

• **OCCURRENCE** S.E. Asia. In humid, tropical forest, swampy areas, and meadows.

Social unit Solitary	Gestation 390–407 days	Young 1	Diet

EVEN-TOED UNGULATES

WILD PIGS

PIGS, HOGS, WARTHOGS, and boars tend to be omnivores, rooting for food in soil and leaves with elongated, blunt-ended snouts that are tough yet sensitive. The typical pig has a large head, prominent ears, small eyes, canine teeth forming tusks, and a heavy, barrel-shaped body with thin legs.

About 14 species make up the family Suidae, including domesticated pigs, which are descended from the Wild Boar. Being even-toed ungulates, each foot, or trotter, has two main hooves, with a smaller lateral hoof on either side. Various species of pigs occur naturally in forest, swamp, and grassland across Africa and Eurasia. Some have been introduced to the Americas, Australia, and New Zealand.

Most species live in sounders, which consist of a sow and her offspring; boars (males) join during the mating season.

Family SUIDAE	Species *Sus scrofa*	Status Locally common

WILD BOAR

One of the most widely distributed terrestrial mammals, the Wild Boar or Eurasian Wild Pig is the ancestor of the domestic pig. It has bristly dark grey to black or brown hair, and a mane of longer hair along its spine. The male is larger than the female and has larger tusks. The piglets are pale brown with paler stripes along the back and sides, providing camouflage when in their nest of grass, moss, and leaves in dense thicket. Females, which usually band together in groups of 20, are very protective of their young. Occupying a wide range of habitats, this pig eats almost anything, runs swiftly, and is a good swimmer.

• **SIZE** Body length: 0.9–1.8 m (3–6 ft). Tail: 30 cm (12 in).

• **OCCURRENCE** Europe, Asia, and N. Africa. In tropical and temperate forest, and wetland.

Social unit Variable	Gestation 100–120 days	Young 4–6	Diet

Family SUIDAE	Species *Sus salvanius*	Status Critically endangered

PYGMY HOG

This stocky, short-tailed species is the smallest member of the pig family, and has a sharply tapering head and snout to push through undergrowth. Dark brown in colour, it has paler undersides. The male's upper canines protrude slightly from the sides of its mouth. Both sexes make nests out of hollows lined with grass. Males are solitary, while females live in groups of 4–6. The Pygmy Hog is the carrier of a unique parasite, the Pygmy Hog Louse.

- **SIZE** Body length: 50–71 cm (20–28 in). Tail: 3 cm (1 ¼ in).
- **OCCURRENCE** S. Asia. In riverine grassland.
- **REMARK** A legally protected species, it is still under threat from poaching and the loss of habitat.

Social unit Social	Gestation 100 days	Young 2–6	Diet

Family SUIDAE	Species *Hylochoerus meinertzhageni*	Status Endangered*

GIANT FOREST HOG

The largest of the African pigs, this hog has a massive head with two large, wart-like skin growths below and behind each eye, and tusks that grow out horizontally from each jaw. The blackish grey skin is covered with coarse black hair, which becomes sparser with age. The lighter coloured piglets darken to brown or black as they mature. Unlike other pigs, this species does not root, but grazes and browses on grasses, sedges, shrubs, or cultivated crops.

- **SIZE** Body length: 1.3–2.1 m (4 ¼–7 ft). Tail: 30–45 cm (12–18 in).
- **OCCURRENCE** W., C., and E. Africa. In subalpine areas, bamboo groves, swamp forest, and savanna.

Social unit Social	Gestation 151 days	Young 2–11	Diet

Family SUIDAE	Species *Potamochoerus porcus*	Status Locally common

BUSH PIG

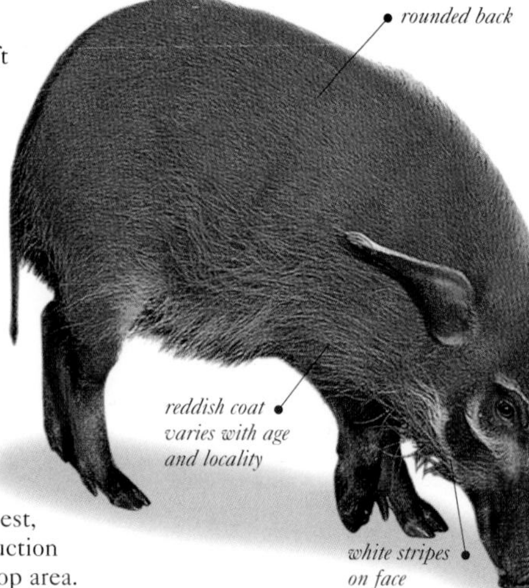

Sometimes called the Red River Hog, this swift runner and agile swimmer is bright reddish in colour and has a distinctive white stripe on its back. Its long, pointed ears have prominent tufts, and there are white stripes on its face. Highly social, the male lives with his harem of females and their young, defending them from intruders. Aggressive in combat, Bush Pigs erect the crest along their backs and circle each other in ritualized displays of strength. The female makes a shallow burrow, lined with grass, for her spotted piglets.

• **SIZE** Body length: 1–1.5 m (3½–5 ft). Tail: 30–43 cm (12–17 in).

• **OCCURRENCE** W. and C. Africa. In tropical forest.

• **REMARK** An agricultural pest, it has benefitted from the reduction of predators and increase in crop area.

AFRICA

Social unit Social	Gestation 4 months	Young 1–8	Diet

Family SUIDAE	Species *Phacochoerus africanus*	Status Locally common

WARTHOG

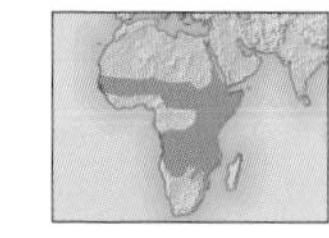

AFRICA

A long-legged, large-headed pig, with prominent facial warts, this species has a long, dark mane extending from the nape of its neck to the middle of its back, where there is a gap, before it continues to the rump. Its tufted tail is held straight and upright when running. The Warthog usually kneels down on its padded "wrists" to feed on newly growing grass tips, using its lips or long incisor teeth. In the dry season, it roots around for underground stems or rhizomes. Burrows, natural or excavated by Aardvarks, are used for shelter, to rear young, or to sleep.

• **SIZE** Body length: 0.9–1.5 m (3–5 ft). Tail: 23–50 cm (10–20 in).

• **OCCURRENCE** Sub-Saharan Africa. In grassland and light mountain forest.

• **REMARK** This is the only pig adapted to grazing in savanna and grassland.

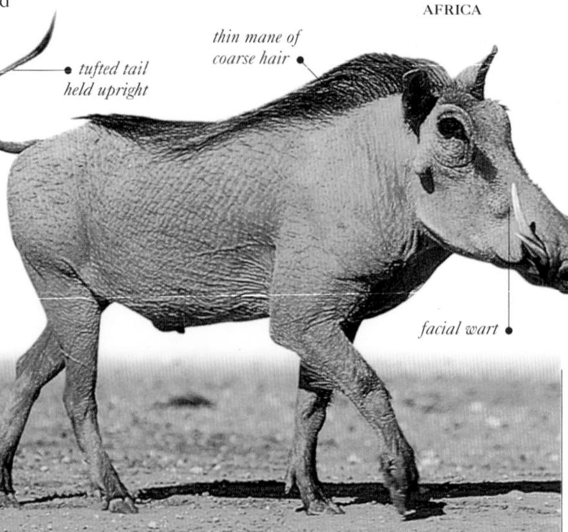

Social unit Social	Gestation 150–175 days	Young 1–8	Diet

Family SUIDAE	Species *Babyrousa babyrussa*	Status Vulnerable

BABIRUSA

ASIA

With a grey to brown, almost hairless body supported by long, thin legs, the Babirusa is unusual in that the male has distinctive upper tusks that grow through the muzzle and then curve back towards the face. Up to 30 cm (12 in) long, they are brittle and loose in their sockets. The lower dagger-like tusks are used by males for fighting and are kept sharp by shearing against the upper tusks and trunks of trees. Females and their young form roving groups of up to eight individuals, whereas males are usually solitary. A good swimmer, the Babirusa sometimes crosses narrow stretches of ocean to reach offshore islands.

- **SIZE** Body length: 0.9–1.1 m (6 ¼ ft). Tail: 27–32 cm (10 ½–12 ½ in).
- **OCCURRENCE** Sulawesi, Togian and Mangole islands. In rainforest and on the shores of rivers and lakes.

Social unit Variable	Gestation 155–158 days	Young 1–2	Diet

Family SUIDAE	Species *Pecari tajacu*	Status Locally common

COLLARED PECCARY

N., C., & S. AMERICA

The smallest of the three peccary species, this pig is often referred to as the Javelina. With a barrel-shaped body and slim legs, it is usually dark grey with a whitish neck collar, and has small, downward-curving tusks. The young are reddish and have a narrow black band along the back. Extremely sociable, peccaries form groups of up to 15, of mixed age and sex, which cooperate to repel predators, usually Coyotes, Pumas, or Jaguars. Social bonding is important, and members of a group may be found grooming each other.

- **SIZE** Body length: 75–100 cm (30–39 in). Tail: 1.5–5.5 cm (½–2 ¼ in).
- **OCCURRENCE** S.W. USA to S. South America. In desert and tropical forest.
- **REMARK** The Collared Peccary is threatened by hunting, and the destruction and fragmentation of its habitat.

Social unit Social	Gestation 145 days	Young 1–4	Diet

EVEN-TOED UNGULATES

HIPPOPOTAMUSES

THE FAMILY Hippopotamidae is made up of only two species: the Common Hippopotamus and the Pygmy Hippopotamus. The former eats mainly grass and is widespread, or even abundant in places, along African rivers and lakes. The latter has a more mixed diet, and is rare, restricted only to the forests and swamps of west Africa, where it is threatened by habitat destruction and hunting.

The hippopotamus has a huge head, an enormous mouth that gapes wide to reveal its tusk-like canine teeth, and very thick, fatty, almost hairless skin. The bulky body is supported by short, slim legs, each ending in four toes. Adaptations for life in water include webbed toes, and eyes, ears, and nostrils sited high on the head, so that the hippopotamus can remain almost totally submerged for long periods.

Family HIPPOPOTAMIDAE	Species *Hexaprotodon liberiensis*	Status Vulnerable

PYGMY HIPPOPOTAMUS

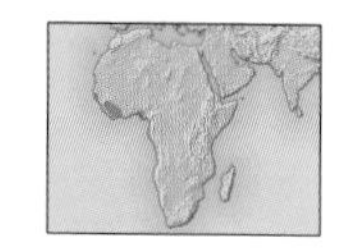

AFRICA

Merely one-fifth the weight of its huge cousin, the Hippopotamus (see opposite), the Pygmy Hippopotamus has a relatively small head, sloping forequarters, and narrower feet with reduced webbing on the toes, all adaptations to a more terrestrial lifestyle. It also feeds on a wider range of plant material including shrubs, ferns, and fruits. A night forager, it follows well-worn trails, spending the day hidden in swamps or in a riverside den, often enlarged from the burrow of some other animal.

- **SIZE** Length: 1.4–1.6 m (4½–5 ft). Weight: 245–275 kg (540–610 lb).
- **OCCURRENCE** W. Africa. In tropical forest and wetland.
- **REMARK** Evidence suggests the Pygmy Hippopotamus has always been rare. Although protected, it is hunted widely for bushmeat.

Social unit Solitary	Gestation 196–201 days	Young 1	Diet

Family HIPPOPOTAMIDAE	Species *Hippopotamus amphibius*	Status Common

HIPPOPOTAMUS

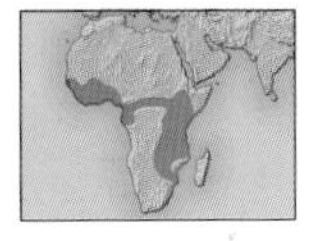

Despite its massive bulk, the Common Hippopotamus is a truly amphibious creature, equally agile on land and in water. The density of its body is slightly greater than that of water and it sinks slowly, walking lightly across the bottom. However, it can stay afloat by inflating its lungs when breathing at the surface, the extra air reducing its body density. The thin, outer layer of its skin dries easily and, despite mucus-producing glands, the hide soon cracks unless regularly moistened in water or mud. During summer, Hippopotamuses are often found wallowing in pools of water in large, temporary groups. Nocturnal by nature, the Hippopotamus chiefly feeds on grass, but has also been observed eating small ungulates or scavenging. The dominant male mates with females in his territory, and the single calf is born underwater. Fiercely protected by its mother, it remains with her until almost mature.

• **SIZE** Length: 2.7 m (9 ft).
Weight: 1.4–1.5 tonnes (1 3/8 –1 1/2 tons).

• **OCCURRENCE** E. and S. sub-Saharan Africa. In grassland and wetland; in water by day.

• **REMARK** DNA analyses reveal that Hippopotamuses are probably more closely related to whales than to other even-toed ungulates.

YAWNING
The Hippopotamus' yawn is actually a threat display. It uses its canine and incisor teeth to defend itself, and may make unprovoked charges on land and in water.

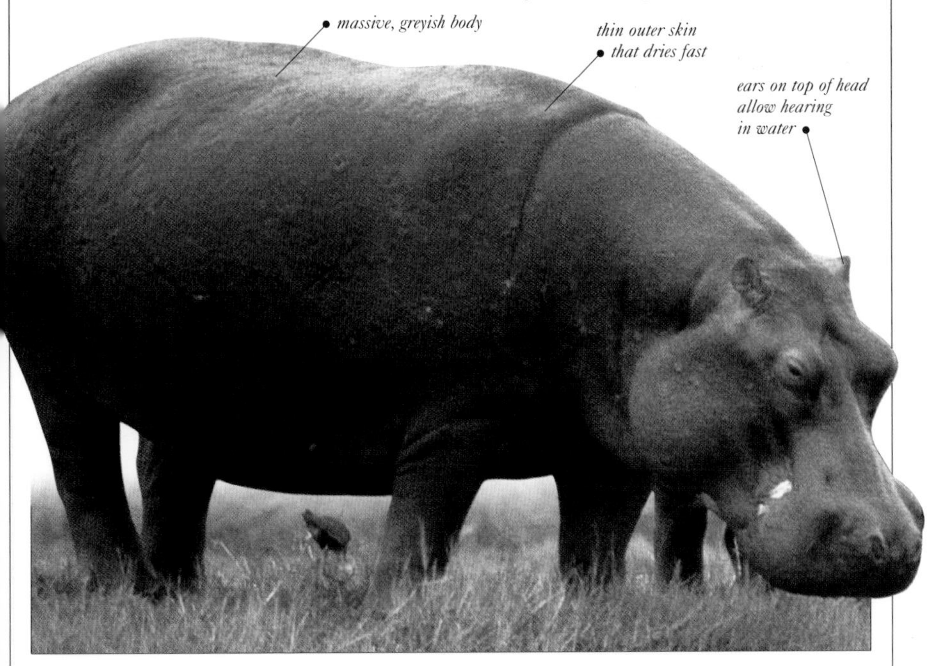

Social unit Solitary	Gestation 240 days	Young 1	Diet

EVEN-TOED UNGULATES

CAMELS AND RELATIVES

THE DROMEDARY (one-humped camel), originally from Africa and West Asia, and the Bactrian Camel (two-humped) of Central Asia, give this family its name, Camelidae. Almost all camels today are either domesticated, or have descended from domestic stock and now roam free.

The other group of camelids is from South America and comprises the wild Guanaco and Vicuña, and their domestic descendants, Llamas and Alpacas.

All camelids have a small head, a split upper lip, long neck, and long legs. They have a "pacing" gait, in which the front and back legs on one side move together, producing a rocking motion. Unlike other even-toed ungulates, the weight is carried not on the hooves, but on fatty pads under the toes.

Family CAMELIDAE	Species *Vicugna vicugna*	Status Endangered

VICUÑA

A selective grazer of perennial grasses, the Vicuña has a prehensile, cleft upper lip and permanently growing incisors to facilitate feeding. It has a uniform light to dark cinnamon coat, with a variable white "bib" and white undersides. Its long legs are placed midway on its body and its hindquarters are contracted. Unlike the camels (see pp. 332–333), it must drink water every day. Groups of one male, 5–10 females, and their young occupy territories marked by dung and urine.

- **SIZE** Length:1.5 (5 ft). Weight: 40–55 kg (88–120 lb).
- **OCCURRENCE** W. South America, in the Andes. In alpine tundra.
- **REMARK** In the 1950s it was discovered that fewer than 10,000 Vicuñas survived in Peru, probably less than 1 per cent of the total population that existed 500 years ago, when the Spanish colonizers first arrived in South America.

S. AMERICA

long, pointed, and erect ears

small head

long and thin neck

ALPACA
Domesticated in the high Andes 4,000 years ago, for its fine wool, the Alpaca was once thought to be a descendant of the Guanaco. DNA tests have now traced its ancestry to the Vicuña.

Social unit Social	Gestation 342–345 days	Young 1	Diet

Family CAMELIDAE	Species *Lama guanicoe*	Status Vulnerable

GUANACO

S. AMERICA

Typically pale to dark brown in colour, with a whitish chest, underside, and inner legs, the Guanaco has a grey to black head with white-fringed eyes, lips, and ears, and a long neck. Preferring cold habitats, it is found from sea level up to 4,000 m (13,000 ft). It feeds on a variety of grasses and shrubs, as well as on lichens and fungi. Family groups consist of one male and around 4–7 females with their young. In the extreme south of the Guanaco's range, where there is significant winter snow accumulation, females and young migrate to more favourable areas in search of food. However, the male remains behind to guard his territory.

- **SIZE** Length: 0.9–2.1m (3–7 ft). Weight: 96–130 kg (210–290 lb).
- **OCCURRENCE** W. to S. South America. In grassland, desert, temperate forest, and mountains.
- **REMARK** Hunted by humans, and threatened by overgrazing of their natural habitat, the four subspecies of Guanaco have been classified as vulnerable. The northernmost subspecies, regarded as the ancestor of the domestic Llama, is critically endangered.

small head

distinctive, long, pointed ears

prehensile, cleft upper lip

pale to dark brown body

neck set low on torso

long legs

fleshy pads on feet, with nails on upper surface

thick, woolly coat

LLAMA
The Llama is a domestic species, first bred from the Guanaco by the Inca civilization, around 6,000 years ago. It is the traditional pack animal of the Andes, and is farmed for its wool and meat (in countries outside South America as well).

Social unit Social	Gestation 345–360 days	Young 1	Diet 

Family CAMELIDAE	Species *Camelus dromedarius*	Status Common

DROMEDARY

The domesticated one-humped camel, now extinct in the wild, is adapted to desert life in various ways. Using the fat reserves in its hump, it loses up to 40 per cent of its body weight when food and water are scarce, and can survive without drinking water longer than any other domestic animal (six or seven months in colder weather). Its body temperature rises in hot weather and sweating is reduced to conserve moisture. It feeds on a huge variety of plants, including salty and thorny species, using its prehensile, split upper lip to browse; it also scavenges on dried out carcasses. Found in small herds of one male and several females and young, the male defends its group by jumping and biting, and pushing away intruders. The Dromedary ranges in colour from cream through almost every shade of brown to black.

• **SIZE** Length: 2.2–3.4 m (7 ¼–11 ft).
Weight: 450–550 kg (990–1,210 lb).
• **OCCURRENCE** N. and E. Africa, W. and S. Asia. In desert.
• **REMARK** The Dromedary has been wholly domesticated, with no record of any individuals existing in the wild since prehistoric times (3000 BC).

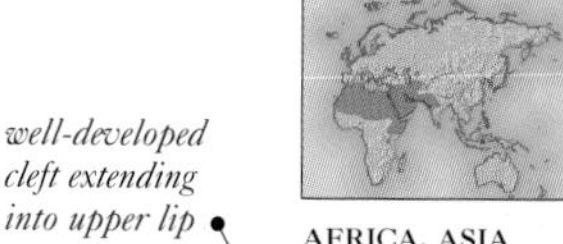

AFRICA, ASIA

FACING THE HEAT
The Dromedary has nostrils that moisten and cool the air it breathes. The heavy eyebrows and double row of curly lashes protect its eyes from desert sand.

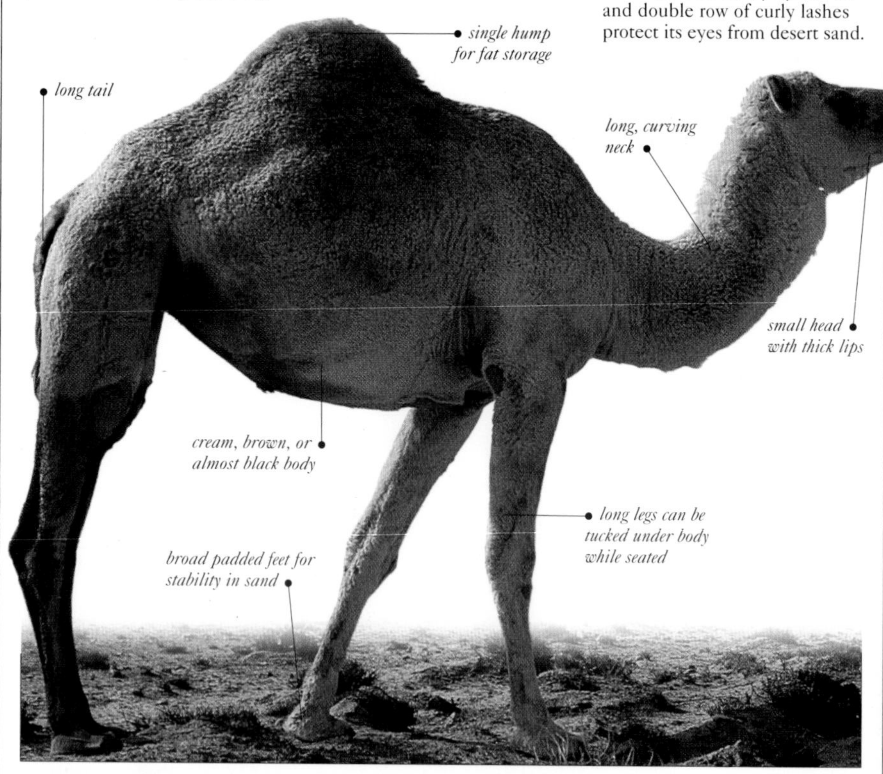

Social unit Social	Gestation 370 days	Young 1	Diet

Family CAMELIDAE	Species *Camelus bactrianus*	Status Endangered

BACTRIAN CAMEL

The two-humped Bactrian Camel is the only Old World camelid found in the wild. It can withstand temperatures from -29°C (-20°F) to 38°C (100°F). After a drought, it can drink up to 110 litres (24 gallons) of water in about 10 minutes. Uniformly light beige to dark brown in colour, it has a shaggy coat in winter that it sheds in summer. The humps, composed of fat, are erect when the animal is well fed, but become flaccid as resources are used up. During the rut, the male puffs out its cheeks, extrudes a red, balloon-like sac from its mouth, grinds its teeth, and paces alongside its competitor in a dramatic, ritualistic display. The dominant male acquires a harem of 6–30 females and their offspring. This family group is not territorial and travels long distances in search of food and water.

• **SIZE** Length: 2.5–3 m (8½–9¾ft). Weight: 450–690 kg (990–1,520 kg).

• **OCCURRENCE** Wild populations in the Gobi Desert; domestic populations in C. Asia. In cold steppe and desert.

• **REMARK** First domesticated 3,500 years ago, this camel is an important beast of burden. Its milk and flesh are also consumed.

ASIA

FLAT-TOED
The two-toed, broad feet of this camel provide stability in sand and snow.

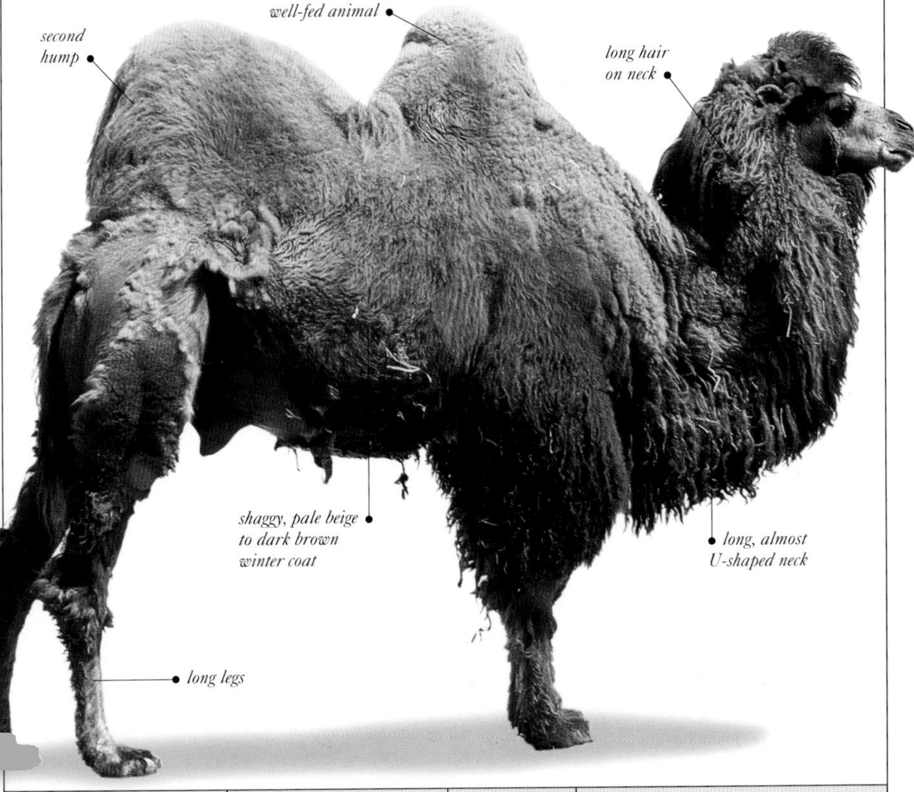

Social unit Social	Gestation 406 days	Young 1	Diet

EVEN-TOED UNGULATES

DEER

SOME 45 SPECIES of deer make up the family Cervidae. They are artiodactyls (even-toed ungulates), with four visible toes on each foot. The first toe is practically lost, and the second and fifth toes are small, leaving the third and fourth toes to bear the body weight. The deer's large eyes, ears, and muzzle provide it with excellent senses to detect the presence of predators.

Deer resemble antelopes, but their antlers are branched and regrow each year. Only male deer have antlers, except reindeer, where females also grow them.

Most species are browsers, at home in woodland and forest, but some also graze in grassland, swamp, or semi-desert. Various cervids are found naturally on all continents, except Australia and Africa (barring the Mediterranean fringe). Many have been introduced to new regions over the centuries.

Family TRAGULIDAE	Species *Tragulus meminna*	Status Unconfirmed

INDIAN SPOTTED CHEVROTAIN

In common with other species of chevrotain (mouse deer), the Indian Spotted Chevrotain, or Spotted Mouse Deer, has four fully developed toes on each foot, unlike "true" deer which have just two. It has a spotted back and white stripes on its throat and flanks. Males use their tusk-like upper canine teeth to fight off opponents.

• **SIZE** Body length: 50–58 cm (20–23 in). Tail: 3 cm (1 ¼ in).

ASIA

• **OCCURRENCE** S. India and Sri Lanka. In tropical rainforest, especially rocky terrain up to 700 m (2,300ft).

Social unit Solitary	Gestation 5 months	Young 1	Diet

Family MOSCHIDAE	Species *Moschus chrysogaster*	Status Lower risk

ALPINE MUSK DEER

Found on rocky, forested mountain slopes, this stocky species has well-developed side toes, enabling it to climb rocks, and to move through snow. It has a dark brown coat, mottled light grey, and a whitish chin and ear fringes.

• **SIZE** Body length: 70–100 cm (28–39 in). Tail: 2–6 cm (¾–2 ¼ in).

• **OCCURRENCE** Afghanistan to C. China. In temperate and coniferous forest, as well as mountains.

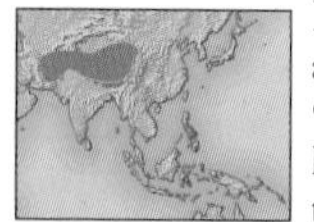

ASIA

• **REMARK** This deer is widely hunted for its musk, a secretion from the navel region, which is highly prized in the perfume industry.

stocky brown body

well-developed side toes help in climbing

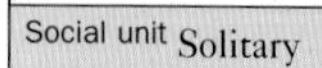
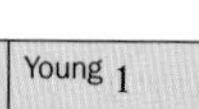

Social unit Solitary	Gestation 185–195 days	Young 1	Diet

Family CERVIDAE	Species *Dama dama*	Status Locally common

FALLOW DEER

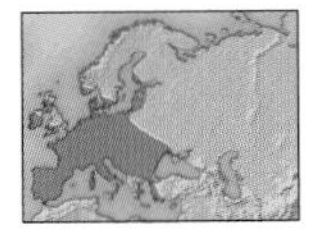
EUROPE, ASIA

Usually brown with white spots, this deer may also be blackish or white. Domesticated for its graceful looks and meat, it has been introduced to the Americas and Africa, as well as Australia. During rut, bucks establish a small patch of territory, where they mate. The Fallow Deer feeds at twilight on a variety of vegetation. The subspecies, Mesopotamian Fallow Deer, is endangered.

• **SIZE** Body length: 1.4–1.9 m (4 ½–6 ¼ ft). Tail: 14–25 cm (5 ½–10 in).

• **OCCURRENCE** Europe and W. Asia. In grassland and forest.

Social unit Social	Gestation 229 days	Young 1	Diet

Family CERVIDAE	Species *Axis axis*	Status Locally common

CHITAL

Also known as the Spotted or Axis Deer, the Chital is found in large herds of 100 or more, grazing in grassland and browsing in open woodland, often below troops of noisy monkeys, which drop fruits on the ground. Swift and agile, the Axis Deer dashes for cover when threatened, reaching a speed of 65 kph (40 mph). With a bright chestnut, white-spotted coat, it has a distinctive white throat patch, and creamy white undersides. The stag's antlers have a brow tine (prong), and the main beam spilts into two, sweeping gracefully backwards.

• **SIZE** Body length: 1–1.5 m (3 ¼–5 ft). Tail: 10–25 cm (4–10 in).

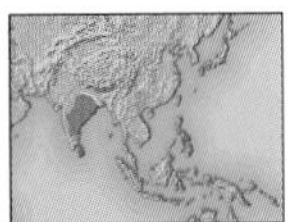
ASIA

• **OCCURRENCE** S. Nepal, India, and Sri Lanka. In open grassland, near cover.

Social unit Social	Gestation 225–230 days	Young 1	Diet

Family CERVIDAE	Species *Cervus elaphus*	Status Locally common

RED DEER

Reddish brown in summer, sometimes with a dark line along its back and neck and indistinct flank spots, the Red Deer turns a dull brown in winter. Its tail and rump patch are straw-coloured. There is great variation in appearance among the 28 or so subspecies of this deer, which include the Wapitis of China and North America. Only the stag has antlers, and the larger they are, the more females he attracts. The antlers are typically a fork or cup of points, and the number of points keep increasing with age, until the male passes his prime. Females (hinds) form herds led by a dominant hind, and the males form separate groups, except during the autumn rut. At this time, stags battle each other and serious fights may take place between evenly matched animals. Using part display and part physical contest, the stags walk parallel to each other as they assess whether to fight. Once the contest begins, they interlock antlers, pushing, twisting, and shoving each other; the victor wins possession of a harem. The diet of this day-and-night grazer varies with the seasons and consists of grasses, sedges, rushes, heather, and a variety of other plants, which it sometimes reaches by standing up on its hindfeet.

- **SIZE** Body length: 1.5–2 m (5–6 ½ ft). Tail: 12 cm (4 ¾ in).
- **OCCURRENCE** Europe to E. Asia. Introduced in Australia and New Zealand, USA, and South America. In grassland, and temperate and coniferous forest.
- **REMARK** This is the most widespread deer in the world, but nine of the subspecies on different continents are endangered, facing the threat of poaching as well as habitat loss and fragmentation.

fully grown antlers

red-brown summer coat turns dull brown in winter

white spots on coat

"LYING UP"
The newly born fawn, reddish brown with white spots, "lies up" hidden in vegetation for the first two weeks, as with all deer young. The mother calls loudly to locate her fawn, taking care of it alone.

Social unit Social	Gestation 225–245 days	Young 1	Diet

BRED IN CAPTIVITY
The Red Deer is adapted to a wide range of climates. It is now widely farmed for its meat, hide, and antler velvet, which is used in oriental medicines. Herds of Red Deer are also kept in urban parks.

MATING SIGNALS
The male Red Deer develops a mane as it enters rut in autumn. During this time it roars, thrashes bushes, and wallows in the mud. It stops feeding and loses a lot of weight.

Family CERVIDAE	Species *Cervus nippon*	Status Critically endangered

SIKA

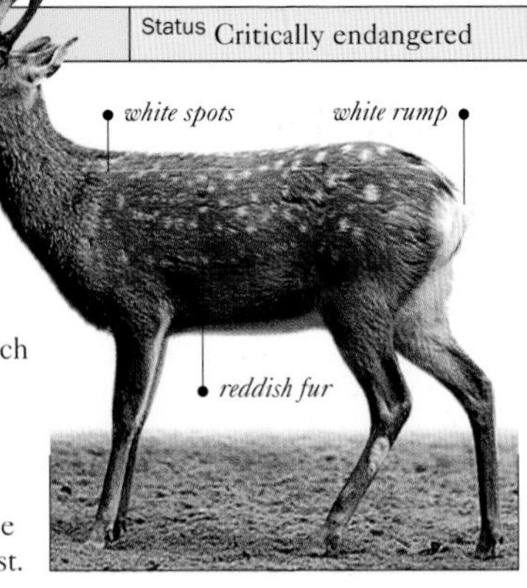

Kept in parks and farmed for centuries, the Sika has also been introduced in many countries. Its appearance varies among the 14 subspecies. However, it is generally rich red-brown with white spots in summer, and turns almost black in winter. Males have a dark chevron, or wavy pattern, on their forehead, and each antler has a maximum of four points.

ASIA

• **SIZE** Body Length: 1.5–2 m (5–6½ ft). Tail: 12–20 cm (4¾–8 in).

• **OCCURRENCE** Vietnam, Taiwan, China, and Japan. Introduced in Europe and New Zealand. In grassland and forest.

Social unit Social	Gestation 220 days	Young 1	Diet

Family CERVIDAE	Species *Cervus unicolor*	Status Locally common

SAMBAR

Uniformly dark brown, except for rusty hues on its chin, inner legs, and underside of its black-tipped tail, the Sambar has a thick mane that is more prominent in rutting males. During this season the male develops a hairless sore spot on the sides of the neck, and stamps on the ground, wallows in mud, and strips bark off trees. This nocturnal browser eats a variety of vegetation and is usually solitary, except for female–fawn pairs.

three-pronged antlers in male

thick mane on neck

uniformly dark brown coat

ASIA

• **SIZE** Body length: 2–2.5 m (6½–8¼ ft). Tail: 15–20 cm (6–8 in).

• **OCCURRENCE** S. and S.E. Asia. In light woodland.

Social unit Solitary	Gestation 240 days	Young 1	Diet

Family CERVIDAE	Species *Elaphurus davidianus*	Status Critically endangered

PÉRE DAVID'S DEER

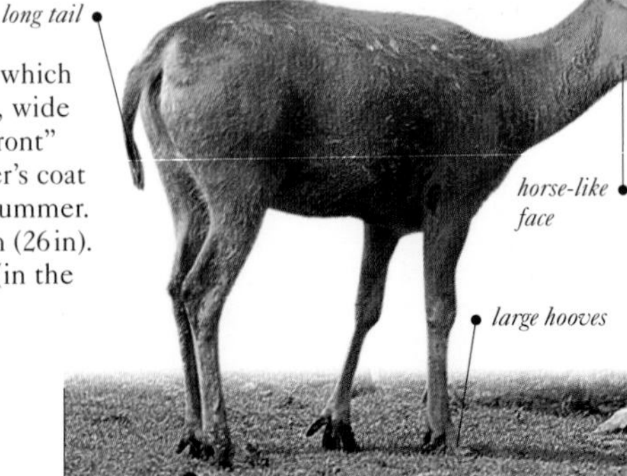

Quite unlike a deer in form, this species (of which a female is shown here) has a horse-like face, wide hooves, and a long tail. The stag's "back to front" antlers are another unusual feature. This deer's coat is greyish fawn in winter, and red-brown in summer.

• **SIZE** Body length: 2.2 m (7¼ ft). Tail: 66 cm (26 in).

• **OCCURRENCE** Reintroduced in China (in the 1980s). In open grassland.

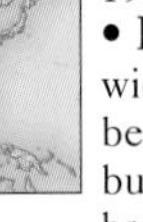

ASIA

• **REMARK** Once widespread in China, it became extinct in the wild, but was saved by captive breeding in the UK.

Social unit Social	Gestation 283 days	Young 1	Diet

Family CERVIDAE	Species *Capreolus capreolus*	Status Locally common

ROE DEER

A black muzzle band and variable white chin and throat patches characterize this deer. Red-brown in summer, it moults to a densely furred, grey coat in winter. Both sexes have a white rump patch with a tiny hidden tail. However, in females the patch is an inverted heart-shape, while in males it is kidney-shaped. The fur on this patch is fluffed up in times of danger, as is the Sika's (shown opposite). The male has rough-surfaced, three-point antlers that it sheds and regrows each winter. Most active at dawn and dusk, the Roe Deer feeds on a variety of vegetation such as grasses, herbs, shrubs, ivy, and fruits.

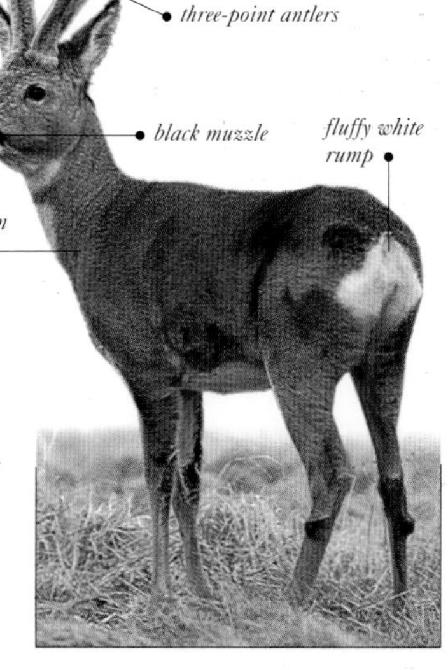

- **SIZE** Body length:1–1.3 m (3¼–4¼ ft). Tail: 5 cm (2 in).

- **OCCURRENCE** Europe to Asia Minor. In forest near glades, or woodland surrounded by pasture.
- **REMARK** A close relative of the Roe Deer is the Siberian Roe (*Capreolus pygargus*).

Social unit Variable	Gestation 300 days	Young 1–3	Diet

Family CERVIDAE	Species *Odocoileus hemionus*	Status Lower risk

MULE DEER

Widely distributed in many habitats, the Mule Deer is recorded as eating hundreds of plant species. Rusty brown in summer, and grey-brown in winter, it has variable white patches on its face and throat, and black bands on its forehead. The tail is black on the upper surface, and white below, giving this species the alternative name of Black-tailed Deer.

As with many other deer, rutting season is from September to November and the young are usually born in June.

- **SIZE** Body length: 0.85–2.1 m (2¾–7 ft). Tail:10–35 cm (4–14 in).
- **OCCURRENCE** W. North America. In grassland and cultivated land to woodland edges, and sometimes urban areas.

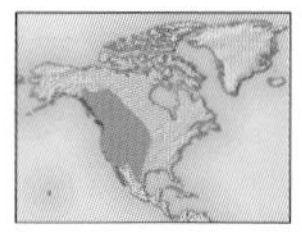

Social unit Social	Gestation 203 days	Young 1–2	Diet

Family CERVIDAE	Species *Blastocerus dichotomus*	Status Vulnerable

MARSH DEER

The largest South American deer, this species is reddish brown in summer and darker in winter, and has a pale face, black lips and nose, and black lower limbs. Long legs with large hooves enable it to walk over marshy or waterlogged ground. It feeds on grasses, reeds, water plants, and shrubs, and lives alone or in groups of two or three, moving seasonally between marshes and higher ground according to the water level.

- **SIZE** Length: Up to 2 m (6½ ft). Weight: 100–140 kg (220–310 lb).
- **OCCURRENCE** S. South America. In marshes, floodplains, and forest edges.
- **REMARK** The Marsh Deer faces a serious threat from habitat loss caused by agriculture and water pollution.

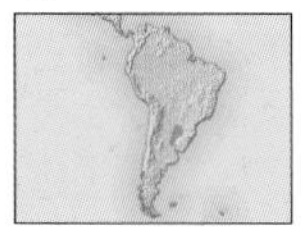

S. AMERICA

large ears, white inside

reddish brown coat turns darker in winter

distinctive long legs, black below, with large hooves

Social unit Variable	Gestation 9 months	Young 1	Diet

Family CERVIDAE	Species *Pudu pudu*	Status Vulnerable

SOUTHERN PUDU

One of the two species of small, stocky, and low-slung pudu, the Southern Pudu has a long-haired, coarse coat of buff to reddish brown, and hardly any tail. The male has antlers that are simple spikes, about 8 cm (3¼ in) long. A solitary species, active both by day and by night, the Southern Pudu takes cover in thickets and undergrowth when threatened.

- **SIZE** Length: 85 cm (34 in). Weight: Up to 15 kg (33 lb).
- **OCCURRENCE** S.W. South America. In humid forest from sea level to a height of 1,700 m. (5,600 ft).

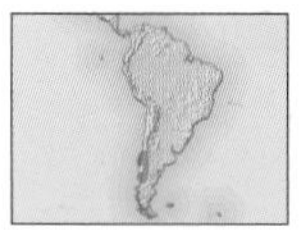

S. AMERICA

Social unit Solitary	Gestation 202 days	Young 1	Diet

Family CERVIDAE	Species *Rangifer tarandus*	Status Endangered*

REINDEER

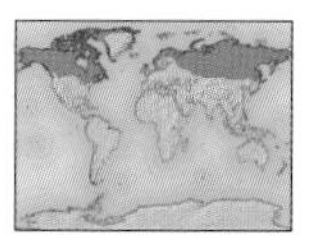

N. AMERICA, GREENLAND, EURASIA

Called Caribou in North America, the Reindeer is the only species of deer in which both sexes have long antlers, with a unique shovel-like brow tine on one side. American forms have chiefly brown coats with darker legs, whereas Eurasian forms are greyer. Adapted to its cold habitat in various ways, the Reindeer has a dense coat that effectively insulates it from freezing water or icy winds, while its large antlers and hooves are used to scrape the snow for lichen. Bulls lose their antlers in spring or early winter, unlike females which retain theirs until the calves are born in May. Pregnant females use their antlers to fight males in winter when food is scarce. Reindeer feed on grasses, sedges, and herbs in summer, and on lichen and fungi in winter.

MIGRATORY PATTERNS
Some Reindeer travel 15–65 km (9–40 miles) daily within the same area; others migrate 1,200 km (750 miles) twice a year in large herds.

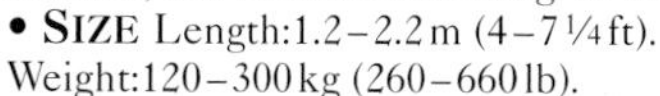

- **SIZE** Length: 1.2–2.2 m (4–7 1/4 ft). Weight: 120–300 kg (260–660 lb).
- **OCCURRENCE** N. North America, Greenland, N. Europe to E. Asia. In mountains and coniferous forest.
- **REMARK** The economy of some peoples in the Arctic depends on this deer for meat, skin, and milk; around two million are semi-domesticated.

large, palmate antlers of male

furry coat to protect from cold

greyish coat of Eurasian Reindeer

Social unit Social	Gestation 210–240 days	Young 1	Diet

Family CERVIDAE	Species *Alces alces*	Status Unconfirmed

ELK

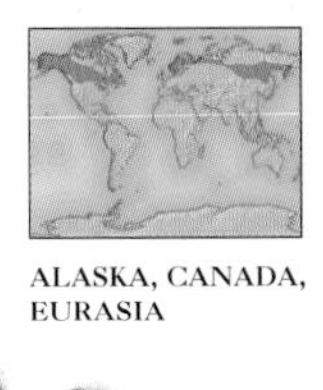

ALASKA, CANADA, EURASIA

The largest member of the deer family, the Elk has several subspecies. In the North American Elk, known as the Moose, the male (shown here), usually has a much larger dewlap than that found in Europe. All Elks have large, blunt muzzles, and wide, palm-shaped antlers in the male, spanning 2 m (6 ½ ft), each with around 20 points. Brownish grey in summer, its greyish winter coat has longer guard-hairs with woollier underfur to withstand severe cold. Its long legs are lighter in colour, and have wide hooves for wading in water and mud, and walking on snow. The Elk inhabits woodland close to swamps and lakes. In summer, it may stand submerged with only its eyes and nostrils above the surface, feeding on the roots of water-lily and other aquatic plants. It also feeds on sedges, horsetails, and other leafy vegetation, although in winter its diet mainly comprises twigs of willow and poplar. The Elk's broad muzzle and flexible lips help it to grasp water plants and strip leaves from twigs. It is usually solitary or found in small family groups, but males may form bachelor groups in winter. Rutting takes place in autumn when the usually silent male grunts and conducts head-to-head battles with other males to establish its dominance.

• **SIZE** Length: 2.5–3.5 m (8 ¼–11 ft). Weight: 500–700 kg (1,100–1,540 lb).

• **OCCURRENCE** Alaska, Canada, and N. Europe to N. Asia. In marshy forest and taiga, close to water.

• **REMARK** The Elk is hunted for its meat and skin, and as a sporting trophy.

wide, splayed antlers

large, blunt muzzle

large dewlap in male

FEMALE GRAZING WITH CALF
The Elk calf is born in May or June, and follows its mother within one or two days of its birth. It grows rapidly, fed on the mother's very rich milk; and she protects her calf from predators, using her hooves.

Social unit Variable	Gestation 242–250 days	Young 1–2	Diet

MARSHY HABITAT
Usually found in areas close to water and marshland, Elks are most active at dawn and dusk. In the cold taiga, they may be active in daylight, because of the long hours of darkness in winter and the reverse in summer.

EVEN-TOED UNGULATES
PRONGHORN

NAMED AFTER the forward-facing "prong" on each horn, this deer-like mammal is the only species in its family, Antilocapridae.

It has several features intermediate between those of deer and antelopes. Its horns consist of a horny sheath over a bony core, as in antelopes, yet they are forked, and are shed and regrown annually, like deer antlers. All males have horns; some females do too, but these are smaller than the male's, and usually do not have prongs. Each foot has two hoof-capped toes.

Pronghorns are restricted to the plains and deserts of North America. They are known to feed on a wide variety of grasses, leaves, and shrubs, as well as cacti and thornbushes. In arid places they can go for days without drinking any water, surviving on the moisture derived from their food.

Family ANTILOCAPRIDAE	Species *Antilopcapra americana*	Status Locally common

PRONGHORN

Locally known as the Prairie Ghost, the Pronghorn can run at a speed of more than 65 kph (40 mph) and vanish from sight within a few seconds. Reddish brown to tan, it has a black mane and underparts, a white rump, and two white stripes across its neck. The female has horns that seldom exceed the length of its ears, whereas in the male the horns grow up to 25 cm (10 in) long, curving backwards at the tip, and with a forward-pointing "prong" arising from the upper half. The white rump hair is raised as a visual warning in times of danger and can be seen from a long distance. In winter, the Pronghorn may form herds of around 1,000 of mixed age and sex, which split into smaller groups in summer.

- **SIZE** Length: 1–1.5 m (3 ¼–5 ft). Weight: 36–70 kg (79–155 lb).
- **OCCURRENCE** S. Canada to W. and C. USA. In grassland and desert.

forward-pointing prong from upper half of horn

two white stripes on neck

white rump hair raised as visual warning in danger

N. AMERICA

Social unit Variable	Gestation 252 days	Young 1–2	Diet

EVEN-TOED UNGULATES

GIRAFFE AND OKAPI

ONLY TWO SPECIES – the Giraffe and the Okapi – make up the family Giraffidae. They are both from Africa, and are similar in many respects.

Almost all parts of their anatomy are slim and elongated – head, tongue, neck, body, legs, and tail; this tendency is most exaggerated in the Giraffe. Both species have horn-like, cartilage-based ossicones, or short conical protrusions, on the head (although these are present only in male Okapis), and both browse with their prehensile tongues, using their peculiar, lobed canine teeth as leaf-rakes.

However, in ecology and habits, the species are different. Giraffes frequent open savanna, and although they have individual home ranges, often gather in loose herds. The Okapi prefers thick forest, is solitary or found in pairs, and only males maintain territories.

Family GIRAFFIDAE	Species *Okapia johnstoni*	Status Lower risk

OKAPI

Shy and secretive, the Okapi feeds in dense forest, relying more on sound than sight to move around, and emitting a characteristic "chuff" sound on meeting another of its kind. Its sleek coat appears deep red, purple, brown, or black, depending on the angle of the light, and its face is white from the middle portion to the neck. Zebra-like stripes on the rump and upper parts of the legs lend a distinctive stamp to its two-toned appearance. The female is taller and heavier than the male, and the male has a pair of short horns on his forehead. The Okapi curls its long, black prehensile tongue around vegetation to feed. The female also uses her tongue to groom herself and her young. The bond between mother and calf, however, is not as strong as it is in many other hooved animals.

- **SIZE** Length: 2–2.2 m (6 ½–7 ¼ ft). Weight: 200–350 kg (440–770 lb).
- **OCCURRENCE** N. and E.C. Zaire. In dense, damp equatorial forest.
- **REMARK** Once called the "forest zebra", the Okapi was not identified as a separate species until 1900.

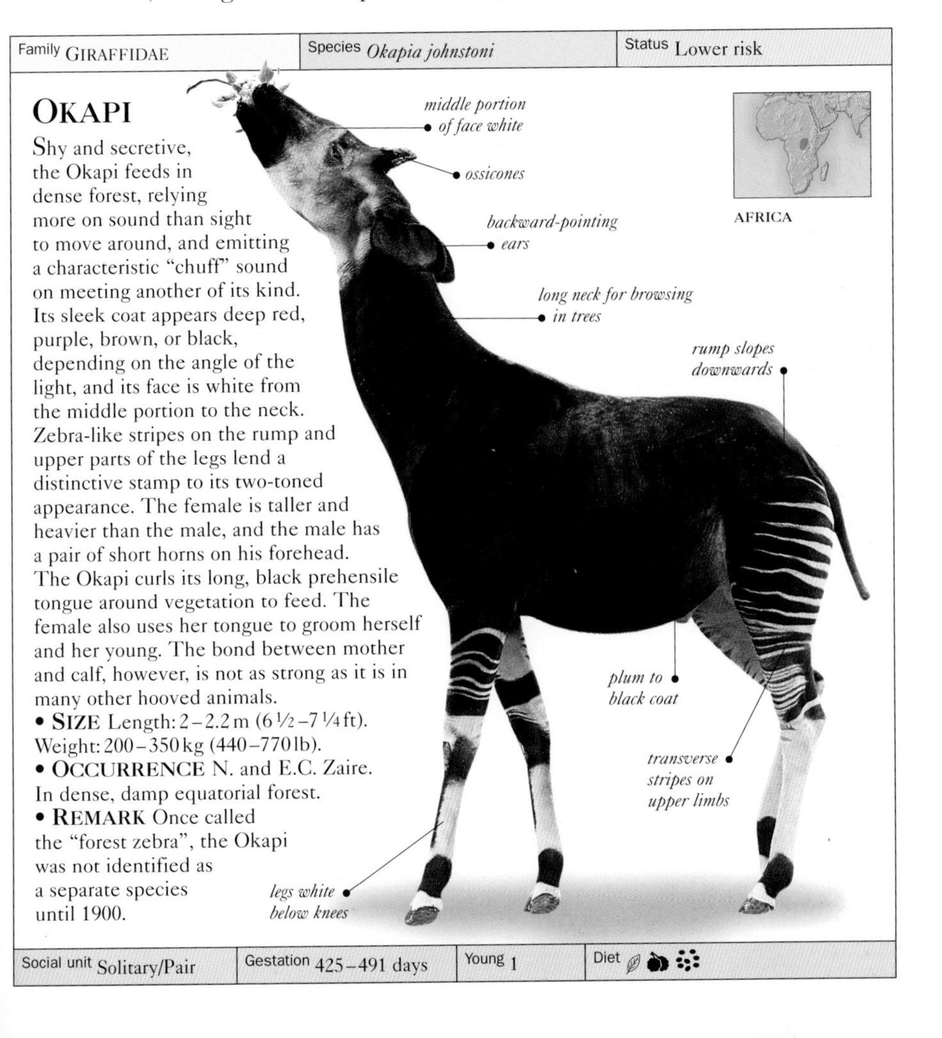

Social unit Solitary/Pair	Gestation 425–491 days	Young 1	Diet

Family GIRAFFIDAE	Species *Giraffa camelopardalis*	Status Lower risk

GIRAFFE

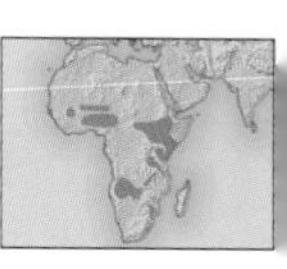

The tallest terrestrial animal, known to grow as tall as 5.5 m (18 ft), the Giraffe is adapted in several ways to browse high among trees. The combined height of its front legs, shoulders and neck, together with its elongated tongue and skull, provide it with great reach. Among the Giraffe's other distinctive features are large eyes and ears, 2–4 horns in both sexes, a back that slopes sharply from shoulders to rump, stilt-like legs with heavy feet, and a thin, tufted tail for whisking away flies. The Giraffe feeds and drinks in the morning and evening, chews cud during the hottest part of the day, and rests (on its feet) at night. Male hierarchy is established by "necking", where two males stand parallel, swinging their necks at each other and striking with the side of the head, in a slow ritualized display. Cows mate with the local dominant bulls. The newborn Giraffe calf stands 2 m (6½ ft) tall and is up on its feet within about 20 minutes of birth. It is weaned at 13 months and stays 2–5 months longer with the mother.

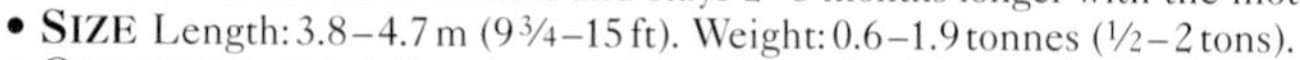

- **SIZE** Length: 3.8–4.7 m (9¾–15 ft). Weight: 0.6–1.9 tonnes (½–2 tons).
- **OCCURRENCE** Africa, south of the Sahara. In dry savanna and open woodland – associated with scattered acacia.

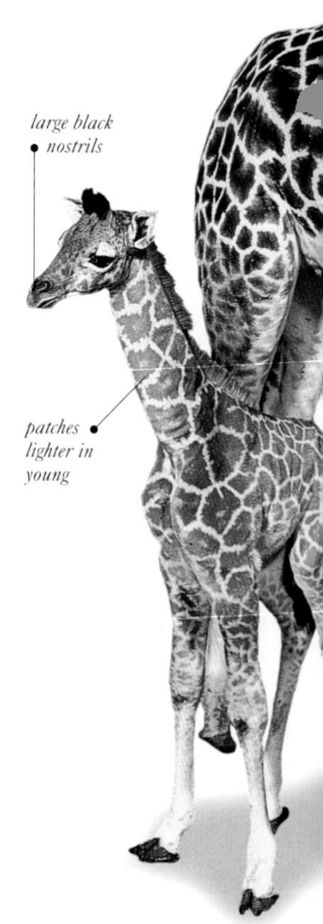

HOW A GIRAFFE DRINKS
The Giraffe splays out its front legs and bends its neck in order to drink water. Normally its heart pumps blood all the way up to its brain at enormous pressure. When it bends, several one-way valves control the force of blood to prevent brain damage.

Social unit Variable	Gestation 457 days	Young 1	Diet

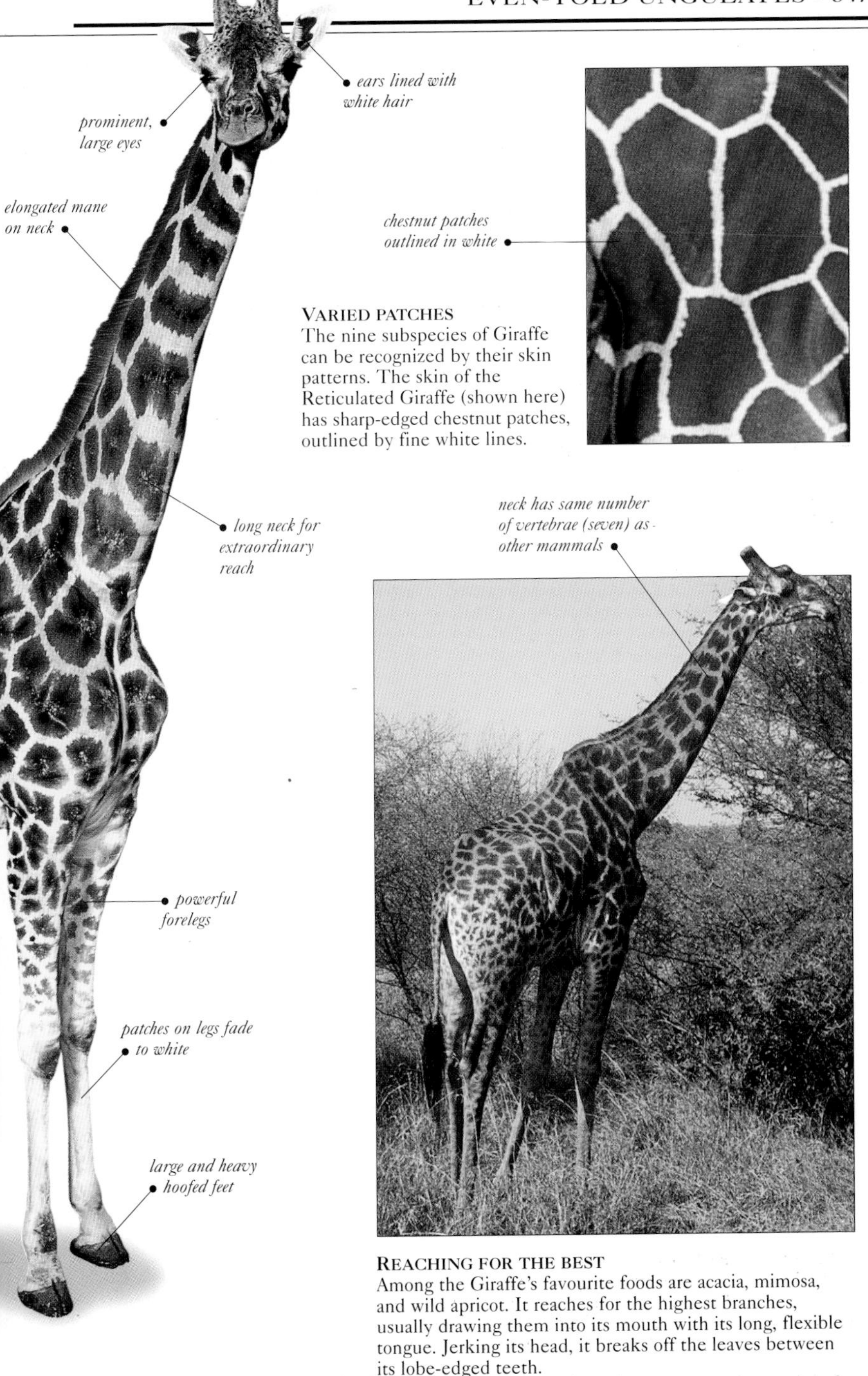

VARIED PATCHES
The nine subspecies of Giraffe can be recognized by their skin patterns. The skin of the Reticulated Giraffe (shown here) has sharp-edged chestnut patches, outlined by fine white lines.

REACHING FOR THE BEST
Among the Giraffe's favourite foods are acacia, mimosa, and wild apricot. It reaches for the highest branches, usually drawing them into its mouth with its long, flexible tongue. Jerking its head, it breaks off the leaves between its lobe-edged teeth.

EVEN-TOED UNGULATES

CATTLE, ANTELOPE AND RELATIVES

THE FAMILY Bovidae, with 140 species, is extremely diverse. Members range from the huge Bison to the small Duiker, and include antelopes such as Wildebeest and Impala, as well as cattle, sheep, and goats.

However, bovids share many features. They have two large, hoof-capped toes on each foot, forming the characteristic cloven hoof of artiodactyls (even-toed ungulates); there are also usually two smaller toes, one on each side. In most bovids, both sexes grow bone-cored horns which are unbranched – although they may be long or short, straight or curled, ridged or smooth.

Many species occur in Africa, with others in Eurasia and North America. Several domesticated bovids, including familiar farmyard cows, sheep, and goats, have been introduced worldwide and are of vast economic importance.

Family BOVIDAE	Species *Tragelaphus spekei*	Status Lower risk

SITATUNGA

Amphibious by nature, the Sitatunga has flexible foot joints, and pointed, widely splayed hooves, allowing it to walk in swamps and marshes. Adult males are grey- to chocolate-brown in colour, whereas females are bright chestnut brown. Both sexes have a white wavy pattern between the eyes and white patches on the cheeks and body. However, only the male has ridged, spiralling horns. The Sitatunga browses on vegetation that is in flower. When faced by a predator, it readily takes refuge in water. The female may be found alone or in groups of around three. The adult male is solitary and barks (usually at night) to keep other males away. On meeting, males display aggressive posturing and rub their horns on the ground.

- **SIZE** Length: 1.2–1.7m (4–5 ½ ft). Weight: 50–125 kg (110–280 lb).
- **OCCURRENCE** W. and C. Africa. In rainforest and wetter regions of the African savanna.

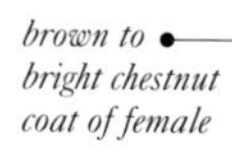

IMMERSED IN WATER
The Sitatunga browses in the morning, evening, and sometimes at night, but rests in swamps by day. It may feed submerged in water, and may rear up on its hindfeet to reach tall sedges and weeds.

Social unit Variable	Gestation 247 days	Young 1	Diet

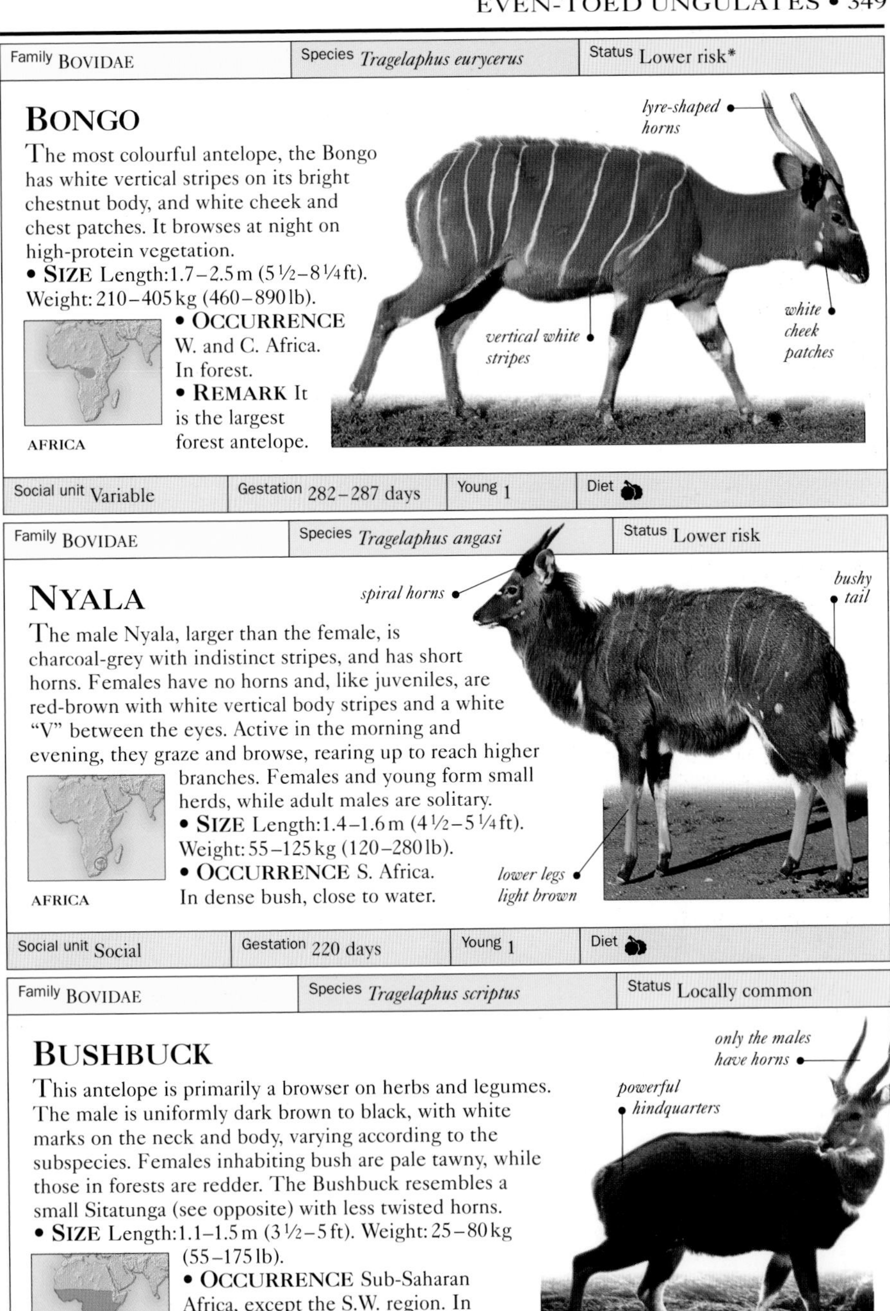

Family BOVIDAE	Species *Tragelaphus eurycerus*	Status Lower risk*

BONGO

The most colourful antelope, the Bongo has white vertical stripes on its bright chestnut body, and white cheek and chest patches. It browses at night on high-protein vegetation.

- **SIZE** Length: 1.7–2.5 m (5½–8¼ ft). Weight: 210–405 kg (460–890 lb).
- **OCCURRENCE** W. and C. Africa. In forest.
- **REMARK** It is the largest forest antelope.

Social unit Variable	Gestation 282–287 days	Young 1	Diet

Family BOVIDAE	Species *Tragelaphus angasi*	Status Lower risk

NYALA

The male Nyala, larger than the female, is charcoal-grey with indistinct stripes, and has short horns. Females have no horns and, like juveniles, are red-brown with white vertical body stripes and a white "V" between the eyes. Active in the morning and evening, they graze and browse, rearing up to reach higher branches. Females and young form small herds, while adult males are solitary.

- **SIZE** Length: 1.4–1.6 m (4½–5¼ ft). Weight: 55–125 kg (120–280 lb).
- **OCCURRENCE** S. Africa. In dense bush, close to water.

Social unit Social	Gestation 220 days	Young 1	Diet

Family BOVIDAE	Species *Tragelaphus scriptus*	Status Locally common

BUSHBUCK

This antelope is primarily a browser on herbs and legumes. The male is uniformly dark brown to black, with white marks on the neck and body, varying according to the subspecies. Females inhabiting bush are pale tawny, while those in forests are redder. The Bushbuck resembles a small Sitatunga (see opposite) with less twisted horns.

- **SIZE** Length: 1.1–1.5 m (3½–5 ft). Weight: 25–80 kg (55–175 lb).
- **OCCURRENCE** Sub-Saharan Africa, except the S.W. region. In forest and bush, near water.
- **REMARK** It is one of the most common antelopes.

Social unit Solitary	Gestation 6 months	Young 1	Diet

Family BOVIDAE	Species *Tragelaphus strepsiceros*	Status Lower risk*

GREATER KUDU

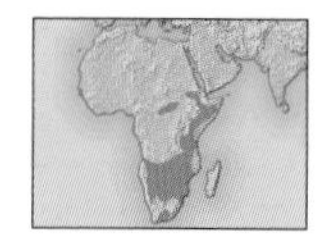

One of the tallest antelopes, the Greater Kudu has a predominantly grey, white-striped coat, with a prominent white chevron (wavy pattern) on its nose, and small spots on either side of its face. It has funnel-shaped ears and a long throat fringe. Only the male has horns – at 1.7 m (5½ ft) they are the longest of any species of antelope. Active by day and by night, this antelope browses on leaves, herbs, tubers, flowers, and fruits. Females form non-hierarchical groups of five or six; males associate in bachelor groups of 2–10. During the rutting season, rival males link horns, shoving and twisting in an effort to overthrow each other. This antelope's acute sense of hearing enables it to locate predators. At times of danger, it stands completely still, or moves away quietly, easily jumping over obstacles of 2 m (6½ ft) or more.

- **SIZE** Length: 2–2.5 m (6½–8¼ ft). Weight: 120–315 kg (260–690 lb).
- **OCCURRENCE** E. to S. Africa. In light forest or fairly thick bush, often in rocky or mountainous country.
- **REMARK** Threatened by habitat loss, the Greater Kudu is also hunted as a sporting trophy and for its meat.

WATER-DEPENDENT
Usually found in forest or hilly bush country, the Greater Kudu prefers to remain close to water, resting during the afternoon to avoid the heat.

Social unit Social	Gestation 9 months	Young 1	Diet

Family BOVIDAE	Species *Taurotragus oryx*	Status Lower risk

COMMON ELAND

Tight spiralling horns are a distinctive feature of both sexes of the Common Eland – the largest antelope. It has 2–15 cream vertical stripes on its upper body and the male has a brownish black "topknot" of matted hair on its head. Females band together to protect their young in the manner of cattle. Like the camel, this antelope allows its body temperature to rise by up to 7°C (13°F) during drought to avoid losing water as sweat.

- **SIZE** Length: 2.1–3.5 m (7–11 ft). Weight: 300–1,000 kg (660–2,210 lb).
- **OCCURRENCE** C., E., and S. Africa. In open plains, dry savanna, montane grassland, heath, and highland forest.
- **REMARK** These docile antelopes are ranched for meat, milk, and hides in Africa. They have been exported to Asia.

Social unit Variable	Gestation 254–277 days	Young 1	Diet

Family BOVIDAE	Species *Boselaphus tragocamelus*	Status Lower risk

NILGAI

Neither cow nor antelope, the Nilgai along with the Chousingha (see p. 352) is a member of the subfamily Boselaphinae. Although the Nilgai is called the Bluebuck or Blue Bull, only the male is grey or bluish grey, while the female is tawny.

- **SIZE** Length: 1.8–2.1 m (6–7 ft). Weight: Up to 300 kg (660 lb).
- **OCCURRENCE** S. Asia. In thinly wooded country, low jungle, and open plains.

Social unit Social	Gestation 243–247 days	Young 1–2	Diet

Family BOVIDAE	Species *Tetracerus quadricornus*	Status Vulnerable

CHOUSINGHA

ASIA

Also called the Four-horned Antelope, this species is unique among bovids in that the male has two pairs of horns. It has a black muzzle and outer ears, and its brownish coat has a dark stripe on the front of each leg. This small, reclusive, and fast-moving antelope usually grazes near water; it uses a low whistle for identification, barking for an alarm call.

- **SIZE** Length: 80–100 cm (32–39 in). Weight: 17–21 kg (37–46 lb).
- **OCCURRENCE** India and Nepal. In wooded hillsides, near water.

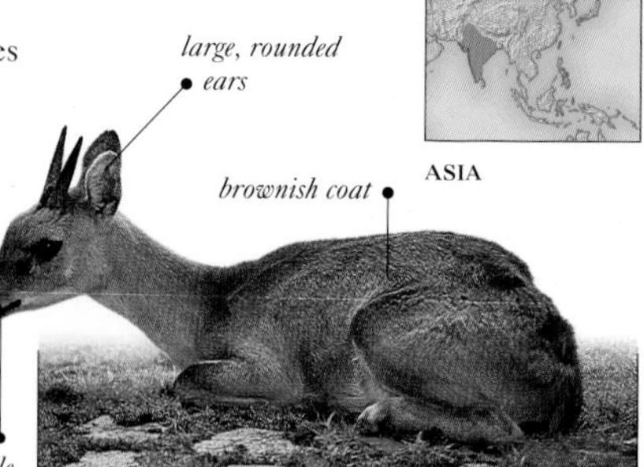

Social unit Solitary	Gestation 7–8 months	Young 1–2	Diet

Family BOVIDAE	Species *Bubalus depressicornis*	Status Endangered

LOWLAND ANOA

This small bovid is dark brown to black, with a pale throat "bib", and facial and leg patches. Its short horns sweep diagonally backwards, enabling it to push through dense, swampy forest. It feeds in the morning on fruits, leaves, ferns, and twigs.

- **SIZE** Length: 1.6–1.7 m (5 1/4–5 1/2 ft). Weight: 150–300 kg (330–660 lb).

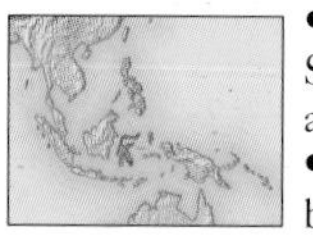

ASIA

- **OCCURRENCE** Sulawesi. In lowland and swampy forest.
- **REMARK** This bovid is one of the smallest wild cattle.

plump, stocky body
flattened horns
short legs

Social unit Solitary	Gestation 9–10 months	Young 1	Diet

Family BOVIDAE	Species *Bubalus arnee*	Status Endangered

ASIAN WATER BUFFALO

Domesticated for centuries and spread around the world, the Asian Water Buffalo is found wild only in small, scattered populations. Also known as the Arni, this massive animal lives in stable clans of females, ruled by a matriarch. Dominant males move into the clan area in the wet season to mate. This buffalo feeds on lush vegetation, and wallows during midday.

ASIA

- **SIZE** Length: 2.4–3 m (7 3/4–9 3/4 ft). Weight: Up to 1.2 tonnes (1 ton).
- **OCCURRENCE** India, Nepal, and possibly Thailand. In wetland.
- **REMARK** It has the largest horns of any bovid: over 2 m (6 1/2 ft) long.

Social unit Social	Gestation 300–340 days	Young 1	Diet 

Family BOVIDAE	Species *Syncerus caffer*	Status Locally common*

AFRICAN BUFFALO

Africa's only cattle-like animal, this buffalo is found in a variety of habitats, but never further than 15 km (9 miles) from water. The male is much larger than the female and has heavier horns that meet at a boss or protuberance on the forehead. It also has a thicker neck, a hump, and small fringe on its dewlap. Extremely gregarious, the nocturnal African Buffalo gathers in herds of 2,000 when food is plentiful. In the dry season herds split into smaller groups of females and young, bachelor bands of males, and solitary older males.

• **SIZE** Length: 2.1–3.4 m (7–11 ft). Weight: Up to 685 kg (1,510 lb).

• **OCCURRENCE** W., C., E., and S. Africa. In primary and secondary forest, savanna, swamps, and grassy plains to mountains.

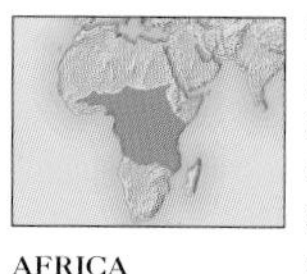
AFRICA

Social unit Social	Gestation 340 days	Young 1	Diet

Family BOVIDAE	Species *Bos javanicus*	Status Endangered

BANTENG

This bovid resembles the domestic cow in appearance. The male is blackish brown to dark chestnut, whereas the female and juvenile are reddish brown; all have white undersides, legs, and rump patches. The male's horns are angled out and then inwards, whereas the female has smaller, crescent-shaped horns. Feeding on grass in the dry season, the Banteng moves to the hills in the monsoon, to eat bamboo and herbs. Females and young live in herds of 2–40 with a single male, while bachelors form their own herds.

• **SIZE** Length: 1.8–2.3 m (6–7½ ft). Weight: 400–900 kg (880–1,990 lb).

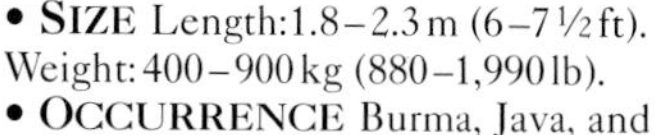

• **OCCURRENCE** Burma, Java, and Borneo. In forest and thicket with open glades.

• **REMARK** The wild population of Banteng is scarce and its habitat is fast diminishing.

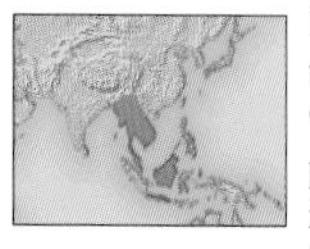
ASIA

Social unit Social	Gestation 9½ months	Young 1–2	Diet

Family BOVIDAE	Species *Bos grunniens*	Status Vulnerable

YAK

ASIA

Brownish black with long, shaggy hair on either side that almost reaches the ground, the Yak has long been domesticated by people of Asia for wool, meat, milk, leather, and as a means of transport. The wild Yak, larger and extremely rare, is confined to desolate, bitterly cold steppe. The dense, soft undercoat of closely matted hair protects it from the extreme cold. Grazing on grasses, herbs, and even lichen, the Yak crunches up frozen ice or snow as a source of water. Females and young gather in large herds, whereas adult males are usually solitary or rove in bachelor herds. During the mating season, which begins in September and lasts several weeks, males battle each other for access to females. A single calf is born every other year and reared by the mother, becoming independent only after it is about a year old. Hunted by Tibetan wolves, Yaks flee speedily when threatened, galloping across large distances with their tails held erect.

- **SIZE** Length: Up to 3.3 m (11 ft). Weight: Up to 525 kg (1,160 lb).
- **OCCURRENCE** Kashmir (India) east to Tibet and Qinghai (China). In desolate steppe at elevations up to 6,000 m (19,800 ft).

DIFFERENT HUES
Domestic Yak can be mottled, black, brown, or red. In spring, Yak moult into a shorter summer coat that is more variable in coloration.

Social unit Variable	Gestation 258 days	Young 1	Diet

Family BOVIDAE	Species *Bos gaurus*	Status Vulnerable*

GAUR

The largest of the wild cattle, the Gaur, also called the Seladang or Indian Bison, has a massive, humped body in shades of red, brown, and black, and white "stockings" on its lower limbs. Both sexes have upward-curving horns up to 1.1 m (3½ ft) long. Males in rut "sing" with a series of low bellows that carry over long distances.

- **SIZE** Length: 2.5–3.3 m (8¼–11 ft). Weight: 650–1,000 kg (1,430–2,210 lb).
- **OCCURRENCE** S. to S.E. Asia. In evergreen and deciduous, forested hills.

Social unit Social	Gestation 270–280 days	Young 1	Diet

Family BOVIDAE	Species *Bison bonasus*	Status Endangered

EUROPEAN BISON

Formerly extinct in the wild, the European Bison or Wisent has been bred in parks and reintroduced in the Bialowieza Forest (on the border of Poland and Belarus). Latest genetic research reveals that it is probably the same species as the American Bison (see pp. 356–357). It has a lighter and shorter coat, but resembles its American cousin in habits and social behaviour. Both sexes have sharp, upturned horns. Mainly a browser, this bison also grazes.

- **SIZE** Length: 2.1–3.4 m (7–11 ft). Weight: 300–920 kg (660–2,030 lb).
- **OCCURRENCE** E. Europe. In woodland, grassland, and coniferous forest.

EUROPE

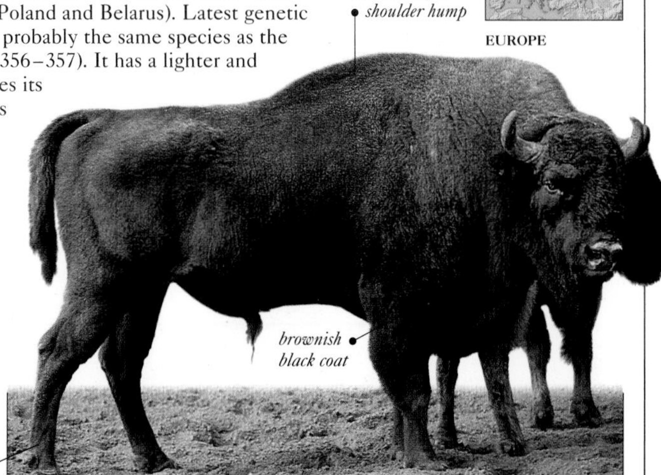

Social unit Social	Gestation 260–270 days	Young 1	Diet

Family BOVIDAE	Species *Bison bison*	Status Lower risk

AMERICAN BISON

Sometimes called the American Buffalo, this massively built bison looks deceptively tall because of its well-developed shoulder hump, which gives it a height of 2 m (6 ½ ft). It has long, shaggy, brownish black hair on its head, neck, shoulders, and forelegs, but the rest of its body is covered in shorter, paler hair. The large, heavy head has a broad forehead with short, upturned horns and a straggly beard. The male is significantly larger than the female. The American Bison is a swift runner despite its huge bulk, and can reach a speed of up to 60 kph (37 mph). Its sense of smell and hearing are excellent and essential for detecting danger. Found in loose herds, females and their young form hierarchical groups, whereas males form bachelor herds and join the females only during the mating season. Males fight fiercely for a mate, usually by head-to-head ramming and threat displays. In the past, when large numbers of bison roamed all over the USA, Canada, and Alaska, herds would make annual migrations of hundreds of kilometres moving along traditional routes.

- **SIZE** Length: 2.1–3.5 m (7–11 ft). Weight: 350–1,000 kg (770–2,200 lb).
- **OCCURRENCE** Yellowstone National Park (USA) and Wood Buffalo National Park (Canada). In grassland, mountains, and open forest.
- **REMARK** Recent genetic evidence indicates that the European Bison (see p. 355), is more similar to its American counterpart than was previously believed.

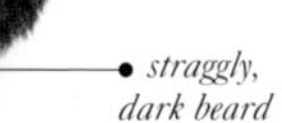

CONSERVATION-DEPENDENT
Once numbering around 50 million, the American Bison was widely hunted and is now virtually extinct in the wild. Although conservation efforts have led to an increase in numbers, most bison are either captive or come from captive stock.

Social unit Social	Gestation 285 days	Young 1	Diet

WINTER MIGRATION
The American Bison migrates south in winter, in search of food. It is, however, not territorial by nature.

N. AMERICA

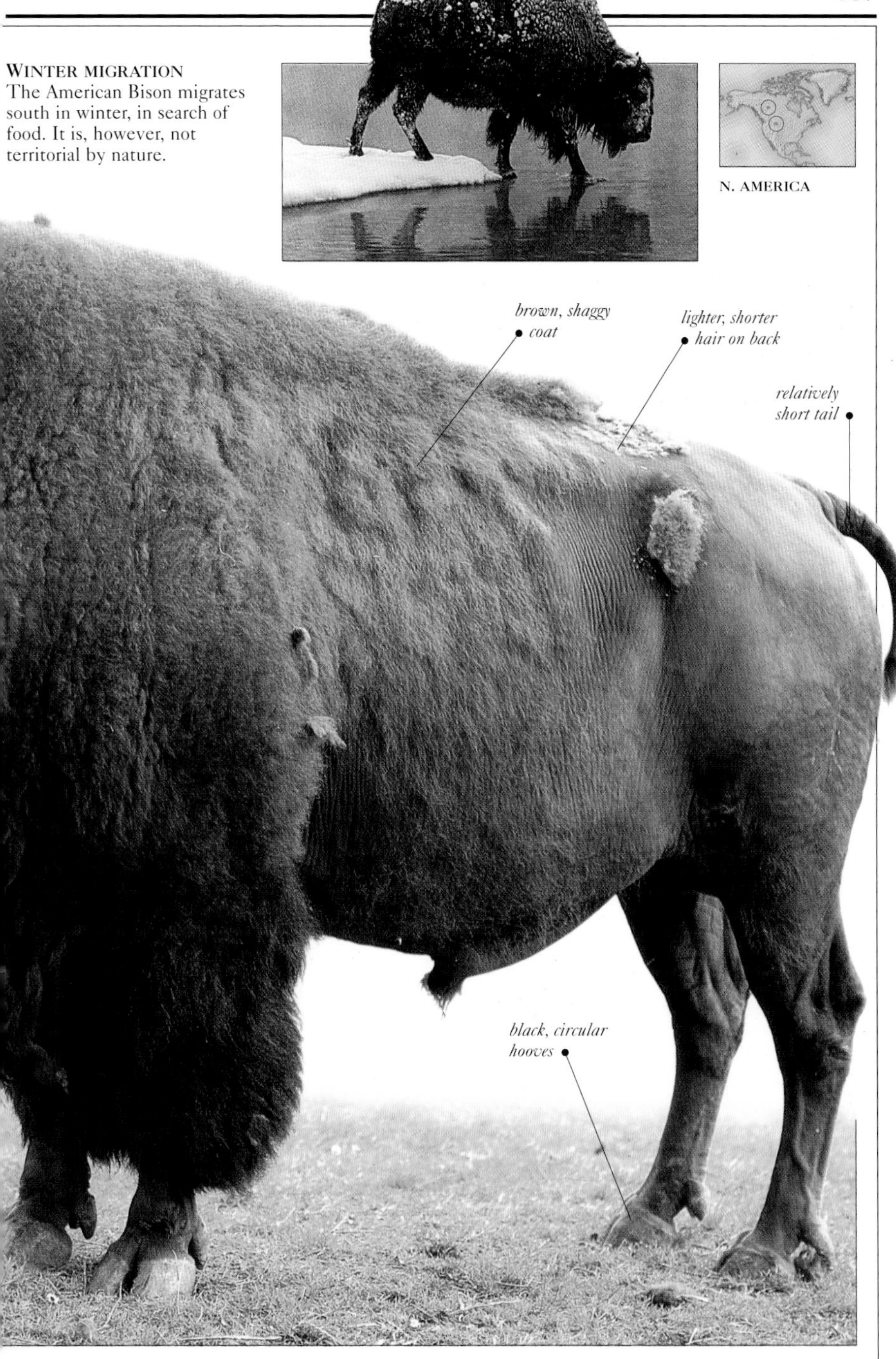

Family BOVIDAE	Species *Cephalophus natalensis*	Status Lower risk

RED FOREST DUIKER

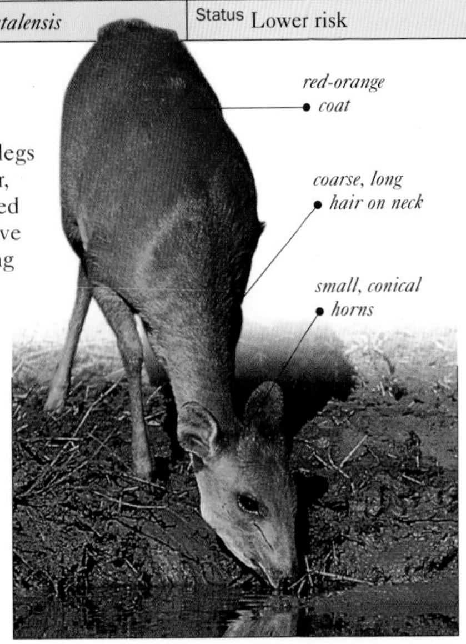

One of 18 or so duiker species, the Red Duiker is a small, arch-backed antelope with longer hindlegs than forelegs. Red-orange to dark brown in colour, it has a tail that is reddish at the base, with a tufted tip that is black-and-white. Both sexes usually have short, backward-pointing horns with a crest of long hair in between, and large scent glands beneath each eye. Duikers freeze mid-stride or sink into the undergrowth when they spot a predator, and dart away at great speed if discovered. They stamp their hindfeet as an alarm call, instead of the forefeet as in other savanna antelope.
• **SIZE** Length: 70–100 cm (28–39 in). Weight: 13 kg (29 lb).

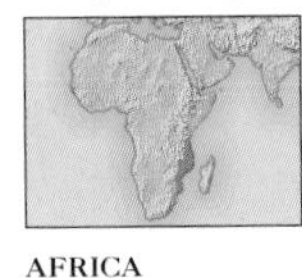
AFRICA

• **OCCURRENCE** Zaire and S. Tanzania to South Africa. In dense bush and forest.
• **REMARK** Duikers are unusual in eating small birds, insects, and carrion.

Social unit Pair	Gestation 120 days	Young 1	Diet

Family BOVIDAE	Species *Sylvicapra grimmia*	Status Locally common

COMMON DUIKER

A tufted forehead, dark nose stripe, and large, pointed ears characterize the Common Duiker. Grey to red-yellow above, it has whitish underparts, and the male has sharp horns, around 11 cm (4 1/4 in) long. Female duikers are usually bigger and heavier than males. A nocturnal browser, the Common Duiker eats leaves, fruits, flowers, tubers, insects, frogs, birds, small mammals, and even carrion. It can go without water for long periods and does not drink in the rainy season, obtaining fluids from fruits, which it often scavenges from beneath trees where monkeys are feeding. Sometimes it reaches for food standing on its hindlegs. The success of this antelope stems from its ability to adapt to a wide spectrum of habitats and food.

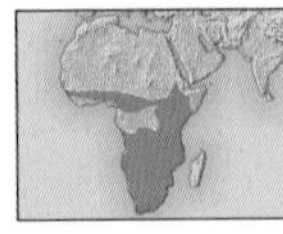
AFRICA

• **SIZE** Length: 0.7–1.2 m (2 1/4–4 ft). Weight: 12–25 kg (26–55 lb).
• **OCCURRENCE** Senegal to Ethiopia and S. Africa. In savanna and hilly areas.

Social unit Solitary/Pair	Gestation 191 days	Young 1	Diet

Family BOVIDAE	Species *Kobus ellipsiprymnus*	Status Lower risk

WATERBUCK

One of the heaviest antelopes, the Waterbuck has a long neck and body, and short legs. Ranging from grey to red-brown, its coat darkens with age. White patches mark its eyebrows, throat, muzzle, underparts, rump, and top of the hooves. Special skin glands secrete an oil that emits a musky odour and waterproofs its fur. Only the males have horns which are ringed and grow up to 1 m (3 ½ ft) long. When threatened, the Waterbuck either fights with its horns and hooves, or takes refuge underwater, exposing only its nostrils.

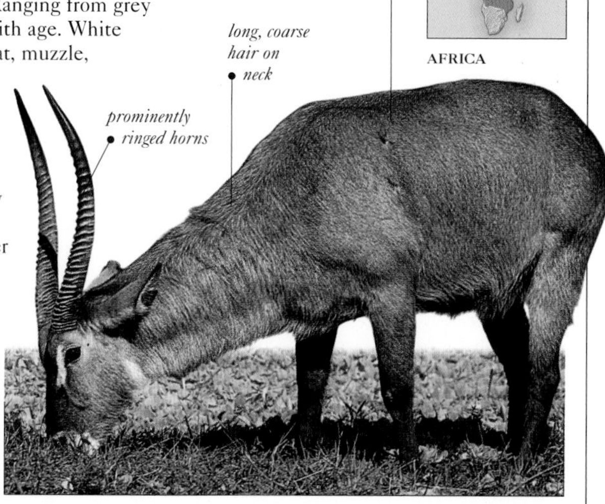

- **SIZE** Length: 1.3–2.4 m (4 ¼–7 ¾ ft). Weight: 50–300 kg (110–660 lb).
- **OCCURRENCE** W., C., and E. Africa. In savanna with patches of woodland and standing water.

Social unit Variable	Gestation 9 months	Young 1	Diet

Family BOVIDAE	Species *Kobus leche*	Status Lower risk

LECHWE

Also known as the Marsh Antelope, the Lechwe lives on floodplains and in swamps, feeding on grass and water plants that are exposed with the seasonal change in water levels. Adapted to wading and swimming, with strong hindquarters and elongated hooves, it moves over shallow water in a series of leaps, and only rests and calves on dry land. The chestnut to black coat contrasts with the white underparts and black leg stripes. Only males have thin, lyre-shaped horns. The Lechwe forms large, loose herds, and where numbers are high follows the "lek" breeding system, like the Kob (see p. 360).

- **SIZE** Length: 1.3–2.4 m (4 ½–7 ¾ ft). Weight: 79–103 kg (175–230 lb).

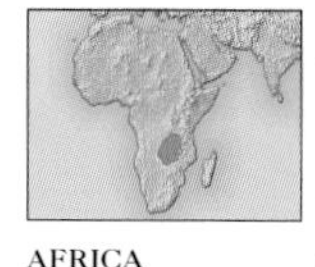

- **OCCURRENCE** C. to S. Africa. In wetland.
- **REMARK** The Lechwe is known to be the fastest moving antelope in water.

Social unit Social	Gestation 7–8 months	Young 1	Diet

Family BOVIDAE	Species *Kobus kob*	Status Lower risk

KOB

Strong, but gracefully built, this antelope is pale cinnamon to brown-black, with white patches on its face and throat, and black leg stripes and feet. The male has distinctive ringed, lyre-shaped horns. Where kobs are found in very dense populations, males gather on an elevated patch of ground (known as a lek) which may be only 15 m (50 ft) across. Usually the heaviest male on the lek gains mating rights over several females through a ritualized display that rarely involves fighting. However, where the density of Kobs is low, males maintain a standard territory and stop females from leaving it by herding them. Active during the morning and evening, the Kob needs to drink water daily. It makes a high bleating sound in times of danger and leaps towards the nearest reedbed to escape.

- **SIZE** Length: 1.3–2.4 m (4 1/4–7 3/4 ft). Weight: 50–300 kg (110–660 lb).
- **OCCURRENCE** W. to E. Africa. In moist savanna, floodplains, and margins of woodland with permanent water.
- **REMARK** The Kob makes up the largest antelope population in Africa after the Wildebeeste.

AFRICA

lyre-shaped, ringed horns in male
white around eyes
short, glossy coat
black markings on front of legs

Social unit Social	Gestation 261–271 days	Young 1	Diet

Family BOVIDAE	Species *Kobus vardonii*	Status Lower risk

PUKU

A uniform golden yellow in colour, the Puku has a whitish area around its eyes, and on its muzzle and throat, and its legs are reddish brown. The male is usually larger than the female, with curved horns about 50 cm (20 in) long. Mostly found in small herds of 3–15, the Puku uses body postures to communicate. It resembles the Kob (see above) in its breeding system, having leks at high population densities and territories at lower ones. It feeds on grasses, and occasionally acacia, and like all plains antelopes, flees rapidly when in danger.

- **SIZE** Length: 1.3–1.8 m (4 1/4–6 ft). Weight: 66–77 kg (145–170 lb).
- **OCCURRENCE** W. to E. Africa. In savanna, near marshes or riverbanks.
- **REMARK** The Puku could be regarded as a southern savanna version of the Kob. However, there is no interbreeding between the two species of antelope.

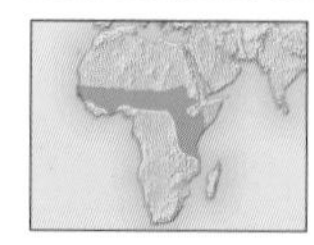
AFRICA

short horns
whitish hue on muzzle

uniformly reddish legs

Social unit Variable	Gestation 8 months	Young 1	Diet

Family BOVIDAE	Species *Redunca redunca*	Status Lower risk

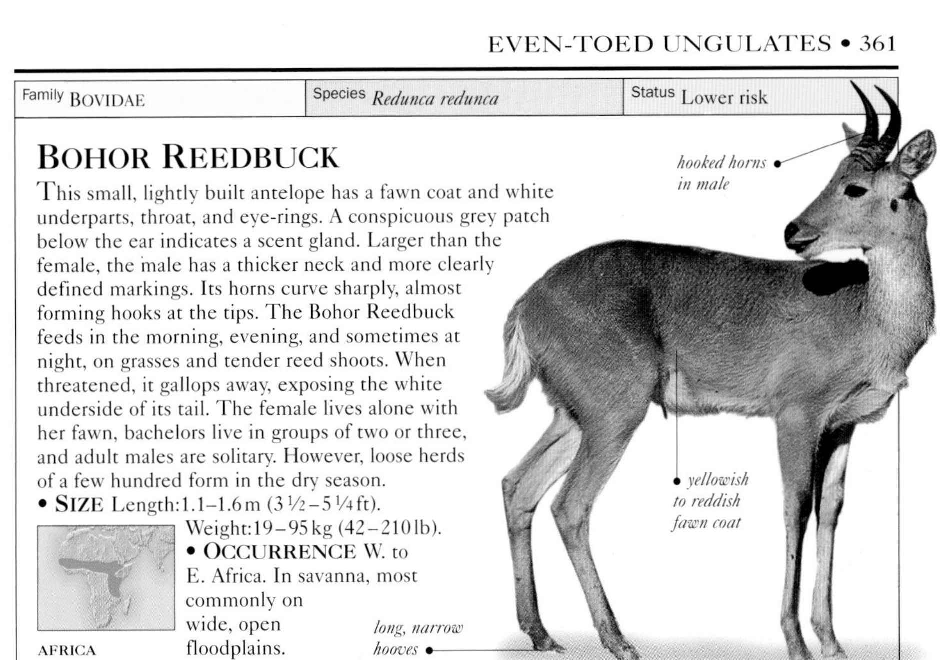

BOHOR REEDBUCK

This small, lightly built antelope has a fawn coat and white underparts, throat, and eye-rings. A conspicuous grey patch below the ear indicates a scent gland. Larger than the female, the male has a thicker neck and more clearly defined markings. Its horns curve sharply, almost forming hooks at the tips. The Bohor Reedbuck feeds in the morning, evening, and sometimes at night, on grasses and tender reed shoots. When threatened, it gallops away, exposing the white underside of its tail. The female lives alone with her fawn, bachelors live in groups of two or three, and adult males are solitary. However, loose herds of a few hundred form in the dry season.

- **SIZE** Length:1.1–1.6 m (3 1/2–5 1/4 ft). Weight:19–95 kg (42–210 lb).
- **OCCURRENCE** W. to E. Africa. In savanna, most commonly on wide, open floodplains.

Social unit Variable	Gestation 7–7 1/2 months	Young 1	Diet

Family BOVIDAE	Species *Hippotragus equinus*	Status Lower risk

ROAN ANTELOPE

Red to brown above with white underparts, the Roan Antelope has black-and-white markings on its face. It is the fourth largest antelope and both sexes have stout, heavily ringed horns, and an upright, black-edged mane. It can survive on very little grass, drinking water two or three times daily. Herds consist of 12–15 females and young with one dominant male, or younger bachelor males. Rival males fight on their knees with violent backward sweeps of their curved horns.

- **SIZE** Length:1.9–2.7 m (6 1/4–8 3/4 ft). Weight:150–300 kg (330–660 lb).
- **OCCURRENCE** W., C., and E. Africa. In savanna.

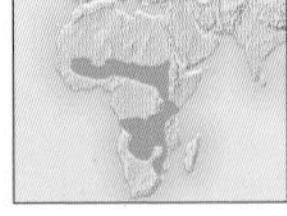

erect mane

pale reddish brown body

black blaze on face

black-and-white markings on face

Social unit Variable	Gestation 268–280 days	Young 1	Diet

Family BOVIDAE	Species *Hippotragus niger*	Status Lower risk

SABLE ANTELOPE

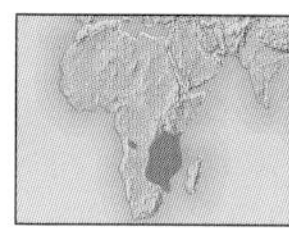

Distinctive facial markings, white with a central black blaze and cheek stripes, characterize this large antelope, which is similar to the Roan Antelope in many respects (see p. 361). Mature males are black, whereas females and juveniles are yellowish brown to rich chestnut, with the same facial pattern as males. Both sexes have stout and heavily ringed horns and a well-developed, erect mane from the top of the neck to the withers, with shorter hair on the throat. This antelope gathers in herds of around 100 during the dry season, splitting into smaller groups in the rainy season, both sexes following a hierarchy. Young males form bachelor groups of 2–12; dominant males vigorously defend their own territories and mate with the females residing there. This antelope is usually active in the early mornings and afternoons, although some herds may be nocturnal.

- **SIZE** Length:1.9–2.7 m (6¼–8¾ ft). Weight:150–300 kg (330–660 lb).
- **OCCURRENCE** E. to S.E. Africa. In lightly wooded grassland.
- **REMARK** Once hunted for trophy heads, political unrest in Africa has also affected this animal and a subspecies, the Giant Sable of Angola, is critically endangered.

FINDING FOOD
Primarily a grazer, the Sable Antelope also browses extensively during the dry season. It always lives within 2–4 km (1¼–2½ miles) of water.

Social unit Social	Gestation 261–281 days	Young 1	Diet

Family BOVIDAE	Species *Oryx gazella*	Status Lower risk

GEMSBOK

With a body in contrasting shades of grey, black, and white, the Gemsbok, or Southern Oryx as it is also known, has a fawn-tinged, grey body with a black longitudinal stripe along its belly, black upper legs and tail, and a distinctive black-and-white facial pattern. The Gemsbok is well-adapted to life in arid regions: it does not pant or sweat until its body temperature reaches 45° C (113° F), and produces concentrated urine and dry faeces. Feeding at night, in the early morning, as well as late afternoon, this antelope supplements its main diet of grass and shrubs with melons and cucumbers for water.

• **SIZE** Length: 1.6–2.4 m (5 ¼–7 ¾ ft). Weight: 100–210 kg (220–460 lb).

• **OCCURRENCE** Namibia and W. South Africa. In arid scrub and desert.

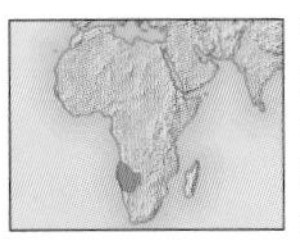

• **REMARK** Almost perfectly adapted to desert life, it lives in dry conditions where few other ungulates would survive.

Social unit Social	Gestation 260–300 days	Young 1	Diet

Family BOVIDAE	Species *Oryx dammah*	Status Critically endangered

SCIMITAR-HORNED ORYX

Pale tan in colour, the Scimitar-horned Oryx has a brownish patch on its forehead, face, and across its eyes, and a russet neck and chest; both sexes have long horns that curve back in a wide arc. The Oryx shares many features with the Gemsbok (see above) to survive in arid regions; it also has enlarged hooves to bear the weight of its stocky body on the sand. Feeding on grasses, legumes, acacia pods, succulents, and fruits, in the early morning, evening, and on moonlit nights, this oryx rests in the shade by day. Mixed herds wander over great distances in search of grazing. Males fight fiercely over a mate, displaying and stabbing with horns.

horns longer and more slender in female

ruddy chest

tufted brown tail

body washed with russet

large, broad hooves

• **SIZE** Length: 1.5–2.4 m (5 ¼–7 ¾ ft). Weight: 100–210 kg (220–460 lb).

• **OCCURRENCE** Chad. On arid, rocky plains.

• **REMARK** Hunted almost to extinction, this oryx is found in a reserve in N.C. Chad. It was reintroduced in Tunisia in 1991, and further releases have been planned.

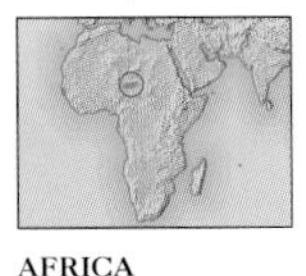

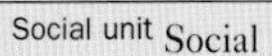

Social unit Social	Gestation 222–253 days	Young 1	Diet

Family BOVIDAE	Species *Addax nasomaculatus*	Status Critically endangered

ADDAX

Found in remote desert areas, this rare, little-known antelope is adapted to its arid habitat in various ways: its short legs and widely splayed hooves enable it to move easily across sand, and it rarely drinks water, getting it instead from the succulent plants it eats. Active early in the morning, evening, and first half of the night, this nomadic animal travels great distances in search of almost any desert vegetation, following rainfall. Formerly living in herds of 5–20, led by an older male, these antelopes are today found alone or in isolated groups of 2–4. The Addax has a white face patch and a chestnut forehead tuft; its sandy to white summer coat turns greyish brown in winter. Its spiral horns have up to three turns.

• **SIZE** Length: 1.5–1.7 m (5–5½ ft).
Weight: 60–125 kg (130–280 lb).

• **OCCURRENCE** N.W. Africa. In desert as well as in semi-desert.

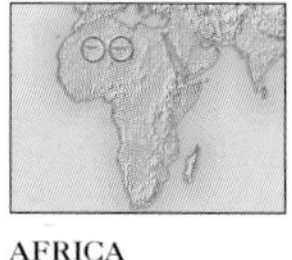
AFRICA

• **REMARK** Not capable of great speed, the Addax is an easy prey for humans. In recent years, it has faced the additional threat of severe drought.

spiralled horns
chestnut forehead tuft
sandy summer coat
short legs
white from hip to hoof

Social unit Variable	Gestation 257–264 days	Young 1	Diet

Family BOVIDAE	Species *Damaliscus dorcas*	Status Vulnerable

BONTEBOK

Rich purplish brown with a broad white blaze on its long muzzle, the Bontebok, or Blesbok as it is also called, has well-developed, lyre-shaped, ringed horns. Adult males use their horns to posture and spar – rarely engaging in actual physical fighting – to gain a territory. The herd of females and young is led by a dominant male, who keeps them together and initiates their travels. The female gives birth to its single offspring at traditional calving grounds, without isolating or concealing its young, unlike other similar antelopes. The newborn can walk within about 5 minutes, and soon follows its mother; it is weaned at about 6 months.

• **SIZE** Length: 1.2–2.1 m (4–7 ft).
Weight: 68–155 kg (150–340 lb).

• **OCCURRENCE** S. Africa. In grassland and sparsely wooded regions; now also in macchia scrub reserves.

AFRICA

• **REMARK** Almost extinct in the wild by the 1830s (only 17 were left), several herds of Bontebok were preserved in parks and reserves. It remains Africa's rarest antelope.

horns have rings for most of length
brown coat with purplish sheen
contrasting white lower legs

Social unit Social	Gestation 8 months	Young 1	Diet

Family BOVIDAE	Species *Damaliscus lunatus*	Status Lower risk

TOPI

Also called the Tsessebe, this antelope has an attractive, glossy reddish brown coat with darker purple patches on the upperparts of its legs, hips, and thighs. It has a long head, a shoulder hump, and a back that slopes downwards. Both sexes have ringed, L-shaped horns which curve inwards at the tips; the female may have a paler coat and smaller horns. Often found in association with Wildebeest, zebras, and Ostriches, the Topi lives in seasonally flooded grassland and uses its long, narrow muzzle and mobile lips to feed on grasses and other vegetation. It has two breeding systems according to habitat and ecology – mature males hold territories and harems among resident populations, but form smaller territories or "leks"among populations that are nomadic. Sentinels standing on termite mounds or higher ground warn others of danger, sending them galloping off with a rocking motion.

• **SIZE** Length:1.2–2.1 m (4–7 ft). Weight: 68–155 kg (150–340 lb).

• **OCCURRENCE** W., E., and S. Africa. In open country and sparsely wooded areas, preferring knee-high grassland.

AFRICA

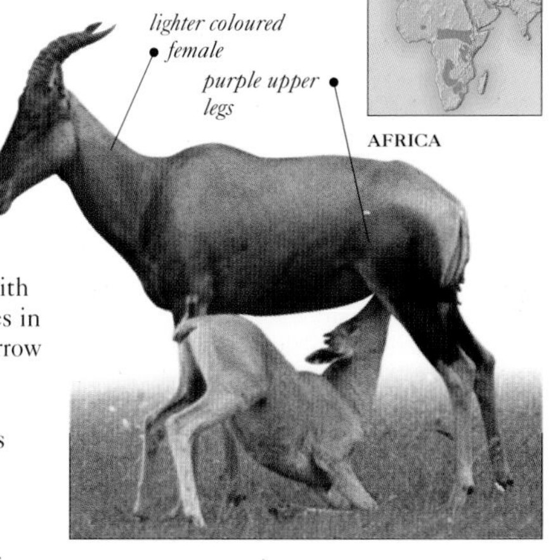

MOTHER AND CALF
The female Topi may mate with one or more males in the breeding season. The single offspring may be kept hidden or follow its mother around.

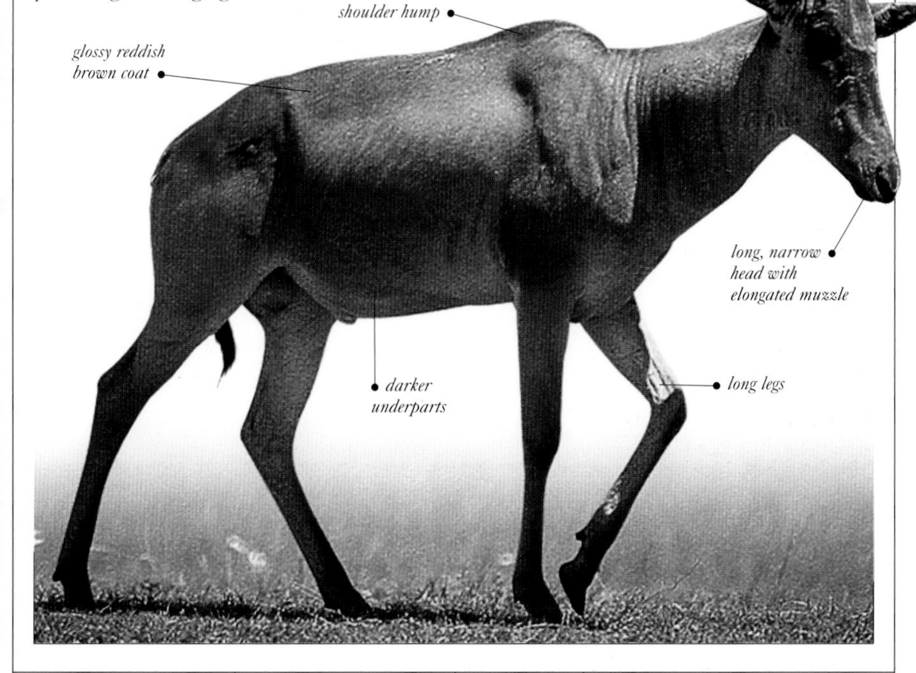

Social unit Social	Gestation 7 1/2–8 months	Young 1	Diet

Family BOVIDAE	Species *Connochaetes taurinus*	Status Lower risk

WILDEBEEST

This large antelope with an unmistakable clumsy appearance, is also known as the Brindled Gnu or Blue Wildebeest. It has high shoulders and a large head and muzzle, cow-like horns, and a copious black mane that spills over its forehead, neck, and shoulders. It is greyish silver with brownish bands on its neck, shoulders, and foreparts, and its beard is either black or white, depending on the subspecies. Both sexes have black, unridged horns that are horizontal at the base, and then curve upwards and inwards. Feeding mainly on grass, as well as succulents, the Wildebeest grazes in the early mornings and late afternoons, and rests during the heat of the day. Females and young form herds of 10–1,000, whereas males form bachelor groups. At the age of three or four years they try to form their individual territory with ritual posturing, pushing, and a typical "ge-nuu" call. Only adult males with territories can mate. Hunted by African lions, Leopards, Cheetahs, hyenas, and hunting dogs, the Wildebeest prances about and paws the ground when startled, galloping off with its head held low and tail waving. If cornered, the Wildebeest is known to fight viciously against its opponents.

• **SIZE** Length: 1.5–2.4 m (5–7¾ ft). Weight: 120–275 kg (260–610 lb).

• **OCCURRENCE** S. Kenya and S. Angola to N. South Africa. In open, grassy plains and acacia savanna, usually near water.

MIGRATING HERDS
Large groups of Wildebeest gather in the dry season, most travelling hundreds of kilometres in search of grazing. When crossing rivers, they are especially vulnerable to crocodile attacks.

Social unit Social	Gestation 8–9 months	Young 1	Diet

FOOD AND DRINK
Wildebeest inhabit grassland close to water. Although most migrate in search of fresh pastures, they become sedentary once they find a plentiful source of food and water.

Family BOVIDAE	Species *Alcelaphus buselaphus*	Status Lower risk

HARTEBEEST

This antelope resembles the Topi (see p. 365) in diet and appearance. Fawn to brownish grey in colour, it has white hip patches and is black on the forehead, muzzle, shoulders, and thighs. The Hartebeest has spiral or lyre-shaped horns, depending on the subspecies (of which there are seven). Active in the early morning and late afternoon, this antelope is often found in association with Ostriches, Wildebeest, or Zebras. It forms large, highly organized herds of up to 300, dominated by territorial males, which may contain smaller herds of non-territorial males or females and young.

- **SIZE** Length: 1.5–2.5 m (5–8 ¼ ft). Weight: 100–225 kg (220–500 lb).
- **OCCURRENCE** W., E., and S. Africa. In dry savanna grassland.

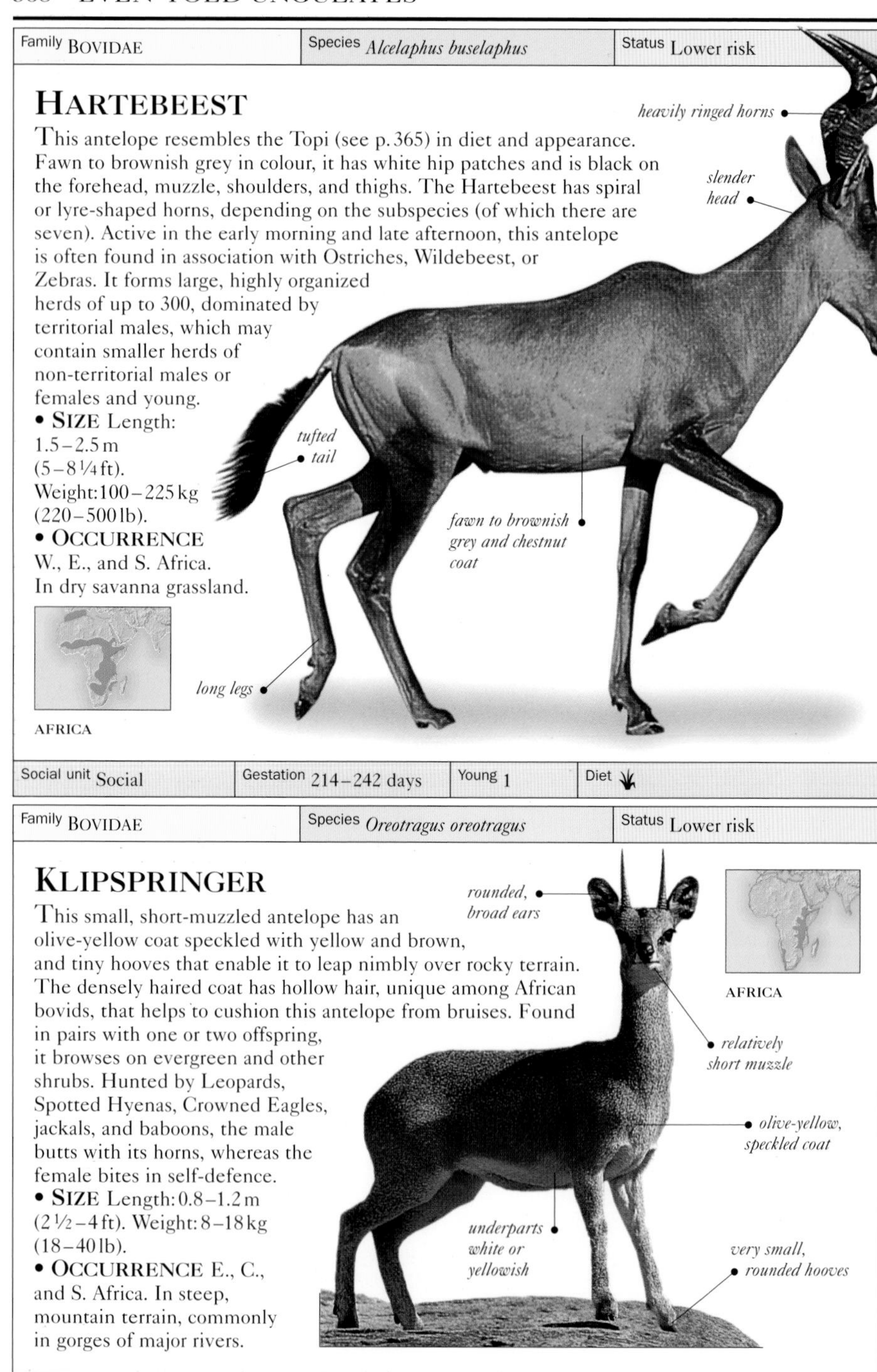

Social unit Social	Gestation 214–242 days	Young 1	Diet

Family BOVIDAE	Species *Oreotragus oreotragus*	Status Lower risk

KLIPSPRINGER

This small, short-muzzled antelope has an olive-yellow coat speckled with yellow and brown, and tiny hooves that enable it to leap nimbly over rocky terrain. The densely haired coat has hollow hair, unique among African bovids, that helps to cushion this antelope from bruises. Found in pairs with one or two offspring, it browses on evergreen and other shrubs. Hunted by Leopards, Spotted Hyenas, Crowned Eagles, jackals, and baboons, the male butts with its horns, whereas the female bites in self-defence.

- **SIZE** Length: 0.8–1.2 m (2 ½–4 ft). Weight: 8–18 kg (18–40 lb).
- **OCCURRENCE** E., C., and S. Africa. In steep, mountain terrain, commonly in gorges of major rivers.

Social unit Pair	Gestation 214–225 days	Young 1–2	Diet

Family BOVIDAE	Species *Ourebia ourebi*	Status Lower risk

ORIBI

A small, slender, long-necked antelope, the Oribi has a silky coat, sandy to rufous above, white below and on the chin and rump, with longer hair on its knees. The male has spiky, ringed horns; the female is usually larger with a dark crown patch. Hunted by Leopards, Caracals, and pythons, this antelope hides in tall grass, stotting or leaping up vertically, if alarmed, and bounding off at high speed. The Oribi is found in male–female pairs, or groups of seven or eight, with two or three adult males, and is active by day as well as by night. The male sometimes helps to groom and guard the offspring.

• **SIZE** Length: 0.9–1.4 m (3–4½ ft). Weight: 14–21 kg (31–46 lb).

• **OCCURRENCE** W. to E. and S. Africa. In savanna grassland as well as wooded areas near water, up to an elevation of 3,000 m (9,900 ft).

Social unit Variable	Gestation 200–210 days	Young 1	Diet

Family BOVIDAE	Species *Raphicerus campestris*	Status Locally common

STEENBOK

Also called the Steinbuck, this antelope is bright rufous-fawn, sometimes tinged silver-grey, and paler below. Found alone or in pairs with largely separate lives, this species marks its territory with scent and dung. It browses and grazes, and digs up roots with its feet. The Steenbok is active by day and night.

• **SIZE** Length: 61–95 cm (24–37 in). Weight: 7–16 kg (15–35 lb).

• **OCCURRENCE** E. and S. Africa. In habitats with dense cover, mainly lightly wooded savanna.

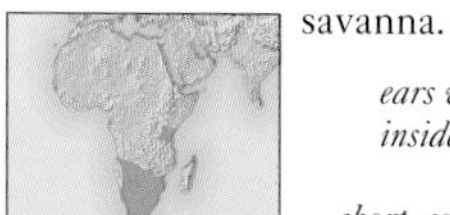

Social unit Solitary/Pair	Gestation 166–177 days	Young 1	Diet

Family BOVIDAE	Species *Madoqua kirkii*	Status Locally common

KIRK'S DIK-DIK

This dwarf antelope has a soft, lank coat, grizzled grey to brown, and more reddish on the head, neck, and shoulders. Its hooves have rubbery pads to grip rocky surfaces, and only the male has antlers. It feeds mainly in the morning and late afternoon, and sometimes at night – its small body and its extremely narrow muzzle enabling it to reach the smallest food items. Along with leaves, buds, flowers, and fruits, salt also forms an important part of its diet. Kirk's Dik-diks live in close male–female pairs with their offspring, within territories.

• **SIZE** Length: 52–72 cm (20 ½–28 in). Weight: 3–7 kg (6 ½–15 lb).

• **OCCURRENCE** E. and S.W. Africa. On arid, stony slopes and in sandy bushland.

• **REMARK** The four species of dik-dik are named after the alarm call they give as they dart away in an erratic zig-zag pattern when faced with danger.

AFRICA

large ears
grizzled grey-brown coat
hooves adapted to walking on rocks

Social unit Pair	Gestation 169–174 days	Young 1	Diet

Family BOVIDAE	Species *Antilope cervicapra*	Status Vulnerable

BLACKBUCK

The male Blackbuck has a rich coffee-brown head, back, sides, and outer legs (these are black in the dominant male of the herd), whereas the female is yellowish fawn on its head and back. Both sexes have pale rings around the eyes, a sheep-like muzzle, white underparts, and short tail. Only the male has horns, which are tightly spiralled, sometimes with up to five turns. A single, dominant male lives with a harem and their offspring, in groups of 5–50. Dominance is established through a visual display with horns and threatening gestures. When alarmed, a single female bounds into the air, followed by others, until the entire herd is in motion. Grazing almost exclusively on grass, Blackbucks also eat cultivated grains.

• **SIZE** Length: 1.2 m (4 ft). Weight: 32–43 kg (71–95 lb).

• **OCCURRENCE** Pakistan, India, and Nepal. In open plains and thorny forest.

• **REMARK** Hunted for its meat and as a sporting trophy, the Blackbuck is also threatened by loss of habitat. However, it receives a certain amount of protection from the local Bishnoi community, which reveres animals.

ASIA

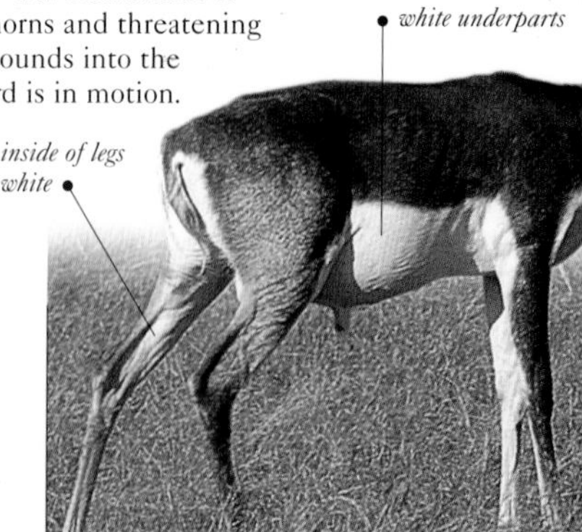

Social unit Social	Gestation 6 months	Young 1	Diet

Family BOVIDAE	Species *Aepyceros melampus*	Status Lower risk

IMPALA

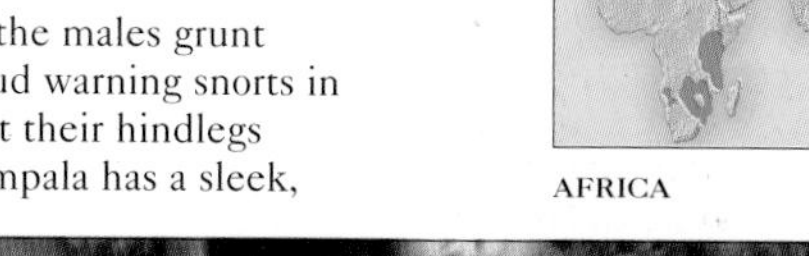

This species is among the noisiest of antelopes: the males grunt loudly while rutting, calves bleat, and all emit loud warning snorts in times of danger, when they leap away kicking out their hindlegs horizontally and landing on their forelegs. The Impala has a sleek, glossy, fawn to reddish coat with lighter thighs and legs. The white ears are tipped black, and upper lip, chin, and underparts are also white; a vertical black streak appears on each side of the hindquarters and on the tail. This adaptable grazer and browser feeds on grass and leaves, and drinks at least twice a day. It forms large, mixed herds in the dry season. However, at other times females and young form herds of 10–100. The dominant male signals its status through scent secretions from glands on its forehead. It also flicks its tongue, a signal at which single females bunch up, and other males either flee or respond to the challenge thrown at them.

• **SIZE** Length: 1.1–1.5 m (3½–5 ft). Weight: 40–65 kg (88–145 lb).

• **OCCURRENCE** Kenya and S. Angola to N. South Africa. In open woodland, and acacia savanna.

FIGHTING FOR POWER
Only male Impalas have horns and they fight to establish dominance. The highest ranking male takes over a territory and mates with the females.

Social unit Social	Gestation 6–7 months	Young 1	Diet

Family BOVIDAE | Species *Litocranius walleri* | Status Lower risk

GERENUK

Also known as the Giraffe Gazelle, the Gerenuk has a slender, elongated neck and long legs. Reddish fawn in colour, it has a broad, dark band along the back and uppersides. Its neck, head, and wedge-shaped muzzle are extremely narrow and allow it to probe into acacia and other thorny vegetation, and it uses its pointed tongue, mobile lips, and sharp incisors to pluck the smallest leaves and shoots. To reach its food it may stand upright on its hindlegs for long periods, curving its spine in a S shape. This allows it to browse higher than most similar-sized herbivores, in open woodland and bush country. Gerenuks are usually found in pairs or small family groups, but young males may form bachelor herds or wander alone, whereas mature males claim their own territories. Chiefly preyed upon by the big cats, the Gerenuk stands motionless to avoid detection.

• **SIZE** Length: 1.4–1.6 m (4½–5¼ ft). Weight: 28–52 kg (62–115 lb).

• **OCCURRENCE** E. Africa. In dry areas with a light covering of bush.

• **REMARK** This poorly known species is hunted by humans for its skin.

AFRICA

Social unit Variable | Gestation 6½–7 months | Young 1 | Diet

Family BOVIDAE | Species *Gazella thomsonii* | Status Lower risk

THOMSON'S GAZELLE

This small, graceful gazelle has a black flank band separating its sandy fawn upperparts from its white underside. Its reddish brown head has a dark blaze, and white eye-rings that extend along the muzzle, above black cheek stripes. The most common gazelle in its region and the staple diet of big cats, it sometimes forms herds with other gazelles. Small herds of females and young join bachelor bands and males, to migrate between grassland (in the rains) and bush (during the dry season).

• **SIZE** Length: 0.9–1.2 m (3–4 ft). Weight: 15–30 kg (33–66 lb).

• **OCCURRENCE** S.E. Sudan, Kenya, and Tanzania. In open country or bushland up to 5,750 m (19,000 ft).

• **REMARK** Thomson's Gazelle is one of the few gazelles that breed twice yearly.

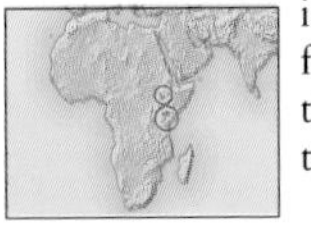

AFRICA

Social unit Social | Gestation 160–180 days | Young 1 | Diet

Family BOVIDAE	Species *Antidorcas marsupialis*	Status Lower risk

SPRINGBOK

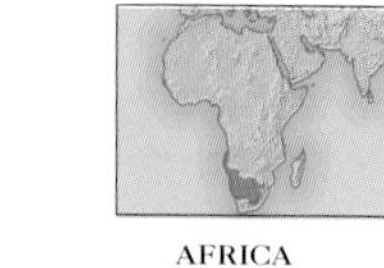

Cinnamon-fawn above with a white underside, the Springbok, or Springbuck, as it is also called, has a dark, reddish brown horizontal band extending from the top of its foreleg to the hip. It has a fold of skin from the middle of its back to the base of its tail, which, when opened, displays a white crest, perhaps helping to confuse predators or serving to warn other members of its herd. Both sexes have short, ringed, black horns. The Springbok is among several bovids that "stott" or "pronk" – leap high and energetically on stiff legs, as if bouncing – behaviour that may deter predators. A highly social herbivore, the Springbok forms large, migratory herds, which split into smaller groups in the dry season. Once millions strong, a "large" migratory herd now consists of only about 1,500 individuals.

• **SIZE** Length: 1.2–1.4 m (4–4½ ft). Weight: 30–48 kg (66–105 lb).

• **OCCURRENCE** S. Africa. In open, dry savanna as well as grassland.

• **REMARK** Previously hunted in large numbers as herds destroyed crops, the Springbok is also killed for its meat.

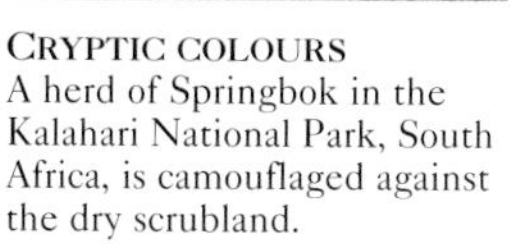

CRYPTIC COLOURS
A herd of Springbok in the Kalahari National Park, South Africa, is camouflaged against the dry scrubland.

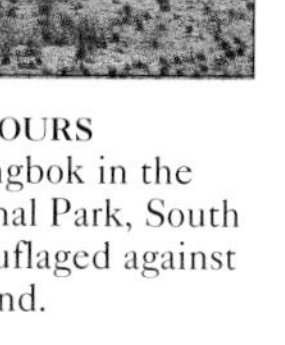

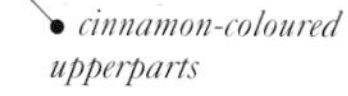

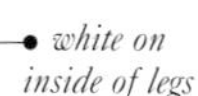

Social unit Social	Gestation 24 weeks	Young 1	Diet 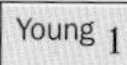 

Family BOVIDAE	Species *Saiga tatarica*	Status Vulnerable

SAIGA

Characterized by its large, "Roman" nose and downward-pointing nostrils, this goat-antelope has a woolly, cinnamon buff coat with a grizzled crown and rump. In winter this becomes whiter and grows thicker by 70 per cent. Small breeding groups congregate in herds of almost 2,000, migrating to feeding grounds in winter. Males form separate herds which travel ahead of females.

- **SIZE** Length: 1–1.4 m (3 ¼–4 ½ ft). Weight: 26–69 kg (57–150 lb).
- **OCCURRENCE** Russia, west of the Caspian Sea, Kazakhstan, and Mongolia. In grassy plains, often including arid areas.

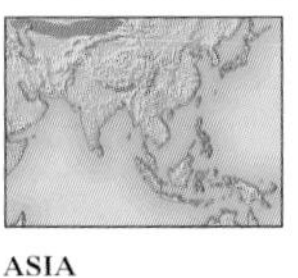

ASIA

Social unit Social	Gestation 139–152 days	Young 1–2	Diet

Family BOVIDAE	Species *Oreamnos americanus*	Status Locally common

MOUNTAIN GOAT

Surviving in extreme cold among ice, snow, and glaciers, the woolly Mountain Goat is yellow-white, with dense underfur to conserve body heat. Its large hooves have hard rims and soft inner pads to grip slippery slopes. Feeding by day and night on grass, moss, lichens, and twigs, it seeks out, and sometimes fights over, salt licks. Groups rarely exceed four in summer, but grow larger in winter; males tend to be solitary.

- **SIZE** Length: 1.2–1.6 m (4–5 ¼ ft). Weight: 46–140 kg (100–310 lb).
- **OCCURRENCE** W. Canada, and N. and W. USA. In alpine tundra and arid slopes.

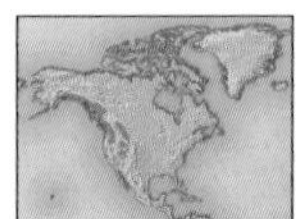

N. AMERICA

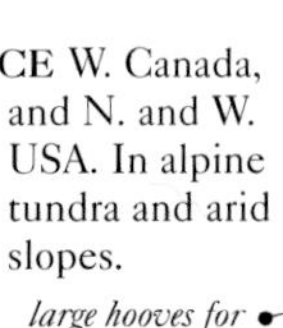

Social unit Variable	Gestation 186 days	Young 1	Diet

Family BOVIDAE	Species *Rupicapra rupicapra*	Status Critically endangered

CHAMOIS

An agile mountain-climber, the Chamois can jump 2 m (6 ½ ft) high, leap 6 m (20 ft) across, and run at 50 kph (31 mph), with its flexible hooves providing a sure grip on slippery terrain. Its stiff and coarse hair is tawny brown in summer, and grows much thicker and blackish brown in winter. A black stripe runs from eye to muzzle and it has white markings on its head and throat. Both sexes have slender, close-set, black horns that grow vertically and bend abruptly at the tips to form hooks. It is active in the early morning and evening, feeding on herbs and flowers in summer, and lichen, moss, and pine shoots in winter. Females and young form groups of 15–30, and "sentinels" warn of danger by foot-stamping and high-pitched calls.

- **SIZE** Length: 0.9–1.3 m (3–4 ¼ ft). Weight: 24–50 kg (53–110 lb).
- **OCCURRENCE** Alps and mountains of S.C. Europe, Balkans, Asia Minor, and Caucasus. In rocky areas and alpine meadows.
- **REMARK** Numbers have been greatly reduced by excessive hunting, habitat loss, and competition with livestock.

EUROPE, ASIA

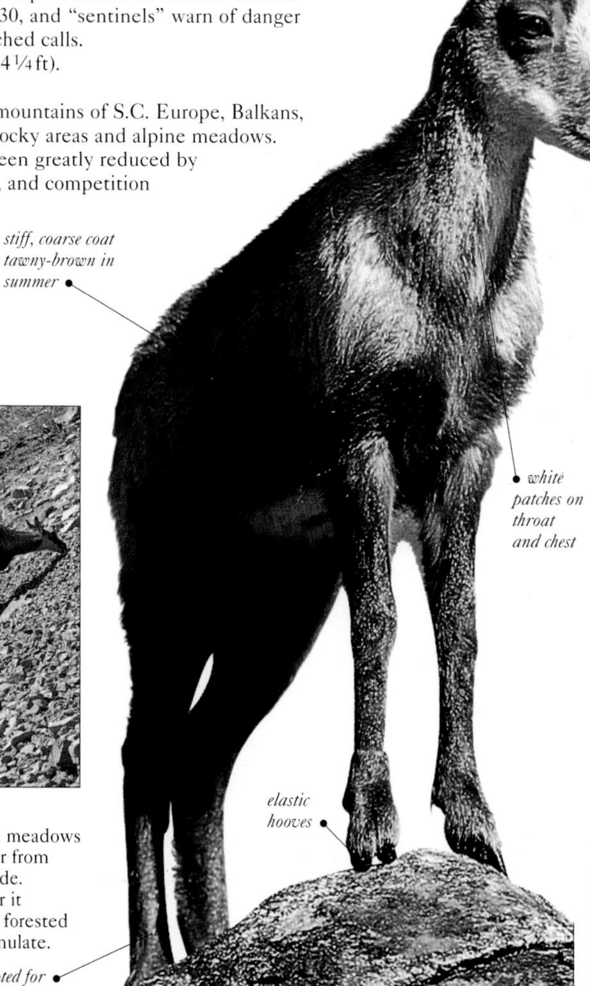

STEEP TERRAIN
In summer, the Chamois stays in meadows above 1,800 m (6,500 ft), never far from rocky outcrops in which it can hide. Although not migratory, in winter it sometimes moves down to steep forested slopes, where snow cannot accumulate.

Social unit Variable	Gestation 170 days	Young 1	Diet

Family BOVIDAE	Species *Ovibos moschatus*	Status Locally common

MUSKOX

N. AMERICA, GREENLAND

The Muskox is so called because of the strong odour emitted by rutting males, who charge at high speed and ram each other with their horns during the mating season. Both sexes have broad horns, which nearly meet at a central protuberance, and curve down and then up at the tips. This massive-bodied bovid has an outer coat of dark brown guard hair that slopes downwards to shed rain and snow. The undercoat of soft, pale brown hair insulates it from the Arctic cold. During summer, the Muskox grazes in river valleys and meadows. In winter, it moves to higher areas, where strong winds keep the ground clear of snow. Adult Muskoxen form a circle around the young to shield them from predators, sometimes charging at the enemy.

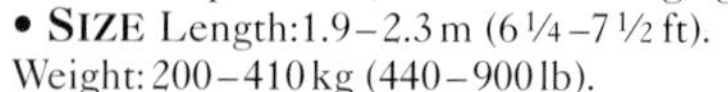

- **SIZE** Length: 1.9–2.3 m (6¼–7½ ft). Weight: 200–410 kg (440–900 lb).
- **OCCURRENCE** Canada and Greenland. In Arctic tundra.
- **REMARK** Effective reintroduction programmes have saved the Muskox from extinction.

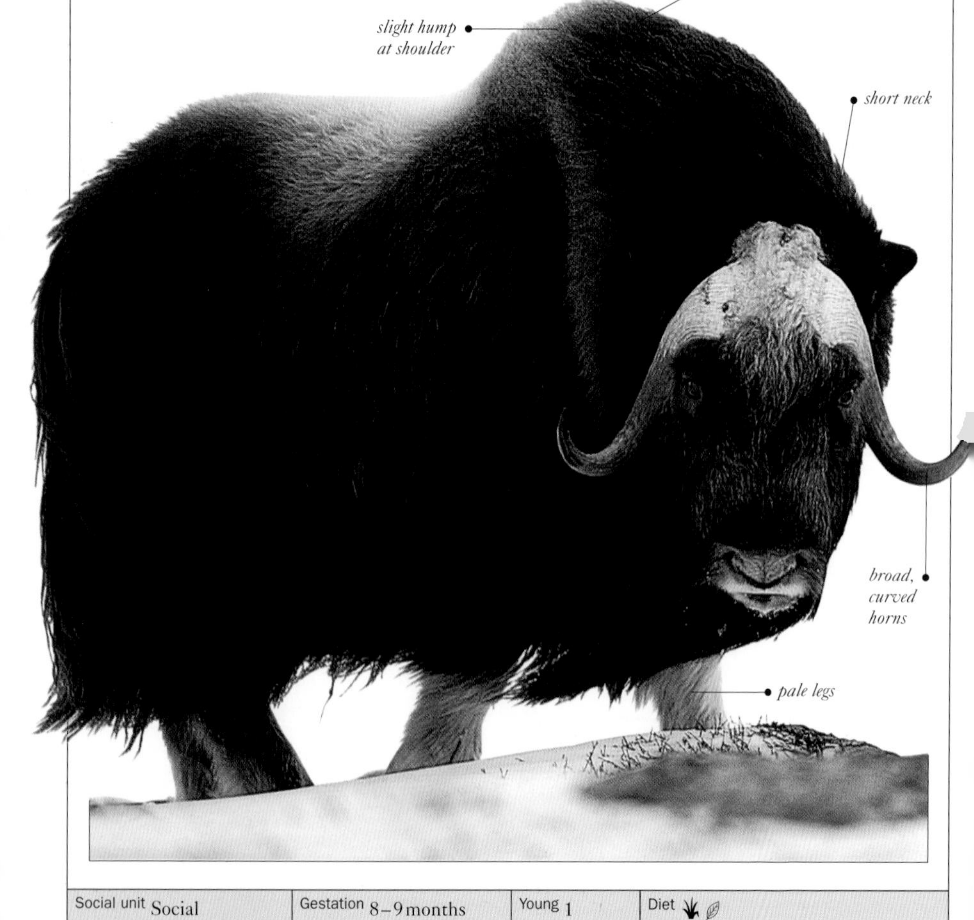

Social unit Social	Gestation 8–9 months	Young 1	Diet

Family BOVIDAE	Species *Hemitragus jemlahicus*	Status Vulnerable

TAHR

flattened horns

reddish brown coat

very short tail

The sure-footed Tahr has a shaggy mane on its neck and shoulders that reaches down to its knees; the fur on its face and head is contrastingly short. The flattened horns, 40 cm (16 in) in males, are twice as long as in females. The male Tahr seldom grazes in open areas, except in the evenings, whereas the female is often found in clearings. During the breeding season, rutting males lock horns and try to throw each other off balance. Like many mountain mammals, the Tahr migrates in spring, moving high into the Himalayas and returning to low temperate forest in autumn. When threatened by predators, this goat clambers rapidly over the rocks, much more able to cross the difficult terrain than the pursuer.

• **SIZE** Length: 0.9–1.4 m (3–4½ ft). Weight: 50–100 kg (110–220 lb).

• **OCCURRENCE** S. Asia. In temperate to subalpine forest up to the treeline between 2,500–5,000 m (8,250–16,500 ft).

ASIA

Social unit Social	Gestation 5–6 months	Young 1	Diet

Family BOVIDAE	Species *Capra ibex*	Status Endangered*

IBEX

Found at or above the treeline, up to 6,700 m (22,000 ft), this goat has a heavy body and relatively short, sturdy legs. The female sports a golden tan coat in summer, which turns grey-brown in winter, while the male has a rich brown coat with yellow-white patches on the back and rump. The enormous curved horns of the male Ibex reach up to 1.4 m (4½ ft), but are a quarter the size in females.

• **SIZE** Length: 1.2–1.7 m (4–5½ ft). Weight: 35–150 kg (77–330 lb).

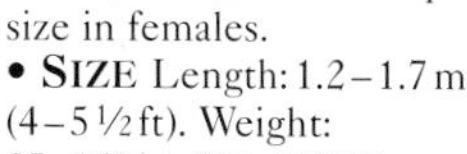

• **OCCURRENCE** S. Europe, W. and S. Asia, and N. Africa. In mountains.

• **REMARK** The Ibex population has been largely destroyed by hunting.

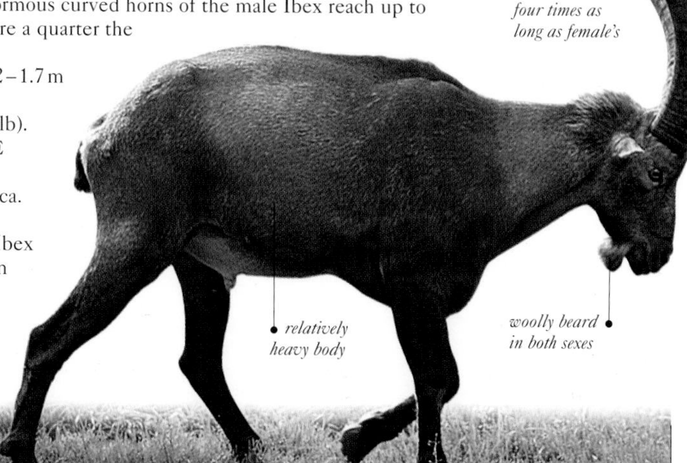

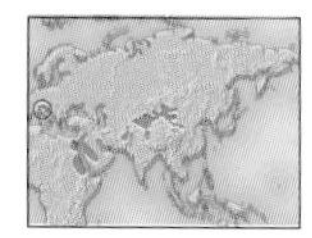

EUROPE, ASIA, AFRICA

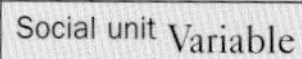
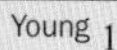

Social unit Variable	Gestation 150–180 days	Young 1	Diet

Family BOVIDAE	Species *Capra aegagrus*	Status Vulnerable

WILD GOAT

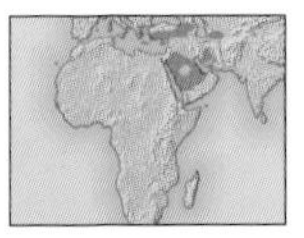

In this species, also known as the Bezoar Goat, females and young males (shown below) are red-grey or yellow-brown, whereas the adult male is silver-grey with dark markings and has a beard. Both sexes have horns. The Wild Goat has broad, rubbery soles on its hooves to provide a better grip on mountain slopes, and it flees to barely accessible cliffs when hunted by predators. Although usually active during the morning and late afternoon, in summer, this goat becomes a night grazer, resting during the day. Females are usually found in groups of 5–25; males fight for mates, using displays with their horns.

• **SIZE** Length: 1.2–1.6 m (4–5 ¼ ft). Weight: 25–95 kg (55–210 lb).

• **OCCURRENCE** W. Asia. In a variety of habitats, from arid scrub to alpine pasture, at elevations up to 4,200 m (13,800 ft).

WILD GENES
The domestic goat is the descendant of *Capra aegagrus* and now poses a threat to the Wild Goat by competing for pasture and by interbreeding.

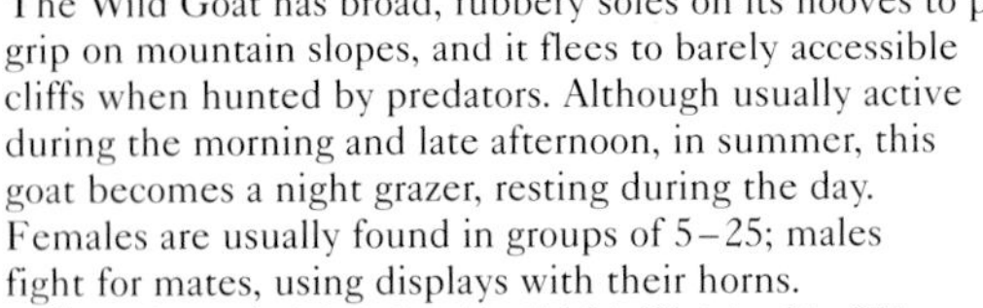

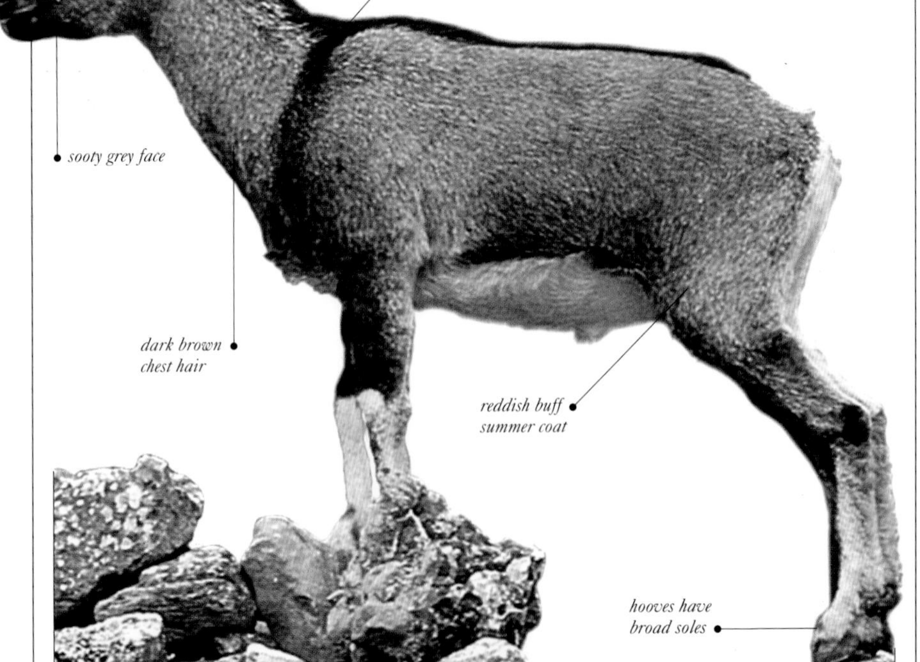

Social unit Social	Gestation 150–170 days	Young 1–3	Diet

Family BOVIDAE	Species *Capra falconeri*	Status Endangered

MARKHOR

This reddish grey goat has a short, smooth coat in summer, which turns longer and greyer in winter. The full grown, black-bearded male is almost twice as heavy as the female, and has a shaggy mane of dark hair extending from the neck to its feet. Its spectacular spiralled horns reach 1.6 m (5 ¼ ft), but are only 25 cm (10 in) long in the female. Active in the early morning and late afternoon, the Markhor rarely ventures above the snow line.

• **SIZE** Length: 1.4–1.8 m (4 ½–6 ft). Weight: 32–110 kg (70–245 lb).

• **OCCURRENCE** C. and S. Asia. In mountains, at a height of 700–4,000 m (2,300–13,200 ft).

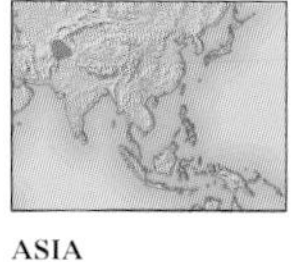
ASIA

• **REMARK** It is hunted for its horns, meat, hide, and body parts (used in local medicines).

Social unit Variable	Gestation 135–170 days	Young 1–2	Diet

Family BOVIDAE	Species *Pseudois nayaur*	Status Lower risk

BLUE SHEEP

Locally called the Bharal, the Blue Sheep is well camouflaged to survive in rocky, icy, mountainous areas. The male is brownish grey, tinged slate-blue, with black flanks and leg stripes; its smooth, rounded horns splay outwards. The female is smaller, with shorter horns, and lacks most of the male's black markings. This sheep freezes when alarmed, its cryptic coloration helping it merge with the background, and takes refuge in the most inaccessible rocky places to escape from predators.

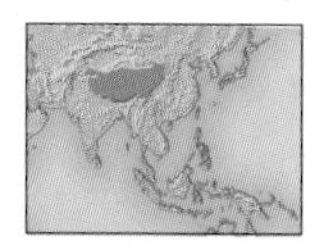
ASIA

• **SIZE** Length: 1.2–1.7 m (4–5 ½ ft). Weight: 25–80 kg (55–175 lb).

• **OCCURRENCE** S. to E. Asia. In the alpine zone between the tree and snow line.

Social unit Social	Gestation 160 days	Young 1–2	Diet

Family BOVIDAE	Species *Ovis canadensis*	Status Lower risk

BIGHORN SHEEP

This sheep's glossy brown coat, made up of outer brittle guard hairs and short, crimped grey underwool, fades to a lighter shade in winter. The male's horns curl almost into a circle and may weigh as much as the rest of the skeleton– up to 14 kg (31 lb). The female's horns, only slightly curved, are smaller. Groups of ewes and young usually consist of 8–10 individuals, whereas males form bachelor herds or remain solitary. During the rut, males walk away from each other, turn to advance with a threatening jump, and lunge to head-butt with tremendous force. The contest goes on for hours until one antagonist gives up. This sheep's thickened skull bones protect its head.

• **SIZE** Length: 1.5–1.8 m (5–6 ft). Weight: 55–125 kg (120–280 lb).

• **OCCURRENCE** S.W. Canada, W. and C. USA, and N. Mexico. In mountainous areas – alpine meadows and grassy mountain slopes close to rocky cliffs.

• **REMARK** The Bighorn Sheep is threatened by human encroachment into its habitat. It is also hunted as a sporting trophy, which removes dominant males from the population.

N. AMERICA

SEASONAL PATHWAYS
Young Bighorn Sheep learn about seasonal pathways from adults. They use their gripping feet and keen eyesight to pick their way over difficult rocky terrain.

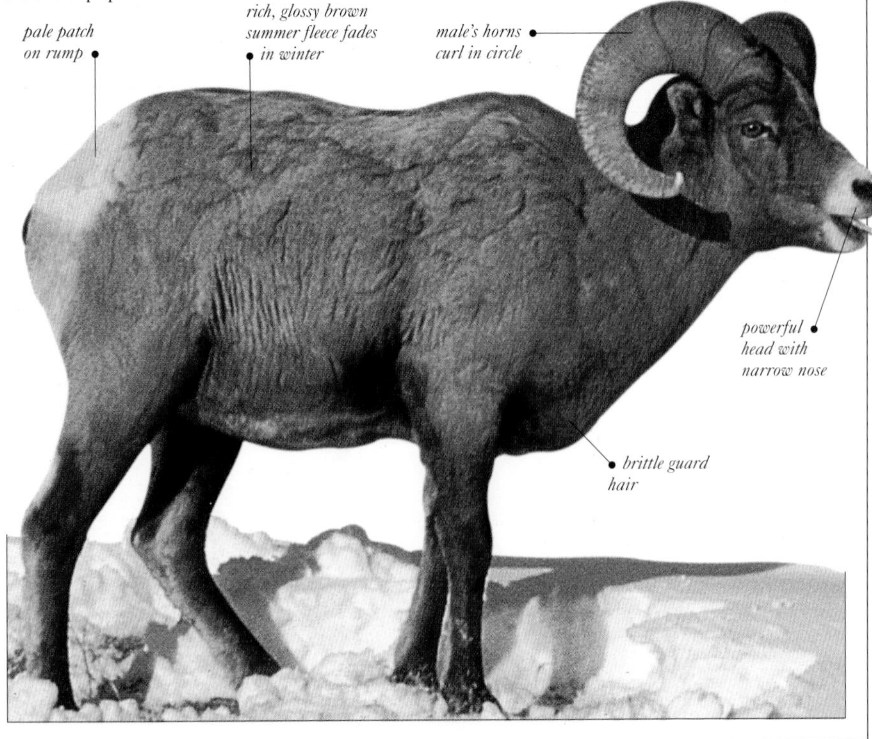

Social unit Variable	Gestation 150–180 days	Young 1–4	Diet

Family BOVIDAE	Species *Ovis orientalis*	Status Endangered

ASIATIC MOUFLON

Probably the ancestor of all domestic sheep breeds, the Asiatic Mouflon is the smallest wild sheep and is reddish brown with a dark central back stripe and paler saddle patch. Its face becomes lighter with age and it has a short dark tail and light underparts. In common with most wild sheep, ewes are found in small herds with their young, whereas males are solitary or roam in bachelor bands. There is a strict hierarchy among males based on age, strength, and size of horns; most do not begin to breed until they are six or seven years old.

• **SIZE** Length: 1.1–1.3 m (3 ½–4 ¼ ft). Weight: 25–55 kg (55–120 lb).

• **OCCURRENCE** W. Asia. In open woodland and on low mountain slopes.

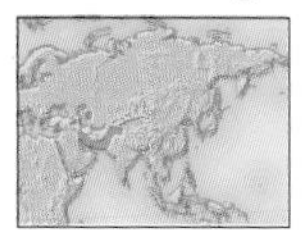

ASIA

Social unit Variable	Gestation 5 months	Young 1–2	Diet

Family BOVIDAE	Species *Ammotragus lervia*	Status Vulnerable

BARBARY SHEEP

Also called the Aoudad, this bovid is intermediate between a sheep and a goat. It has a reddish brown, tawny coat with a short, upright mane on its neck and shoulders, and a much longer one of soft hair on the throat, chest, and upper forelegs. In the female, the mane and the horns are less developed. Grazing in the early morning or late evening, it also browses on herbaceous plants and stunted bushes.

• **SIZE** Length: 1.3–1.7 m (4 ¼–5 ½ ft). Weight: 40–145 kg (88–320 lb).

• **OCCURRENCE** N. Africa. In semi-desert and desert highland.

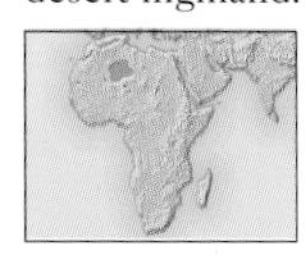

AFRICA

Social unit Variable	Gestation 154–161 days	Young 1–3	Diet

GLOSSARY

• ARBOREAL
Living exclusively or mainly in trees, such as many squirrels and monkeys. Compare Aquatic, Terrestrial.

• AQUATIC
Living exclusively or mainly in water, such as whales and dolphins. Compare Arboreal, Terrestrial.

• ARTIODACTYL
A hoofed mammal (ungulate) with an even number of toes on all or most feet, such as camels, deer, and cattle. Compare Perissodactyl.

• BALEEN / BALEEN PLATES
Comb-like, bristly plates or flaps of a gristly substance hanging from the upper jaw of many large whales. Used to strain small prey, such as krill, from sea water. Also known as "whalebone".

• BLOW
Cloud of moisture-laden air exhaled by whales, dolphins, and porpoises.

• BLOWHOLE
Nostril or nasal breathing opening on the top of the head in whales, dolphins, and porpoises.

• BOSS
A dome-shaped protuberance on the forehead of certain bovid species, where the horns meet.

• BOVID
An even-toed hoofed mammal that is a member of the large family Bovidae; includes wild and domesticated deer, giraffes, antelopes, gazelles, cattle, sheep, and goats.

• BOW-RIDE
To swim or "surf" in the wave created by the bow of a ship, or at the front of a large aquatic animal such as a whale.

• BRACHIATION
Hanging by the arms and swinging hand-over-hand, through the branches, like a gibbon.

• BREACHING
To leap clear of the water and crash back in with a large splash, as carried out by many whales, dolphins, and porpoises.

• BROWSE
To eat the vegation of trees, bushes, and shrubs, which is above ground level. Compare Graze.

• CANID
A member of the mammalian group Canidae, all domesticated and wild dogs, wolves, foxes, and jackals.

• CANINE
1. A usually large, long, pointed, slightly curved tooth found near the front of the mouth, just behind the incisor teeth. Sometimes called the "eye tooth". Well developed in meat-eating mammals such as cats and dogs. 2. A member of the dog group. See Canid.

• CANOPY FOREST
Forest with trees of approximately the same height, with their branches intermeshing to form a dense, roof-like canopy high above the ground.

• CARNIVORE
1. A mammal or other animal that eats mainly flesh or meat; a hunter or predator. Compare Herbivore, Omnivore. 2. A member of the mammalian group Carnivora, which includes cats, dogs, bears, weasels, and civets.

• CERVID
A member of the mammalian group Cervidae, the deer.

• CETACEAN
A member of the mammalian group Cetacea, the whales, dolphins, and porpoises.

• CHANNEL COUNTRY
Area north of the Great Artesian Basin, Australia, which is criss-crossed by rivers.

• CHEVRON
A V- or arrow-shape, especially as a marking on an animal, or within the shape of a horn or antler.

• CIRCUMPOLAR
Spread or distributed all around the polar region, at all longitudes. Applies to North or South Pole.

• COLONY
A group of animals living together and often sharing the tasks necessary for survival, such as foraging for food.

• CORAL RAG SCRUB
Scrub growing among limestone formed from coral.

• CRUSTACEANS
A large group of mainly sea-dwelling invertebrate animals, including lobsters, crabs, prawns, shrimps, barnacles and, on land, woodlice.

• DEWLAP
A fleshy, furry flap hanging from the throat and/or neck region of a mammal.

• DIGIT
A finger, toe, or equivalent structure, including the bones within this. The basic mammalian condition is five digits on each of the four limbs.

• DIGITIGRADE
Standing and moving on the toes (digits), as in horses and deer, rather than on the soles or central parts of the feet. Compare Plantigrade.

• DIURNAL
Active mainly during the day. Compare Nocturnal.

• DOMESTICATED
A mammal or other animal which has been bred to be, in various ways, easily kept and used by people, especially being less aggressive or more tame.

• DORSAL FIN
The fin on the back or upper surface of an aquatic mammal, such as a whale or dolphin; or a fish.

• DREY
The nest, den, or sheltering place of a squirrel.

• ECHOLOCATION
A method of sensing the location of objects in the surroundings, by sending out sounds (usually high-pitched pulses), which bounce off objects, and analyzing the returning echoes. Also known as Sonar.

• EQUID
A member of the mammalian group Equidae, which includes wild and domestic horses, asses, and zebras.

• **Even-toed ungulate**
See Artiodactyl.

• **Ewe**
A female sheep or similar type of mammal.

• **Falcate**
A curved, sickle- or scythe-like shape with one convex and one concave edge, like the back-curving dorsal fin of the Common Dolphin.

• **Felid**
A member of the mammalian group Felidae, including all wild and domesticated cats.

• **Feral**
Refers to animals originally from domesticated stock, but which subsequently take up life in the wild. Common examples include cats, dogs, horses, and pigs.

• **Flipper**
A limb with a broad, flattened shape, effective for moving through water by swimming, rowing, or paddling.

• **Flukes**
The tail of a whale, dolphin, or porpoise, with a broad, muscular, horizontal surface (but no limb bones inside) which is swept up and down for swimming.

• **Forage**
To search for food.

• **Forelimbs**
The front or forward limbs, as opposed to the rear or back limbs (hind limbs).

• **Fossa**
A pit or depression, usually in a bone such as the skull.

• **Gestation**
The period of pregnancy, the time when a baby develops inside the mother's body, before birth.

• **Graze**
To eat the leaves, stems, or other parts of grasses and similar low-growing plants, at or just above ground level. Compare Browse.

• **Guard hairs**
Longer, usually thicker hairs (fur) that form a mammal's outer coat, the guard coat. This shows over and protects the inner coat or under fur which has shorter, softer, usually more dense hairs.

• **Harem**
A group of females that breed with one male and are defended and protected by him against rival males.

• **Herbivore**
An animal that eats mainly leaves, fruits, or other plant matter. Compare Carnivore, Omnivore.

• **Hibernation**
A period of dormancy in winter. During hibernation an animal's body processes drop to a low level.

• **Hind**
A female deer or similar mammal.

• **Hindlimbs**
The back or rear limbs and their parts, as opposed to the front limbs.

• **Home range**
An area used regularly by an animal or group of animals, for feeding, shelter, breeding, and other needs, but not necessarily defended against others. Compare Territory.

• **Incisor**
A usually chisel- or spade-like tooth found at the front of the mouth. Incisors are used for biting, nibbling, and gnawing, and are especially well-developed in rodents.

• **Insectivore**
1. An animal that eats insects and similar smallish prey items such as worms. 2. A member of the mammalian group Insectivora, which includes, shrews, moles, and hedgehogs.

• **Introduced species**
A species that has been brought, usually by people, to a new area where it did not formerly occur naturally. Compare Native.

• **Invertebrate**
An animal without a backbone (vertebral column).

• **Keel**
A distinctive bulge or ridge at the base of the tail of whales and dolphins, just in front of the flukes.

• **Keratin**
A hard, tough body protein that forms many mammalian structures, such as skin, nails, hooves, claws, and horns.

• **Krill**
Small, shrimp-like crustaceans that form the major food of various marine mammals, including baleen whales and certain seals.

• **Lagomorph**
A member of the mammalian group Lagomorpha, which includes rabbits, hares, and pikas.

• **Lek**
A generally small territory held and defended by the male of a species, such as deer, during breeding to attract females.

• **Lodge**
The large, strong, stick-built nest or den for a group of beavers or similar animals.

• **Marine**
An organism living in the sea or in salty water.

• **Marsupial**
A mammal whose young are born at a very early stage of development, and which suckle and develop further in a pouch, the marsupium, on the female's front or underside.

• **Melanism**
Extra amounts of the brown-black colouring substance melanin, which makes the skin and fur very dark.

• **Melon**
Bulbous forehead of many toothed whales, dolphins, and porpoises. It is believed to focus the outgoing and incoming sound pulses used in echolocation.

• **Molar**
A usually broad, flat, or ridged tooth found near the back of the mouth, often called a cheek tooth. Molars are used for crushing and chewing.

• **Molluscs**
A large group of invertebrates in which the body is surrounded by a large fleshy "hood" or "cloak", and perhaps a hard, protective shell. The group includes oysters, mussels, clams, and similar "shellfish", as well as slugs, snails, octopus, cuttlefish, and squid.

• **Monotreme**
A mammal that reproduces by laying eggs that hatch into live young; as opposed to giving birth to formed young, which is the case in most mammals.

• **Moulting**
Shedding fur, layers of skin, feathers, scales, or similar external structures, usually in a regular seasonal fashion.

• **Musk**
A pungent-scented substance produced by various animals, especially males, such as elephants, deer, cattle, and carnivores. It usually signifies readiness to breed.

• **Mustelid**
A member of the mammalian group Mustelidae, which includes stoats, weasels, polecats, otters, mink, sables, and badgers.

• **Native**
Having been born in an area, or having naturally occurred there for a very long time. Compare Introduced.

• **Nictitating membrane**
"Third eyelid" – a cover or flap-like lid that can be moved across the eye for reasons such as protection or to reduce illumination.

• **Nocturnal**
Active mainly at night or during darkness. Compare Diurnal.

• **Nose leaf**
A flap-like structure around the nose of certain animals, especially some bats. It helps to direct the sound pulses for echolocation.

• **Odd-toed ungulate**
See Perissodactyl.

• **Omnivore**
An animal that eats all types of food, including flesh and plant matter.

• **Opportunistic feeder**
An animal that takes advantage of many and varied foods, as and when they become available.

• **Pack**
A group of usually predatory animals, such as wolves.

• **Patagium**
The thin, leathery, elastic, skin-like flight membrane of a bat or other flying or gliding mammal.

• **Perissodactyl**
A hoofed mammal (ungulate) with an odd number of toes on all or most feet, including horses, tapirs, and rhinos. Compare Artiodactyl.

• **Pinniped**
A member of the mammalian group Pinnipedia ("flipper-feet"), which includes seals, sea lions, and walruses.

• **Plankton**
Tiny plants and animals that drift in oceans and lakes.

• **Plantigrade**
Standing and moving on the soles or central part of the feet, as in bears and humans. Compare Digitigrade.

• **Pod**
A coordinated group of members of the Cetacea, especially great whales. See also School.

• **Predator**
A hunter; an animal that actively catches and kills other animals that form its prey.

• **Prehensile**
Able to grasp. Many New World Monkeys have prehensile tails.

• **Prey**
An animal which is hunted as food by another.

• **Primary forest**
Forest that has been undisturbed (especially by human activity) for a very long time, and has reached a stable, mature end-point of development.

• **Primate**
A member of the mammalian group Primates, which includes lemurs, galagos, lorises, pottos, monkeys, apes, and humans.

• **Proboscis**
A flexible, elongated, or enlarged nose or snout, like the elephant's long trunk.

• **Pronk**
To move in a stiff-legged, high-leaping, bounding, or "bouncing" manner, as in the Springbok. Also called "stotting".

• **Prosimian**
A member of the mammalian group Primates that is not a monkey or ape – a lemur, galago, loris, potto, or similar type.

• **RANGE**
1. To roam or travel long distances in search of food, shelter, or mates. 2. See Home range.
• **REGURGITATE**
To "cough up" or bring back swallowed food from the stomach, crop, or similar digestive storage part.
• **REINTRODUCE**
To restore or put back a type of living thing, into an area where it once occurred, but from which it has become locally extinct.
• **RETRACTABLE (RETRACTILE) CLAWS**
Claws that can be pulled back or withdrawn into fleshy pocket-like sheaths at the ends of the toes. Found especially in the cat family.
• **RODENT**
A member of the very large mammalian group Rodentia, which includes rats, mice, voles, squirrels, gerbils, beavers, and porcupines.
• **ROOST**
To rest or sleep, usually above the ground such as in a tree hole, on a tree branch, or in the roof of a cave.
• **RORQUAL**
Strictly speaking, a baleen whale of the genus Balaenoptera; however, many experts include the Humpback Whale in this group.
• **ROSTRUM**
An elongated or enlarged snout or forehead area, especially the upper jaw region of the skull, as in the "beak" of a dolphin.
• **RUTTING**
When male animals come together to display and battle each other, for breeding access to females.
• **SAVANNA**
A grassland habitat, with scattered trees and bushes, and a long dry season. Often applied to the open grassland of E. and S. Africa.
• **SCATTER-HOARDER**
An animal that hides or stores (hoards) food items in many different locations, such as a squirrel burying nuts at various sites.
• **SCAVENGE**
Eating leftovers, or dying, dead, or rotting carcasses and similar remains of animals.
• **SCHOOL**
A coordinated group of members of the Cetacea, especially dolphins. See also Pod.
• **SECONDARY FOREST**
Forest which has been disturbed by human activity, wildfire, flood, etc., and is in the process of re-developing its mature trees and animal life.
SEMI-RETRACTABLE CLAWS
Claws which can partly be drawn into sheaths. See Retractable claws.
• **SETT**
The large, complex underground home, nest site, or den of a group of badgers or similar mammals.
• **SIMIANS**
Monkeys (including marmosets and tamarins) and apes; members of the mammalian group Primates that are not Prosimians.
• **SIRENIANS**
Members of the group Sirenia, including Dugongs and manatees.
• **SONAR**
Sound Navigation And Ranging. See Echolocation.
• **SPYHOPPING**
Raising the head vertically out of the water, then sinking below the surface without much splash.
• **STAG**
A male deer or similar mammal.
• **STEPPE**
A predominantly grassland habitat, with no or few trees and bushes, and a long dry season. Often applied to the open grasslands of Asia.
• **STOTT**
See Pronk.
• **SUB-SAHARAN**
The region of Africa to the south of the Sahara Desert.
• **SUB-ANTARCTIC**
The islands and ocean to the north of the southern polar region of Antarctica, often includes southern tips of South America and Africa.
• **SUBMONTANE**
Hilly regions, above the low plains and foothills, but below the true mountains.
• **SYMBIOTIC**
A relationship shared by two individuals or two different species, in which both gain.
• **TAIGA**
Vast region of mainly evergreen forest and woodland across northern Asia and North America.
• **TERRESTRIAL**
Living exclusively or mainly on the ground. Compare Arboreal, Aquatic.
• **TERRITORY**
An area used regularly by an animal or group of animals, for feeding, shelter, and breeding, which is usually defended against others. Compare Home range.
• **TINE**
The end or point of a branch on a deer's antler.
• **TRAGUS**
A flap or fleshy projection in front of a bat's ear.
• **TUBERCLES**
Wart-like lumps, circular bumps, or similar knobbly structures.
• **TUNDRA**
Open, treeless region of low-growing plants in the far north, covered in snow and ice for part of the year.
• **UNDER FUR**
See Guard hairs.
• **UNGULATE**
Mammal with hooves. See Artiodactyl, Perissodactyl.
• **VENTRAL**
Relating to the belly or underside of the body.
• **VERTEBRATE**
An animal with a backbone.
• **VESTIGIAL** Relating to an organ that is atrophied or non-functional.
• **VIVERRID**
A member of the mammalian group Viverridae, which includes civets, genets, linsangs, and mongooses.
• **WARREN**
The underground nest or den of a rabbit or similar mammal, especially those that live in colonies.
• **WEANING**
In mammals, the period when the mother gradually ceases to provide milk for her young.
• **WILD**
Not domesticated or tamed; natural and unaffected by humans.

INDEX

A

Aardvark 311
Aardwolf 273
Acinonyx jubatus 295
Acomys minous 168
Acrobates pygmaeus 69
Acrobatidae 69
Addax 364
Addax nasomaculatus 364
Aepyceros melampus 371
African Buffalo 353
African Clawless Otter 264
African Elephant 308
African Forest Elephant 309
African Golden Cat 285
African Polecat *see* Zorilla
African Striped Weasel 251
African Wild Ass 315
African Wild Dog 233
Agouti paca 178
Agoutidae 178
Ailuridae 242
Ailuropoda melanoleuca 240
Ailurus fulgens 242
Alaskan Bear 236
Alcelaphus buselaphus 368
Alces alces 342
Allactaga tetradactyla 173
Alopex lagopus 220
Alouatta pigra 109
Alouatta seniculus 109
alpacas 330
Alpine Musk Deer 334
Amami Rabbit 137
Amazon River Dolphin 184
Amazonian Manatee 313
American Badger 259
American Beaver 154
American Bison 356
American Black Bear 234
American Buffalo
  see American Bison
American Harvest Mouse 156
American Mink 250
Ammotragus lervia 381
Anathana ellioti 96
anatomy 16–19
  bone 16
  ear 18
  cetaceans 17
  flying mammals 17
Andean Bear
  see Spectacled Bear
Andean Cat 277
Angolan Free-tailed Bat 93
Angwantibo 100
Anoa, Lowland 352
Anoura geoffroyi 90
Ant-bear *see* Aardvark
Ant-pig *see* Aardvark
Anteater
  Giant 132
  Lesser *see* Southern Tamandua
  Silky 132
anteaters 52, 130
Antechinomys laniger 59
Antelope
  Four-horned *see* Chousingha
  Marsh *see* Lechwe
  Roan 361
  Sable 362
antelopes 348
Antidorcas marsupialis 373
Antilocapridae 53, 344
Antilopcapra americana 344
Antilope cervicapra 370
Aonyx capensis 264
Aonyx cinerea 264
Aotidae 111
Aotus lemurinus 111
Aoudad *see* Barbary Sheep
apes 52, 122
Aplodontia rufa 143
Aplodontidae 143
Apodemus flavicollis 166
Apodemus sylvaticus 166
aquatic mammals 15, 42–43
  body shape
  breathing
  movement
  propulsion
  temperature control
arboreal mammals 15
Arctic Fox 220
Arctic Hare 141
Arctictis binturong 268
Arctocebus calabarensis 100
Arctocephalus pusillus 297
Arctonyx collaris 260
Armadillo
  Giant 133
  Hairy 133
  Nine-banded 134
  Northern Naked-toed 134
armadillos 130
Armoured Shrew 81
Arni *see* Asian Water Buffalo
artiodactyls 334
Arvicola terrestris 164
Asian Elephant 348
Asian Water Buffalo 352
Asian Wild Ass *see* Onager
Asian Wild Dog *see* Dhole
Asiatic Black Bear 238
Asiatic Mouflon 381
Asiatic Two-horned Rhinoceros *see* Sumatran Rhinoceros
Ateles chamek 108
Ateles geoffroyi 108
Atelidae 107, 108
Atelocynus microtis 225
Atlantic Spotted Dolphin 187
Australian False Vampire Bat 84
Australian Fur Seal
  see Cape Fur Seal
Australian Water Rat 171
Axis axis 335
Axis Deer *see* Chital
Aye-aye 106
Azara's Agouti 177

B

Babirusa 327
Baboon
  Guinea 115
  Olive 116
baboons 23

Babyrousa babyrussa 327
Bactrian Camel 333
Badger
American 259
Burmese Ferret 259
Eurasian 261
Hog- 260
Honey 260
Baikal Seal 304
Baird's Tapir 323
Balaena mysticetus 204
Balaenidae 206
Balaenoptera acutorostrata 214
Balaenoptera musculus 212
Balaenoptera physalus 210
Balaenopteridae 210
Bald Uakari 110
baleen 31
baleen whales 53, 204
Banded Mongoose 270
Bandicoot 63
Eastern Barred 63
New Guinean Spiny 63
Bank Vole 163
Banteng 353
Barbary Sheep 381
Bassaricyon gabbii 245
Bassariscus astutus 243
Bat-eared Fox 225
Bathyergidae 181
Bat
Angolan Free-tailed 93
Australian False Vampire 84
Brown Long-eared 93
Common Noctule 92
Daubenton's 93
Davy's Naked-backed 89
Egyptian Fruit 88
Egyptian Rousette *see* Egyptian Fruit Bat
False Vampire 90
Franquet's Epauletted 86
Geoffroy's Hairy-legged *see* Geoffroy's Tail-less Bat
Geoffroy's Tail-less 90
Ghost *see* Australian False Vampire Bat
Greater Bulldog 88
Lesser Horseshoe 88
Lesser Mouse-tailed 88
Linnaeus' False Vampire *see* False Vampire Bat
Mauritian Tomb 84
Mexican Funnel-eared 89
Bat cont.
Proboscis 85
Tent-building 90
Vampire 91
bats 29, 52, 84
Beaked Whale, Cuvier's 198
Bear
American Black 234
Asiatic Black 238
Brown 236
Dog *see* Sun Bear
Polar 235
Sloth 239
Spectacled 239
Sun 238
bears 53, 234
Beaver
American 154
Eurasian 153
Mountain 143
Beech Marten 252
Beluga 196
Bengal Tiger 292
Bettong, Brush-tailed 71
Bettongia penicillata 71
Bezoar Goat *see* Wild Goat
Bi-coloured French Lop 139
Bighorn Sheep 380
Bilby, Greater 63
Binturong 268
bird-shaped mammals 11
Bison
American 356
European 355
Indian *see* Gaur
Bison bison 356
Bison bonasus 355
Black Bear 31
Black Lemur 103
Black Panther 291
Black Rat 171
Black Rhinoceros 320
Black Spider Monkey 108
Black Wallaby *see* Swamp Wallaby
Black-backed Jackal 229
Black-footed Ferret 247
Black-handed Spider Monkey 108
Black-lipped Pika 137
Black-tailed Deer *see* Mule Deer
Black-tailed Jackrabbit 142
Black-tailed Prairie Dog 145
Blackbuck 370
Blanford's Fox 217
Blarina brevicauda 82
Blastocerus dichotomus 340
Blesbok *see* Bontebok
blow hole 17
blubber 43
Blue Buck *see* Nilgai
Blue Bull *see* Nilgai
Blue Hare *see* Mountain Hare
Blue Sheep 379
Blue Whale 14, 212
Blue Wildebeest *see* Wildebeest
Bobcat 282
Bohor Reedbuck 361
Bolivian Squirrel Monkey 112
Bongo 349
Bonobo *see* Pygmy Chimpanzee
Bontebok 364
Bornean Orangutan 124
Bos gaurus 355
Bos grunniens 354
Bos javanicus 353
Boselaphinae 351
Boto *see* Amazon River Dolphin
Botta's Pocket Gopher 152
Bottlenose Dolphin 189
Bovidae 53, 348
bovids 348
Bowhead Whale 204
brachiation 29
Brachylagus aquaticus 138
Brachylagus floridanus 138
Brachylagus idahoensis 138
Brachyteles arachnoides 108
Bradypodidae 130
Bradypus torquatus 131
breeding season 20
Brindled Gnu *see* Wildebeest
Brown Bear 236
Eurasian 236
Brown Capuchin 112
Brown Hare 140
Brown Hyena 273
Brown Lemming 165
Brown Lemur 103
Brown Long-eared Bat 93
Brown Rat 170
Brush-tailed Bettong 71
Brush-tailed Rock Wallaby 71
Bubalus arnee 352

Bubalus depressicornis 352
Buffalo
African 353
Asian Water 352
Bulldog Bat, Greater 88
Bumblebee Bat 14
Burchell's Zebra 315
Burmese Ferret Badger 259
Burramyidae 70
Bushbaby, Lesser *see* South African Galago
Bush Dog 232
Bush Elephant *see* African Elephant
Bush Pig 326
Bushbuck 349
Bushy-tailed Olingo 245

C

Cabassous centralis 134
Cacajao calvus 110
caching 32
Cacomistle *see* Ringtail
Californian Sea Lion 296
Callicebus moloch 111
Callimico goeldii 113

Callithrix argentata 115
Callithrix pygmaea 114
Callitrichidae 107
Callorhinus ursinus 298
Callosciurus prevostii 151
Camel, Bactrian 333
Camelidae 53, 330
camelids 330
camels 53, 330
Camelus bactrianus 333
Camelus dromedarius 332
Canid 216
Canidae 53, 216
Canis adustus 229
Canis aureus 228
Canis dingo 227
Canis latrans 227
Canis lupus 230
Canis mesomelas 229
Canis simensis 228
Cape Fur Seal 297
Cape Ground Squirrel 145
Cape Porcupine 174
Capra aegagrus 378
Capra falconeri 379
Capra ibex 377
Capreolus capreolus 339
Capromyidae 178
Capromys pilorides 178
Capuchin, Brown 112
Capybara 177
Caracal 284
Caribou *see* Reindeer
Carnivora 53, 296
carnivores 30, 53, 216
Castor canadensis 154
Castor fiber 153
Castoridae 153
Cat
African Golden 285
Andean 277
Cat cont.
Fishing 279
Flat-headed 279
Jungle 276
Marbled 280
Sand 277
Wild 276
cats 50, 53, 276
cattle 50, 53, 348
Cavia aperea 176
Caviidae 176
Cavy *see* Guinea Pig
Cebidae 107, 112
Cebus apella 112
Celebes Black Ape
see Celebes Macaque
Celebes Macaque 119
Cenozioc Era 13
Cephalophus natalensis 358
Cephalorhynchus commersonii 191
Cephalorhynchus hectori 191
Ceratotherium simum 318
Cercartetus lepidus 70
Cercocebus torquatus 118
Cercopithecidae 107
Cercopithecus neglectus 118
Cerdocyon thous 224
Cervidae 53, 334
cervids 334
Cervus elaphus 336
Cervus nippon 338
Cervus unicolor 338
Cetacea 53
cetaceans 17, 53, 182
Chaetophractus villosus 133
Chamois 375
Cheetah 30, 295
Cheirogaleidae 101
Cheirogaleus medius 101
Chevrotain, Indian Spotted 334
Chimpanzee 126
Pygmy 127
Chinchilla 179
Chinchilla lanigera 179
Chinchillidae 178
Chinese Pangolin 135
Chinese River Dolphin 184
Chinese Tiger 292
Chipmunk, Eastern 146
Chironectes minimus 57
Chiroptera 52
Chital 335
Choloepus didactylus 130
Chousingha 352
Chrysochloridae 80
Chrysocyon brachyurus 222
Civet
Malayan 267
Palm 268
civets 53
classification 52–53
Clethrionomys glareolus 163
Clouded Leopard 288
Coati, Southern Ring-tailed 244
Coatimundi *see* Southern Ring-tailed Coati
Coendou prehensilis 175
Collared Peccary 327
Colobus, Eastern Black-and-White 120
Colobus guereza 120
Columbian Ground Squirrel 148
Commerson's Dolphin 191
Common Brush-tailed Possum 66
Common Cuscus 66
Common Dolphin 189
Common Dormouse
see Hazel Dormouse
Common Duiker 358
Common Eland 351
Common Genet *see* Small Spotted Genet
Common Hamster 159
Common Mouse Opossum 57
Common Noctule Bat 92
Common Pipistrelle 91
Common Raccoon 243
Common Rat *see* Brown Rat
Common Ringtail 68
Common Seal 305
Common Tenrec 80
Common Vole 163
Common Wombat 64
Common Zebra
see Burchell's Zebra
communication 26
Condylura cristata 82
Conepatus humboldti 258
Connochaetes taurinus 366
conservation 48–49
convergence 15
Cottontail, Eastern 138
Cougar *see* Puma
courtship 20, 27
Coyote 227
Coypu 180
Crab-eating Fox 224
Crab-eating Macaque 119
Crab-eating Raccoon 244
Crabeater Seal 300
Crested Porcupine
see Cape Porcupine
Cretaceous Period 12
Cricetomys gambianus 169
Cricetus cricetus 159
critically endangered mammals 46
Crocuta crocuta 274
Cryptoprocta ferox 270
Culpeo Fox 224
Cuon alpinus 232
Cuscus, Common 66
Cuvier's Beaked Whale 198
Cyclopes didactylus 132
Cynictis penicillata 269
Cynocephalidae 95
Cynocephalus variegatus 95
Cynomys ludovicianus 145
Cystophora cristata 303

D

Dactylopsila trivirgata 67
Dall's Porpoise 183
Dama dama 335
Damaliscus dorcas 364
Damaliscus lunatus 365

Dasycercus byrnei 60
Dasypodidae 133, 134
Dasyprocta azarae 177
Dasyproctidae 177
Dasypus novemcinctus 134
Dasyuridae 58
Dasyurus viverrinus 61
Daubenton's Bat 93
Daubentonia madagascariensis 106
Daubentoniidae 106
Davy's Naked-backed Bat 89
De Brazza's Monkey 118
Deer 53, 334
  Alpine Musk 334
  Axis *see* Chital
  Black-tailed *see* Mule Deer
  Fallow 335
  Marsh 340
  Mule 339
  Pére David's 338
  Red 336
  Roe 339
deforestation 47
Degu 181
Deinotherium 12
Delphinapterus leucas 196
Delphinidae 185
Delphinus delphis 189
Dendrohyrax arboreus 310
Dendrolagus dorianus 77
Dermoptera 52
Desert Hamster 159
Desert Hedgehog, Long-eared 79
Desert Jerboa 173
Desert Lynx *see* Caracal
desert mammals 34–35
  food storage
  temperature control
  temporary food surplus
  water balance
  water conservation
  water storage
Desmana moschata 83
Desmarest's Hutia 178
Desmodus rotundus 91
Dhole 232
Dibbler 59
Dicerorhinus sumatrensis 321
Diceros bicornis 320
Didelphidae 56
Didelphis virginiana 56
digitigrade 28
Dik-dik, Kirk's 370
Dingo 51, 227
Dipodidae 173
Dipodomys merriami 152
Distoechurus pennatus 69
divergence 15
diversity 14–15
Dog
  African Wild 233
  Bush 232
  Hunting *see* African Wild Dog
  Raccoon 226
  Small-eared 225
dogs 53, 216
Dog Bear *see* Sun Bear
Dolichotis patagonum 176
Dolphin
  Amazon River 184
  Atlantic Spotted 187
  Bottlenose 189
  Chinese River 184
  Commerson's 191
  Common 189
  Dusky 185
  Ganges River 185
  Hector's 191
  Irrawaddy 190
  Northern Right-whale 190
  Pacific White-sided 186
  Pantropical Spotted 188
  Risso's 186
  Spinner 187
  Striped 188
  White-beaked 186
dolphins 17, 182
domestic dog 231
domestic ferret 246
domestic goat 378
domestic yak 354
domestication 50
Doria's Tree-kangaroo 77
Dormouse 172
  Common *see* Hazel Dormouse
  Edible 172
  Fat *see* Edible Dormouse
  Hazel 172
Douroucouli *see* Night Monkey
Dromedary 332
Duck-billed Platypus 55
Dugong 53, 312
Dugong dugon 312
Dugongidae 312
Duiker 358
  Common 358
  Red Forest 358
Dunnart, Fat-tailed 58
Dusky Dolphin 185
Dusky Titi Monkey 111
Dwarf Hamster *see* Desert Hamster
Dwarf Lemur, Fat-tailed 101
Dwarf Mongoose 269

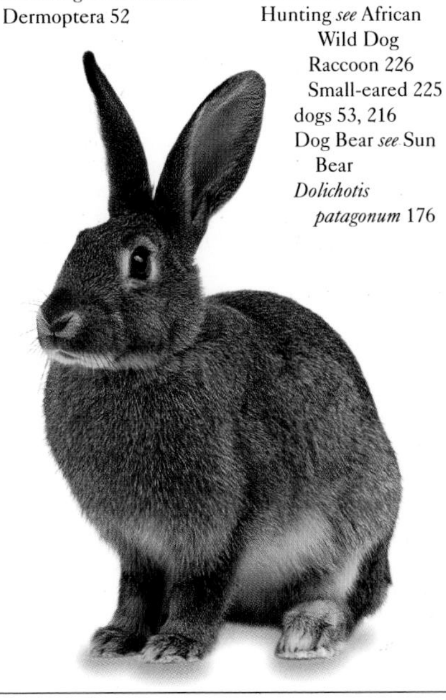

E

Eastern Barred Bandicoot 63
Eastern Black-and-white Colobus 120
Eastern Chipmunk 146
Eastern Cottontail 138
Eastern Gorilla 128
Eastern Grey Squirrel 149
Eastern Quoll 61
Eastern Spotted Skunk 258
Echidna
  Long-nosed 55
  Short-nosed 54
Echinosorex gymnura 79
echolocation 26
Echymipera kalubu 63
Edentata 130
Edible Dormouse 172
egg-laying mammals 11, 21, 52, 54
Egyptian Fruit Bat 88
Egyptian Rousette Bat *see* Egyptian Fruit Bat
Elaphurus davidianus 338
Elephant
  African 308
  Asian 306
  Bush *see* African Elephant
elephants 13, 32, 53, 306
Elephant-shrew
  Golden-rumped 94
  Rufous 94
elephant-shrews 52, 94
Elephantidae 308
Elephantulus rufescens 94
Elephas maximus 348
Elk 342
  North American 342
Emballonuridae 84
Emperor Tamarin 114
endangered mammals 46
endoskeleton 16
Enhydra lutris 265
Epomops franqueti 86
Equidae 53, 314
Equus africanus 315
Equus burchelli 315
Equus ferus przewalskii 314
Equus grevyi 316
Equus hemionus 315
Eremitalpa granti 80
Erethizon dorsatum 175
Erethizontidae 175
Erinaceidae 78
Erinaceus europaeus 78
Ermine *see* Stoat
Erythrocebus patas 117
Eschrichtiidae 208
Eschrichtius robustus 208
Ethiopian Wolf 228

Eubalaena glacialis 206
Eurasian Badger 261
Eurasian Beaver 153
Eurasian Brown Bear 236
Eurasian Harvest Mouse 168
Eurasian Lynx 283
Eurasian Red Squirrel 148
Eurasian River Otter *see* European Otter
Eurasian Water Shrew 81
Eurasian Wild Pig *see* Wild Boar
Euro *see* Wallaroo
European Bison 355
European Mink 251
European Mole 15, 83
European Otter 262
European Polecat 246
European Rabbit 139
European Water Vole 164
even-toed ungulates 330
evolution 12–13

F

Fallow Deer 335
 Mesopotamian 335
False Killer Whale 192
False Vampire Bat 90
Fat Dormouse *see* Edible Dormouse
Fat-tailed Dunnart 35, 58
Fat-tailed Dwarf Lemur 101
Fat-tailed Gerbil 162
Fat-tailed Pseudantechinus *see* Red-eared Antechinus
Feather-tailed Glider *see* Pygmy Glider
Feather-tailed Possum 69
feeding 30–33
 carnivore
 herbivore
 omnivore
Felidae 53, 276
Felis 276
Felis aurata 285
Felis caracal 284
Felis chaus 276
Felis jacobita 277
Felis lynx 283
Felis margarita 277
Felis marmorata 280
Felis pardalis 281
Felis planiceps 279
Felis rufus 282
Felis serval 278
Felis silvestris 276
Felis viverrinus 279
Felis wiedi 280
Fennec Fox 35, 216
feral mammals 51
Ferret, Black-footed 247
filter feeder 31
Fin Whale 210
Finless Porpoise 183
Fisher 254
Fishing Cat 279
Flat-headed Cat 279
fliers 15
Flying Lemur, Malayan 95
flying lemurs 52, 95
flying mammals 17
Flying Squirrel, Giant 151
foetus 22
forest mammals 38–39
 arboreal
 camouflage
 canopy dwellers
 fliers
 gliders
 nest builders
 terrestrial
Fossa 270
Four-horned Antelope *see* Chousingha
Four-toed Jerboa 35, 173
Fox
 Arctic 220
 Bat-eared 225
 Blanford's 217
 Crab-eating 224
 Culpeo 224
 Fennec 216
 Grey 221
 Red 218
 Rueppell's 220
 Swift 217
foxes 216
Franquet's Epauletted Bat 86
Fruit Bat, Egyptian 88
fruit bats 17
Fur Seal
 Australian *see* Cape Fur Seal
 Cape 297
 Northern 298

G

Galago
 South African 101
 Thick-tailed 100
Galago crassicaudatus 100
Galago moholi 101
Galagonidae 100
Ganges River Dolphin 185
Gaur 355
Gazella thomsonii 372
Gelada 117
Gemsbok 363
Genet
 Common *see* Small Spotted Genet
 Small spotted 266
Genetta genetta 266
Geoffroy's Hairy-legged Bat *see* Geoffroy's Tail-less Bat
Geoffroy's Long-nosed Bat *see* Geoffroy's Tail-less Bat
Geoffroy's Spider Monkey *see* Black-handed Spider Monkey
Geoffroy's Tail-less Bat 90
Geomyidae 152
Gerbil
 Fat-tailed 162
 Mongolian 160
Gerenuk 372
Ghost Bat *see* Australian False Vampire Bat
Giant African Mole Rat 163
Giant Anteater 132
Giant Armadillo 133
Giant Flying Squirrel 151
Giant Forest Hog *see* Wild Boar
Giant Otter 265
Giant Panda 48, 240
Giant Pouched Rat 169
Giant Rat, Malagasy 162
Giant Squirrel, Indian 150
gibbons 38
Giraffa camelopardalis 346
Giraffe 346
Giraffe Gazelle *see* Gerenuk
Giraffidae 53, 345
Glider
 Feather-tailed *see* Pygmy Glider
 Greater 68
 Pygmy 69
 Squirrel 67

gliders 15, 29
Gliridae 172
Glis glis 172
Globicephala macrorhynchus 192
Glutton *see* Wolverine
Goat
 Bezoar *see* Wild Goat
 Mountain 374
 Wild 378
goats 348
Goeldi's Marmoset 113
Golden Hamster 158
Golden Jackal 228
Golden Lion Tamarin 49, 113
Golden Snub-nosed Monkey 121
Golden-rumped Elephant-shrew 94
Gomphotherium 12
Goosebeak *see* Cuvier's Beaked Whale
Gopher, Botta's Pocket 152
Gorilla beringei 128
Grampus griseus 186
Grant's Golden Mole 80
grassland mammals 36–37
 burrowing
 herds
 limbs
 migration
 movement
 pack
 speed
Gray Whale 208
Greater Bilby 63
Greater Bulldog Bat 88
Greater Glider 68
Greater Grison 254
Greater Kudu 350
Grevy's Zebra 316
Grey Fox 221
Grey Gentle Lemur 104
Grey Seal 303
Grey Squirrel, Eastern 149
Grey Wolf 230
Grey Woolly Monkey 107
Grey-bellied Night Monkey 111
Grison, Greater 254
Grizzly Bear 236
Ground Squirrel
 Cape 145
 Columbian 148
Groundhog *see* Woodchuck
Guanaco 330, 331
Guatemalan Black Howler Monkey *see* Mexican Howler Monkey
Guinea Baboon 115
Guinea Pig 176
Gulf of California Porpoise 183
Gulo gulo 255
Gymnobelideus leadbeateri 67
gymnures 78

H

habitat 15
 habitat destruction 47, 48
 habitat protection 48
hair 19
Hairy Armadillo 133
Halichoerus grypus 303
Hamster
 Common 159
 Desert 159
 Golden 158
Hanuman Langur 120
Hapalemur griseus 104
Haplorhini 52, 107
Harbour Porpoise 182
Harbour Seal *see* Common Seal
Hare
 Arctic 141
 Blue *see* Mountain Hare
 Brown 140
 Mountain 142
hares 52, 136
Harp Seal 304
Hartebeest 368
Harvest Mouse, Eurasian 168
hearing 26
Hector's Dolphin 191
Hedgehog
 Long-eared 79
 West European 78
Helarctos malayanus 238
Heliosciurus gambianus 150
Helogale parvula 269
Hemiechinus auritus 79
Hemitragus jemlahicus 377
herbivores 32
herd 25
Hero Shrew *see* Armoured Shrew
Herpestidae 269
Heterocephalus glaber 181
heterodont dentition 18
Heteromyidae 152
Hexaprotodon liberiensis 328
Hill Kangaroo *see* Wallaroo
Hippopotamidae 53, 328
Hippopotamus
 Common 329
 Pygmy 328
hippopotamuses 53, 328
Hippopotamus amphibius 329
Hippotragus equinus 361
Hippotragus niger 362
Hispaniolan Solenodon 79
Hispid Cotton Rat 157
Hog 325
 Giant Forest 325
 Pygmy 325
Hog-badger 260
Hog-nosed Bat 14
Hokkaido Bear 236
home range 24
Hominidae 126
homeothermic 10
Honey Badger 260
Honey Bear *see* Sun Bear
Honey Possum 69
Hooded Seal 303
Hooker's Sea Lion *see* New Zealand Sea Lion
horses 50, 53, 314
House Mouse 167
Howler Monkey
 Guatemalan Black *see* Mexican Howler Monkey
 Mexican 109
howler monkeys 27
Humpback Whale 27, 214
hunting 47
Hutia, Desmarest's 178
Hunting Dog *see* African Wild Dog
Hyaena hyaena 272
Hyaenidae 53, 272
Hydrochaeridae 177
Hydrochaerus hydrochaeris 177
Hydromys chrysogaster 171
Hydrurga leptonyx 301
Hyena
 Brown 273
 Spotted 274
 Striped 272
hyenas 53, 272
Hylobates lar 122
Hylobates syndactylus 123
Hylobatidae 122
Hylochoerus meinertzhageni 324
Hyperoodon ampullatus 198
Hypogeomys antimena 162
Hypsiprymnodon moschatus 70
Hyracoidea 53, 310
Hyrax
 Rock 310
 Tree 310
hyraxes 53, 310
Hystricidae 174
Hystrix africaeaustralis 174

I

Iberian Lynx 283
Ibex 377
Ictonyx striatus 255
Impala 371
Indian Bison *see* Gaur
Indian Giant Squirrel 150
Indian Rhinoceros 321
Indian Spotted Chevrotain 334
Indian Tree Shrew 96
Indo-Chinese Tiger 292

Indri 106
Indri indri 106
Indriidae 105
Inia geoffrensis 184
Iniidae 184
Insectivora 52
insectivores 52, 78
internal fertilization 20
introduced species 51
Irrawaddy Dolphin 190
IUCN Red List 46

J

Jackal
  Black-backed 229
  Golden 228
  Side-striped 229
Jaculus jaculus 173
Jaguar 289
Javan Rhinoceros 46, 321
Javelina *see* Collared Peccary
jaws 18
Jerboa
  Desert 173
  Four-toed 173
joey 21
Jungle Cat 276
Jurassic Period 12

K

Kangaroo
  Hill *see* Wallaroo
  Red 77
  Western Grey 74
Kangaroo Rat, Merriam's 152
Killer Whale 194
  False 192
Kinkajou 245
Kirk's Dik-dik 370
Klipspringer 368
Koala 39, 65
Kob 360
Kobus ellipsiprymnus 359
Kobus kob 360
Kobus leche 359
Kobus vardonii 360
Kodiak Bear 236
Kowari 60
Kultarr 59

L

lactation 23
Lagenorhynchus albirostris 186
Lagenorhynchus obliquidens 186
Lagenorhynchus obscurus 185
Lagomorpha 52, 136
Lagomorphs 136
Lagorchestes conspicillatus 72
Lagostomus maximus 178
Lagothrix cana 107
Lagurus lagurus 165
Lama guanicoe 331
Langur, Hanuman 120
Lar Gibbon 122
Leadbeater's Possum 67
Lechwe 359
Lemming
  Brown 165
  Steppe 165
Lemmus sibericus 165
Lemniscomys striatus 165
Lemur
  Black 103
  Brown 103
  Fat-tailed Dwarf 101
  Grey Gentle 104
  Ring-tailed 102
  Ruffed 104
  Weasel 105
Lemur catta 102
Lemur fulvus 103
Lemur macaco 103
Lemuridae 102
Leontopithecus rosalia 113
Leopard 290
  Clouded 288
  Snow 288
Leopard Seal 301
Lepilemur mustelinus 105
Leporidae 137
Leptonychotes weddelli 301
Lepus arcticu 141
Lepus californicus 142
Lepus europaeus 140
Lepus timidus 142
Lesser Anteater *see* Southern Tamandua
Lesser Bushbaby *see* South African Galago
Lesser Horseshoe Bat 89
Lesser Mouse-tailed Bat 88
Lesser Panda 242
Lesser Tree Shrew 96
Lime's Two-toed Sloth 130
Linnaeus' False Vampire Bat *see* False Vampire Bat
Lion 294
Lipotes vexillifer 184
Lipotidae 184
Lissodelphis borealis 190
Litocranius walleri 372
Little Pygmy-possum 70
Llama 331
llamas 330, 331
Lobodon carcinophagus 300
locomotion 28–29
  aquatic mammals 29
  bipedal locomotion 29
  quadrupedal locomotion 28
Long-clawed Marsupial Mouse 61
Long-eared Desert Hedgehog 79
Long-footed Potoroo 70
Long-nosed Echidna 55
Lontra canadensis 264
Loridae 100
Loris tardigradus 98
Loris
  Slender 98
  Slow 99
Lorisidae 98
Lowland Anoa 352
Loxodonta africana 308
Lutra lutra 262
Lycaon pictus 233
Lynx
  Desert *see* Caracal
  Eurasian 283
  Iberian 283

M

Macaca fascicularis 119
Macaca nigra 119
Macaque
  Celebes 119
  Crab-eating 119
Macroderma gigas 84
Macropodidae 71
Macropus fuliginosus 74
Macropus parma 76
Macropus robustus 73
Macropus rufus 77
Macroscelidea 52, 94
Macroscelididae 94
Macrotis lagotis 63
Madoqua kirkii 370
Madras Tree Shrew *see* Indian Tree Shrew
Malagasy Giant Rat 162
Malayan Civet 267
Malayan Flying Lemur 95
Malayan Tapir 323
mammal-watching 44–45
  equipment 45
  safety precautions 44, 45
mammary glands 10
Manchurian Bear 236
Mandrill 116
Mandrillus sphinx 116
Maned Three-toed Sloth 131
Maned Wolf 37, 222
Manidae 135
Manis pentadactyla 135
Manis temminckii 135

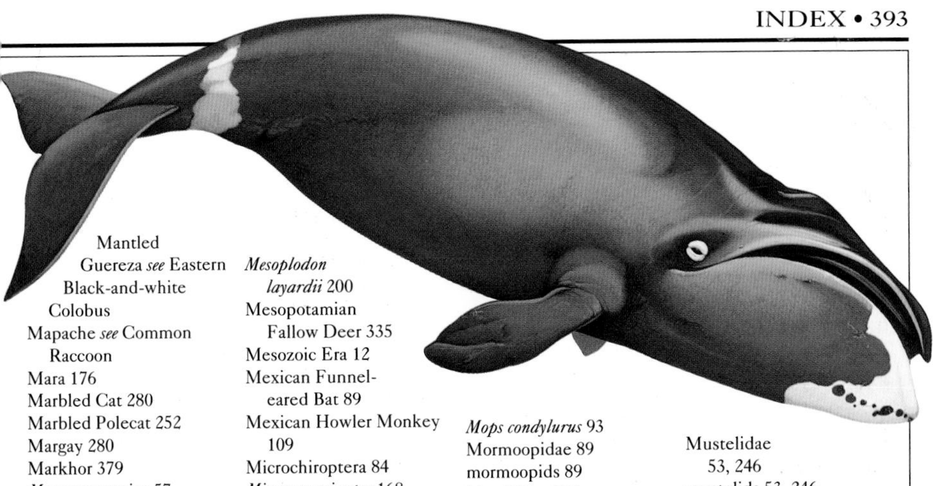

Mantled Guereza *see* Eastern Black-and-white Colobus
Mapache *see* Common Raccoon
Mara 176
Marbled Cat 280
Marbled Polecat 252
Margay 280
Markhor 379
Marmosa murina 57
Marmoset
Goeldi's 113
Pygmy 114
Silvery 115
Marmota flaviventris 144
Marmota monax 144
Marsh Antelope *see* Lechwe
Marsh Deer 340
Marsupial Mole 15, 64
Marsupial Mouse, Long-clawed 61
Marsupialia 52
marsupials 11, 21, 52, 56
Marten
Beech 252
Yellow-throated 253
Martes flavigula 253
Martes foina 252
Martes pennanti 254
Martes zibellina 253
mating 20
Mauritian Tomb Bat 84
Mediterranean Monk Seal 300
Meerkat 35, 271
Megachiroptera 84
Megadermatidae 84
Megaptera novaeangliae 214
Megazostrodon 12
Meles meles 261
Mellivora capensis 260
Melogale personata 259
Melursus ursinus 239
Mephitis mephitis 256
Meriones unguiculatus 160
Merriam's Kangaroo Rat 152
Mesocricetus auratus 158
Mesoplodon bidens 200
Mesoplodon layardii 200
Mesopotamian Fallow Deer 335
Mesozoic Era 12
Mexican Funnel-eared Bat 89
Mexican Howler Monkey 109
Microchiroptera 84
Micromys minutus 168
Microtus arvalis 163
Minke Whale 214
Mink
American 250
European 251
Mirounga leonin 302
Moeritherium Phiomia 12
Mole
European 83
Grant's Golden 80
Marsupial 64
Star-nosed 82
Mole Rat
Giant African 163
Naked 181
Monachus monachus 300
Mongolian Gerbil 160
Mongolian Wild Horse *see* Przewalski's Wild Horse
Mongoose
Banded 270
Dwarf 269
Yellow 269
Monk Seal, Mediterranean 300
Monkey
Bolivian Squirrel 112
De Brazza's 118
Dusky Titi 111
Golden Snub-nosed 121
Grey Woolly 107
Night 111
Patas 117
Proboscis 121
monkeys 52, 107
Monodon monoceros 196
Monodontidae 196
Monotremata 52
monotremes 11, 21, 54
Moonrat 79
Mops condylurus 93
Mormoopidae 89
mormoopids 89
Moschidae 334
Moschus chrysogaster 334
Mountain Beaver 143
Mountain Cat *see* Andean Cat
Mountain Goat 41, 374
Mountain Hare 142
Mountain Lion *see* Puma
mountain mammals 40–41
heat retention 40
insulation 40
moulting 41
Mountain Tapir 322
Mouse
American Harvest 156
Eurasian Harvest 168
House 167
Spinifex Hopping 169
Spiny 168
Striped Grass 165
White-footed 156
Wood 166
Yellow-necked Field 166
Mouse Opossum, Common 57
Mule Deer 339
Mungos mungo 270
Muridae 156
Muriqui 108
Mus musculus 167
Muscardinus avellanarius 172
Muskox 40, 376
Muskrat 164
Musky Rat-kangaroo 70
Mustela erminea 247
Mustela lutreola 251
Mustela nigripes 247
Mustela nivalis 248
Mustela putorious 246
Mustela vison 250
Mustelidae 53, 246
mustelids 53, 246
mutual grooming 27
Myocastor coypus 180
Myocastoridae 180
Myotis daubentonii 93
Myrmecobiidae 62
Myrmecobius fasciatus 62
Myrmecophaga tridactyla 132
Myrmecophagidae 131
Mysticeti 53

N

Naked Mole Rat 181
Narwhal 196
Nasalis larvatus 121
Nasua nasua 244
Natalidae 89
Natalus stramineus 89
Neofelis nebulosa 288
Neomys fodiens 81
Neophascogale lorentzii 61
Neophocaena phocaenoides 183
New Guinean Spiny Bandicoot 63
New Zealand Sea Lion 297
Night Monkey 111
Grey-bellied 111
Nilgai 351
Nine-banded Armadillo 134
Ningaui ridei 58
Noctilio leporinus 88
Noctule Bat, Common 92
Noctilionidae 88
North American Elk 342
North American Grey Squirrel 51

North American Pika 136
North American Porcupine 175
North American River Otter 264
North American Short-tailed Shrew 82
Northern Bottlenose Whale 198
Northern Fur Seal 298
Northern Naked-toed Armadillo 134
Northern Right Whale 206
Northern Right-whale Dolphin 190
Northern White Rhinoceros 318
Norway Rat *see* Brown Rat
Notomys alexis 169
Notoryctes typhlops 64
Notoryctidae 64
Numbat 62
Nyala 349
Nyctalus noctula 92
Nyctereutes procyonoides 226
Nycticebus coucang 99
Nyctomys sumichrasti 157

O

Ocelot 281
Ochotona curzoniae 137
Ochotona princeps 136
Ochotonidae 136
Octodon degus 181
Octodontidae 181
odd-toed ungulates 314
Odobenidae 296
Odobenus rosmarus 299
Odocoileus hemionus 339
Odontoceti 53
Okapi 53, 345
Okapia johnstoni 345
Olingo, Bushy-tailed 245
Olive Baboon 116
Ommatophoca rossii 303
omnivores 31
Onager 315
Ondatra zibethicus 164
Opossum
  Common Mouse 57
  Virginia 56
  Water 57
Orangutan
  Bornean 124
  Sumatran 124
orangutans 49
Orcaella brevirostris 190
Orcinus orca 194
Oreamnos americanus 374
Oreotragus oreotragus 368
Oribi 369
Oriental Linsang 267
Oriental Short-clawed Otter 264
Ornithorhynchidae 55
Ornithorhynchus anatinus 55
Orycteropodidae 311
Orycteropus afer 311
Oryctolagus cuniculus 139
Oryx dammah 363
Oryx gazella 363
Otaria byronia 298
Otariidae 296
Otocyon megalotis 225
Otter
  African Clawless 264
  Eurasian River *see* European Otter
Otter cont.
  European 262
  Giant 265
  North American River 264
  Oriental Short-clawed 264
  Sea 265
Ourebia ourebi 369
Ovibos moschatus 376
Ovis canadensis 380
Ovis orientalis 381

P

Paca 178
Pachyuromys duprasi 162
Pacific White-sided Dolphin 186
packs 36
Pagophilus groenlandicus 304
Palm Civet 268
Pan paniscus 127
Pan troglodytes 126
Panda
  Giant 240
  Lesser 242
Pangolin
  Chinese 135
  Temminck's 135
pangolins 52, 135
Panther *see* Puma
Panthera leo 294
Panthera onca 289
Panthera pardus 290
Panthera tigris 292
Pantropical Spotted Dolphin 188
Papio anubis 116
Papio papio 115
Paradoxurus hermaphroditus 268
Parahyaena brunnea 273
Parantechinus apicalis 59
parental care 23
parks and zoos 44
Parma Wallaby 76
patagium 17
Patagonian Cavy *see* Mara
Patagonian Hog-nosed Skunk 258
Patas Monkey 117
Pecari tajacu 327
Peccary, Collared 327
Pedetes capensis 153
Pedetidae 153
Pentalagus furnessi 137
Perameles gunnii 63
Peramelidae 63
Pére David's Deer 51, 338
Perissodactyla 314
perissodactyls 322
Perodicticus potto 99
Peromyscus leucopus 156
Peroryctidae 63
Petauridae 67
Petaurista elegans 151
Petauroides volans 68
Petaurus norfolcensis 67
Petrogale penicillata 71
Phacochoerus africanus 326
Phalanger orientalis 66
Phalangeridae 66
Phascolarctidae 65
Phascolarctos cinereus 65
Philodata 52, 135
Phoca sibirica 304
Phoca vitulina 305
Phocarctos hookeri 297
Phocidae 296
Phocoena phocoena 182
Phocoena sinus 183
Phocoenidae 182
Phocoenoides dalli 183
Phodopus roborovskii 159
Phyllostomidae 90
Physeter macrocephalus 202
Physeteridae 202
Pichi 134
Pig, Bush 326
pigs 53
Pika
  Black-lipped 137
  North American 136
pikas 52, 136
Pinnipedia 53, 296
Pinnipeds 296
Pipistrelle, Common 91
Pipistrellus pipistrellus 91
Pithecia pithecia 110
Pitheciidae 110
Pizote *see* Southern Ring-tailed Coati

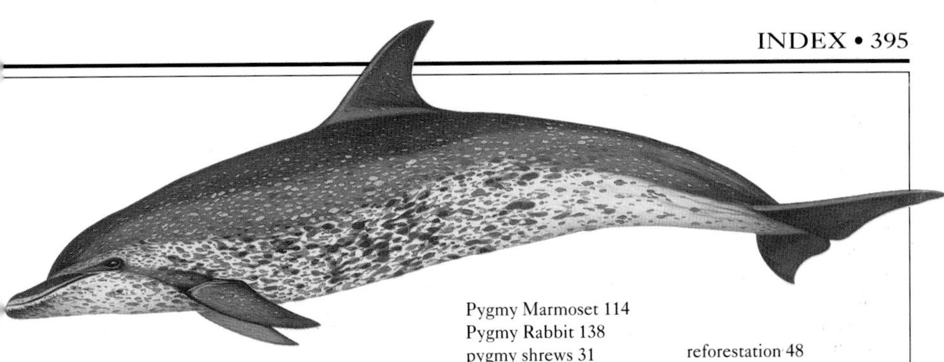

placenta 20
placental mammals
  11, 21, 22
Plains Prairie Dog *see*
  Black-tailed Prairie Dog
Plains Viscacha 178
plantigrade 28
Platanista gangetica 185
Platanistidae 185
Plateau Pika
  see Black-lipped Pika
platypus 21
Plecotus auritus 93
pod 25
Poecilogale albinucha 251
Polar Bear 40, 235
polar mammals 40–41
  heat retention
  insulation
  moulting
Polecat
  European 246
  Marbled 252
pollution 47, 48
Pongidae 124
Pongo pygmaeus 124
Porcupine
  Cape 174
  Crested *see* Cape
    Porcupine
  North American 175
  Prehensile-tailed 175
Porpoise
  Dall's 183
  Finless 183
  Gulf of California 183
  Harbour 182
Possum
  Common Brush-tailed 66
  Feather-tailed 69
  Honey 69
  Leadbeater's 67
  Striped 67
Potamochoerus porcus 326
Potoroidae 70
Potoroo, Long-footed 70
Potorous longipes 70
Potos flavus 245
Potto 99
pouched mammals 11, 21
Prairie Dog, Black-tailed
  145
prairie dogs 37
Prehensile-tailed
  Porcupine 175
Prevost's Squirrel 151
primates 52, 97
Priodontes maximus 133
Prionodon pardicolor 267
Proboscidea 53, 348
Proboscis Bat 85
Proboscis Monkey 121
Procavia capensis 310
Procaviidae 310
Procyon cancrivorus 244
Procyon lotor 243
Procyonidae 242, 244
Procyonids 242
Pronghorn 53, 344
Propithecus verreauxi 105
prosimians 52, 97
protected parks 48
Proteles cristatus 273
Przewalski's Wild Horse
  50, 314
Pseudalopex culpaeus 224
Pseudantechinus
  macdonnellensis 60
Pseudocheiridae 68
Pseudocheirus peregrinus
  68
Pseudois nayaur 379
Pseudorca crassidens 192
Pteroneura brasiliensis 265
Pteronotus davyi 89
Pteropodidae 86, 88
Pudu pudu 340
Puku 360
Puma 286
Puma concolor 286
Pygmy Chimpanzee 127
Pygmy Glider 69
Pygmy Hippopotamus
  328
Pygmy Hog *see* Wild Boar
Pygmy Marmoset 114
Pygmy Rabbit 138
pygmy shrews 31
Pygmy White-toothed
  Shrew 14, 82

Q

Quarternary Period 13
Quokka 73
Quoll, Eastern 61

R

Rabbit-eared Bandicoot
  see Greater Bilby
Rabbit
  Amami 137
  European 139
  Pygmy 138
  Swamp 138
  Volcano 140
rabbits 52, 136
Raccoon
  Common 243
  Crab-eating 244
Raccoon Dog 226
raccoons 18, 242
Rangifer tarandus 341
Raphicerus campestris 369
Rat
  Australian Water 171
  Black 171
  Brown 170
  Common *see* Brown Rat
  Giant African Mole
    163
  Giant Pouched 169
  Hispid Cotton 157
  Malagasy Giant 162
  Merriam's Kangaroo 152
  Naked Mole 181
  Sumichrast's Vesper 157
Ratel *see* Honey Badger
Rattus norvegicus 170
Rattus rattus 171
Ratufa indica 150
Razorback *see* Fin Whale
reforestation 48
reintroduced species 51
reproduction 20–23
  egg-laying mammals
  placental mammals
  pouched mammals
reproductive system
  20–23
  egg-laying mammals
  placental mammals
  pouched mammals
Red Bald Uakari 110
Red Deer 336
Red Forest Duiker 358
Red Fox 218
Red Howler Monkey 109
Red Kangaroo 77
Red Panda
  see Lesser Panda
Red River Hog
  see Bush Pig
Red Squirrel, Eurasian
  148
Red-eared Antechinus 60
Red-legged Pademelon 72
Redunca redunca 361
Reed Cat *see* Jungle Cat
Reindeer 341
Reithrodontomys raviventris
  156
Reticulated Giraffe 347
Rhinoceros sondaicus 321
Rhinoceros unicornis 321
Rhinoceros
  Black 320
  Indian 321
  Javan 321
  Sumatran 321
  White 318
rhinoceroses 49, 318
Rhinocerotidae 53, 318
Rhinolophidae 89
Rhinolophus hipposideros 89
Rhinopithecus roxellana 121
Rhinopoma hardwickei 88
Rhinopomatidae 88
Rhynchocyon chrysopygus 94
Rhynchonycteris naso 85
Ring-tailed Lemur 102

Ringtail 243
Common 68
Ringtailed Cat *see* Ringtail
Risso's Dolphin 186
River Dolphin
Amazon 184
Chinese 184
Ganges 185
River Otter, North American 264
Roan Antelope 361
Rock Dassie *see* Rock Hyrax
Rock Hyrax 310
Rock Wallaby, Brush-tailed 71
Rodentia 52
rodents 16, 52, 143
Roe Deer 339
Romerolagus diazi 140
Ross Seal 303
Rousettus egyptiacus 88
Rueppell's Fox 220
Ruffed Lemur 104
Rufous Elephant-shrew 94
ruminants 32
Rupicapra rupicapra 375
Russian Desman 83
Ryukyu Rabbit *see* Amami Rabbit

S

Sable 253
Sable Antelope 362
Saguinus imperator 114
Saiga 374
Saiga tatarica 374
Saimiri boliviensis 112
Sambar 338
Sand Cat 277
Sand Fox *see* Rueppel's Fox
Sarcophilus harrisii 62
Savanna Elephant *see* African Elephant
Scandentia 52, 96
scavengers 30
scent 27
Scimitar-horned Oryx 363
Sciuridae 144
Sciurus carolinensis 149
Sciurus vulgaris 148
Scutisorex somereni 81
Sea Lion
Californian 296
Hooker's *see* New Zealand Sea Lion
New Zealand 297
South American 298
sea lions 40, 53, 296
Sea Otter 43, 265
Seal
Baikal 304
Cape Fur 297
Common 305
Crabeater 300
Grey 303
Harbour *see* Common Seal
Harp 304
Hooded 303
Leopard 301
Mediterranean Monk 300
Northern Fur 298
Ross 303
Weddell 301
seals 40, 53, 296
Seladang *see* Gaur
Semnopithecus entellus 120
senses 26
Serval 278
Setonix brachyurus 73
Sharp-nosed Bat *see* Proboscis bat
Sheep
Barbary 381
Bighorn 380
Blue 379
sheep 348
Ship Rat *see* Black Rat
Short-finned Pilot Whale 192
Short-nosed Echidna 54
Shrew
Armoured 81
Eurasian Water 81
Hero *see* Armoured Shrew
North American Short-tailed 82
Pygmy White-toothed 82
Siamang 123
Siberian Bear 236
Siberian Tiger 292, 293
Side-striped Jackal 229
sight 26
Sigmodon hispidus 157
Sika 338
Silky Anteater 132
Silver-backed Jackal *see* Black-backed Jackal
Silvery Marmoset 115
Simian Jackal *see* Ethiopian Wolf
simians 97
Sirenia 53, 312
sirenians 312
Sitatunga 348
skeleton 16
skin 19
Skunk
Eastern Spotted 258
Patagonian Hog-nosed 258
Striped 256
Slender Loris 98
Sloth
Lime's Two-toed 130
Maned Three-toed 131
Sloth Bear 239
sloths 39, 130
Slow Loris 99
Small Spotted Genet 266
Small-eared Dog 225
Small-eared Zorro *see* Small-eared Dog
smell 26
Sminthopsis crassicaudata 58
Snow Leopard 288
social groups 24–25
Solenodon, Hispaniolan 79
Solenodon paradoxus 79
Solenodontidae 79
Soricidae 81
South African Galago 101
South American Sea Lion 298
South American Tapir 323
Southern Elephant-seal 302
Southern Oryx *see* Gemsbok
Southern Pudu 340
Southern Right Whale 206
Southern Ring-tailed Coati 244
Southern Tamandua 131
Southern White Rhinoceros 318
Sowerby's Beaked Whale 200
specialized mammals 11
Speckled Dasyure *see* Long-clawed Marsupial Mouse
Spectacled Bear 239
Spectacled Hare Wallaby 72
Speothos venaticus 232
Sperm Whale 43, 202
Spermophilus columbianus 148
Spider Monkey
Black 108
Black-handed 108
Geoffroy's *see* Black-handed Spider Monkey
Spilogale putorius 258
Spinifex Hopping Mouse 169

Spinner Dolphin 187
Spiny Anteater *see* Short-nosed Echidna
Spiny Mouse 168
Spotted Deer *see* Chital
Spotted Hyena 274
Spotted Mouse Deer *see* Indian Spotted Chevrotain
Spotted Native Cat *see* Eastern Quoll
Springbok 373
Springbuck *see* Springbok
Springhaas *see* Springhare
Springhare 153
Square-lipped Rhinoceros *see* White Rhinoceros
Squirrel
- Eastern Grey 149
- Eurasian Red 148
- Giant Flying 151
- Indian Giant 150
- Prevost's 151
- Sun 150

Squirrel Glider 67
Squirrel Monkey, Bolivian 112
Star-nosed Mole 26, 82
Steenbok 369
Steinbuck *see* Steenbok
Stenella coeruleoalba 188
Stenella frontalis 187
Stenella longirostris 187
Steppe Lemming 165
Stinker *see* Swamp Wallaby
Stoat 247
Stone Marten *see* Beech Marten
Strap-toothed Whale 200
Strepsirhini 52, 97
Striped Dolphin 188
Striped Grass Mouse 165
Striped Hyena 272
Striped Polecat *see* Zorilla
Striped Possum 67
Striped Skunk 256
Suidae 53, 324
Sumatran Orangutan 124
Sumatran Rhinoceros 321
Sumatran Tiger 292, 293
Sumichrast's Vesper Rat 157
Sun Bear 238
Sun Squirrel 150
Suncus etruscus 82
Suricata suricatta 271
Suricate *see* Meerkat
Sus salvanius see Wild Boar
Sus scrofa 324
Swamp Cat *see* Jungle Cat
Swamp Rabbit 138
Swamp Wallaby 73
Swift Fox 217
Sylvicapra grimmia 358
Syncerus caffer 353
Syrian Bear 236

T

Tachyglossidae 54
Tachyglossus aculeatus 54
Tachyoryctes macrocephalus 163
Tahr 377
Talpa euoropaea 83
Talpidae 82
Tamandua, Southern 131
Tamandua tetradactyla 131
Tamarin
- Emperor 114
- Golden Lion 113

Tamias striatus 146
Tammar Wallaby 21
Taphozous mauritianus 84
Tapir
- Baird's 323
- Malayan 323
- Mountain 322
- South American 323

tapirs 53, 322
Tapiridae 53, 322
Tapirus bairdii 323
Tapirus indicus 323
Tapirus pinchaque 322
Tapirus terrestris 323
Tarsiidae 97
Tarsipedidae 69
Tarsipes rostratus 69
Tarsius bancanus 97
Tasmanian Devil 62
Taurotragus oryx 351
Taxidea taxus 259
teeth 18
Tertiary Period 13
Tejón *see* Southern Ring-tailed Coati
Temminck's Pangolin 135
temperature control 19
Tenella attenuata 188
Tenrec ecaudatus 80
Tenrecidae 80
Tent-building Bat 90
Tetracerus quadricornus 352
Theropithecus gelada 117
Thick-tailed Galago 100
Thick-tailed Greater Bushbaby *see* Thick-tailed Galago
Thomomys bottae 152
Thomson's Gazelle 372
threats 46–47
Thylogale stigmatica 72
Tiger 46, 292
- Bengal 292
- Chinese 292
- Indo-Chinese 292

Titi Monkey, Dusky 111
toothed whales 53, 182
Topi 365
touch 26
Tragelaphus angasi 349
Tragelaphus eurycerus 349
Tragelaphus scriptus 349
Tragelaphus spekei 348
Tragelaphus strepsiceros 350
Tragulidae 334
Tragulus meminna 334
Tree Dassie *see* Tree Hyrax
Tree Fox *see* Grey Fox
Tree Hyrax 310
Tree Shrew
- Indian 96
- Lesser 96

tree shrews 52, 96
Tremarctos ornatus 239
Triassic Period 12
Trichechidae 313
Trichechus manatus 313
Trichosurus vulpecula 66
Tsessebe *see* Topi
Tubulidentata 53, 311
Tufted Capuchin *see* Brown Capuchin
Tupaia minor 96
Tupaiidae 96
Tursiops truncatus 189

U

Uakari, Bald 110
umbilical cord 22
Uncia uncia 288
underground colonies 37
underwater birth 22
unguligrade 28
Urocyon cinereoargenteus 221
Uroderma bilobatum 90
Ursidae 53, 234
Ursids 234
Ursus americanus 234
Ursus arctos 236
Ursus maritimus 235
Ursus thibetanus 238
uterus 22

V

Vampire Bat 30, 91
  Australian False 84
  False 90
Vampyrum spectrum 90
Vaquita *see* Gulf of California Porpoise
Varecia variegata 104
Verreaux's Sifaka 105
vertebrates 18
Vespertilionidae 92
Vicugna vicugna 330
Vicuña 41, 330
Virginia Opossum 23, 56
Viverra tangalunga 267
Viverridae 53, 266
viverrids 266
Volcano Rabbit 140
Vole
  Bank 163
  Common 163
  European Water 164
Vombatidae 64
Vombatus ursinus 64
Vormela peregusna 252
Vulpes cana 217
Vulpes rueppelli 220
Vulpes velox 217
Vulpes vulpes 218
Vulpes zerda 216

W

Wallabia bicolor 73
Wallaby
  Black *see* Swamp Wallaby 73
  Brush-tailed Rock 71
  Parma 76
Wallaby cont.
  Spectacled Hare 72
  Swamp 73
Wallaroo 73
Walrus 43, 299
warm-blooded 10
warren 25
Warthog 326
Water Bat
  see Daubenton's Bat
Water Buffalo, Asian 352
Water Opossum 57
water pollution 46
Water Rat, Australian 171
Water Shrew, Eurasian 81
Water Vole, European 164
Waterbuck 359
Weasel Lemur 105
Weddell Seal 301
West African Manatee 313
West European Hedgehog 78
West Indian Manatee 313
Western Gorilla 128
Western Grey Kangaroo 74
Western Tarsier 97
Whale
  Blue 212
  Bowhead 204
  Cuvier's Beaked 198
  False Killer 192
  Fin 210
  Gray 208
  Humpback 214
  Killer 194
Whale cont.
  Minke 214
  Northern Bottlenose 198
  Northern Right 206
  Short-finned Pilot 192
  Sowerby's Beaked 200
  Sperm 202
  Strap-toothed 200
whales 49
whaling 47
White Rhinoceros 318
White-beaked Dolphin 186
White-collared Mangabey 118
White-faced Saki 110
White-footed Mouse 156
Whitefin Dolphin *see* Chinese River Dolphin
Wild Boar 324
Wild Cat 276
Wild Goat 378
wild mammals 44
Wild Pig, Eurasian
  see Wild Boar
wild pigs 324
Wildebeest 366
Wisent
  see European Bison
Wolf
  Ethiopian 228
  Grey 230
  Maned 222
Wolverine 255
Wombat, Common 64
Wongai Ningaui
  see Inland Ningaui
Wood Mouse 166
Woodchuck 144
Woolly Monkey, Grey 107
Woolly Spider Monkey
  see Muriqui
World Wide Fund for Nature (WWF) 48

X

Xenarthra 52, 130
Xerus inauris 145

Y

Yak 40, 354
Yapok *see* Water Opossum
Yellow Mongoose 269
Yellow-bellied Marmot 144
Yellow-necked Field Mouse 166
Yellow-throated Marten 253

Z

Zaedyus pichiy 134
Zaglossus bartoni 55
Zalophus californianus 296
Zebra
  Burchell's 315
  Common *see* Burchell's
  Grevy's 316
Ziphiidae 182
Ziphius cavirostris 198
Zorilla 255

ACKNOWLEDGMENTS

DORLING KINDERSLEY would like to thank Kim Bryan for her invaluable editorial consultancy and Merrol Parker for her help with research. Thanks also to Suresh Kumar for his cartographic work and to Arun P. for the line artworks used in the end-papers.

Picture Researchers: Cheryl Dubyk-Yates, Sean Hunter
Picture Librarian: Richard Dabb

The publisher would like to thank the following for their kind permission to reproduce their photographs: (Abbreviations key:
t=top, b=bottom, r=right, l=left, c=centre)

Animals Animals/Earth Scenes: Anthony Bannister 174tr; David J Boyle 238br; Dani Jeske 354b; **Ardea London Ltd:** Tony Beamish 29c; Hans & Judy Beste 76cr; M. D. England 360tr; Jean Paul Ferrero 283tr, 334cr; Kenneth W. Fink 131b, 151tr, 152t, 374tr; Francois Gohier 25cr, 171br; Nick Gordon 244br, 262t, 262c; Joanna Van Gruisen 268tr; Chris Harvey 34c; Chris Knights 275t; Keith & Liz Laidler 244t, 265t; P. Morris 79br, 96br, 163tr; B. Moose Peterson 156tr; Starin 348br; Peter Steyn 80tr; M. Watson 8bl, 106tr, 255b, 288tr; Alan Weaving 169br; **Auscape:** Jean-Paul Ferrero 66tr, 67br; T. Shivanandappa 98tr; **Robert E Barber:** 250bl, 254tr, 298tr, 298br; **David Barnes:** 297br, 301br, 302b; **Bat Conservation International:** Merlin D. Tuttle 89cr, 90cr, 90br, 93br; **Fred Bavendam:** 313b, 338cr; **BBC Natural History Unit:** Doug Allan 304br; Peter Blackwell 37cr; John Cancalosi 21bl; Jim Clare 111tr; Alain Compost 105tr; Bruce Davidson 36c; Georgette Douwma 21tl; Jeff Foott 304t, 313tr, 325br, 356bl; Charlie Hamilton James 228tr; Martha Holmes 205tr; Kevin J Keatley 261tr; Thomas D Mangelsen 43cr, 235b; Dietmar Nill 252b; Pete Oxford 103br; Colin Preston 163br; Anup Shah 116br, 124r, 233tr; Tom Vezo 56br; Doc White 110br, 213t; Niall Benvie: 376b; William Bernard Photography: 134tr, 243t, 299b;
BIOS Photo: J. Alcalay 77tr; Jany Sauvanet 132br; R. Seitre 79cr, 83b, 252tr, 259bt; **C. K. Bryan:** 227br, 351br; **John Cancalosi:** 60br, 71tr, 75cr, 108tr; **Bruce Coleman Ltd:** Erwin & Peggy Bauer 222cr; John Cancalosi 337b; Mark Carwardine 128tr; Bruce Coleman 151b, 273cr; Alain Compost 253tr; M & P Fogden 135c; Jeff Foott 256b; C & D Frith 260br; Janos Jurka 339tr; Joe Macdonald 357t; Mary Plague 321cr; Hans Reinhard 257tl; John Shaw 15c; Jorg & Petra Wegner 39c; Staffan Widstrand 129tr; Rod Williams 99br, 118tr, 270br, 280tr; Bill Wood 95cr, 95bl; Konrad Wothe 101br; Gunter Ziesler 277b, 280b; **Bruce Coleman Inc:** 293cr; **Wendy Conway:** 341b, 341b; **Peter Cross:** 71br, 73tr, 127b, 148tr, 175tr, 177br, 264br, 297tr, 315br, 320b, 336c, 360br, 366c, 369br, 373b, 381br; **Dennis Cullinane:** 81tr; **Nigel Dennis:** 97br, 102b, 103tr, 105br, 106br, 135br, 225tr, 229tr, 273br, 311b, 320tr, 326br, 358tr, 358br, 359tr, 364br, 368tr, 368br, 369tr, 370tr, 371b; **Brock Fenton:** 84br, 88cr, 89br, 90tr; **Fotomedia:** R. Dev 355tr; Joanna Van Gruisen 334br; Neeraj Mishra 99tr; Otto Pfister 335br; E. Hanumantha Rao 98b, 232b, 315cr; Vivek R. Sinha 321tr; **Foto Natura:** Frans Lanting 38c; S. Maslowski 155b; **Brian Gibbs:** 260tr; **Michael P. Gillingham:** 138br; **Francois Gohier:** 210b; **Derek Harvey:** 115tr; **Dr C Andrew Henley-Larus:** 58tr, 58br, 59tr, 62br, 68br, 69br, 73cr; **ImageState:** ImageState 236bl; National Geographic 220br; **Jacana Hoa-Qui:** Gunter Ziesler 133br, 134cr; **F. Jack Jackson:** 268br; **Mike Jordan:** 79tr, 81br, 137tr, 150b, 152b, 157br, 159tr, 159br, 162tr, 164tr, 165cr, 168br, 169tr, 172tr, 172br, 173b, 173t, 379tr; **Hiromitsu Katsu:** 137b; **Saul Kitchener:** 225b; **Mark Kostich Photography:** 267tr; **FLPA - Images of nature:** Rolf Bender 251tr; Brake/Sunset 131tr; Di Domerico 300tr; Free Pictures 22br; David Hosking 27cl; E & D Hosking 35cl; S. Jonasson 303cr; Gerard Laci 323tr; Frank W. Lane 153tr; Leeson/Sunset 109cr; K. Maslowski 156br; S. Maslowski/Foto Natura 155b; Meinderts/Foto Natura 176tr; Mikhail 253br; Minden Pictures 27cr; Philip Perry 319tl; L Lee Rue 366bl; R. Van Nostrand 121br; Terry Whittaker 24br, 222b, 273tr, 285tr; **Vanessa**

Latford: 349br; **Lincoln Park Zoo:** Saul Kitchener 225br; **The Mammal Images Library:** P. Myers 57t; **Chris Mattison Nature Photographics:** Geoff Trinder 144br, 335t; Martin Withers 66br, 69tr, 72br, 84cr, 136b, 166b, 247b, 344b, 361tr, 372tr, 372br, 380b; **National Geographic Society:** Richard T. Nowitz 44l; **Natural History Museum, London:** 12t; Natural Visions: Heather Angel 180b; **N.H.P.A.:** A.N.T. 15bc, 59br, 64tr; Anthony Bannister 45tr, 255tr, 49l; Mark Bowler 125br; Laurie Campbell 142t, 258tr; David Currey 47bl; Stephen Dalton 38tr, 44tr, 88br; Nigel J. Dennis 251br; Pavel German 55tr; Daniel Heuclin 63br, 82cr, 178cr; Ralph & Daphne Keller 67cr; Stephen Krasemann 22bl, 217b; Gerard Lacz 76tr; Michael Leach 246cr; Haroldo Palo Jr 57b, 132c; Steve Robinson 7tr; Andy Rouse 229br, 235tr, 250r, 257tr; Jany Sauvanet 175br, 224br; Kevin Schafer 108br, 224tr, 239br; John Shaw 138tr, 142br, 310br; Eric Soder 375bl; Morten Strange 267br; Norbert Wu 40bl; **Oxford Scientific Films:** AnimalsAnimals/Rick Edwards 48cl, Animals-Animals/Richard K. LaVal 134br; Kathie Atkinson 60tr; Eyal Bartov 217t; Joel Bennett 43tfr; Niall Benvie 142tr; Joe Blossom 325tr; Scott Camazine 82tr; Daniel J. Cox 49br; Mark Deeble & Victoria Stone 264cr; Ajay Desai 120br; Michael Dick 121tr; Michael Fogden 80br, 157tr, 245br; Jeff Foott 133cr; Michael Habicht 26tr; Howard Hall 42c, 203tl; Mike Hill 35bl; Mark Jones 303tr; Isaac Kehimka 150tr; Breck P. Kent 82br; Keith & Liz Laidler 241c; Raymond A. Mendez 25br; Stan Osolinski 348bl; Andrew Plumptre 129b; Partridge Productions 100tr; Dieter and Mary Plage 46tr; Norbert Rosing 40c; Edwin Sadd 317t; David Shale 143br; Chris Sharp 48tr; Survival Anglia 31cl, Survival Angila/John Harris 140tr; Tom Ulrich 164br, 258br; **Otto Pfister:** 377tr, 379br; **Mark Picard:** 138cr, 342c; **Planet Earth Pictures:** Tom Brakefield 293t; Jim Brandenburg 141tr; M & C Denis-Hout 309cr; Robert Franz 339br; Doug Perrine 312b; Tom Walker 9c, 141b; John Waters 178br; Andrew Zvoznikov 165tr; **Galen B. Rathbun:** 94cr, 94br; **Wendy Shattil:** Wendy Shattil & Bob Rozinski 178tr, 247t, 259tr, 285br; **Still Pictures:** Fred Bruemmer 299tr; Mark Edwards 47tr; Al Grillo 47cl; Michel Gunther 126b; Roland Seitre 240bl, 279br, 341t; **Michael P. Turco:** 265b; **Andre Van Huizen:** 351tr; **Colin Varndell:** 148br; **Judith Wakelam:** 119br; **Dave Watts:** 55br, 61br, 63tr, 63cr, 67tr, 70tr, 70cr, 70br, 72tr, 73br, 104br, 117tr, 301t, 375br, 381tr; **Jim Winkley:** 374b; **Winfried Wisniewski:** 195tl; **Art Wolfe:** 347tr; **P. A. Woolley and D. Walsh:** 61tr, 69c.

DK PICTURE LIBRARY: American Museum of Natural History 14tr; British Museum 8c; Paignton Zoo ; Parc Zoologique de Paris 13c; Philip Dowell 289cl, 289 br, 290b; Oxford University Museum of Natural History 123bl; Jerry Young: 10bl, 11tr, 19t, 30b, 45 br, 49tr, 50cr, 51tl, 52 tr, 54 b, 86 bl, 88tr, 91 c, 108 cr, 111 br, 112 br, 117 br, 118 br, 122 br, 130b, 174 b, 216 br, 218-219 c, 220 tr, 221 b, 221 tr, 222cr, 223 c, 226 b, 226 tr, 227 tr, 228 br, 230c, 232 tr, 233 b, 235 b, 272 cr, 276 cr, 284 b, 286 bl, 316 bl, 318 b, 332b, 332 tr, 388 tl, 394 bl

Additional Photography: Max Alexander, Peter Anderson, Irv Beckman, Jon Bouchier, Geoff Brightling, Jane Burton, Peter Bush, Martin Camm, Peter Chen, Andy Crawford, Peter Cross, Geoff Dann, Philip Dowell, Alistair Duncan, Mike Dunning, Ken Findlay, Neil Fletcher, Christopher and Sally Gable, Frank Greenaway, Steve Gordon, Kit Houghton, Colin Keates, Dave King, Bob Langrish, Jane Miller, Gary Ombler, Brian Pitkin, Susanna Price, Rob Reichenfeld, Tim Ridley, Guy Ryecart, Tim Shepard, Steve Shott, Harry Taylor, Kim Taylor, Linda Whitwam, Alex Wilson.

FRONT JACKET: Oxford Scientific Films: Theo Allofs tl; John Cancalosi cr; K & L Laidler tp.

INSIDE FLAP FRONT: William Bernard Photography.

BACK JACKET: Planet Earth Pictures: Doug Perrine cr.